Guido

INDUSTRIAL ELECTRONICS

James T. Humphries

Leslie P. Sheets

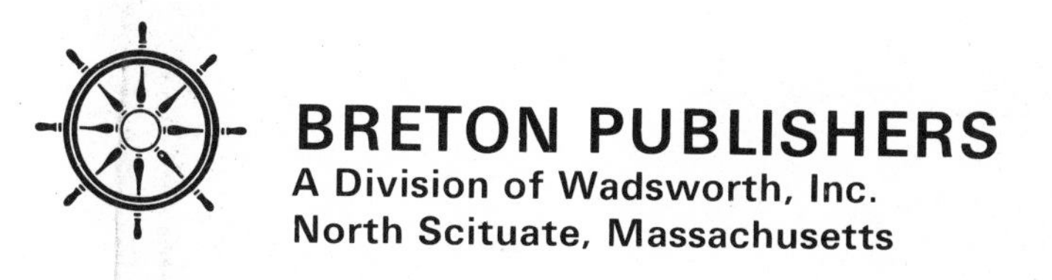

To our wives, Marg and Joyce

Breton Publishers
Division of Wadsworth, Inc.

Library of Congress Cataloging in Publication Data

Humphries, James T., 1946–
Industrial electronics.

Bibliography: p.
Includes Index.
1. Industrial electronics. I. Sheets, Leslie P., 1941– . II. Title.
TK7881.H85 1983 621.381 82-14784
ISBN 0-534-01415-1

Printed in the United States of America
1 2 3 4 5 6 7 8 9—87 86 85 84 83

Industrial Electronics was prepared for publication by the following people:

Sponsoring editor: George Horesta
Production editor: Linda F. Kluz
Art coordinator: Mary S. Mowrey
Copy editor: Carol Beal
Interior designer: Amato Prudente
Cover designer: Stephen Wm. Snider

Illustrations were done by Arvak. The book was typeset in Times Roman and Univers by Syntax International; it was printed and bound by Halliday Lithograph, Inc. The cover photo was provided by American Iron and Steel Institute.

The schematic in Figure 1.7 is reprinted with permission from Signetics Corporation; copyright © 1982 Signetics Corporation.

The schematics in Figures 2.48 and 6.14 are redrawn with permission from National Semiconductor Corporation.

The schematics in Figures 2.57 and 2.58 are redrawn with permission from RCA Solid State Division.

The charts in Figures 3.5 and 11.6 are reprinted with permission from EDN, 1978 © Cahners Publishing Company.

The schematic in Figure 5.39 is redrawn with permission from Hamlin, Inc.

The chart in Figure 5.41 is reprinted with permission from *IEEE Spectrum*, Vol. 4, No. 8, pp. 102–11, August 1967.

The schematics in Figures 6.4, 6.5, 6.11, 6.15, 6.17, 6.18, and 6.19 are copyrighted by and reprinted with permission from General Electric Company.

The schematic in Figure 6.6 is redrawn with permission from RCA Solid State Division.

The schematics in Figure 7.10 and 7.16 are redrawn with permission from Omega Engineering, Inc.

The schematics in Figure 7.14 are redrawn with permission from Fenwal Electronics, Division of Kidde, Inc.

The schematic in Figure 7.18 is redrawn with permission from Analog Devices, Inc.

The schematics in Figure 7.19 are redrawn with permission from Hewlett-Packard Company.

The art in Figure 8.17 is redrawn with permission from C.A. Norgren Company.

The art in Figures 10.1, 10.3, 10.5, 10.7, 10.8, 10.9, 10.10, and 10.11 are adapted with permission from IRIG 106-80, *Telemetry Standards*, written by the Telemetry Group, Range Commanders Council, and published by Range Commanders Council, Secretariat, White Sands Missile Range, NM.

Contents

CHAPTER 3 DC MOTORS AND GENERATORS 55

CHAPTER 4 AC MOTORS 95

Preface

The goal of *Industrial Electronics* is to provide students with an understanding of the basic components and systems used in industrial electronics. The book is intended for use in associate degree, electronics technician programs offered in postsecondary technical and community colleges.

The text avoids design questions. Instead, the text focuses on the underlying concepts and the operation of electronic devices, circuits, and systems. We feel that if concepts are understood, designing circuits in most cases is not a problem. We definitely do not subscribe to the notion that the best way to understand electronic circuits is to design them.

The text is comprehensive. Experience has shown that a course in industrial electronics requires the coverage of a large number of topics. But how does a teacher cover these topics in one course? One solution is to use several textbooks for the course. Another is to supplement one text extensively with instructor-prepared materials. A third approach is to modify the course and teach only those topics covered in the textbook. However, we feel that all of these alternatives are unacceptable. Thus, we have purposely written a comprehensive text and have tried to include most of the topics you will wish to discuss.

Alternatively, we have also written the book so that its presentation is as flexible as possible, should you not wish to present all the topics included. Although we cover most of the topics in the book in a one-semester course, some of the topics can be treated in other courses. For example, we teach the material in the first two chapters (op amps) in our course on electronic circuits. Thus, the material in this book could easily be expanded to two semesters or two quarters, depending on your depth of treatment for each topic.

We require all our electronics students at the School of Technical Careers at Southern Illinois University, Carbondale, to take this course. We feel that every graduate of a two-year or four-year electronics program should have a basic understanding of both digital and analog circuitry. Many of our graduates report that they have found their preparation in analog circuits and systems invaluable. They typically find themselves acting as interfaces between analog applications and people who have a predominantly digital background.

The text is not overly mathematical. We expect students to have a mathematics background no higher than algebra and trigonometry. There is evidence from industry that mathematics preparation beyond this level is not necessary for the technician, and we have avoided any higher level of mathematics in our presentation. We realize, however, that mathematics is a concise way of representing concepts. Thus, we have included it where we feel that an adequate grasp of the concept demands a mathematical treatment.

We also expect students to have some background in basic digital gates and logic gained from an introductory course in digital electronics earlier in their electronics education. It may also be helpful, but not essential, for the student to have completed a technical physics course.

Learning aids have been built into the presentation. We have included questions at the end of every chapter that give immediate feedback on student understanding. Where appropriate, problems are also included. In addition, at the end of the text, bibliographies are included for each chapter. You may wish to use these references as supplemental reading or study assignments if you cover certain topics in more depth.

We hope you will find this textbook a useful teaching aid in your program. An ancillary instructor's guide is available without charge from the publisher for the convenience of instructors who adopt this book for classroom use.

Publishing a textbook is the result of a coordinated effort that involves many people. We would like to express our thanks to the following: to Bob White and his students for their excellent photographic work; to Arthur Barrett, David Longobardi, Thomas Dunsmore, and Eddy E. Pollock, who reviewed the manuscript and offered many helpful suggestions; to the editorial staff at Breton Publishers, and especially to George J. Horesta for words of encouragement at just the right moments and to Linda Kluz and Carol Beal for the almost Herculean task of making our writing readable; to our colleagues here at Southern Illinois University for their support; to our wives, Marg and Joyce, for typing the manuscript; and to our children for understanding why we had to be away from home so many evenings and weekends.

James T. Humphries
Leslie P. Sheets

Acknowledgments

The authors would like to express their thanks to the following manufacturers and photographers who provided photographs and illustrations for the preparation of *Industrial Electronics*:

Allen-Bradley Company
American Iron and Steel Institute
Analog Devices
EDN, a Cahners Publication
Fenwal Electronics, Division of Kidde, Inc.
G. M. Ferree
General Electric Company
Hamlin, Inc.
Hersey Products, Inc., Industrial Measurement Group
Hewlett-Packard
IEEE Spectrum
Instruments & Control Systems, Chilton Company
Intel Corporation
IRG Telemetry
Metritape, Inc.
Micro Switch, a Honeywell Division
Nanmac Corporation
National Aeronautics and Space Administration
National Semiconductor Corporation
C.A. Norgren Company
Omega Engineering, Inc.
Powell
RCA Solid State Division
Reliance Electric Company
Signetics Corporation
Simpson Electric Company
Taylor Instrument Company, Division of Sybron Corporation
United States Navy
Bob White

Introduction

Industrial electronics can be defined as the control of industrial machinery and processes through the use of electronic circuits and systems. Each of the topics in this text has been carefully chosen to help you, the future technician, survive in such an environment. We feel that knowledge gained by studying these topics will make you better prepared for entry-level employment as an electronics technician.

Although many topics have been included in this text, there are many additional topics you will need to know to function successfully in industry. As a starting point, we assume that your previous electronics courses have given you a firm grounding in alternating and direct current theory, the functions of electrical and electronic components, and mathematics through algebra and trigonometry.

The topics in this text are built around a very *general* process control system, since we believe the electronics technician should be a generalist. Furthermore, it is not possible to cover, in one text, all the circuits and systems you are likely to encounter in industry. The range of industrial applications is simply too broad. Therefore, we have concentrated on some basic circuit and component concepts with appropriate examples. Once you have mastered the basic circuit and component functions, you will be ready to put these parts together into a functional system. That is, your knowledge of the functions of subparts and subsystems will help you understand how the overall system functions.

An example of a very basic process control system is illustrated in Figure I.1. Each block in the diagram represents a division of the elements

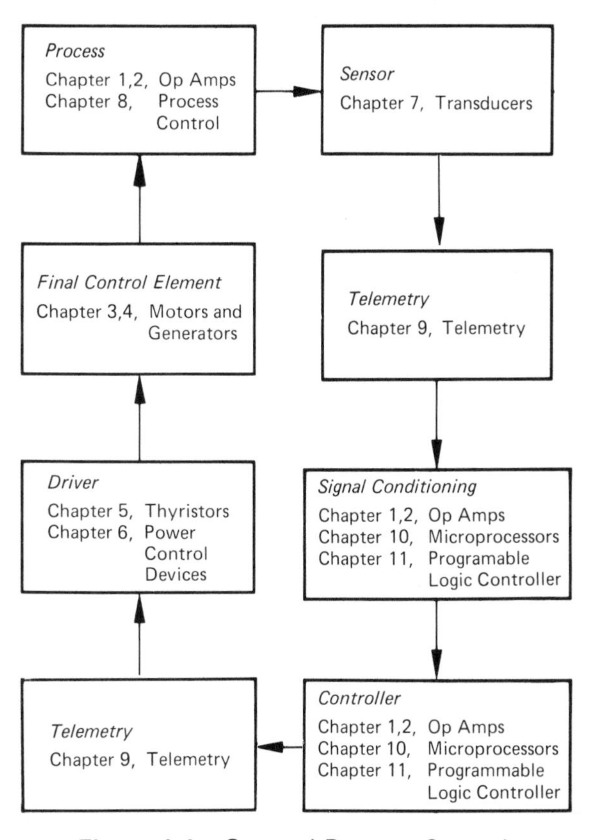

Figure I.1 General Process Control System

within a process control system. Note that it is sometimes difficult to physically separate one block or topic from another in a real system. And, of course, not all blocks will be used in every control system. All, however, are likely to be encountered in industry.

Each block in Figure I.1 contains the associated chapter in which that topic is discussed. The chapters by themselves may seem disconnected from each other, but they obviously are related when considering the complete system.

The process in Figure I.2 is a hypothetical example. It shows a conveyor belt and a motor driver that must keep a constant speed on the belt regardless of the load. This example is related to the system of Figure I.1 in the following manner: The process in Figure I.1 is the transportation of the load from one point to another at a constant speed. As the load on the belt changes, some type of speed sensor is used to detect changes in speed and convert this change into an electric output. If the controlling system is some distance from the belt, telemetering can be used to transport the information from the sensor to the controller.

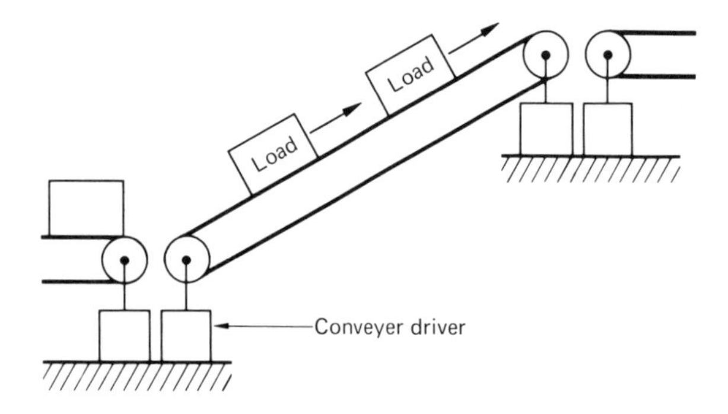

Figure I.2 Conveyer Belt System Used to Illustrate a Simple Process and Its Control

Before the controller can operate on this sensor output, it may require conditioning of some type to get the electric signal in the correct range or form. The controller can then make a decision, on the basis of the results of the incoming data, as to whether the belt should have more or less power than it had previously or the same power. This decision is then telemetered back to the belt drive circuitry, which will apply the appropriate amount of power, depending on the results of the controller's decision. The final control element, the motor, will then be provided with the appropriate power necessary to keep the belt operating at a constant speed.

This simple process illustrates the concept of process control and the basis for the inclusion of the topics in this text.

Operational Amplifiers

CHAPTER 1

Objectives

On completion of this chapter, you should be able to:

- Describe the characteristics of the ideal op amp, and identify which stage produces each characteristic;
- Define the following parameters and describe their effect on circuit performance: input offset voltage, input offset current, input bias current;
- Describe the parameters that affect the frequency response of the op amp;
- Calculate the gain, output voltage, and output current of given inverting and noninverting op amps;
- Compare the characteristics of inverting and noninverting op amps;
- Calculate the output voltage, given the schematic diagrams and input voltages, for the summing and averaging amplifiers;
- Outline the procedure for offset-nulling an inverting op amp.

Introduction

The *operational amplifier* (*op amp*) has a short but eventful history. The concept of the op amp dates back to the late 1940s, when they were used in analog computers. They performed the mathematical operations of addition, subtraction, integration, and differentiation; hence, the name *operational* amplifier. The original op amp was constructed mainly of vacuum tubes. Those early op amps consumed a lot of power and were costly and bulky.

The integrated circuit op amp as we know it today was developed in 1965 by Robert Widlar. Widlar was then working for Fairchild Semiconductor, which marketed the first integrated circuit op amp, the μA709. The 709 was so well designed that it is still in use today. In the intervening years, that one design has been further developed and expanded. Today, it is estimated that there are several thousand types of op amps. Furthermore, over one-third of all linear integrated circuits (ICs) are op amps. In fact, the op amp is one of the most popular active linear devices on the market today. The reason for its popularity is threefold. First, many different kinds of op amps are commercially available, and they are low in cost. Second, as we will see later, the op amp displays enormous versatility in application. Third, op amps are easy to use in designing and developing prototypes.

Just what is an op amp? We can describe an *op amp* very simply as a high-gain, direct-coupled amplifier. Characteristics such as circuit gain and frequency response are established by external components.

In this chapter, we will look at several important features of the op amp, such as their symbols, parameters, and configurations. We will discuss the ideal op amp, with a brief look at how the internal structure affects the performance of the device. After considering the ideal op amp, we will spend the remainder of the chapter dealing with practical op amp circuits and configurations. We will also introduce the two major classifications of op amp circuits: the inverting and noninverting op amp.

Ideal Op Amp

If asked to describe the perfect or ideal amplifier, what characteristics would we choose? For instance, what should the gain, frequency response, and input and output impedances be? If small signals are to be amplified, certainly we would want the voltage gain to be high, ideally infinite ($A_V = \infty$). (Note that we use A to symbolize gain, although G is sometimes used in other textbooks.) Again, if the input signal is very small, we do not want the amplifier to load down (reduce) the signal. So, ideally, the input impedance should be infinite ($Z_i = \infty$). What about frequency response? If an amplifier is to be versatile, it should amplify (ideally) any signal from 0 Hz (DC) to infinity (bandwidth $BW = \infty$). Finally, what is the ideal output impedance? There are basically two conditions we need to think about in choosing an output impedance: maximum transfer of power and maximum transfer of voltage. Since the op amp is not a power device, we do not need to worry about matching impedances for maximum power transfer. If the output impedance of the op amp is zero, we have maximum transfer of voltage, which is what we want. Also, if the output impedance is high, the gain of the amplifier is reduced. The output impedance forms a voltage divider with the load and feedback resistors. Ideally, then, the amplifier's output impedance should be zero ($Z_o = 0\ \Omega$).

We do not normally need to be concerned with the internal workings of the op amp. However, a brief consideration of the op amp's internal circuitry is helpful in understanding how the ideal op amp parameters are so nearly achieved. It also gives some insight into the limitations in testing and measuring op amp performance.

Inside the Op Amp

Basically, the op amp can be broken down internally into three stages: input, intermediate, and output.

Input Stage

The *input stage* gives the op amp its high Z_i and contributes to its high A_V. The basic component in this stage, and the heart of the op amp, is the *differential amplifier*, illustrated in Figure 1.1. Note

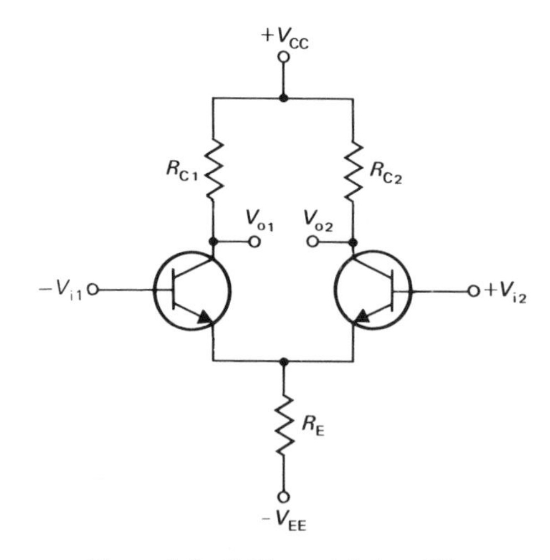

Figure 1.1 Differential Amplifier

that V_{i1} and V_{i2} are the two input voltages to the differential amplifier, and V_{o1} and V_{o2} are the two outputs. If V_{o2} is taken as a reference, then a signal applied at V_{i1} is inverted at the V_{o1} output terminal. This input terminal is usually identified with a minus sign. If a signal is applied at V_{i2}, the signal is unchanged in phase at V_{o1}. This input terminal is called the *noninverting terminal.* It is usually identified with a plus sign.

Recall that the differential amplifier exhibits a very high input impedance, mathematically approximated by the following equation:

$$Z_i \approx 2\beta r'_e \tag{1.1}$$

where β = current gain of transistor (h_{fe})

r'_e = AC resistance of base-emitter junction, approximately 25 mV/I_E

I_E = emitter current

Equation 1.1 is derived from considering the differential amp as a common emitter cascaded to a common base. The common emitter's input impedance is the current gain times r'_e. The common base's input (r'_e) is in series with r'_e from the common emitter. Since r'_e is relatively large owing to small emitter currents, Z_i approaches the range 1–2 MΩ.

Another interesting feature of the differential amplifier is its supply voltages. Note that we must have both a positive and a negative supply. This requirement will be especially important later when we discuss level shifters.

A useful property of the differential amplifier is its ability to reject common-mode signals. *Common-mode rejection* is defined as the amplifier's ability to reject identical signals when they occur on both inputs at the same time. Why is this property so important? Many industrial electronics applications exhibit high noise levels. This unwanted interference is fed equally to both the inputs of the op amp. Since the amplitude and phase of the noise is equal at any instant, the differential amp amplifies the common-mode signals equally. The difference voltage at the output terminals due to the common-mode signal is then zero. The desired signal, however, uses the difference-mode amplifying capabilities of the differential amp. Here, the input signal is the difference between the input terminals, and the amplifier amplifies this difference.

Well-designed differential amplifiers are rarely seen in the form shown in Figure 1.1. One reason is that R_E is typically a large-value resistor and is not well suited to IC fabrication techniques. Transistors generally replace large-value resistors in IC design. Thus, differential amplifiers are more often seen in the form shown in Figure 1.2.

What is the difference between Figure 1.1 and Figure 1.2? Notice that the resistor R_E in Figure 1.1 has been replaced by a *constant-current source,* that is, a transistor Q_3 with constant bias. This procedure improves common-mode rejection characteristics. Good common-mode rejection characteristics come about when the common-mode gain is low. Since the impedance of the constant-current source is high and is in the denominator of the gain equation, the common-mode gain is low.

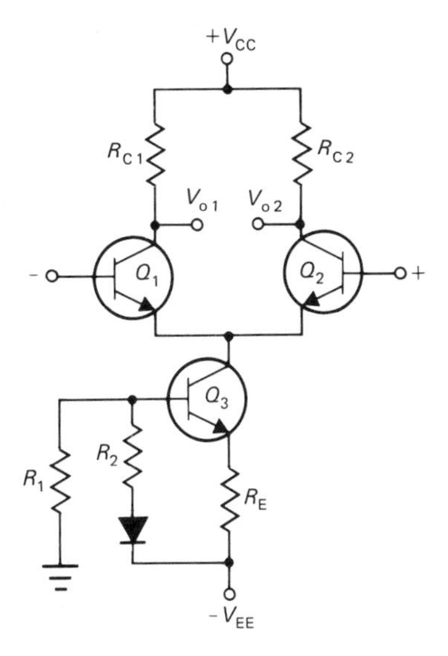

Figure 1.2 Differential Amplifier with Constant-Current Source

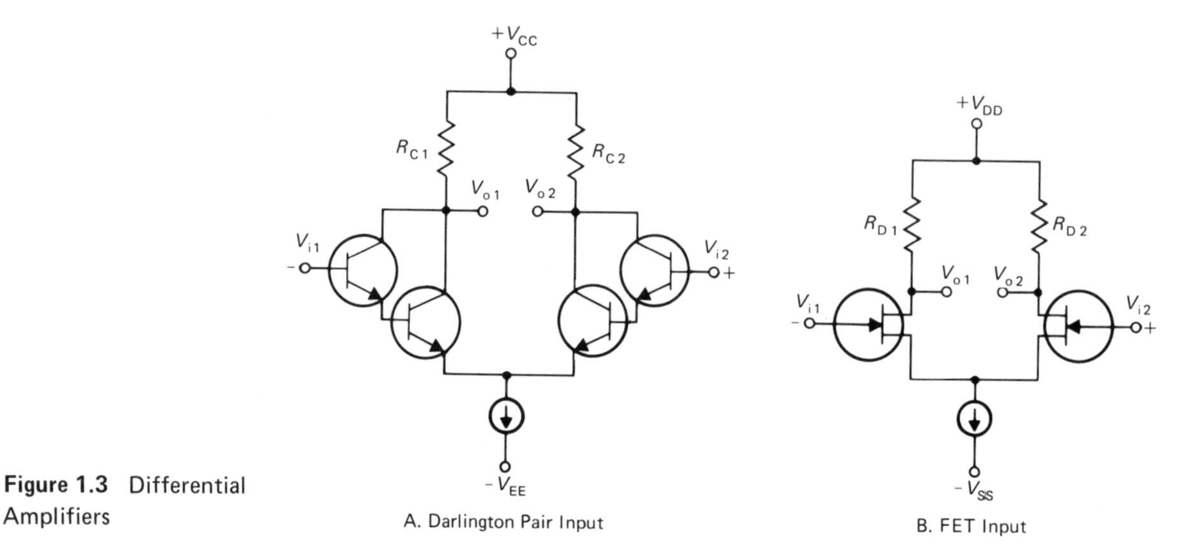

Figure 1.3 Differential Amplifiers

Sometimes, other components and other configurations are used to increase the input impedance of the differential amplifier. Two of the most common configurations are illustrated in Figure 1.3.

The circuit shown in Figure 1.3A has a *Darlington pair input.* The input impedance of the Darlington configuration is approximately $3\beta^2 r'_e$. This value increases the input impedance of the differential amplifier by a factor of 1.5β.

Note that in the circuit in Figure 1.3B, the input devices are field effect transistors (FETs), not bipolar junction transistors (BJTs). The advantage of this configuration is the high input impedance provided by the FETs.

Intermediate Stage

The *intermediate stage* serves two functions. First, it increases the current and voltage gain of the op amp. Second, it provides level shifting. These functions are described next.

A problem encountered in the differential amplifier stage is overcome here in the intermediate stages. The gain of the common emitter is inversely related to input impedance. As the designer seeks to raise input impedance by lowering emitter current, gain is also lowered. The intermediate stage provides the additional voltage gain needed to provide high overall amplifier gain.

Level shifting is a change in DC bias levels owing to direct-coupled cascaded amplifiers. Level shifting is necessary to combat the phenomenon of voltage buildup. This phenomenon occurs in any direct-coupled amplifier using cascaded common-emitter stages. Thus, each succeeding stage requires a higher collector supply to keep the collector at a higher potential than that of the base.

Look at Figure 1.2 again. If the inputs are grounded, there is no difference in potential between the inputs, and we expect 0 V out. We can see that if the bases of Q_1 and Q_2 are grounded, their collectors must be at a higher potential in order to prevent saturation. Thus, the output is not zero, as we desired. But suppose we couple the outputs of the circuit in Figure 1.2 to the inputs of the circuit of Figure 1.4A. In this situation, the collectors of Figure 1.4A are at a lower potential (and closer to 0 V).

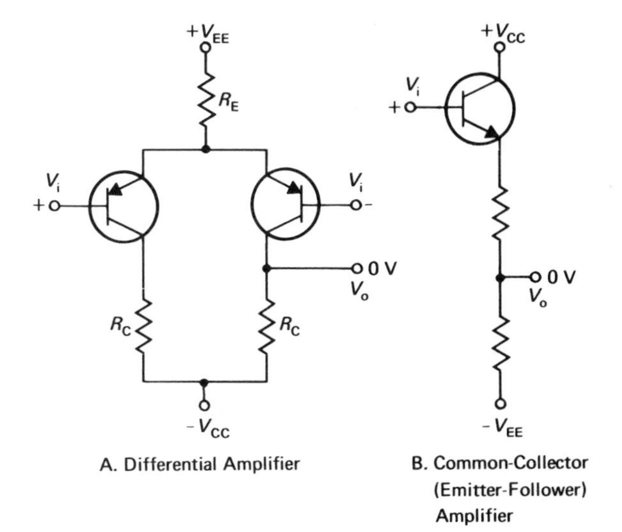

Figure 1.4 Level Shifting

Level shifting and additional voltage and current gain are provided in the circuit shown in Figure 1.4A. Another method for solving the voltage buildup problem is the circuit of Figure 1.4B. The common-collector amplifier in Figure 1.4B only gives level shifting and current gain.

Another circuit generally employed in the intermediate stage is the one shown in Figure 1.5. This stage contains an *internal compensating capacitor* C_{comp} or provides for an external capacitor connection. Among other things, as discussed later, the capacitor C_{comp} helps prevent high-frequency oscillations and instability.

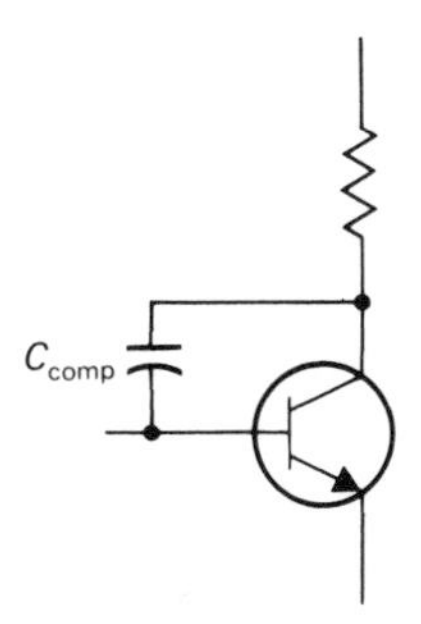

Figure 1.5 Intermediate-Stage Circuit with Compensating Capacitor C_{comp}

Output Stage

The *output stage* provides the power necessary to drive the load while allowing maximum output signal voltage swing and minimum output impedance. The simplest circuit that satisfies these requirements is the *emitter-follower.* A disadvantage of the emitter-follower is that its emitter resistor consumes too much power at high current levels. A more practical substitute for the emitter-follower is the simplified *complementary symmetry power amplifier* shown in Figure 1.6. This amplifier can supply large currents with good power gain and low output impedance.

Figure 1.7 shows a simplified but complete op amp schematic. See if you can identify the components that make up the three stages discussed in this section. Note that all resistor values are typical.

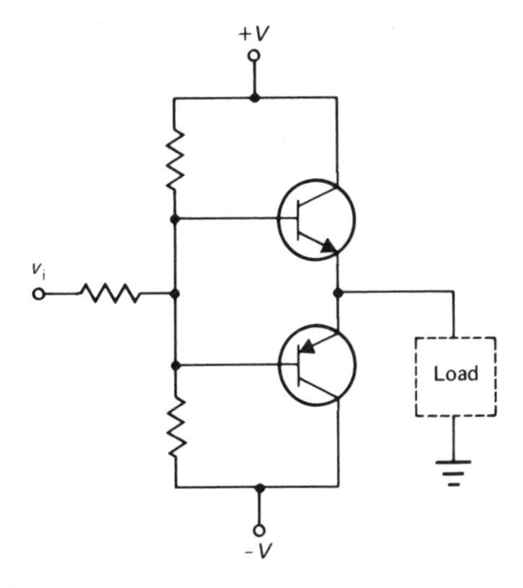

Figure 1.6 Output Stage of Complementary Symmetry Power Amplifier

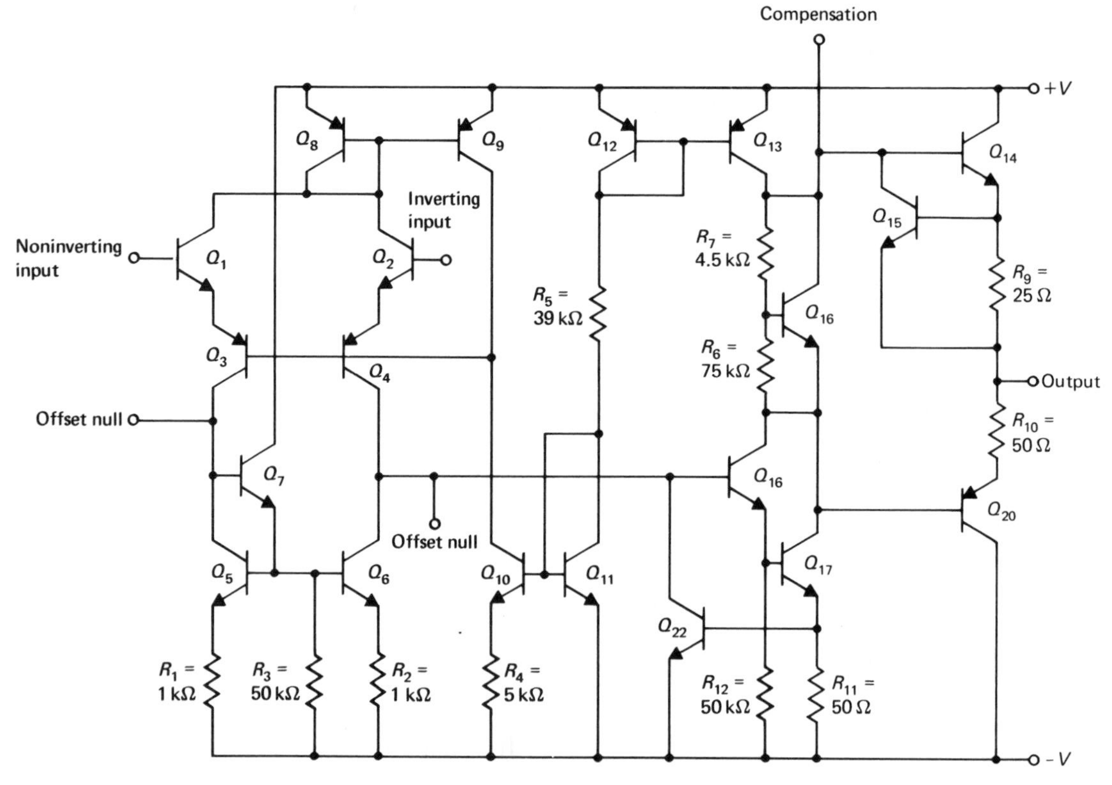

Figure 1.7 Integrated Circuit Op Amp Schematic with Provision for External Compensation

Op Amp Schematic Symbol

The op amp schematic symbol is shown in Figure 1.8. Terminals 2 and 3 are the differential input terminals of the device. Terminal 2 is the inverting terminal. Any signal connected to this terminal will experience a 180° phase shift at the output. Terminal 3 is the noninverting terminal. No phase inversion occurs between this input and the output.

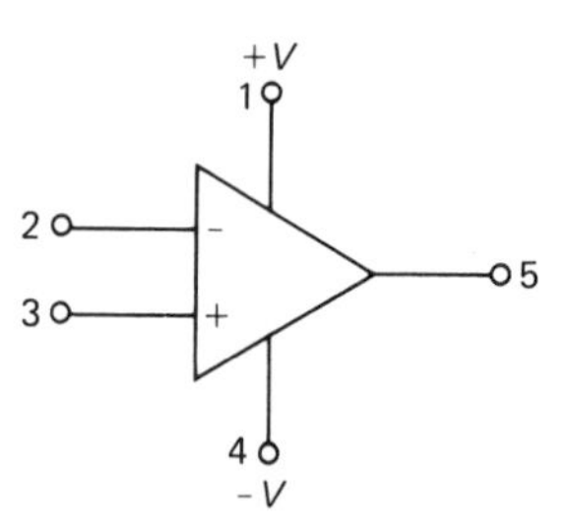

Figure 1.8 Op Amp Schematic Symbol

Terminals 1 and 4 are the power supply terminals. Normally, a positive voltage is connected to terminal 1 and a negative voltage to terminal 4. The op amp may have either one of these terminals grounded. In such a case, the terminal shows a ground schematic symbol attached to the proper terminal. Be cautious of operating the op amp with either power input grounded. This mode is not the normal mode of operation, and special consideration must be taken. For example, the output cannot go to 0 V, and the op amp can be destroyed if the input signal is of the incorrect polarity. Another word of caution: No matter what the power supply voltage is, within its specified range, the input signal should not exceed it. If no power supply connections are shown on the schematic, you can assume that positive and negative voltages are connected as shown.

Terminal 5 is the output terminal. Notice that the output is single-ended. In other words, it has one output terminal that carries the signal with respect to ground.

Op Amp Parameters

Technicians need to be familiar with op amp parameters. The op amp parameters are contained in the data sheets supplied by the device manufacturer. The specifications for the μA741 are contained in the Appendix. Here, we will discuss only the more frequently used op amp parameters.

Input Offset Voltage

In the ideal op amp, the output voltage is 0 V when the input voltage is 0 V. However, it is impossible to construct a perfectly balanced differential amplifier. Therefore, there is an output with no input voltage. The voltage needed at the input to adjust the amplifier for 0 V out is called the *input offset voltage* (V_{io}). It ranges in value from a few millivolts to microvolts. The popular 741 op amp's input offset voltage is typically 1 mV. In general, the lower the input offset voltage, the better the device is.

Input Offset Current

The ideal op amp has equal input currents when the output is 0 V. However, the input currents are not always equal. The difference between the input currents when the output voltage is 0 V is called the *input offset current* (I_{os}). Typically, the input offset current for the 741 is about 20 nA.

Input Bias Current

We noted above that the input currents are not necessarily equal. The *input bias current* is the average of the bias currents flowing into or out of both inputs. This parameter specifies what the approximate input current is. In general, the smaller the input bias current, the better the op amp is. Op amps that use FET front ends generally have much smaller input currents than those with BJT front ends. The 741 has a bias current of approximately 80 nA, while FET op amps have input bias currents that extend down into the picoampere range. Input bias current is symbolized as I_b.

Frequency Response Parameters

There are several parameters that involve the op amp's frequency response. Of all the ideal op amp characteristics we have discussed, frequency response does not approach the ideal very closely. It is important, therefore, that you have a thorough understanding of the op amp's frequency limitations.

Slew Rate

The slew rate limitation is probably the most important frequency limitation since it affects the amplifier's large-signal performance. Basically, the *slew rate parameter* indicates how well an amplifier follows a rapidly changing input signal. The slew rate (SR) limitation is defined mathematically as the change in output voltage (ΔV_o) over a change in time (Δt):

$$\text{SR} = \frac{\Delta V_o}{\Delta t} \qquad \textbf{(1.2)}$$

The slew rate parameter is measured in volts per microsecond. For example, the 741 has a typical slew rate of 0.5 V/μs. Thus, during 1 μs, the output voltage cannot change more than 0.5 V.

Slew rate limiting is caused by the presence of internal or external capacitances, the largest being the compensating capacitor. As we know, it takes a finite amount of time to charge and discharge a capacitor with a fixed amount of current available. The op amp has a number of internal constant-current sources that fix the amount of current the op amp can provide to charge the compensating capacitor. In the 741, this current is about 15 μA, and the compensating capacitor is about 30 pF. Thus, the slew rate is as follows:

$$\begin{aligned} SR &= \frac{\Delta V_o}{\Delta t} = \frac{I_{max} R_L}{R_L C_{comp}} = \frac{I_{max}}{C_{comp}} \\ &= \frac{15\ \mu A}{30\ pF} \\ &= 0.5\ V/\mu s \end{aligned} \quad \textbf{(1.3)}$$

where I_{max} = maximum charging current for C_{comp}

R_L = load resistance

We see from Equations 1.2 and 1.3 that there are several things that can be done to prevent slew rate limiting:

1. Decrease the input signal amplitude (which decreases the transistor's output current requirement).
2. Decrease the input signal frequency (which reduces the slope of the input and thus the output).
3. Increase the current the op amp provides to charge the compensating capacitor or decrease the size of the compensating capacitor.

The most difficult of these solutions is increasing the current the op amp provides. To do so, we must substitute another op amp.

You may be required to measure the slew rate characteristic of an op amp. Slew rate measurements require a function generator and an oscilloscope. A square wave input produces an output like that shown in Figure 1.9A if the frequency is high enough to cause slew rate limiting. Notice that the output signal has risen 5 V in 10 μs. The slew rate is the slope of the line from point *A* to point *B*.

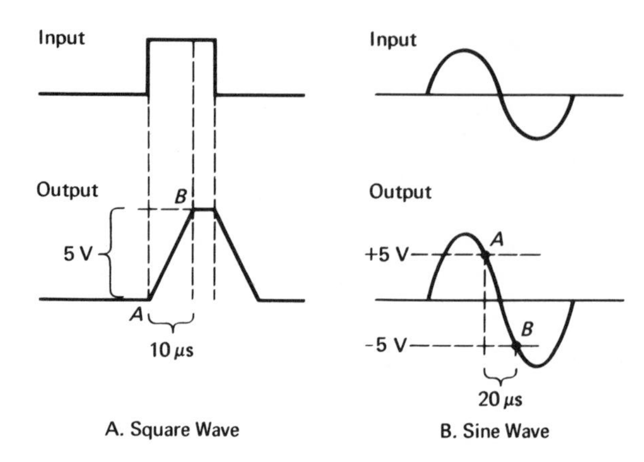

Figure 1.9 Waveforms for Slew Rate Limiting

Therefore, the slew rate is rise divided by run, or 5 V divided by 10 μs, which equals 0.5 V/μs.

Now, look at Figure 1.9B. Notice that the shape of the output waveform is triangular for a sine wave input. It must be a pronounced triangle wave before slew rate limiting can be measured. The slew rate, as before, is the slope of this distorted waveform. In this case, we again choose points *A* and *B* for our measurements. So, the calculation is rise divided by run, or 10 V divided by 20 μs, which equals 0.5 V/μs.

It should be obvious from the previous discussion that the two important factors in slew rate limiting are input frequency and amplitude. A useful equation is derived from an application of the calculus. The equation enables us to calculate the op amp's maximum frequency or amplitude limitations (for sine waves only). The equation is as follows:

$$f_{max} = \frac{SR}{2\pi \times V_{pk}} \quad \textbf{(1.4)}$$

where V_{pk} = peak output voltage of sine wave

f_{max} = maximum undistorted frequency, in hertz, before slewing occurs.

The term f_{max} is sometimes referred to as the *power bandwidth*.

Gain-Bandwidth Product

We have seen that the op amp's frequency response is limited in large-signal applications by the slew rate. In small-signal applications, the op amp's response again centers around circuit capacitances, of which the compensating capacitor is the largest.

Most amplifiers at high frequencies have a problem with feedback. As we know, an amplifier may be made to oscillate if enough positive feedback is applied. All amplifiers experience a significant phase shift between input and output at some high frequency. This result is especially true in multistage amplifiers. Suppose a sufficient number of stages contribute a portion of the phase shift. Then, eventually, some of the output signal will be fed back in phase with the input, and oscillations may occur.

For prevention of these oscillations, some op amps have a built-in capacitor that causes the gain to decrease at high frequencies. The phase shift may still be there. But the decrease in gain ensures that the amount of regenerative feedback will never be enough to sustain oscillations. If the op amp has an internal capacitor, the amplifier is said to be internally compensated. If the capacitor is added on externally, the op amp is said to be externally compensated.

Recall, from basic AC circuit theory, that the reactance X_C of a capacitor varies inversely with frequency f. The equation that expresses this relationship is as follows:

$$X_C = \frac{1}{2\pi f C} \quad \textbf{(1.5)}$$

As frequency increases, the capacitor's reactance decreases, and vice versa. If frequency is increased by a factor of 10 (a decade), then reactance decreases by a factor of 10. Since the compensating capacitor is in the collector load of the differential amplifier, the gain is as follows:

$$A_V = \frac{r_C \parallel X_{C(\text{comp})}}{r'_e} \quad \textbf{(1.6)}$$

where r_C = AC collector resistance
and $\parallel$ indicates a parallel circuit.

According to the equation, as X_C decreases by a factor of 10, the op amp gain also decreases by 10, or -20 dB, where $1\ \text{dB} = 20 \log_{10} V_o/V_i$. This decrease in gain as frequency increases is called *roll-off*. Roll-off is expressed in either of two ways: a number of decibels per decade or a number of decibels per octave. A decade is a tenfold change in frequency, while an octave is a twofold change. The change may be either an increase, expressed as a positive number, or a decrease, expressed as a negative number. The roll-off of a 741 is -20 dB per decade or -6 dB per octave. That is, as the frequency increases by 10, voltage gain decreases by 20 dB. This concept is graphically illustrated in Figure 1.10, which plots gain versus frequency response. This diagram is called a *Bode diagram*, which is a plot of gain and phase angle versus frequency. In Figure 1.10, only the gain-frequency relationship is shown.

From this graph, we see that frequency and gain are inversely related and by factors of 10. Notice that when the gain is 1 (0 dB), the frequency limit is 1 MHz. Data books often call this frequency the small-signal, *unity-gain frequency* (f_T). Sometimes, you will not be able to find this parameter in the data books. You may, instead, see a parameter called *transient-response rise time* (T_R).

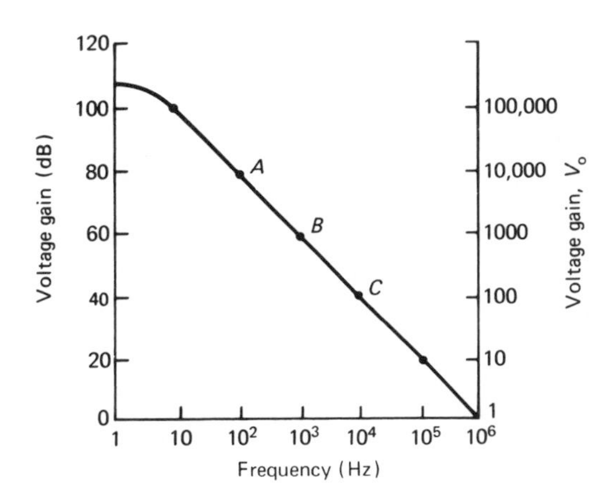

Figure 1.10 Bode Diagram of Frequency Response Characteristics of a Compensated Op Amp (μA 741)

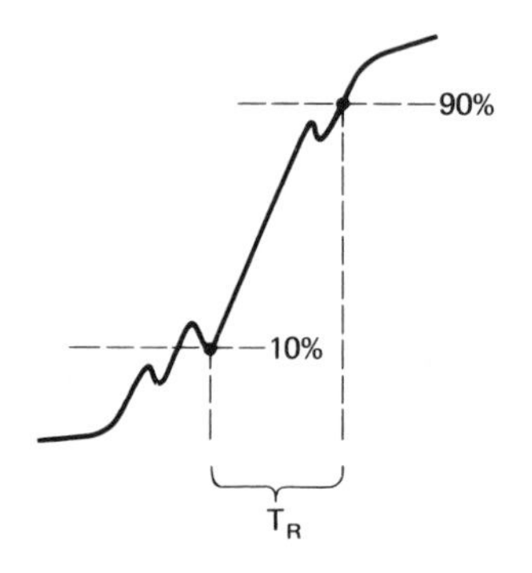

Figure 1.11 Rise Time Measurement

Rise time is defined as the time it takes a waveform to go from 10% to 90% of its final amplitude. Measurement of rise time is shown in Figure 1.11.

Unity-gain frequency bandwidth (*BW*) can be calculated from the rise time T_R by the following equation:

$$BW = \frac{0.35}{T_R} \quad \textbf{(1.7)}$$

A factor intimately related to frequency response is the *gain-bandwidth product* (GBP). The GBP is defined mathematically by the following equation:

$$\text{GBP} = BW \times A_V \quad \textbf{(1.8)}$$

where A_V = closed-loop voltage gain of circuit

This equation is used to find the bandwidth of an op amp at a specific gain. Suppose you are working with an amplifier whose gain is 100 and whose GBP is 1 MHz. What is the bandwidth of this op amp? The bandwidth is calculated as follows:

$$BW = \frac{\text{GBP}}{A_V} = \frac{1\ \text{MHz}}{100} = 10\ \text{kHz} \quad \textbf{(1.9)}$$

One interesting observation about the GBP is that it is always constant. From the graph in Figure 1.10, we see that as gain goes up by 10, frequency goes down by the same amount. Thus, the product is always the same. In practice, it is difficult to measure the GBP for low-gain settings. This limitation is due, in part, to the small voltages and currents involved.

For purposes of comparison, Table 1.1 lists some common parameters for several common types of op amps.

Inverting Op Amp

Op amps are used in two basic configurations: the inverting and the noninverting amplifier. An *inverting op amp* is one that takes a signal at the input and phase-shifts it 180° at the output. The *noninverting op amp* does not phase-shift the signal from the input to the output. Almost all the circuits we discuss later are built on these foundational circuits. In this section, we discuss the inverting op amp. Later in the chapter, we consider the noninverting op amp.

Table 1.1 *Parameters of Various Op Amps*

	Op Amp					
Parameters	*709*	*201*	*301*	*741*	*791*	*Ideal*
Input bias current (nA)	1500	500	250	500	500	0
Input offset voltage (mV)	7.5	7.5	7.5	6.0	6.0	0
Input offset current (nA)	500	200	50	200	200	0
GBP (MHz)	1.0	1	1	1	0.2	∞
Slew rate (V/μs)	3.0	2.0	2.0	0.5	6.0	∞
Input impedance (MΩ)	0.7	4.0	2.0	2.0	2.0	∞

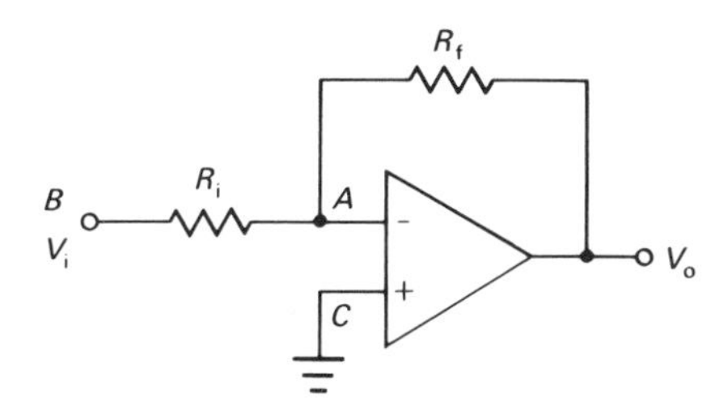

Figure 1.12 Basic Inverting Op Amp Circuit

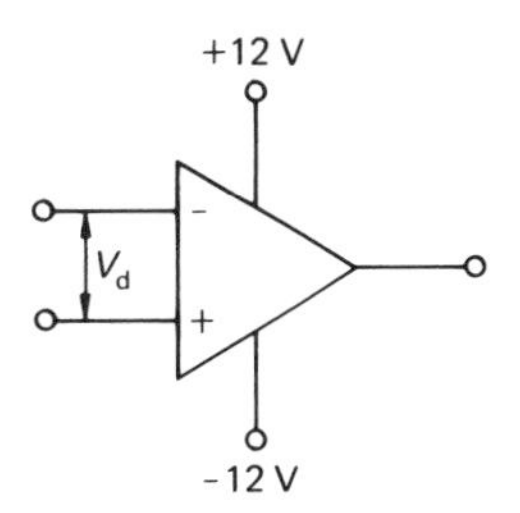

Figure 1.13 Op Amp Circuit with No Feedback

Figure 1.12 shows an op amp connected as an inverting amplifier. Notice that a *feedback resistor* R_F connects the output to the inverting terminal. This connection provides negative, or degenerative, feedback. We would therefore expect this amplifier to have a gain significantly less than the op amp's open-loop gain. *Open-loop gain* is the gain of the op amp when there is no feedback path. For the ideal op amp, the open-loop gain is infinity. However, the practical op amp's gain is somewhat less than infinity. For example, the 741 has an open-loop gain of about 200,000.

Also, notice in Figure 1.12 that the noninverting terminal (C) is grounded and that R_i connects the input (B) to the inverting terminal (A).

Virtual Ground

The concept of the virtual ground is an important one in understanding the behavior of inverting op amps.

Figure 1.13 shows a basic op amp with no feedback. Now, let us suppose that we have a power supply of ± 12 V powering the op amp. Let us also suppose that the maximum output voltage (saturation voltage) can go no higher than $+10$ V and no lower than -10 V. In this case, the output voltage of $+10$ or -10 V represents output saturation voltage, V_{sat}. *Output saturation voltage* occurs when the value of the output voltage of the op amp can get no closer to the value of its supply voltage because a portion of the supply voltage is dropped across the saturated output transistors.

With a gain of 200,000, what is the input voltage that gives an output of 10 V? The voltage gain equation is as follows:

$$A_V = \frac{V_o}{V_i} \tag{1.10}$$

Rearranging this equation, we get the following:

$$V_i = \frac{V_o}{A_V} = \frac{10\text{ V}}{200{,}000} = 50\ \mu\text{V} \tag{1.11}$$

We see, then, that the maximum difference (without signal distortion) in potential (V_d) between the inverting and noninverting terminals is 50 μV. This important fact brings us to the first rule in understanding op amps.

Rule 1.1
The difference in potential between the two input terminals of an op amp is approximately zero.

Since the input voltage is almost always significantly higher than V_d, we can assume that V_d is approximately equal to zero.

Now, if the noninverting terminal is grounded, what is the approximate difference in potential between the inverting terminal and ground? Look at the circuit in Figure 1.14. The noninverting terminal is at ground potential, and there is essentially no difference in potential between the input terminals. Thus, the inverting input can be considered to be 0 V. Or we can say it is *virtually grounded*, from which comes the name virtual ground.

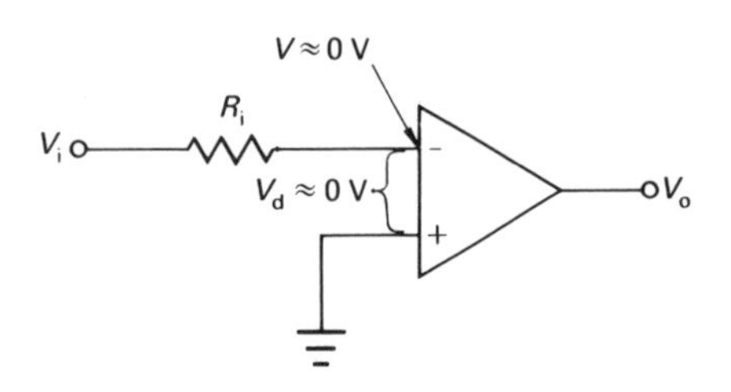

Figure 1.14 Op Amp with Noninverting Terminal Grounded

Input Current

The second rule we need for analyzing the inverting amplifier concerns the input current. The input impedance of a practical op amp is very high; for instance, the 741 input impedance is about 2 MΩ. We then find the maximum input current by using the 50 μV input voltage V_i, the input impedance Z_i, and Ohm's law:

$$I_i = \frac{V_i}{Z_i} = \frac{50\ \mu\text{V}}{2\ \text{M}\Omega} = 25\ \text{pA} \tag{1.12}$$

Notice how small the input current is. This result brings us to our second rule.

Rule 1.2
Essentially no current flows into or out of the input terminals of the op amp.

Since the input current is in the order of 10^{-12} A, we can safely neglect it.

Analyzing the Inverting Amplifier

Armed with these two rules, we can develop equations to help us understand this amplifier. Refer to the schematic in Figure 1.12. If the potential at point A is 0 V, we can say that all of V_i is dropped across R_i. Using Ohm's law, we can then find the current flowing through R_i:

$$I_i = \frac{V_i}{R_i} \tag{1.13}$$

Since no appreciable current flows into or from the inverting terminal, all the current flowing through R_i also flows through R_F. So, we have the following equation:

$$I_i = I_F \tag{1.14}$$

Now, also because of the virtual ground, V_o is dropped across R_F. From Ohm's law, V_o can be expressed as follows:

$$V_o = -I_F R_F \tag{1.15}$$

where I_F = DC feedback current

And from Equation 1.13, V_i is as follows:

$$V_i = I_i R_i \tag{1.16}$$

Therefore, we have the following relationship:

$$\frac{V_o}{V_i} = -\frac{R_F}{R_i} = A_V \tag{1.17}$$

We now have an equation that defines the gain of an inverting op amp circuit. Note that, due to the op amp's high open-loop gain, the circuit gain is entirely determined by components external to the op amp. The negative sign indicates that a 180° phase shift occurs between input and output. Or in the case of a DC input, a change in polarity between input and output occurs. The polarities shown in Figure 1.12 support this statement.

Input Impedance

One interesting feature of the inverting op amp configuration is its input impedance. Look at the op amp circuit in Figure 1.12 again. If point A is always at 0 V, what is the input impedance at point B? Obviously, if all the input voltage is dropped across R_i, the input impedance at point B is simply the resistance of R_i. Within limits, then, the input impedance of this inverting amplifier is changed by changing R_i. It is recommended that R_i be no less than 1 kΩ in order to limit the possible current drain on the op amp.

Output Current

As a technician, you may be called upon to evaluate an op amp's output current. We will calculate the output current for the op amp circuit in Figure 1.15. Notice that the current (I_o) flowing in the output terminal is a combination of the currents flowing through R_F (I_F) and R_L (I_L). Thus, we can equate the output current to the sum of those currents:

$$I_o = I_F + I_L \tag{1.18}$$

Both I_F and I_L can be easily calculated.

Suppose we are using the circuit in Figure 1.15. All the input voltage (5 V) is dropped across R_i. The current through R_i, then, is as follows:

$$I_i = \frac{V_i}{R_i} = \frac{5\text{ V}}{10\text{ k}\Omega} = 0.5\text{ mA} \tag{1.19}$$

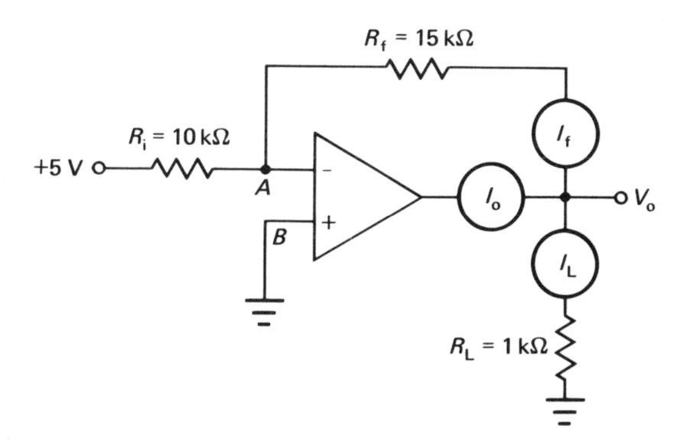

Figure 1.15 Circuit for Output Current Calculation

We showed previously (Equation 1.14) that $I_i = I_F$. So, $I_F = 0.5$ mA. Now, V_F, the voltage drop across R_F, is as follows:

$$V_F = I_F \times R_F = 0.5\text{ mA} \times 15\text{ k}\Omega = 7.5\text{ V} \tag{1.20}$$

Notice that since A and B are at the same potential (ground), 7.5 V must be dropped across R_L as well as R_F. Therefore, we can find I_L as follows:

$$I_L = \frac{V_L}{R_L} = \frac{7.5\text{ V}}{1\text{ k}\Omega} = 7.5\text{ mA} \tag{1.21}$$

The total output current is, then, as follows:

$$I_o = I_L + I_F = 7.5\text{ mA} + 0.5\text{ mA} = 8\text{ mA} \tag{1.22}$$

The maximum output current that an op amp can deliver to a load is found in the data sheets. In no case should the output current demanded by the circuit and the load exceed the maximum allowable output current. Some op amps, including the 741, are protected from excess currents by circuitry that limits the output current to a maximum value. Some op amps, though, are not so protected. If the maximum allowable output current is exceeded, the gain equation (1.17) is no longer valid, since the op amp is operating outside its normal range. If the load requirements exceed the op amp's capabilities (25 mA for the 741), a transistor circuit can be added to the op amp output to boost its current output.

AC Input Signals

Thus far, we have only considered DC input voltages in our examples. The inverting op amp is not limited to DC applications, though. Let us now consider an AC input to the op amp.

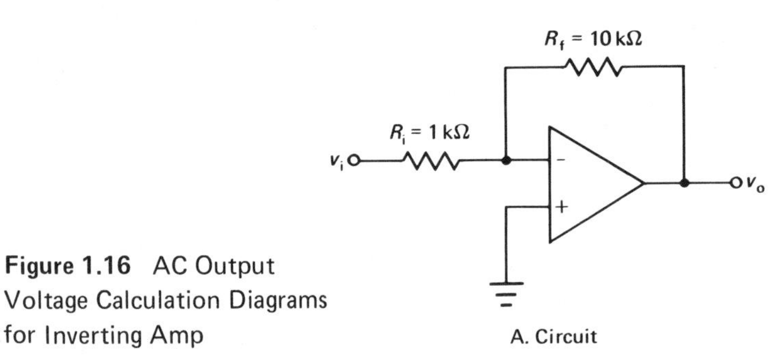

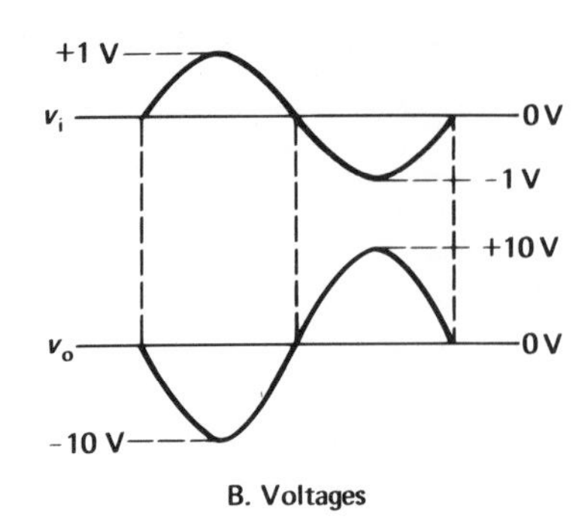

Figure 1.16 AC Output Voltage Calculation Diagrams for Inverting Amp

The op amp circuit in Figure 1.16A has a gain of -10 with an input signal of ± 1 V peak. By looking at the output in Figure 1.16B, we see that the signal has been amplified by a factor of 10, with 180° phase inversion.

DC and AC Input Signals

The op amp circuits you will encounter may have AC signals riding on DC levels. Let's see how the op amp treats such a signal. Figure 1.17A depicts an inverting op amp; the input signal is shown in Figure 1.17B. Notice that the input signal is shifted in phase at the output by 180°. The input DC voltage level is translated from $+3$ V to -6 V.

Algebraically, this change can be calculated by knowing that the gain is -2. We use the following equations:

$$A_V = -2 \qquad \text{and} \qquad A_V = \frac{V_o}{V_i} \tag{1.23}$$

So, if V_i is $+3$ V, then V_o is $+3$ V multiplied by -2. The output DC voltage is then -6 V, as shown in Figure 1.17B. By the same token, the AC output signal is calculated by knowing that the gain is -2 and the input signal is ± 2 V peak.

Summing Op Amp

The *summing op amp* is a special application of the inverting amplifier. The summing op amp performs the mathematical operation of addition. It can be

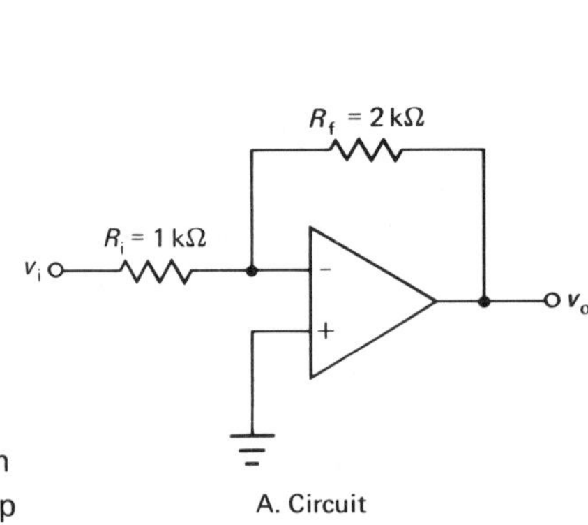

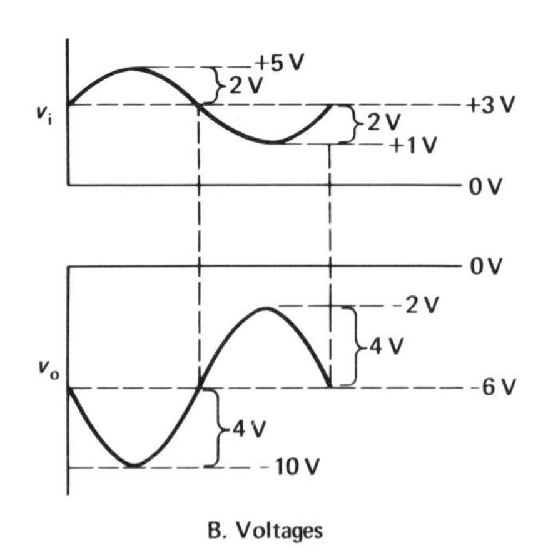

Figure 1.17 AC and DC Output Voltage Calculation Diagrams for Inverting Amp

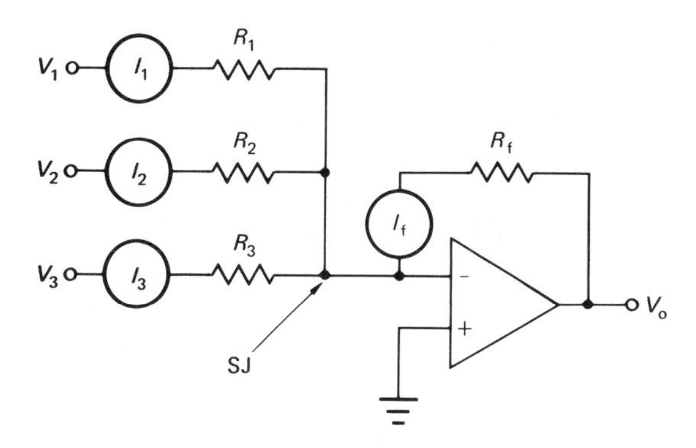

Figure 1.18 Inverting Summing Amp

considered as an inverting op amp with several inputs. A diagram of the summing amp configuration is shown in Figure 1.18.

An important point to notice in this circuit is that each applied voltage is dropped across its respective resistor. Thus, V_1 is dropped across R_1, V_2 across R_2, and V_3 across R_3. Why? Because the voltage at the summing junction (SJ) is 0 V due to the virtual ground. Note also that the currents join at SJ to form the feedback current. (Recall that no current flows into the inverting terminal). So, we have the following equation:

$$I_1 + I_2 + I_3 = I_F \quad \textbf{(1.24)}$$

Also:

$$I_F = -\frac{V_o}{R_F} \quad \textbf{(1.25)}$$

Substituting voltage and resistance for current, we have the following equation:

$$\frac{V_1}{R_1} + \frac{V_2}{R_2} + \frac{V_3}{R_3} = = -\frac{V_o}{R_F} \quad \textbf{(1.26)}$$

Solving for V_o, we obtain the next equation:

$$V_o = -R_F\left(\frac{V_1}{R_1} + \frac{V_2}{R_2} + \frac{V_3}{R_3}\right) \quad \textbf{(1.27)}$$

This equation is the general equation used to find the output voltage of a summing amp. This equation reduces to the following when all resistors are equal, that is, $R_F = R_1 = R_2 = R_3$:

$$V_o = -(V_1 + V_2 + V_3) \quad \textbf{(1.28)}$$

Let us suppose in Figure 1.18 that R_1, R_2, R_3, and R_F are 10, 20, 30, and 60 kΩ, respectively. Further, let V_1, V_2, and V_3 equal +0.1, +0.2, and +0.3 V, respectively. The output voltage is found by using the general equation we developed. Substituting the values above yields the following result:

$$\begin{aligned} V_o &= -60\ \text{k}\Omega\left(\frac{0.1\ \text{V}}{10\ \text{k}\Omega} + \frac{0.2\ \text{V}}{20\ \text{k}\Omega} + \frac{0.3\ \text{V}}{30\ \text{k}\Omega}\right) \\ &= -1.8\ \text{V} \end{aligned} \quad \textbf{(1.29)}$$

The summing amp is used for AC applications as well as DC applications. In fact, the audio microphone mixer shown in Figure 1.19 is an excellent and useful AC application. The advantage of this circuit arises from the fact that each input voltage is dropped across its own input resistor. Therefore, there is no interaction between microphones. Also, provision is made to vary the gain of each input individually (R_1, R_2, R_3) or of all channels at the same time (R_F).

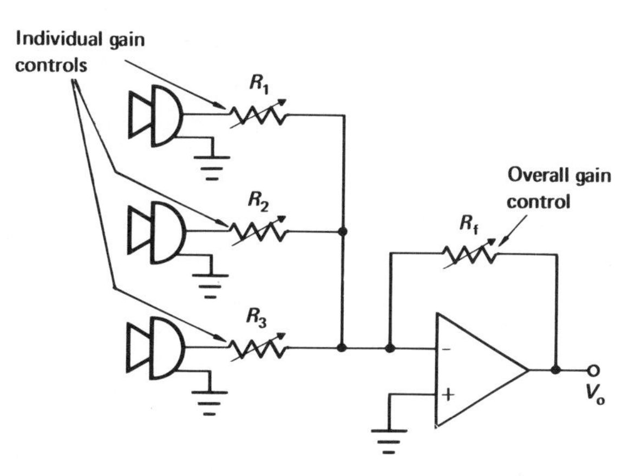

Figure 1.19 Summing Amp Used as Microphone Mixer

Averaging Op Amp

The *averaging op amp* is a special case of the summing amp. Consider Figure 1.18 again. If all the input resistors are of equal value, and R equals the resistance of one input resistor divided by the number of inputs (N), then this circuit arithmetically averages the input voltages. Suppose each input resistor is 30 kΩ. To average the input voltages, the feedback resistor would have to be 10 kΩ. With V_1, V_2, and V_3 of +3, +5, and +2 V, respectively, we can calculate the output voltage by using the general equation:

$$V_o = -10\ \text{k}\Omega\left(\frac{3\ \text{V}}{30\ \text{k}\Omega} + \frac{4\ \text{V}}{30\ \text{k}\Omega} + \frac{2\ \text{V}}{30\ \text{k}\Omega}\right)$$
$$= -3\ \text{V} \qquad \textbf{(1.30)}$$

You can see from this example that this circuit does indeed average voltages. Like the summing amp, it can also be used to average AC input signals. You may also have noticed that the output is inverted. This problem can easily be solved by using another inverting amplifier with a gain of 1 in series with the averaging amp.

Noninverting Op Amp

Up to this time, we have concentrated on inverting op amps. As stated previously, the noninverting amp makes up the second major division of op amp circuits.

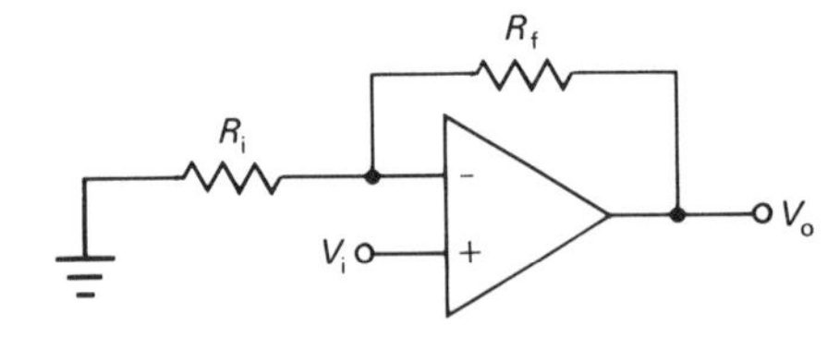

Figure 1.20 Basic Noninverting Op Amp Circuit

Figure 1.20 illustrates the basic noninverting op amp configuration. Note the difference between this amplifier and the inverting amplifier of Figure 1.12. In the noninverting amplifier, the noninverting terminal is connected to the input rather than ground, while R_i is grounded. Going back to op amp Rule 1.1, we can state that V_i appears across R_i because the voltage difference between the inverting and noninverting terminals is essentially zero. Therefore, we have the following equations:

$$I_i = \frac{V_i}{R_i} \quad \text{and} \quad V_F = I_i R_F \qquad \textbf{(1.31)}$$

Substituting, we get the following:

$$V_F = \frac{R_F}{R_i} V_i \qquad \textbf{(1.32)}$$

Looking at the circuit in Figure 1.20, we also see that the output voltage (V_o) is equal to V_F added to V_i: $V_o = V_F + V_i$. Substituting the known factors into this equation gives us the following equation:

$$V_o = \frac{R_F}{R_i} V_i + V_i \qquad \textbf{(1.33)}$$

Factoring this equation, we get the following:

$$V_o = \left(\frac{R_F}{R_i} + 1\right) V_i \qquad \textbf{(1.34)}$$

Rearranging, we arrive at the gain equation:

$$A_V = \frac{V_o}{V_i} = \frac{R_F}{R_i} + 1 \qquad \textbf{(1.35)}$$

From this derivation, we see that the gain equation (1.35) for the noninverting amp differs from the gain equation (1.17) for the inverting amp in two ways: (1) There is no negative sign and therefore no phase inversion, and (2) the gain is never less than unity.

There are other important differences. Recall that the input impedance of the inverting amplifier equalled the input resistor. As we can see in the

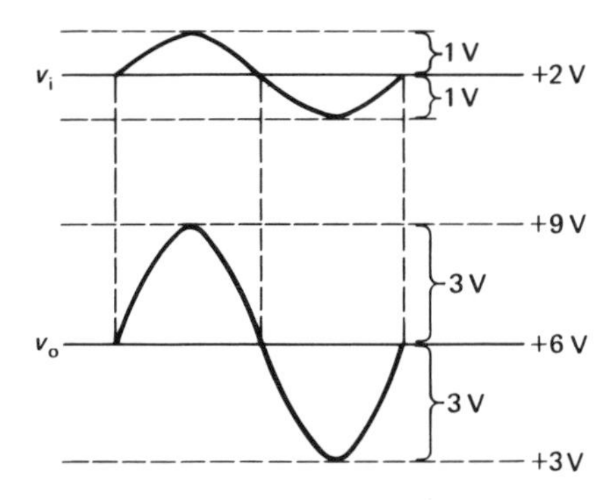

Figure 1.21 AC and DC Input Voltage to Noninverting Amp

noninverting amplifier, the input impedance is that of the device itself—that is, very high. The input impedance of the noninverting amp does not depend on external circuit characteristics.

Let us apply an input to the noninverting amp, as shown in Figure 1.21. The input signal is a 1 V peak sine wave riding on a +2 V DC level, with an input resistor of 10 kΩ and a feedback resistor of 20 kΩ. The gain of the amplifier is as follows:

$$A_V = \frac{R_F}{R_i} + 1 = \frac{20\text{ k}\Omega}{10\text{ k}\Omega} + 1$$
$$= 3 \qquad \textbf{(1.36)}$$

Therefore, the +2 V DC level becomes +6 V, while the 1 V peak sine wave becomes 3 V peak.

Output current is calculated in the same way as for the inverting amplifier.

Offset Nulling

In the ideal amplifier, when the input (or inputs) is grounded, the output is 0 V. This result is not true for the practical amplifier owing to differential amplifier imbalances. Normally, such imbalances produce outputs that can be neglected. However, if the amplifier gain is high, or if it is necessary for the output to approach 0 V as closely as possible, offset nulling may be in order. *Offset nulling* is the procedure by which the output of the op amp circuit is adjusted for 0 V with 0 V input.

The best way to null an op amp circuit uses the built-in circuit provided by the manufacturer for that purpose. The 741 nulling circuit is shown in Figure 1.22.

The nulling procedure for the 741 is as follows:

1. Ground the input (V_i, not pin 2 of the op amp).
2. Place a voltmeter from the output to ground.
3. Adjust the offset, or adjust until 0 V is reached at the output.
4. Unground the input—do not move the offset-adjust potentiometer once it is set.

Note that the amplifier is nulled with feedback. It is virtually impossible to null the amplifier without DC feedback because of capacitance charging associated with high open-loop gain.

If an op amp does not have provisions for nulling, external circuits may be used. Examples of these circuit configurations are shown in Figures 1.23A and 1.23B. In each case, the nulling procedure is the same as previously stated.

Another configuration you may see associated with an inverting amp is illustrated in Figure 1.24. Note the resistor R_b in series with the noninverting terminal. The resistor R_b compensates for the error caused by bias currents of the op amp's input differential amplifier. For best results, the value

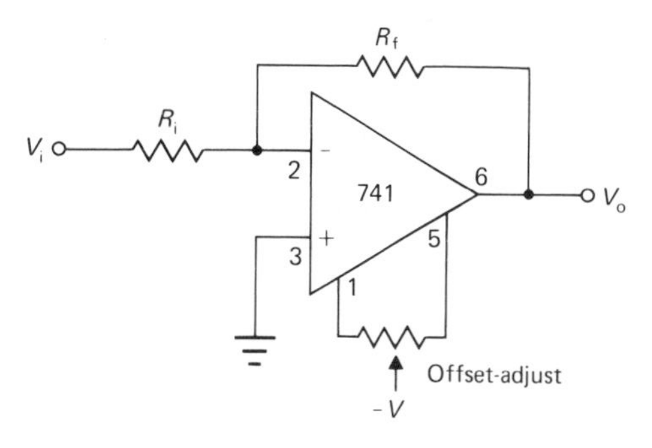

Figure 1.22 Offset-Nulling Circuit for 741 Op Amp

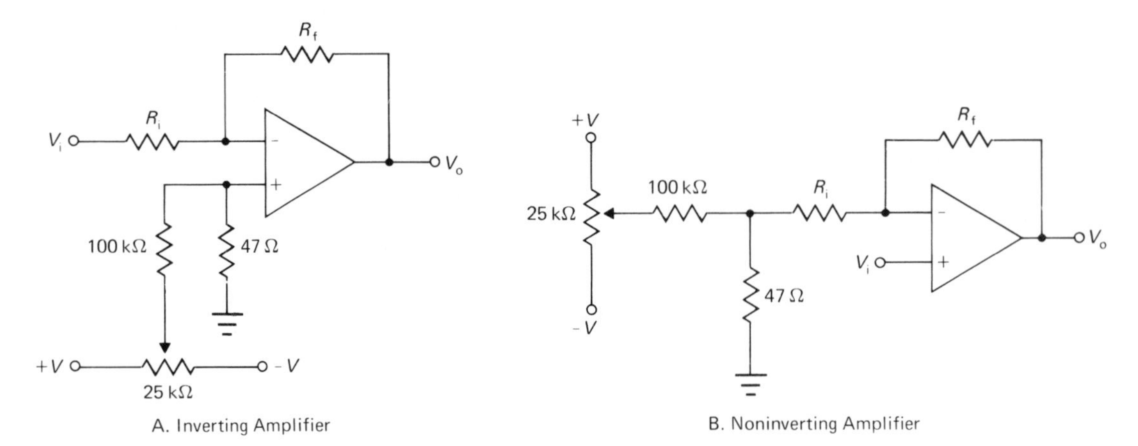

Figure 1.23 Offset-Nulling Circuits

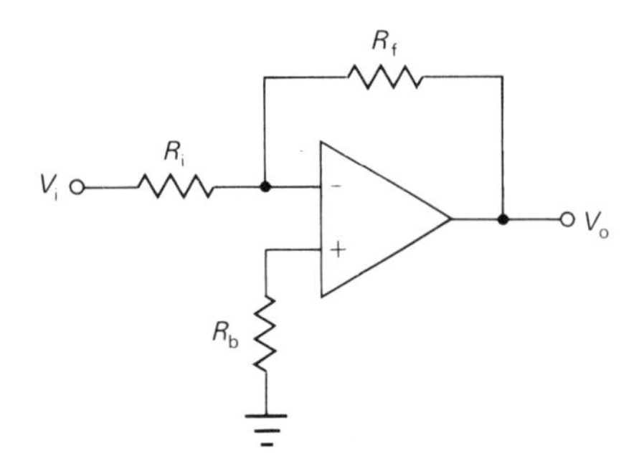

Figure 1.24 Bias Current Compensation

usually selected for R_b is as follows:

$$R_b = R_i \parallel R_F \tag{1.37}$$

(The $\parallel$ symbol indicates that the resistors are electrically connected in parallel.)

Conclusion

We have considered the two basic op amp configurations, the inverting and noninverting amplifier. If you understand these two basic op amp circuits, you should have no trouble understanding other op amp circuits. Most of them use one of these two fundamental configurations.

As we have seen, the op amp is a circuit building block with enormous potential. In the next chapter, we discuss some of the applications of these two configurations.

Questions

1. The op amp is defined as a ______-gain, direct-coupled amplifier. The op amp received its name from its ability to perform the ______ of addition, subtraction, multiplication, and division, among others.
2. An ideal op amp has ______ gain, input impedance, and bandwidth, while the output impedance is ______ Ω.
3. If the ideal op amp has 0 V input, the output should be ______ V. Practically, the output of an op amp is never 0 V out with an input of 0 V. The voltage needed at the input to produce 0 V output is called the input ______ voltage.
4. Of the two types of frequency limitations discussed, ______ ______ limiting causes distortion.
5. The compensating capacitor prevents high-frequency ______.
6. The rules for analyzing op amp operation are that, first, no potential difference exists between the

_____ terminals and, second, no current flows into or out of the _____ terminals of the op amp.

7. The two basic configurations of op amp circuits are _____ and _____ amplifiers.
8. Describe the ideal op amp in terms of input and output impedance, voltage gain, and bandwidth characteristics.
9. List the three basic stages of the op amp, and explain briefly how the ideal parameters relate to each of the three stages.
10. Explain common-mode rejection and how it is achieved in the op amp.
11. Define, in your own words, the following op amp parameters: (a) input offset voltage and current, (b) input bias current, (c) slew rate, and (d) gain-bandwidth product.
12. Describe the concept of virtual ground.
13. Write, in your own words, the two rules for analyzing inverting amplifiers.
14. What is the input impedance of the inverting amp? Of the noninverting amp? Why?
15. Describe the operation of the summing amplifier circuit.
16. Why is there a need for offset nulling? Describe the procedure by which you would offset-null an op amp.
17. Describe the purpose of feedback in an op amp. Why is it necessary?
18. Explain what compensation is in the op amp. Distinguish between internal and external compensation.
19. List and describe the five basic terminals of an op amp.
20. Describe the difference between the summing amp and the averaging amp.

Problems

1. A step function (a suddenly applied DC voltage) is applied to an op amp. The output voltage waveform exhibits slew rate limiting with a slope of 10 V in 0.5 μs. What is the slew rate of the op amp?
2. What is the highest-frequency sine wave that can be applied to an op amp with a slew rate of 0.5 V/μs with a 10 V peak input signal?
3. An op amp has a gain-bandwidth product of 2.5 MHz. The op amp is used in a circuit with a gain of 10. What is the bandwidth of this circuit?
4. Suppose the circuit of Figure 1.15 has the following resistance values: $R_F = 20$ kΩ, $R_i = 5$ kΩ, and $R_L = 1$ kΩ.
   a. Calculate V_o, I_F, and I_L with a $+2$ V DC input voltage.
   b. Draw the output voltage waveform with a 2 V peak input signal riding on a $+1$ V DC reference.
5. Suppose the circuit in Figure 1.18 has the following resistance values: $R_1 = 20$ kΩ, $R_2 = 40$ kΩ, $R_3 = 10$ kΩ, and $R_F = 30$ kΩ. Calculate the DC output voltages with the following inputs: $V_1 = -2$ V, $V_2 = +1$ V, and $V_3 = +3$ V.
6. Describe the resistance values necessary to make the circuit shown in Figure 1.18 an averaging op amp.
7. Suppose the circuit in Figure 1.20 has $R_F = 15$ kΩ, $R_i = 5$ kΩ, and $V_i = -2$ V. Calculate (a) A_V, (b) V_o, and (c) I_o.
8. A 741 op amp has supply voltages equal to ± 15 V. When the op amp is saturated, what will the output voltage be (approximately)? Why?

Applications of Op Amps

CHAPTER

2

Objectives

On completion of this chapter, you should be able to:

- Describe the characteristics and applications of the voltage follower;
- Compare the integrator and differentiator, and indicate how the ideal circuits are made practical;
- Calculate the output voltage of a log amp with a given input;
- List the advantages of the active op amp filter over the passive devices;
- List and explain the Barkhausen criteria;
- Describe the operation of the comparator with and without hysteresis;
- Differentiate among the phase-shift, Wien-bridge, and twin T oscillators;
- Compare the op amp and the CDA;
- Describe the operation of the CDA, and draw the schematic diagram for the CDA;
- Describe the operation of the OTA, and draw the OTA schematic diagram;
- Describe an application for the op amp, current-to-voltage converter, and draw the schematic diagram of that application;
- List the general procedures for troubleshooting op amp circuits.

Introduction

In this chapter, we consider applications of the two basic op amp configurations discussed in Chapter 1. Several of these circuits perform higher-level mathematical functions. We will examine how the op amp can be used to subtract, integrate, differentiate, and generate logarithms. All these circuits are used in analog computers.

A common application of op amps in audio circuits is the filter. We will see in this chapter how op amp filters are superior to other types at certain frequencies. We will also consider the op amp oscillator. Other devices that perform similar functions but are not strictly considered op amps are also included in our presentation.

The versatility of the op amp is seen in the variety of its applications. It is used in signal interfacing, signal conditioning, impedance matching, active filtering, and many other applications.

Voltage Follower

The *voltage follower*, a special case of the noninverting amplifier, is a circuit in which the output duplicates, or follows, the input in phase and

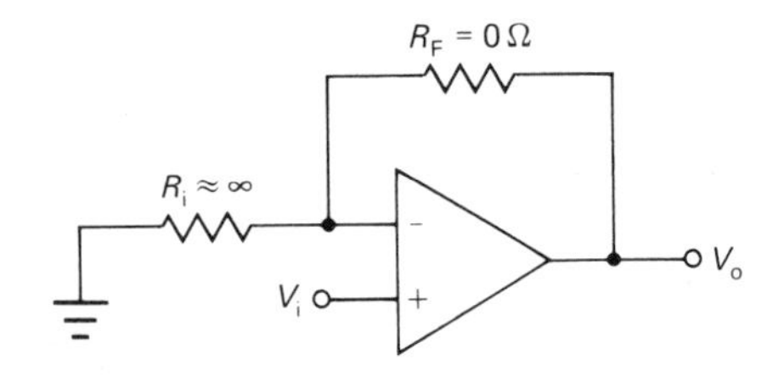

Figure 2.1 Developing the Voltage Follower from the Noninverting Amp

amplitude. Let us take the noninverting amplifier of Figure 1.20, change the input resistor R_i to infinity, and decrease the feedback resistor R_F to $0\ \Omega$, as shown in Figure 2.1.

What is the gain of this circuit? We can use the following equation:

$$A_V = \frac{R_F}{R_i} + 1 \tag{2.1}$$

where A_V = voltage gain

R_i = resistor across which input voltage is developed

In this equation, we substitute zero for R_F and let R_i approach infinity. The resulting gain is 1, as shown in Equation 2.2:

$$A_V = \frac{R_F}{R_i} + 1 = \frac{0}{\approx \infty} + 1$$
$$= 1 \tag{2.2}$$

Now, the circuit can be redrawn as in Figure 2.2, which shows the voltage follower in its most common representation.

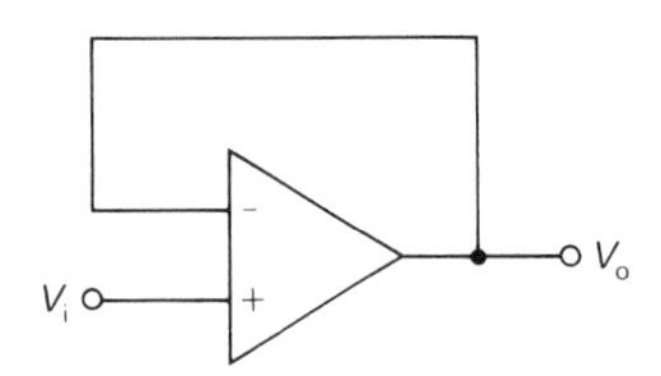

Figure 2.2 Voltage Follower

What are its advantages? The voltage follower is primarily used as a *buffer amplifier*, an amplifier that isolates the preceding stage from the following stage. It has a very high input impedance, even higher than a device with no feedback. Negative feedback causes input impedance to increase and output impedance to decrease. The voltage follower's high impedance ensures that the input circuit is not loaded down. And, of course, the voltage follower has unity gain with no phase shift between input and output. The voltage follower does not change the input in either phase or amplitude. The output follows the input—hence, the name, voltage follower.

Difference Amplifier

If we combine the inverting amplifier and the noninverting amplifier, we have the *difference* (or differencing) *amp*. It is shown in Figure 2.3. The output produced results from a combination of the input sources, V_{i1} and V_{i2}. From the principle of superposition, the contribution of V_{i1} to the output (V_{o1}) can be represented as follows:

$$V_{o1} = -\left(\frac{R_F}{R_i}\right) V_{i1} \tag{2.3}$$

Here (Equation 2.3), we consider V_{i2} shorted to ground; thus, the circuit becomes an inverting op amp. The contribution of V_{i2} to the output (V_{o2}) is

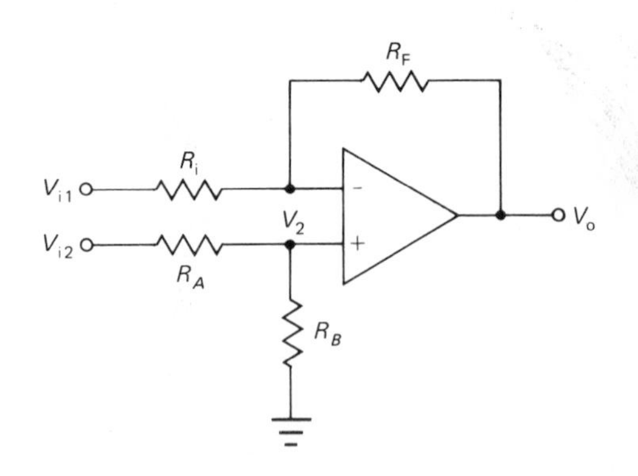

Figure 2.3 Difference Amp

determined by grounding V_{i1}, thus producing a noninverting op amp. The output (V_{o2}) produced by V_{i2} is as follows:

$$V_{o2} = \left(\frac{R_F}{R_i} + 1\right) V_2 \qquad (2.4)$$

For V_{i1} shorted to ground, V_2 is defined (from the voltage divider equation) as follows:

$$V_2 = \left(\frac{R_B}{R_A + R_B}\right) V_{i2} \qquad (2.5)$$

where R_A = resistance between input signal and noninverting terminal

R_B = resistance between noninverting terminal and ground

Suppose we combine all these factors into one equation, that is, by superposition:

$$V_o = V_{o2} + V_{o1} \qquad (2.6)$$

Then:

$$V_o = \left(\frac{R_F}{R_i} + 1\right)\left(\frac{R_B}{R_A + R_B}\right) V_{i2} - \left(\frac{R_F}{R_i}\right) V_{i1} \qquad (2.7)$$

The first expressions on the right side of the equation are for the noninverting amp; the expressions following the minus sign are for the inverting amp.

If all resistors are equal, this rather formidable equation can be simplified as follows:

$$V_o = V_{i2} - V_{i1} \qquad (2.8)$$

Further, if $R_i = R_A$ and $R_F = R_B$, then the output voltage is as follows:

$$V_o = \left(\frac{R_F}{R_i}\right)(V_{i2} - V_{i1}) \qquad (2.9)$$

Instrumentation Amplifier

An interesting application of the difference amp circuit involves the idea that neither one of the op amp inputs is at ground potential. The difference amp can then be used to connect inputs that are floating to a grounded load without grounding either source. The instrumentation amplifier is such an application.

An *instrumentation amp* is an amplifier that is generally used to amplify DC and low-frequency signals produced by a transducer. (Transducers are discussed in Chapter 7.) These signals are typically only a few millivolts in amplitude, while common-mode noise may be several volts. The instrumentation amp has good common-mode rejection characteristics. As we saw in Chapter 1, common-mode rejection is the ability of the amplifier to reject noise common to both inputs. Since noise will be common to both inputs, the amplifier of Figure 2.3 is frequently used when a signal may be masked by noise or when a signal is very small.

One problem inherent in the amplifier of Figure 2.3 is its input impedances. The input impedances to this amplifier may be unequal. And they are relatively low for use in instrumentation applications when the inverting and noninverting gains are not equal. This problem is reduced by making the gains large and equal, that is, $R_F = R_B$ and $R_i = R_A$, with R_F and R_B large. The input impedances then become much higher but may still be unequal. The amplifier can be improved by adding a voltage follower to each input. Such an amplifier is then called an instrumentation amp. Desirable characteristics of very high input impedance and excellent common-mode rejection combine in the instrumentation amp of Figure 2.4.

The gain of this circuit is expressed by the following equation:

$$A_V = \left(\frac{R_3}{R_2}\right)\left(\frac{2R_1 + R_A}{R_A}\right) \qquad (2.10)$$

where R_2 = resistance between voltage follower and output op amp

R_3 = feedback resistance to inverting terminal of output op amp and resistance to ground at noninverting terminal of output op amp

The instrumentation amplifier finds use in industry where signals are masked by noise and

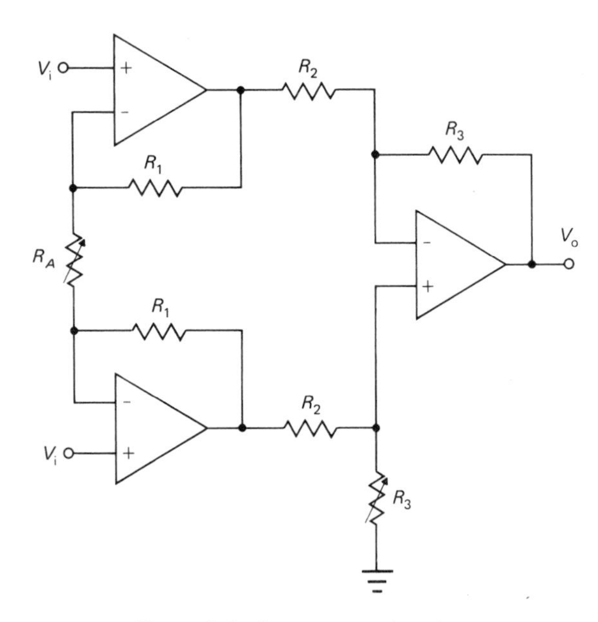

Figure 2.4 Instrumentation Amp

where it is desirable not to load the sensor unduly. Some manufacturers make an instrumentation amp within a single chip. The LM725 is an example. A single chip is desirable because, in that case, the amplifiers and resistors of the instrumentation amp can be more closely matched. Close matching improves instrumentation amp performance.

Integrator

Operational amplifiers can be arranged so as to perform the mathematical function of integration. Such a device is called an *integrator*. *Integration* is the process of summing the input signal over time or, in analytical geometry, of finding the area under a curve. The op amp integrator is illustrated in Figure 2.5. Notice that this circuit looks like the inverting amplifier except that the feedback resistor (R_F) has been replaced by a capacitor (C).

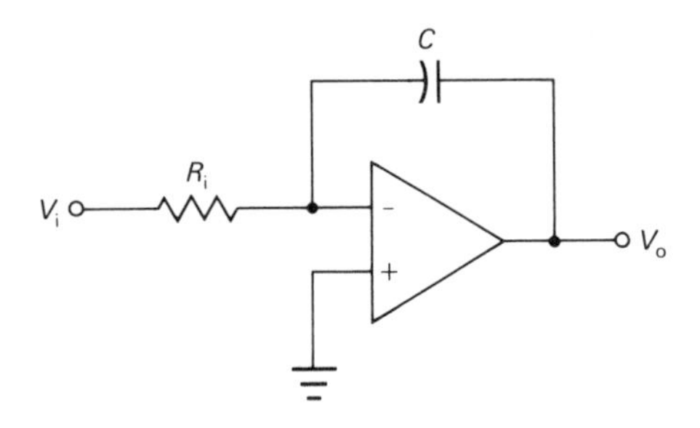

Figure 2.5 Ideal Op Amp Integrator

Integration of DC

To get a better idea of how an integrator works, we start with a DC input. Consider the circuit in Figure 2.6A with switch S_1 closed and switch S_2 open.

Assuming current is constant for the moment, the charge (Q) on any capacitor is related to the

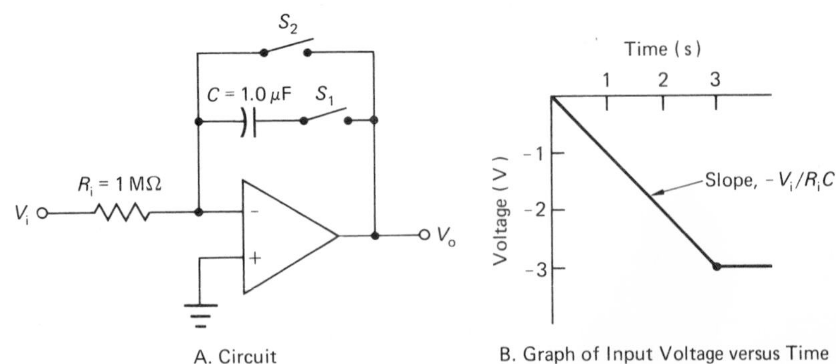

Figure 2.6 Integrator Example

current (I) flowing through it and the time (t) spent charging it:

$$\Delta Q = I\,\Delta t \quad \textbf{(2.11)}$$

where ΔQ = change in charge, measured in coulombs

Δt = change in time, measured in seconds

In a similar fashion, we can say that the change in voltage (ΔV) across the capacitor is a function of its change in charge (ΔQ) and the capacitor's capacitance (C):

$$\Delta V = \frac{\Delta Q}{C} \quad \textbf{(2.12)}$$

Substitution gives us the following equation:

$$\Delta V = \frac{I\,\Delta t}{C} \quad \textbf{(2.13)}$$

Equation 2.13 tells us that if we know the capacitance of the capacitor, the amount of the current flowing through it, and the time of charging, we can predict the change of voltage across it.

Now, recall our discussion about inverting circuits. What factors determined the current flowing through R_i and R_F? We found that input voltage (V_i) and the input resistance (R_i) alone determined this current. So, if we now substitute these factors into Equation 2.13 and simplify, we get the following:

$$\Delta V = -\frac{V_i\,\Delta t}{R_iC} \quad \textbf{(2.14)}$$

Let us now apply these equations to a specific example. Figure 2.6A has values of 1 MΩ for R_i and 1 μF for C. A positive voltage is applied to the input for a certain time interval.

When switch S_1 is closed, current flows through the capacitor at a constant rate. If the capacitor is initially uncharged, how can we determine what the voltage on the capacitor will be after 3 s? We must know how fast the capacitor charges. The rate of change of voltage on the capacitor can be determined by rearranging Equation 2.14 and is measured in volts per second:

$$\begin{aligned}\text{rate of change of voltage} &= \frac{\Delta V_o}{\Delta t} = \frac{-V_i}{R_iC}\\ &= \frac{-1\text{ V}}{(1\text{ M}\Omega)(1\ \mu\text{F})}\\ &= -1\text{ V/s}\end{aligned} \quad \textbf{(2.15)}$$

In our example, the rate of change of output voltage is −1 V/s. When, at the end of 3 s, we open the switch S_1, we should have a voltage of −3 V across the capacitor. This result is graphically illustrated in Figure 2.6B. Notice that the slope of the line in Figure 2.6B is $-V_i/R_iC$. Switch S_2 can be used to discharge the capacitor (C).

An interesting variation of this circuit results when switch S_2 in Figure 2.6A is replaced with silicon-controlled rectifier (SCR). SCRs are discussed in Chapter 5. What we have then is a sawtooth wave generator, as illustrated in Figure 2.7. The reference voltage (V_{ref}) on the SCR gate must be well within the range of 0 V to the $-V$ supply of the op amp.

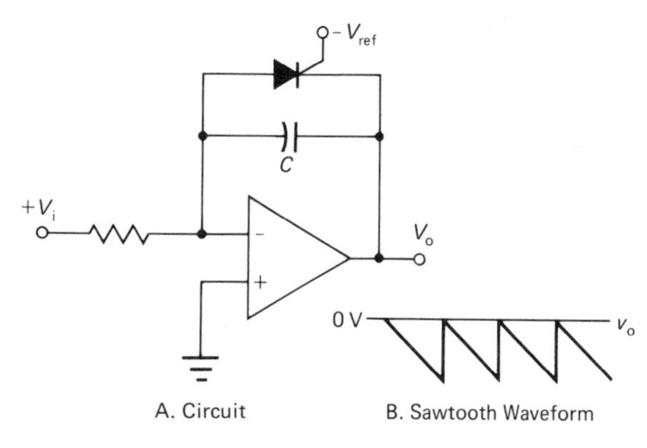

Figure 2.7 Op Amp Integrator Combined with an SCR

Integration of a Square Wave

We have discussed the integrator's operation with a positive DC placed on its input. Now, instead of placing a constant positive DC voltage at the input of the integrator, let us alternate positive and negative DC voltages (otherwise known as a *square wave*). The output, then, instead of ramping negative, ramps positive and negative, a waveform known as a *triangle wave*. The situation is illustrated in Figure 2.8.

In this example, a triangle wave output results from a square wave input. The triangle wave output is established by the following equation:

$$V_{o(p-p)} = -\frac{V_{i(pk)}t}{R_i C} = \frac{1\ V_{pk} \times 100\ \mu s}{(10\ k\Omega)(0.0022\ \mu F)} = 4.54\ V_{(p-p)} \qquad \textbf{(2.16)}$$

where $V_{o(p-p)}$ = peak-to-peak output voltage
$V_{i(pk)}$ = peak input voltage

Figure 2.8 shows a *practical integrator*. Resistor R_A has been added to prevent the amplifier from integrating the DC offset voltage. If this small voltage is integrated over a sufficiently long time, it drives the amplifier into saturation and the amplifier no longer operates as an integrator. Resistance R_A should be approximately 10 times R_i. Another way to see the effect of this resistor is to think of the amp's low-frequency gain as very high because of the capacitor's high reactance X_C. The resistor R_A limits this gain at low frequencies to R_A/R_i. But since the offset voltage is small, the DC output is also small.

Integration of a Sine Wave

The integrator gain equation for a sine wave input is similar to the inverting amplifier gain equation (1.17). Replacing R_F with X_C, we have the following equation:

$$A_V = \frac{X_C}{R_i} \angle 90^\circ \qquad \textbf{(2.17)}$$

where $\angle 90^\circ$ = notation that indicates a 90° phase shift occurs

A phase shift is expected across a capacitor. When X_C divided by the op amp, open-loop gain equals R_i, the phase shift between the input sine wave and the output waveform is approximately 45°. As frequency increases, the phase shift approaches 90°. In this region, where X_C divided by op amp, open-loop gain is much smaller than R_i, the circuit in Figure 2.8 operates as an integrator. The output of

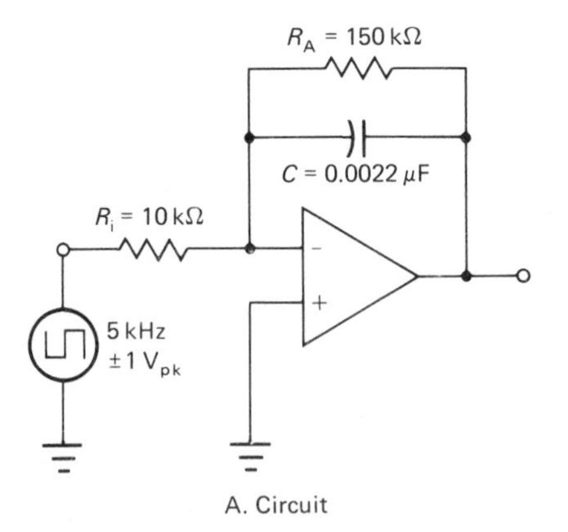

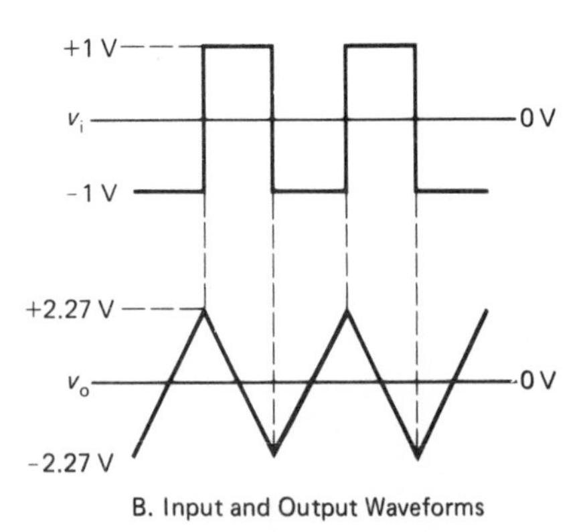

Figure 2.8 Practical Op Amp Integrator with Square Wave Input

the integrator with a sine wave input is a cosine wave, as shown in Figure 2.9.

In calculus, the integral of a sine wave is a negative cosine wave. Thus, the phase relationship between input and output is 90°. But notice, in Figure 2.9, that the output leads the input by 90°. We expect the output to lag the input since the integrator is, in fact, a lag network. A *lag network* is a circuit in which the output voltage lags the input voltage. But the integrator of Figure 2.8 uses the inverting amp (180° lag) and the integrator lag network (90° lag), which work in conjunction to provide a phase-leading relationship, output to input.

In Figure 2.9, to determine the maximum amplitude of the output cosine wave, $V_{o(pk)}$, we expand Equation 2.17, drop the angle notation, and multiply it by the peak of the input sine wave, $V_{i(pk)}$:

$$V_{o(pk)} = \frac{V_{i(pk)}}{2\pi f R_i C} \qquad (2.18)$$

where f = frequency of input sine wave

Equation 2.18 indicates that as the input frequency increases, the output amplitude of the cosine wave decreases. In Figure 2.8, if the input signal is a 1 V peak, 5 kHz sine wave, the output cosine wave amplitude is 1.45 V peak.

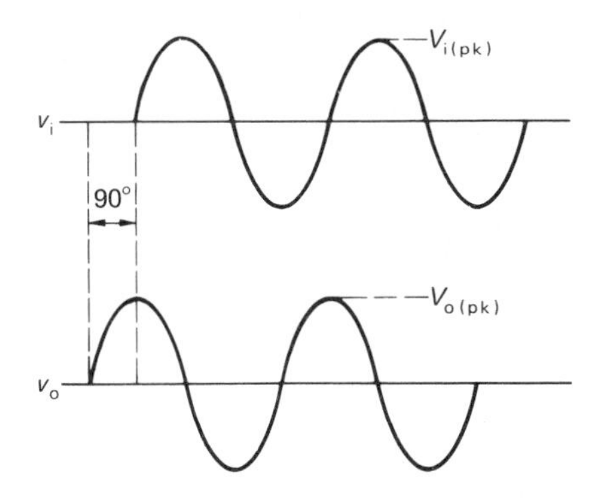

Figure 2.9 Integrator Input and Output Waveforms

Differentiator

Interchanging the resistor and capacitor in the integrator produces a *differentiator*, which is shown in Figure 2.10. Mathematically, the *differentiation function* is the inverse of integration.

The differentiator is not as frequently used as the integrator. We noted that the integrator (for sine waves) was a basic lag network, that is, a low-pass filter. Recall that a *low-pass filter* passes all frequencies below a given frequency. The inverse is true of the differentiator: It is a basic high-pass filter. A *high-pass filter* is a circuit that passes all frequencies above a given frequency. Thus, the differentiator is much more likely to oscillate or become unstable than the integrator is, and, at the same time, it will amplify any noise that is present on the signal.

The mathematical concept of differentiation is simple. Instead of producing an output that is proportional to the input voltage (as the inverting amp does), the differentiator output is proportional to the slope of the input voltage. If the slope of the input voltage is zero, the differentiator output is 0 V. If the slope of the input is constant, the output is constant. If the slope of the input is infinite (as in the edges of a square wave), the output is an infinite voltage (ideally).

Note the circuit in Figure 2.11A. Resistor R_F and capacitor C form a lead network (or high-pass filter). *A lead network* is a circuit in which the output voltage leads the input voltage. Series

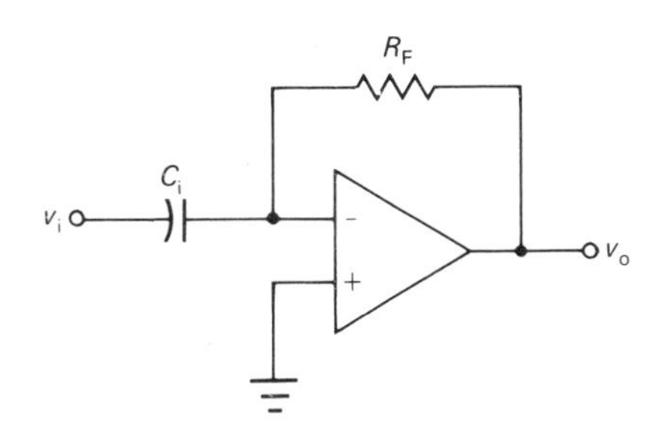

Figure 2.10 Ideal Op Amp Differentiator

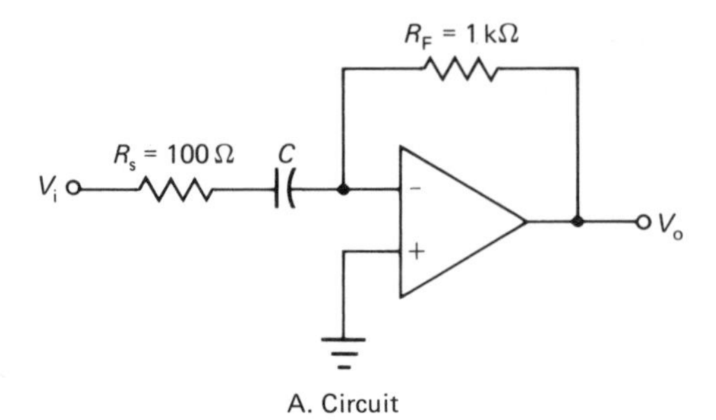

Figure 2.11 Differentiator with a Sine Wave Input and Cosine Wave Output

resistor R_s holds the high-frequency gain to $-R_F/R_s$. The input and output waveforms are shown in Figure 2.11B.

Differentiation of a Sine Wave

As in the integrator, a sine wave input to the differentiator gives a cosine wave output. Mathematically, a differentiated sine wave gives a cosine wave. The difference between integrator and differentiator output due to a sine wave input is in phase angle and amplitude. The differentiator, with its 180° phase inversion, causes the output voltage to lag the input by 90°. This relationship is diagramed in Figure 2.11B.

The output amplitude is expressed by the following equation:

$$V_{o(pk)} = 2\pi f R_F C V_{i(pk)} \tag{2.19}$$

In this equation, as the input frequency increases, so does the output amplitude.

Differentiation of a Square Wave

Notice what happens when we change the input from a sine wave to a square wave, as shown in Figure 2.12. The output of a differentiator with such an input is a series of positive and negative

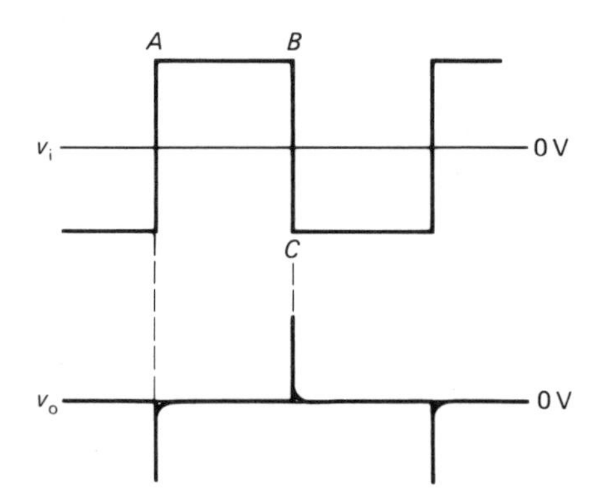

Figure 2.12 Square Wave Input to a Differentiator and Spiked Waveform Output

spikes. Why? Note that the slope of the *AB* portion of the square wave of Figure 2.12 is zero. We expect to get no output. The slope of the *BC* portion is infinity (or close to it). The output quickly rises to its maximum value and then falls back to zero. Thus, the output waveform is a series of positive and negative spikes.

Differentiation of a Triangle Wave

Let us examine what would happen if a triangle wave were the input for the differentiator. In calculus, the operation of differentiation is the

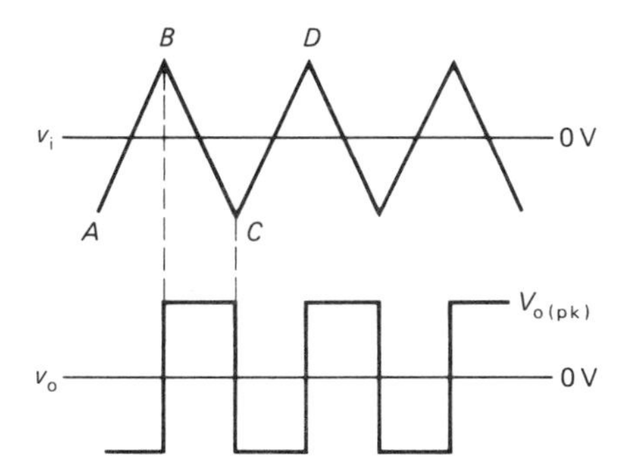

Figure 2.13 Triangle Wave Input to a Differentiator and Square Wave Output

inverse of integration, as we mentioned earlier. Integration produces a triangle output with a square wave input. Thus, the differentiator produces a square wave output with a triangle input.

The input-output relationship with a practical differentiator is shown in Figure 2.13. The slope of the input from A to B and C to D is a constant positive value. The output at those times is a constant negative voltage. This output value changes only if the input slope changes.

To determine the amplitude of the output square wave of Figure 2.13, we use the following equation:

$$V_{o(pk)} = -R_F C\left(\frac{\Delta V_i}{\Delta t}\right) \tag{2.20}$$

The change in input voltage with respect to time ($\Delta V_i/\Delta t$) is the slope of the input waveform.

The verification of the equations in the differentiator section is left for you to do.

Logarithmic Amplifier

A circuit frequently seen in analog computers is the logarithmic amplifier (log amp). In a log amp, the output is proportional to the logarithm of the input. The *log amp* takes advantage of the logarithmic relationship between the current through a semiconductor device and the voltage across it. Two such nonlinear devices are the semiconductor diode and the BJT. The circuit configurations for these log amps are illustrated in Figures 2.14A and 2.14B.

In the circuit shown in Figure 2.14A, the voltage across the diode (V_D) is logarithmically related to the current through the diode (I_D) by the following relationship:

$$V_D = A \log_B \frac{I_D}{I_r} \tag{2.21}$$

where $\log_B$ = logarithm to a specified base (B)
A = constant of proportionality, which depends on the base of the logarithm
I_r = theoretical reverse leakage current of diode

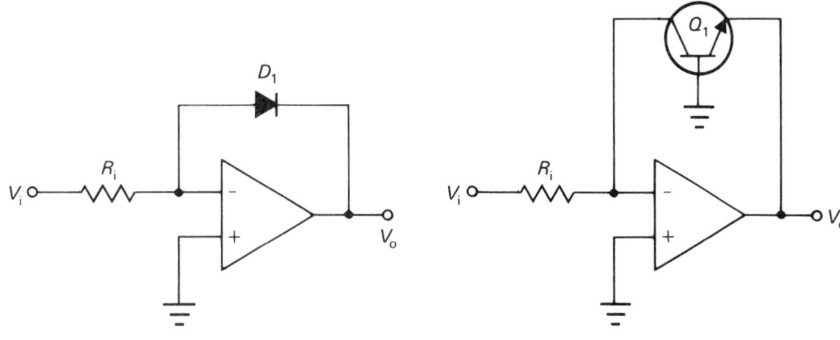

Figure 2.14 Ideal Log Amp

Likewise, in Figure 2.14B, the collector current (I_C) divided by its reverse leakage current (I_{Cr}) is logarithmically related to the voltage from base to emitter (V_{BE}) by the following relationship:

$$V_{BE} = A \log_B \frac{I_C}{I_{Cr}} \qquad \textbf{(2.22)}$$

Practical Log Amp

In actual use of the log amp, a more practical circuit such as the one in Figure 2.15A would be utilized. Here, the more practical equation is as follows:

$$V_{o1} = -V_{BE} = A \log_B(V_i) + K \qquad \textbf{(2.23)}$$

where V_{o1} = output of log amp
V_{BE} = voltage from base to emitter of transistor
A = constant of proportionality
K = offset value

The value of resistance for both R_1's in Figure 2.15A is given by the following expression:

$$\frac{V_{i(max)}}{I_{C(max)}} \leq R_1 \leq \frac{V_{i(min)}}{I_b} \qquad \textbf{(2.24)}$$

where $V_{i(max)}$ = maximum input voltage
$V_{i(min)}$ = minimum input voltage
$I_{C(max)}$ = maximum collector current of transistor used
I_b = input bias current of op amp

For the 741, I_b is approximately 80 nA.

For evaluation of A and K in Equation 2.23, the following equations may be used:

$$A = \frac{V_{o2} - V_{o1}}{\log(V_{i2}) - \log(V_{i1})} \qquad \textbf{(2.25)}$$

$$K = V_{o1} - A \log(V_{i1}) \qquad \textbf{(2.26)}$$

where V_{i1} = first applied input voltage
V_{i2} = second applied input voltage
V_{o1} = output voltage of log amp due to V_{i1}
V_{o2} = output voltage of log amp due to V_{i2}

An example might be helpful at this point to demonstrate the evaluation of A and K. Assume the input voltage applied will range over three

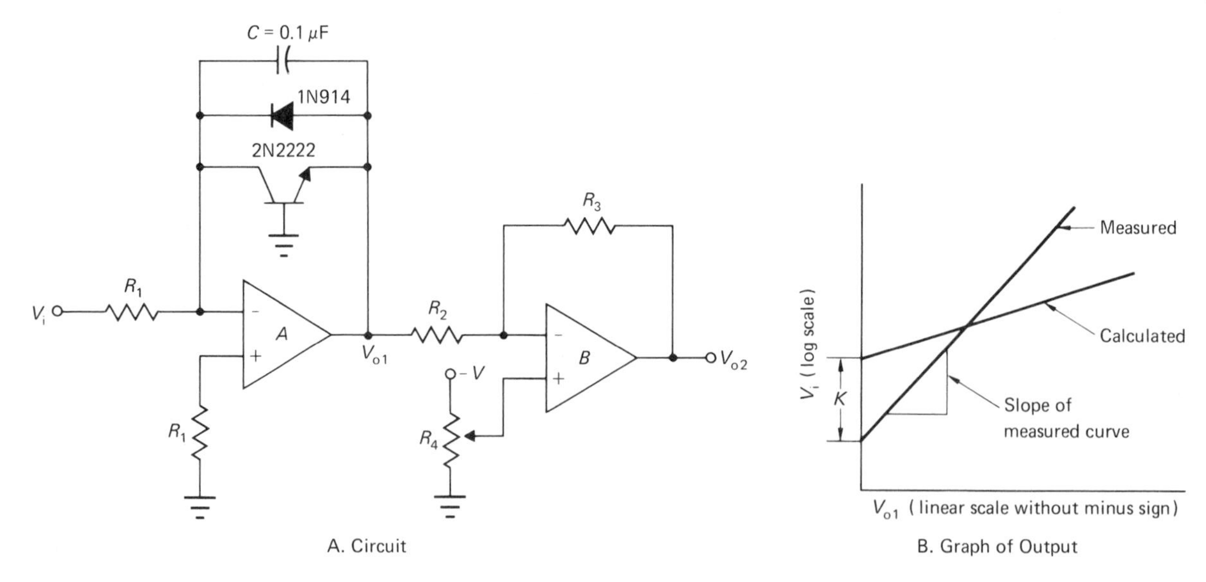

Figure 2.15 Practical Log Amp

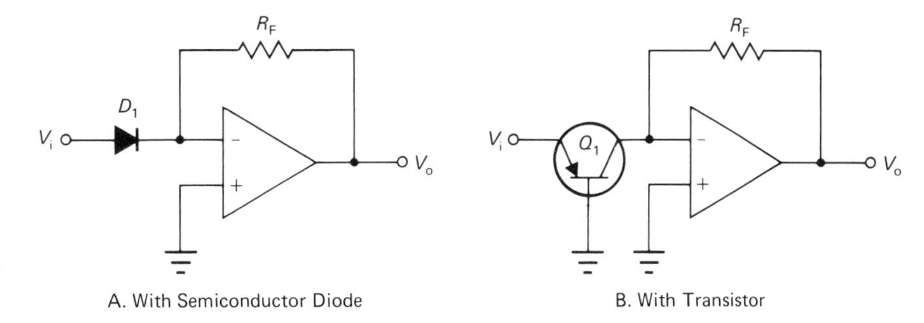

Figure 2.16 Ideal Antilog Amp

different values: 1 V, 2 V, and 3 V. For these three inputs, three output voltages (V_{o1} in Figure 2.15A) result: 0.60 V, 0.62 V, and 0.64 V (hypothetically).

In Equation 2.25, V_{o1} is 0.60 V, V_{o2} is 0.62 V, V_{i1} is 1 V, and V_{i2} is 2 V. (Notice that the negative signs were dropped from the output voltage measurements.) Substituting these values into Equation 2.25 results in a value of 0.0664 for A. In Equation 2.26, V_{o1} is 0.60 V, A is 0.0664, and V_{i1} is 1 V. Substituting these values into Equation 2.26 results in a value of 0.60 for K.

Now, the procedure will be repeated for the second group of values for V_i and V_o. For Equation 2.25 and the second group of values, V_{o1} is now 0.62 V, V_{o2} is 0.64 V, V_{i1} is 2 V, and V_{i2} is 3 V. Substituting these values into Equation 2.25 results in a value of 0.114 for A. In Equation 2.26, V_{o1} is now 0.62 V, A is 0.114, and V_{i1} is 2 V. Substituting these values into Equation 2.26 results in a value of 0.586 for K.

After all values for A and K have been calculated, they can be averaged in order to obtain values that can be used in the amplifier circuit following the log amp in Figure 2.15A.

After A and K are determined, they are used with the second amplifier (B amplifier) in Figure 2.15A. The noninverting terminal of amplifier B is set to the value of K, and the gain of amplifier B is set to the reciprocal of A. When a voltage is input at V_i, the logarithm of that voltage is now output at V_{o2}.

As a precaution for the circuit of Figure 2.15A, be sure to use only positive input voltages in the range for which you determined the R_1 value in Equation 2.24. In addition, the transistor in the feedback circuit of the log amp (2N2222 in Figure 2.15A) can be any small-signal, high-speed transistor. If the transistor's response is a straight line on the graph in Figure 2.15B, the circuit of Figure 2.15A should produce an accurate logarithmic output. Whichever transistor you choose, though, will be heat-sensitive. For best results, either keep the transistor at a constant temperature or use a temperature-stabilized circuit. The bibliography for this chapter given at the back of the text lists good sources for such designs.

Antilog Amp

The *antilog amplifier* performs the inverse function of the log amp. This operation is done by simply reversing the position of the semiconductor and the input resistor. The antilog amp is shown in Figures 2.16A and 2.16B.

Designing and working with log amps can be a tedious and exacting job. For this reason, several manufacturers have designed and built log amps within a single IC chip.

Applications

Applications of log amps are numerous. For example, they are used for compressing large voltage ranges into smaller ones. Also, since many

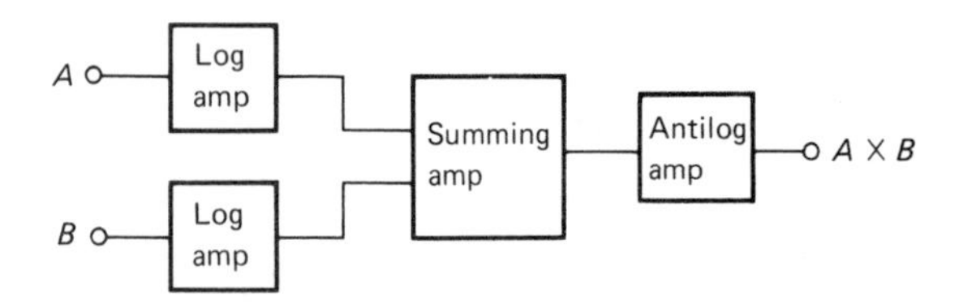

Figure 2.17 Multiplying with Log and Antilog Amps

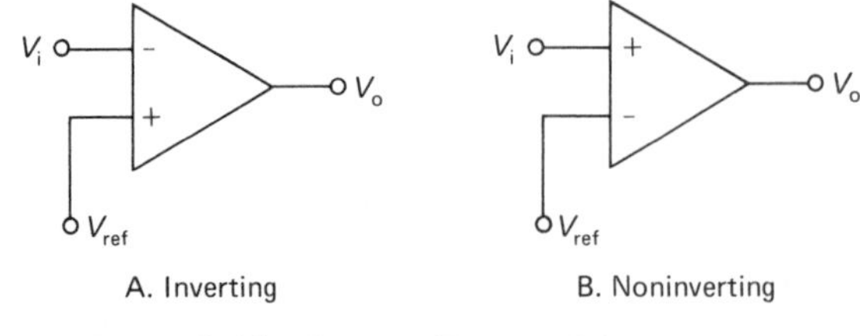

Figure 2.19 General Forms of Comparators

transducers have a logarithmic transfer function, the log amp is used to compensate for this nonlinear response.

Probably the most frequent use of the log amp is in analog computers. Recall that multiplying and dividing may be accomplished by adding and subtracting the logarithms of numbers and then obtaining the antilogarithm. Since log amps perform these operations, they can be used in multipliers and dividers in computers. The block diagram of a multiplier is shown in Figure 2.17.

Comparator

The simplest op amp circuit is the *comparator*, which is an op amp without feedback. With no feedback, the op amp's gain is very high, equal to its open-loop gain. Figure 2.18 shows an example of an inverting and a noninverting comparator.

Since the gain of a comparator is very high, a voltage of only a few microvolts causes the amplifier to saturate. Therefore, the output of this amplifier is normally either $+V_{sat}$ (saturation voltage) or $-V_{sat}$. The op amps in Figure 2.18 compare the input against a reference, which in this case is 0 V. Such comparators are also called *zero-crossing detectors* because the output changes when the input crosses the 0 V point. Whenever the input voltage is above or below 0 V, the amplifier saturates.

A more general form of the inverting and noninverting comparators is shown in Figure 2.19. In these comparators, the op amp compares the input to a reference voltage (V_{ref}). Figure 2.20 shows a sine wave input to a comparator like the one shown in Figure 2.19B. Figure 2.19B (the noninverting comparator) has a 5 V peak sine wave applied with a reference of +3 V. We see that when the input rises above +3 V, the output goes to $+V_{sat}$. When the input falls below +3 V, the output goes to $-V_{sat}$.

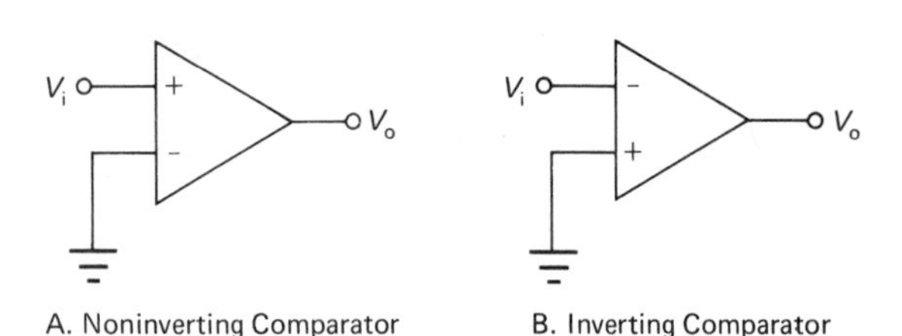

Figure 2.18 Comparators (Zero-Crossing Detector)

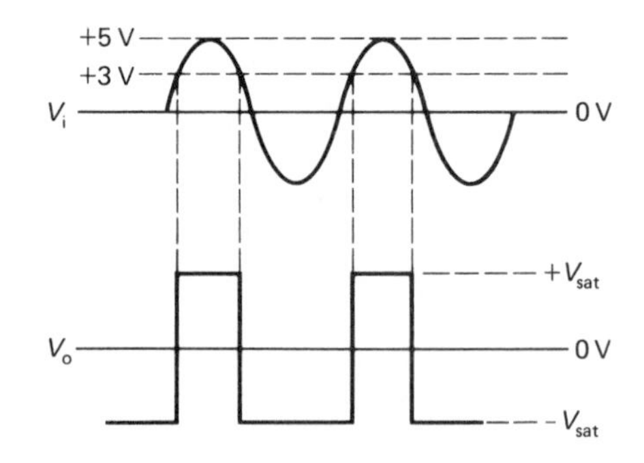

Figure 2.20 Input and Output Waveforms for Noninverting Comparator

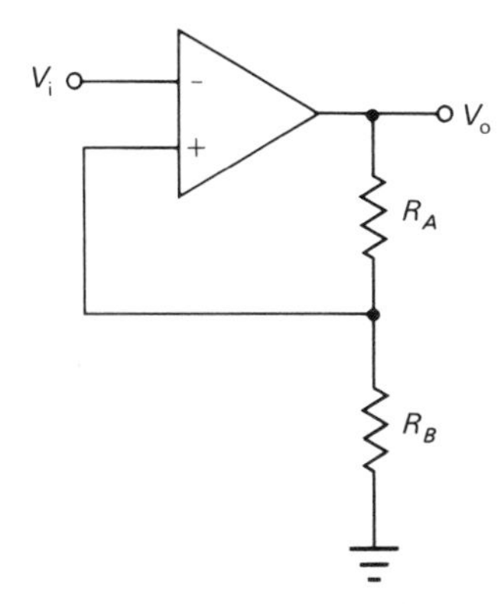

A. Circuit, with Feedback

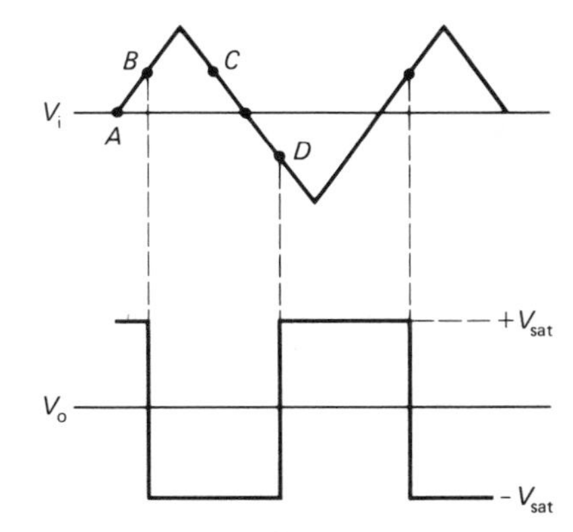

B. Input and Output Waveforms

Figure 2.21 Comparator with Hysteresis

The comparators we have just discussed are susceptible to triggering by noise voltages, alone or in combination with a signal source. The basic comparator can be improved by adding positive feedback, as shown in Figure 2.21A. This circuit is called a comparator with hysteresis. *Hysteresis* is the property by which the output of a device is dependent on the history of the input movement and the present direction of the input movement. Originally applied to magnetic circuits, the term *hysteresis* now has been applied to other systems that have that type of action.

In Figure 2.21A, the noninverting terminal receives a reference from the voltage divider formed by resistors R_A and R_B. Note that from point A to B on the input waveform in Figure 2.21B, the output is at $+V_{sat}$ $(+10 \text{ V})$. When the input voltage rises above $+1$ V, the amplifier's output changes to $-V_{sat}$ (-10 V). At this time, the *trip point*, or reference voltage, at the noninverting terminal is no longer $+1$ V. It is -1 V. Now, the output of Figure 2.21B does not change at point C but, instead, at point D.

The reference voltage (trip point) is calculated from the following equation:

$$V_{ref} = \left(\frac{R_B}{R_A + R_B}\right)(\pm V_{sat}) \qquad \textbf{(2.27)}$$

For example, suppose $R_A = 9 \text{ k}\Omega$ and $R_B = 1 \text{ k}\Omega$. Then, the circuit of Figure 2.21, when at $-V_{sat}$ (-10 V), has the following trip point:

$$V_{ref} = \left(\frac{1 \text{ k}\Omega}{9 \text{ k}\Omega + 1 \text{ k}\Omega}\right)(-10 \text{ V})$$
$$= -1 \text{ V} \qquad \textbf{(2.28)}$$

Active Filter

The operation of many electronic systems requires that certain bands of frequencies be passed and others attenuated. A device that does this operation is called a *filter*.

Classification of filters depends on point of view. If the components from which the filter is made are the basis of classification, then we say filters are either passive or active. A *passive filter* uses capacitors, resistors, and inductors. *Active filters* employ amplifiers along with the components mentioned in the passive category. Inductors are usually not used at audio frequencies since they can be simulated with amplifiers, capacitors, and resistors. Also, inductors are relatively expensive, large, and heavy. Figure 2.22 shows this type of classification of filters and the frequencies at which they are used.

Since an in-depth treatment of active filters would be quite lengthy, only an introduction is given in this text. For more information on active

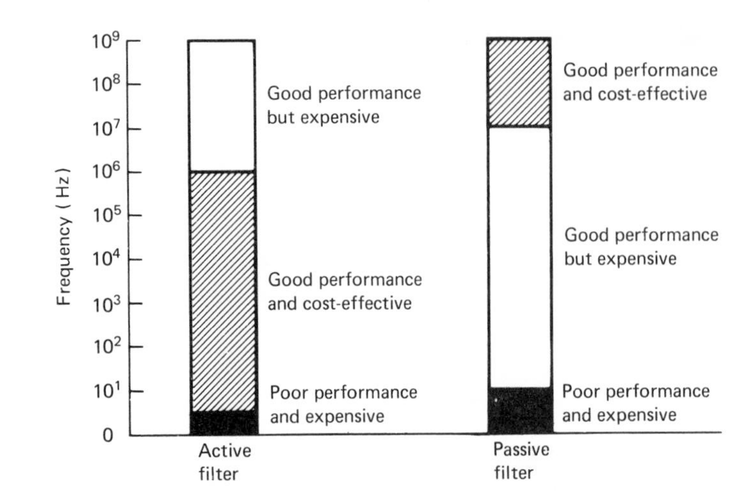

Figure 2.22 Performance of Active and Passive Filters

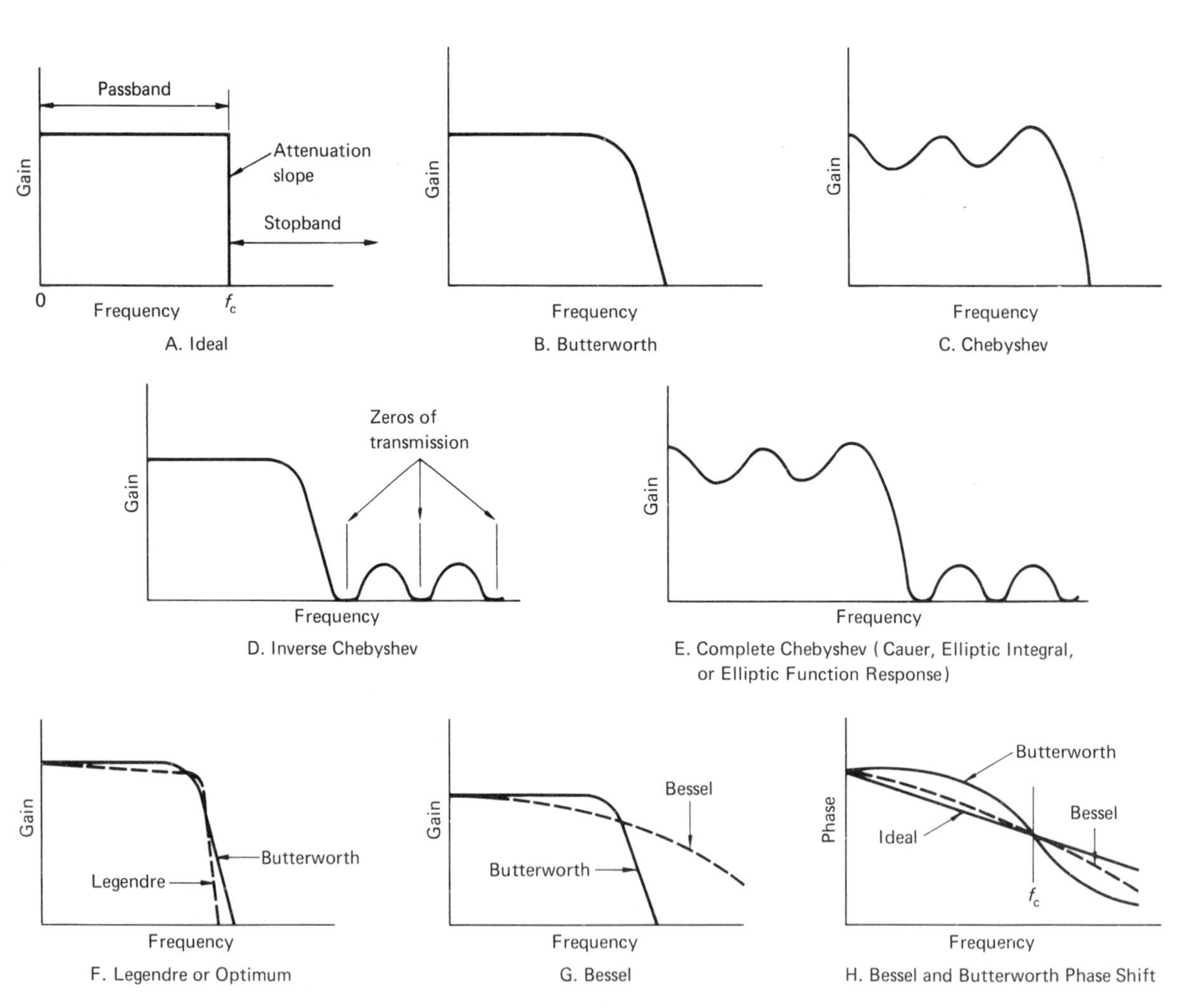

Figure 2.23 Classification of Filters by Shape of Response Curves

Table 2.1 *Comparison of Filter Types*

Name of Filter Type	*Main Distinguishing Characteristic*	*Remarks*
Butterworth	Maximally flat amplitude response	Most popular general-purpose filter
Chebyshev	Equal-amplitude ripples in passband	Attenuation slope steeper than in Butterworth near cutoff
Inverse Chebyshev	Equal-amplitude ripples in stopband	No passband ripple; zeros of transmission in stopband
Complete Chebyshev (also called Cauer, elliptic function, elliptic integral, or Zolatarev)	Equal-amplitude ripples in both passband and stopband	Zeros of transmission in stopband
Legendre	No passband ripple, but steeper attenuation slope than in Butterworth	Not maximally flat
Bessel (also called Thomson)	Phase characteristic nearly linear in pass region, giving maximally flat group delay	Good for pulse circuits because ringing and overshoot minimized; poor attenuation slope

filters, consult the bibliography for this chapter (at the end of the book) or other reference material.

As we will see, the frequency behavior of any given amplifier can be controlled by adding reactive components to it in different configurations. Such active filters are reliable, low in cost, and easy to tune. In addition, when op amps are used, they give not only voltage gain but also high input impedance and low output impedance characteristics.

Filters are also classified according to the shapes of their response curves. Figure 2.23 shows some of these curves. Table 2.1 compares these filter types.

Probably the most common classification of filters is related to how they treat frequencies. These classifications are low-pass, high-pass, bandpass, and band-reject (or notch).

Low-Pass Filter

A *low-pass filter* is defined as a circuit that passes all frequencies below a certain frequency while attenuating all higher frequencies. A simple low-pass Butterworth filter with its bandpass characteristic is shown in Figure 2.24.

As shown in Figure 2.24B, the output of the filter is constant until the *cutoff frequency* (f_c) is

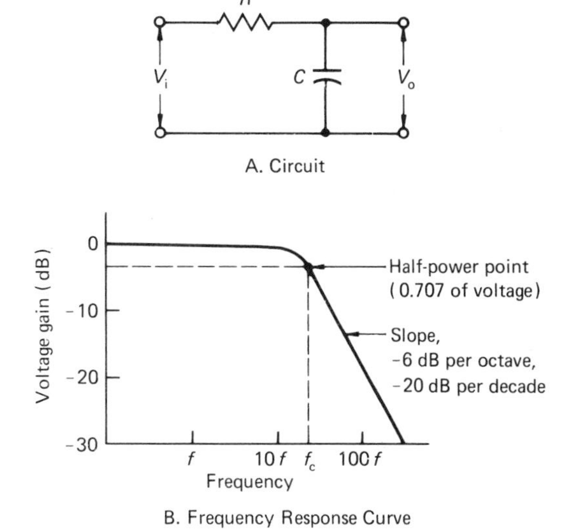

Figure 2.24 Low-Pass Filter

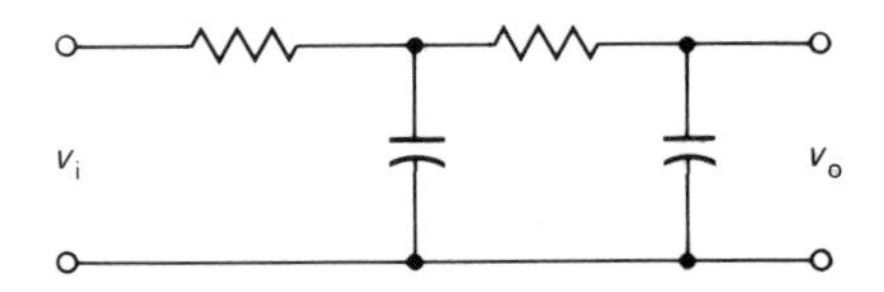

Figure 2.25 Second-Order, Low-Pass, Butterworth Filter

approached. After f_c is passed, the output voltage decreases, or rolls off, at a rate of -6 dB per octave, or -20 dB per decade. (Recall that a decade is a tenfold increase or decrease in frequency, while an octave is a twofold increase or decrease in frequency.) This filter is called a *first-order filter* because its transfer function is a linear equation and because it has only one *RC* filter combination.

Table 2.2 *Phase Shifts at Half-Power Point*

Order of Filter	*Phase Shift at f_o or f_c**	*Roll-Off (dB per Decade)*
1	$-45°$	-20
2	$-90°$	-40
3	$-135°$	-60
4	$-210°$	-80

* f_o = oscillation frequency for oscillators; f_o = cutoff frequency for filters.

Figure 2.25 shows a second-order Butterworth filter. Notice that there are two *RC* networks here. This filter exhibits a roll-off of -12 dB per octave, or -40 dB per decade. A third-order Butterworth filter has three *RC* networks and roll-offs at -18 dB per octave or -60 dB per decade; and so on. Table 2.2 shows the phase shifts at the half-power (0.707) points for these filters.

Let us now examine some low-pass active filters, as shown in Figure 2.26. These filters are first-order, *voltage-controlled, voltage source (VCVS) filters*, recognized by their use of the noninverting amplifier configuration. The only difference between the two amplifiers in Figures 2.26A and 2.26B is the gain.

Figure 2.27 shows a second-order, VCVS, low-pass filter. Components R_1 and C_1 make up one low-pass *RC* network, while R_2 and C_2 make up the other. For the resistors and capacitors as labeled in Figure 2.27, the equations for cutoff frequency f_c and voltage gain A_V are as follows:

$$f_c = \frac{1}{2\pi\sqrt{R_1C_1R_2C_2}}$$

and

$$A_V = \frac{R_F}{R_i} + 1 \qquad \textbf{(2.29)}$$

Thus far, we have shown examples of VCVS filters only. Figure 2.28 gives an example of a second-order, multiple-feedback, low-pass Butter-

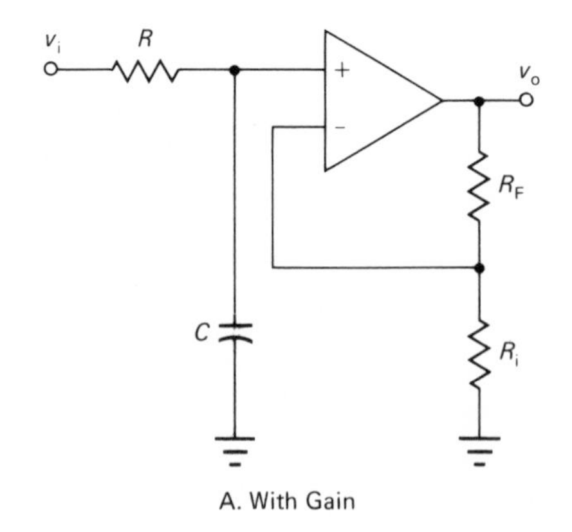

A. With Gain

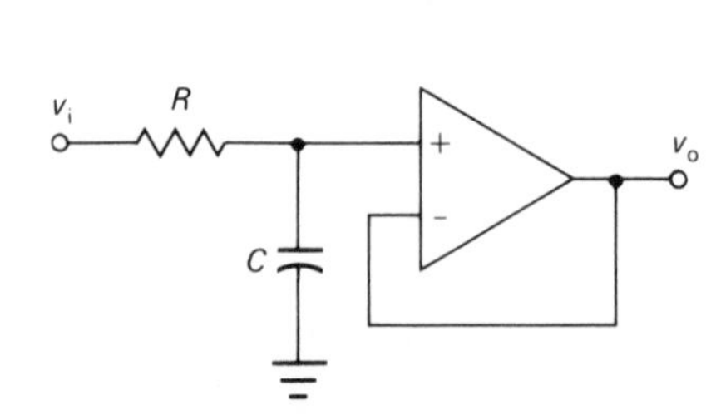

B. With Unity Gain

Figure 2.26 First-Order, VCVS, Low-Pass Filter

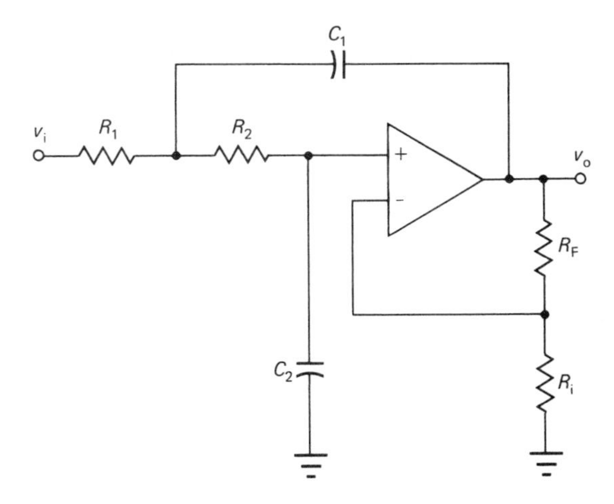

Figure 2.27 Second-Order, VCVS, Low-Pass Filter with Gain

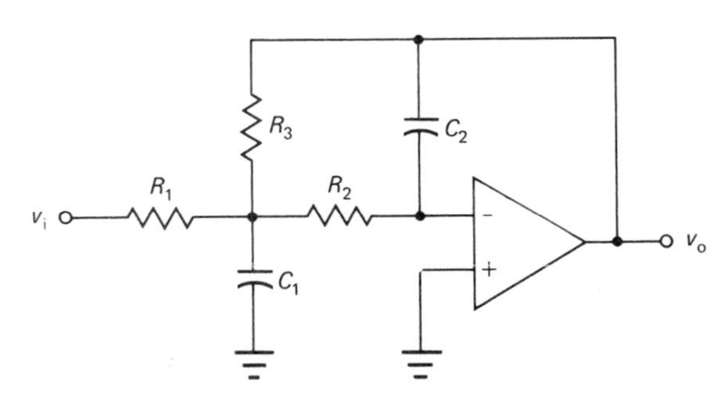

Figure 2.28 Second-Order, Multiple-Feedback, Low-Pass Butterworth Filter

worth filter. In this filter, identified as a *feedback filter*, the gain is fixed by the ratio of R_3 and R_1. The cutoff frequency equation is similar to the one in Equation 2.29. This circuit does have a disadvantage: Its gain and frequency characteristics cannot be separately controlled, as in the VCVS filters.

Higher-order filters (above second-order) may be constructed by cascading lower-order filters. For example, cascading two second-order filters creates a fourth-order filter.

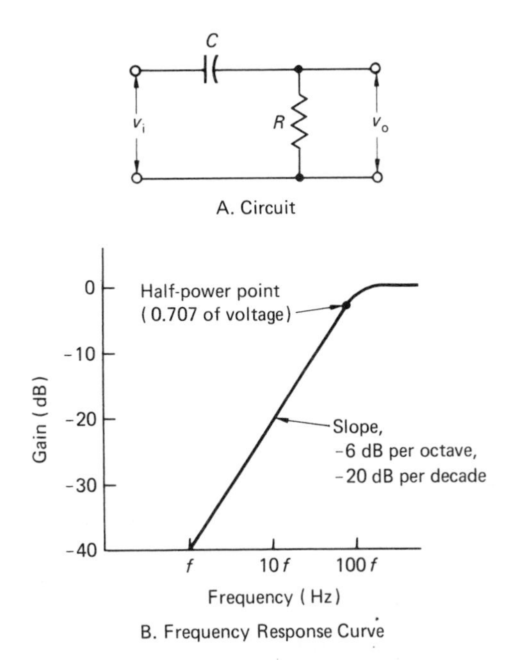

Figure 2.29 High-Pass, First-Order Butterworth Filter

High-Pass Filter

A *high-pass filter* is defined as a circuit that passes all frequencies above a certain frequency while attenuating all lower frequencies. The high-pass, first-order Butterworth filter and its associated graph are shown in Figure 2.29. The graph is similar to that of the low-pass filter. As frequency decreases, the output of the filter is constant until f_c is approached. Once f_c is passed, the output rolls off at -6 dB per octave (-20 dB per decade).

Active high-pass filters resemble low-pass filters, with the positions of capacitors and resistors interchanged. For instance, compare the low-pass filter in Figure 2.28 with the high-pass filter in Figure 2.30. For the resistors and capacitors shown in Figure 2.30, the equations for cutoff frequency f_c and voltage gain A_V are as follows:

$$f_c = \frac{1}{2\pi\sqrt{C_2 C_3 R_1 R_2}} \qquad A_V = \frac{C_3}{C_1} \tag{2.30}$$

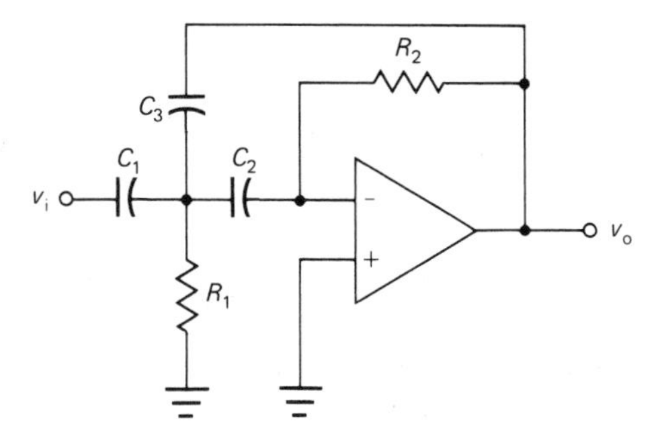

Figure 2.30 Second-Order, Multiple-Feedback, High-Pass Filter

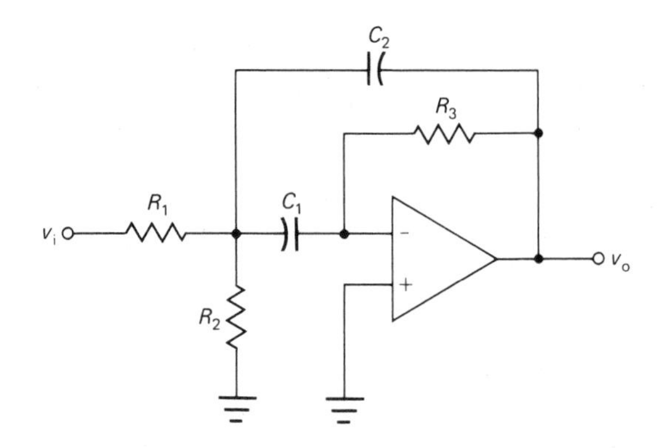

Figure 2.32 Second-Order, Multiple-Feedback Bandpass Filter

Bandpass Filter

A *bandpass filter* is defined as one that passes a band of frequencies while attenuating all others. A simple way to construct a bandpass filter is to cascade a high-pass and a low-pass filter if their frequency response curves are made to overlap. The result of such a union is illustrated in Figure 2.31, with a graph showing gain versus frequency.

An example of a second-order, multiple-feedback bandpass filter is shown in Figure 2.32. For the resistors and capacitors shown in Figure 2.32, the equations for cutoff frequency f_c and voltage gain A_V are as follows:

$$f_c = \frac{1}{2\pi C_1}\sqrt{\frac{R_1 + R_2}{R_1 R_2 R_3}}$$

and $$A_V = \frac{R_3}{2R_1} \quad (\text{at } f_c) \qquad \textbf{(2.31)}$$

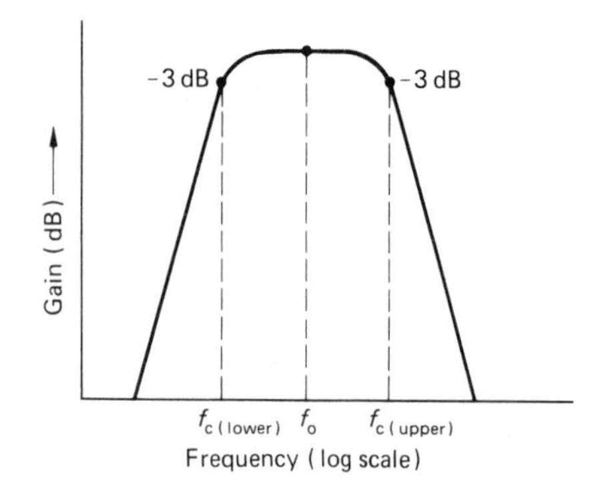

Figure 2.31 Frequency Response Curve for Bandpass Filter

Note the difference between the filter of Figure 2.32 and the high-pass filter of Figure 2.30. The capacitor C_1 of Figure 2.30 has been replaced by resistor R_1, thus making Figure 2.32 a bandpass filter. Generally, the multiple-feedback filter shown in Figure 2.32 is used for narrow-bandwidth applications ($Q > 10$) (recall that Q represents quality factor). Where wider bandwidths are required ($Q < 10$), high-pass and low-pass filters are cascaded.

Band-Reject (Notch) Filter

The *band-reject (notch) filter* is defined as a circuit that attenuates a specified band of frequencies while passing all others. Like the bandpass filter, the band-reject filter is constructed by connecting high-pass and low-pass filters in series (cascade) or in parallel. The parallel connection is most commonly used. It is sometimes called the *twin T filter*.

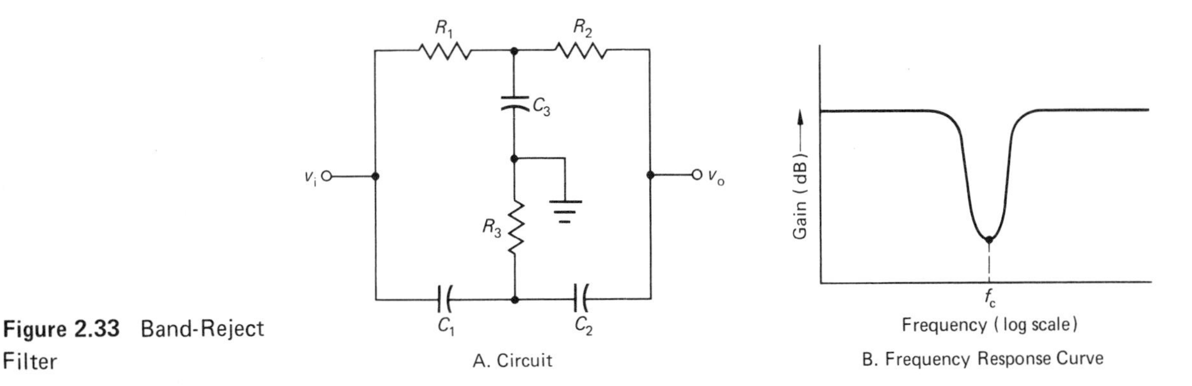

Figure 2.33 Band-Reject Filter

An example of the twin T and its associated bandpass characteristic is shown in Figure 2.33. Note that C_1, C_2, and R_3 form the high-pass section, and R_1, R_2, and C_3 form the low-pass section.

An active version using the twin T filter is shown in Figure 2.34. In this version, $C_1 = C_2$, $C_3 = 2C_1$, $R_1 = R_2$, and $R_3 = (1/2)R_1$. The gain for this filter is 1. For the resistors and capacitors as labeled in Figure 2.34, the equation for the cutoff frequency is as follows:

$$f_c = \frac{1}{2\pi R_2 C_2} \tag{2.32}$$

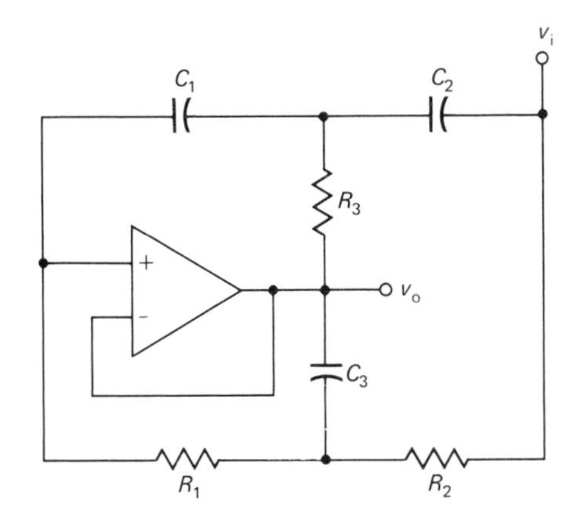

Figure 2.34 Twin T Band-Reject Filter

Gyrator

While not strictly a filter circuit, the gyrator may be classified along with filters because of its unique reactive properties. The *gyrator* acts as a simulated inductor. We have already discussed the disadvantages of inductors at audio frequencies. At frequencies below radio frequencies (RF), the gyrator can simulate the behavior of an inductor. The gyrator is basically a two-terminal (or single-port) device. When a capacitor is connected to one port, the device rotates electrically, or gyrates, the capacitor so that it looks like an inductor at the other port. Two examples of gyrators are shown in Figure 2.35.

Gyrators may be used to replace inductors in low-frequency circuits. However, one terminal of the replaced inductor must be grounded for these circuits to work properly.

Oscillator

An *oscillator* is a circuit that generates an alternating current (AC) waveform with only a direct current (DC) input. Commonly, the frequency of alternation is determined by the oscillator's components.

Oscillators are quite widely used in industrial electronics for both analog and digital applications.

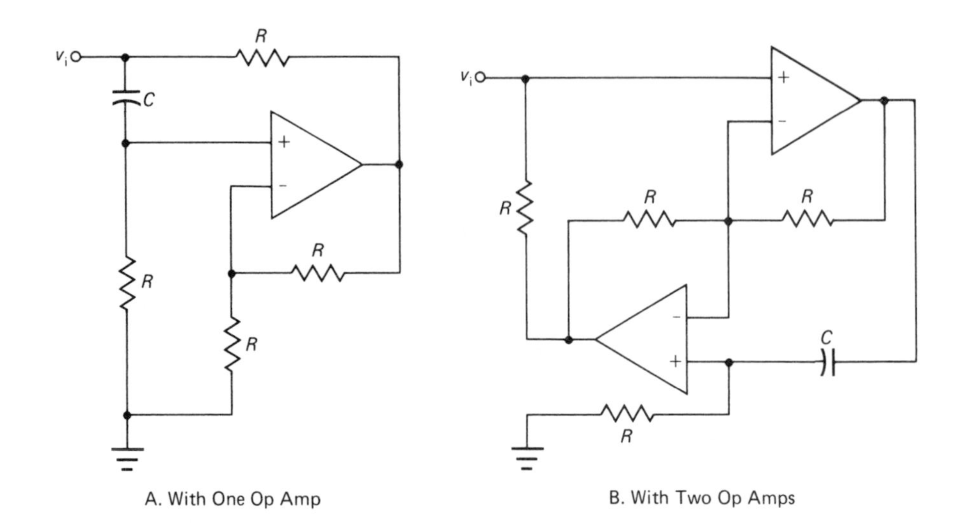

Figure 2.35 Gyrator (Simulated Inductor)

Because most oscillators require an amplification device of some kind, op amps can be used to form basic oscillator circuits.

Sine Wave Oscillator

Sine wave oscillators, whether they use op amps or other amplification devices, must meet certain criteria to sustain oscillations. These criteria are sometimes called the Barkhausen criteria. The *Barkhausen criteria* are as follows:

1. The overall gain of the circuit (called *loop gain*) must be 1 or greater. Thus, losses around the circuit must be compensated for by an amplifying device.
2. The phase shift around the circuit (input to output and back to the input) must be 0° (or 360°).

The three common types of sine wave oscillators are phase-shift, Wien-bridge, and twin T.

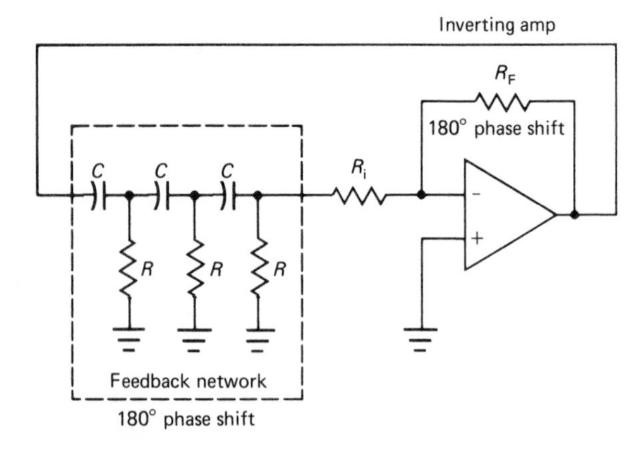

Figure 2.36 Phase-Shift, Sine Wave Oscillator

Phase-Shift Oscillator

To illustrate the Barkhausen criteria, we consider the phase-shift oscillator shown in Figure 2.36. The feedback network of the phase-shift oscillator ensures that its own 180° phase shift, when added to the same shift for the inverting amp, equals 360°. Thus, one part of the Barkhausen criteria is met. The phase-shift network attenuates by a factor of about 30. The amplifier must then have a gain of at least 30 to satisfy the remaining Barkhausen criterion.

With the resistors and capacitors as labeled in Figure 2.36, the equation for the frequency of oscillation (f_o) of this phase-shift oscillator is as follows:

$$f_o = \frac{1}{2\pi RC\sqrt{6}} \tag{2.33}$$

Wien-Bridge Oscillator

Another common sine wave oscillator is the Wien-bridge oscillator, shown in Figure 2.37. The *Wien-bridge oscillator* operates on the balanced bridge principle. When the impedance of the R_1C_1 branch equals the impedance of the R_2C_2 branch, the feedback voltage is in phase with the output voltage. Again, when the gain of the amplifier is sufficient to replace losses around the circuit, both Barkhausen criteria are met.

For the resistors and capacitors as labeled in Figure 2.37, the frequency of oscillation of the Wien-bridge oscillator is as follows:

$$f_o = \frac{1}{2\pi\sqrt{R_1C_1R_2C_2}} \tag{2.34}$$

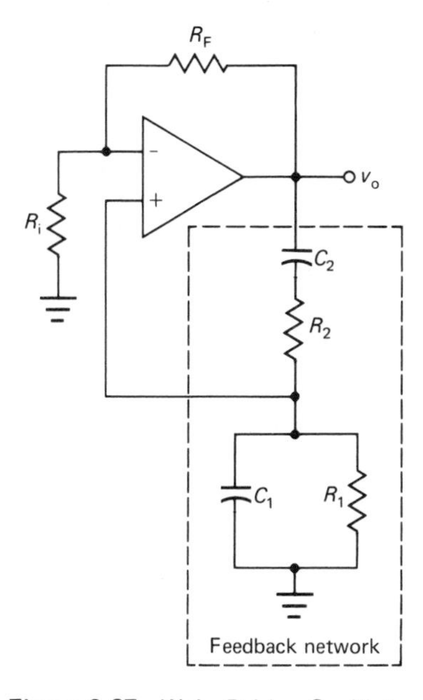

Figure 2.37 Wein-Bridge Oscillator

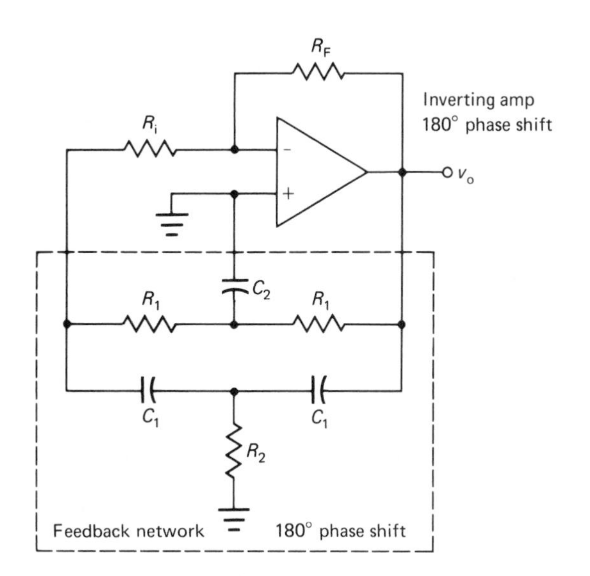

Figure 2.38 Twin T Oscillator

Twin T Oscillator

The third sine wave oscillator is the *twin T oscillator* of Figure 2.38, so named because it uses the twin T configuration discussed in the section on notch filters. The feedback network gives a 180° phase shift only to the frequencies around f_o. The twin T network has losses around 25. Hence, if the amplifier's gain is at least that much, the circuit oscillates. For the resistors and capacitors as labeled in Figure 2.38, the frequency of oscillation of the twin T oscillator is as follows:

$$f_o = \frac{1}{2\pi R_1C_1} \tag{2.35}$$

when $C_1 = 2C_2$ and $R_2 = R_1/2$.

Square Wave Oscillator

Figure 2.39 shows a square wave oscillator. When power is applied to this circuit, a small positive or negative voltage appears across R_A because of the presence of DC offset voltage. Since, at the first

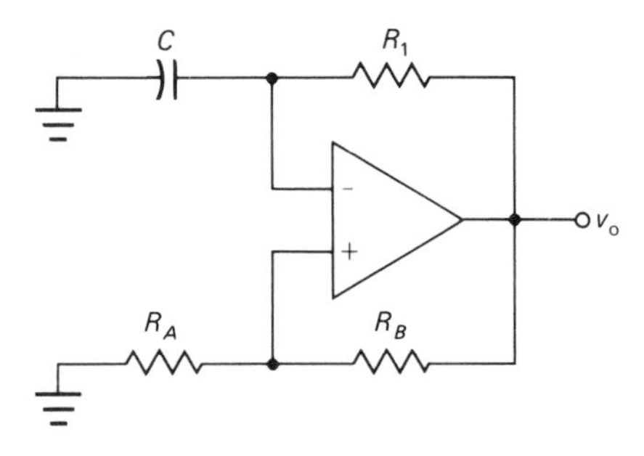

Figure 2.39 Square Wave (Relaxation) Oscillator

instant, capacitor C is uncharged, it applies 0 V to the inverting terminal. The small offset voltage appearing across R_A drives the amplifier into saturation. Capacitor C starts to charge and continues charging until it reaches the voltage across R_A. When this voltage is exceeded, the oscillator output changes from one output saturation voltage state to the other. The capacitor now discharges and then charges in the opposite direction. Again, when the voltage across the capacitor reaches the new reference voltage, it changes state. Thus, the process repeats itself.

As we see, this circuit is no more than a relaxation oscillator with an op amp comparator. A *relaxation oscillator* is an RC circuit with a capacitor that is charged and discharged (relaxed) to provide the basis for the oscillation. The output voltage oscillates between $+V_{sat}$ and $-V_{sat}$—a square wave.

For the resistors and capacitors as labeled in Figure 2.39, and when $R_A = 10R_B$, the frequency of oscillation of the square wave oscillator is as follows:

$$f_o = \frac{5}{R_1 C} \qquad \textbf{(2.36)}$$

The square wave oscillator can be used to produce a triangle wave output by cascading it to an integrator, as shown in Figure 2.40.

Miscellaneous Op Amp Applications

Throughout the text, we show applications of op amps in conjunction with other components and devices. We have chosen seven applications of op amps to discuss here to further show the flexibility and versatility of op amps.

Op Amp Supply

Figure 2.41 illustrates a method by which a dual power supply can be made from a single supply input, one op amp, two transistors, a few resistors, and a few capacitors. The circuit is basically a high-gain amplifier with a very high input impedance. Resistors R_1 and R_2 form a voltage divider that references the noninverting terminal at about half the input voltage. Feedback current from the

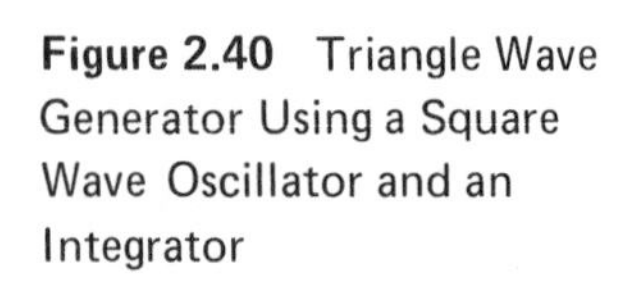
Figure 2.40 Triangle Wave Generator Using a Square Wave Oscillator and an Integrator

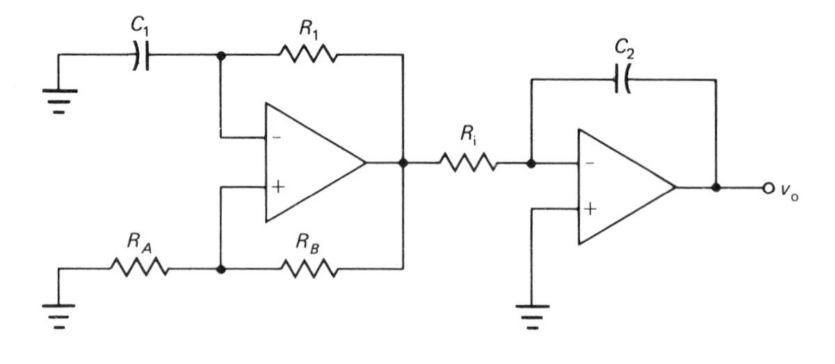

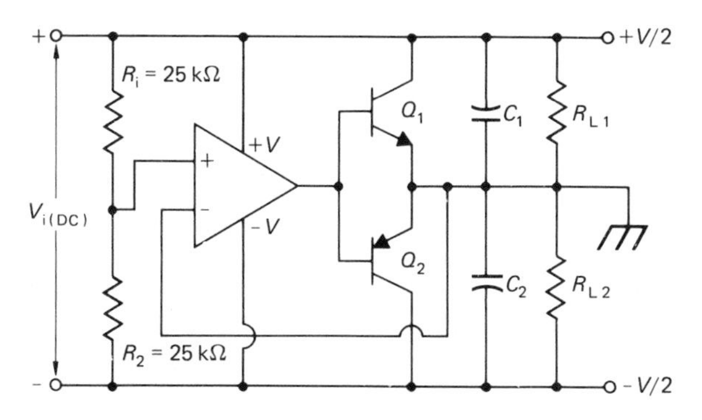

Figure 2.41 Op Amp Power Supply Circuit

junction of the Q_1 and Q_2 emitters to the input of the op amp keeps the output voltage evenly balanced.

LED Overvoltage Indicator

Figure 2.42 shows a circuit that is used to indicate the presence of an overvoltage by turning on an LED (light-emitting diode). By an adjustment of the gain of the op amp (R_F/R_i) and by proper choice of zener diodes, this circuit can be used to show an overvoltage visually. In this case, if the input voltage goes over +3 V, diode D_4 conducts, turning on D_2. If the input voltage goes lower than +3 V, diode D_3 conducts, turning on D_1.

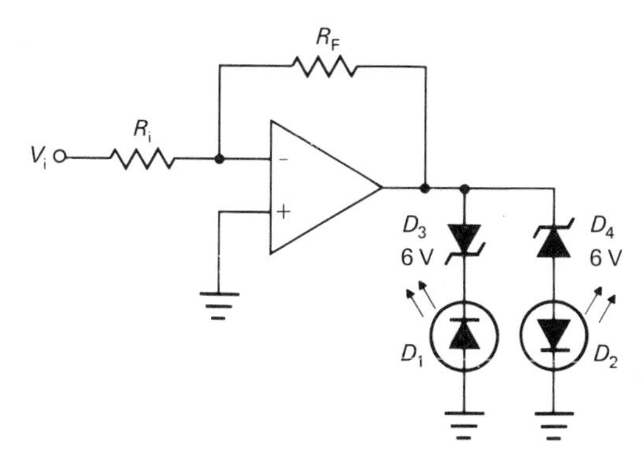

Figure 2.42 Op Amp LED Overvoltage Indicator

Precision Half-Wave Rectifier

Diodes in a precision half-wave rectifier usually do not turn on fully until about 0.7 V is dropped across them. Refer to Figure 2.43. Because the diodes are in the feedback circuit of the op amp, they turn on as soon as the input goes slightly above or below 0 V. They do so because of their high resistance before they turn on. Very high feedback resistance makes gain high. The 0.7 V needed to turn on the diodes occurs with a very small input voltage.

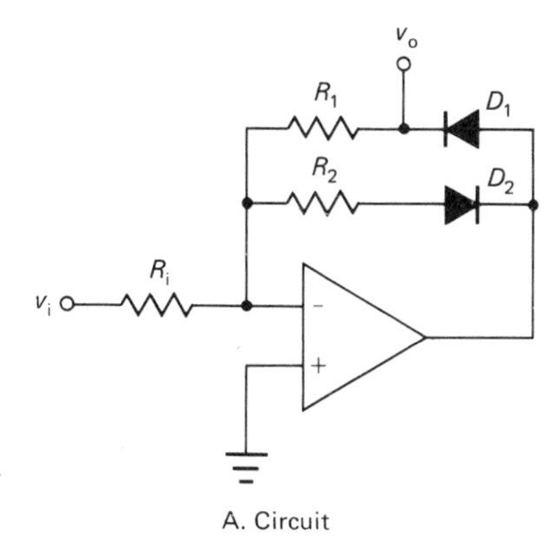

A. Circuit

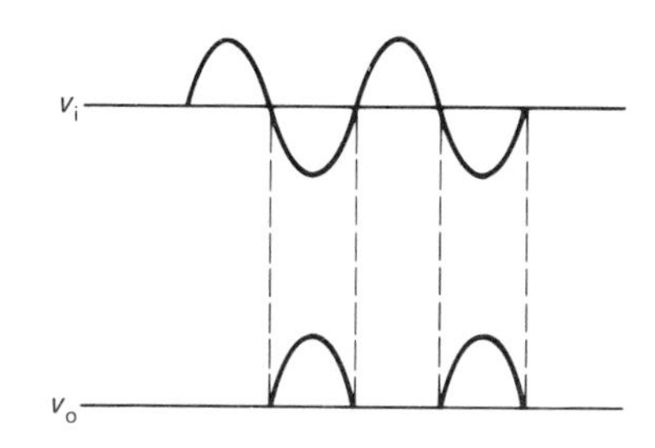

B. Input and Output Waveforms

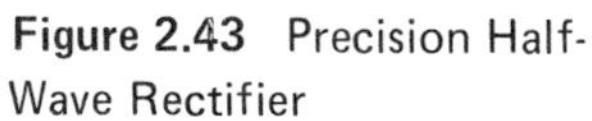

Figure 2.43 Precision Half-Wave Rectifier

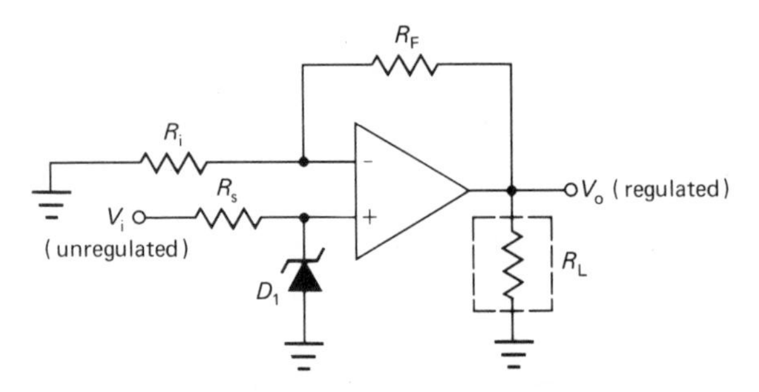

Figure 2.44 Op Amp Regulator

Op Amp Voltage Regulator

The op amp is also used as a low-current voltage regulator. Figure 2.44 illustrates such an application. Changes in the unregulated input result in a very small change across D_1 (provided it is biased correctly). Therefore, the output voltage remains relatively constant with changing load demands. The output voltage is a function of the zener voltage times the gain ($R_F/R_i + 1$). Since zener voltage, R_F, and R_i remain constant, the output remains constant.

Op Amp Voltage-to-Current Converter

Some applications require a current source rather than a voltage source. Examples include analog meter movements, resistance measurement circuits, and driving inductive loads such as relay coils. Figure 2.45 shows a device that provides a current source, the voltage-to-current converter. In such a circuit, the amount of feedback or load current does not depend on the resistance of the load. Load current depends on the input voltage and current. This effect is important, for example, in metering circuits. Only the input voltage being measured should affect the current through the meter. Any changes in the current through the meter because of meter resistance or voltage changes are undesirable.

Op Amp Current-to-Voltage Converter

Some transducers, particularly light- and heat-sensitive ones, give an output current proportional to the input variable (light or heat). To work properly with this current output device, we need an amplifier that can process input current rather than input voltage. The op amp shown in Figure 2.46 converts input current to an output voltage.

In this circuit, almost no current flows in the inverting terminal. So, any input current from the source will flow through the feedback resistance. Because of the virtual ground, the voltage drop produced by the feedback resistor is also the output voltage. The output voltage is then determined by the input current, providing that the feedback resistor stays constant.

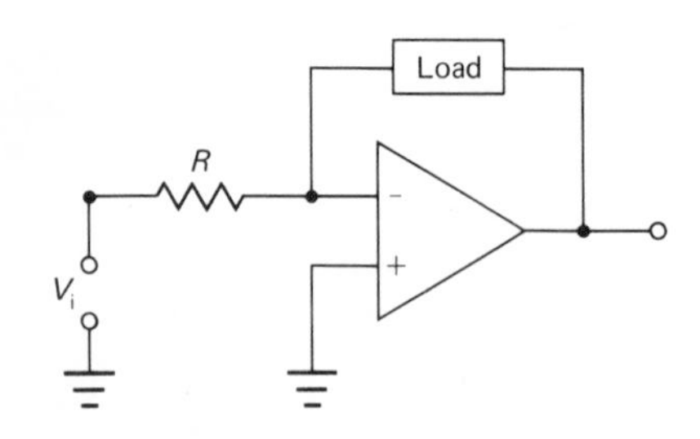

Figure 2.45 Op Amp Voltage-to-Current Converter

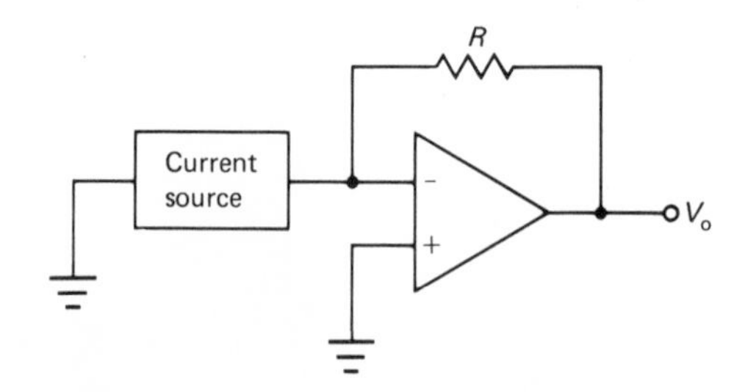

Figure 2.46 Op Amp Current-to-Voltage Converter

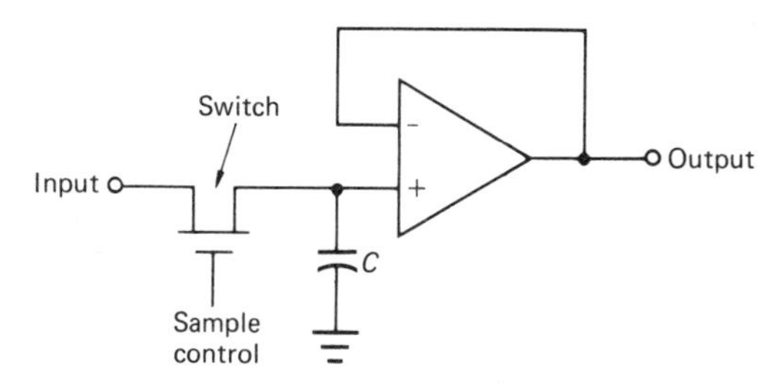

Figure 2.47 Op Amp Sample-and-Hold Circuit

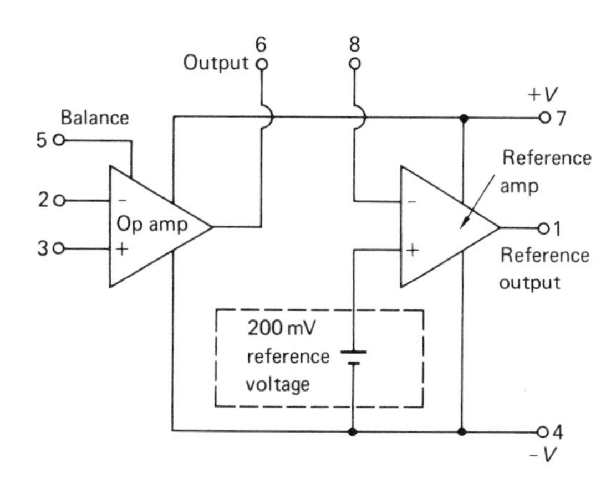

Figure 2.48 Block Diagram of LM10 Low-Voltage Op Amp

Op Amp Sample-and-Hold Circuit

A *sample-and-hold (S/H) circuit*, as its name implies, is used to sample a voltage at a particular time and hold it constant for another specified time interval. Figure 2.47 shows just such a circuit designed with an op amp.

In the S/H circuit, when a sample control voltage is applied to the switch, the switch turns on. The capacitor charges owing to the input voltage applied across it by the switch. The capacitor continues to charge until the switch is turned off. Since there is very little resistance to current flow in the charging circuit, the capacitor charges instantly to the input voltage.

After the switch is turned off, the input voltage stays on the capacitor. The voltage on the capacitor is then a measurement of the voltage at the input at the instant the sample was taken. The time when the switch is on is called the *sample period*. When the switch is off, the time is called the *holding period*.

Low-Voltage Op Amp

In 1979, National Semiconductor Corporation introduced a new device into the op amp world, the LM10. The LM10 is a low-voltage op amp. Its two main advantages are a voltage supply range from as low as 1.1 V to as high as 40 V and the ability to function in a floating or conventional mode. A block diagram of the LM10 is shown in Figure 2.48. Performance characteristics of the LM10 are similar to those of the LM108 op amp. The LM10, though, has a high output drive capability in addition to built-in thermal overload circuitry.

Current-Differencing Amplifier (CDA)

The *current-differencing amplifier* (*CDA*), or Norton amplifier, developed in the early 1970s, answered a need in the linear device field. Industry required a device that would be linear yet compatible with digital circuitry and one that would operate from a single power supply. The CDA meets these needs and is relatively inexpensive as well. Four CDAs are housed within a 14-pin, IC, dual inline package (DIP).

The basic CDA circuit is shown in Figure 2.49. The inverting section of the basic CDA is shown in Figure 2.49A. In Figure 2.49B, the resistors R_C and R_E have been replaced by transistors acting as constant-current sources. Since constant-current sources have very high dynamic resistances, the gain of this circuit reaches 60 dB.

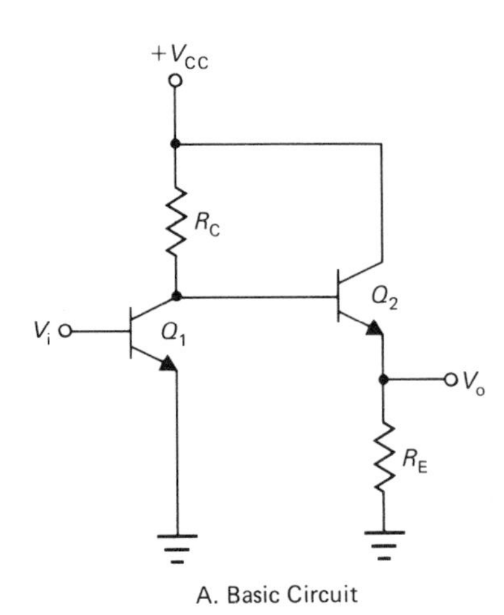

A. Basic Circuit

B. With Constant-Current Sources (Transistors) Replacing Resistors

Figure 2.49 Inverting Section of CDA

Figure 2.50 illustrates the addition of a circuit that provides a noninverting input. This addition makes a differential input device. Transistor Q_5 and diode D_1 form a *current mirror stage.* Since the base-emitter junction of Q_5 and the cathode-anode junction of D_1 are matched, the application of a bias voltage causes both devices to conduct the same amount of current. Or we say that the current in D_1 is *mirrored* in Q_5. The CDA then tries to keep the difference between the two currents equal to zero. The same concept appears in the op amp, which tries to keep the difference in potential between the two inputs equal to zero. The feedback resistor, the only component external to the IC in Figure 2.50, provides a path for current from Q_5 to the output transistor Q_3.

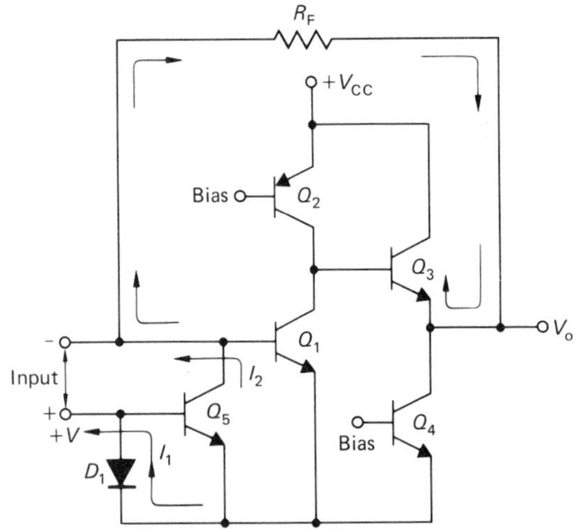

Figure 2.50 Simplified CDA Schematic Showing Inverting and Noninverting Terminal Inputs

Biasing the CDA

Figure 2.51 shows the biasing arrangement used most for CDAs. When $+V$ is connected to R_1, current flows (I_b). The CDA mirrors the same amount of current through R_F as flows through R_1. The current through R_1 can be found from the following equation:

$$I_1 = \frac{+V - 0.7}{R_1} \approx \frac{+V}{R_1} \tag{2.37}$$

The output voltage likewise is as follows:

$$V_{out} = (I_F R_F) + 0.7 \approx I_F R_F \tag{2.38}$$

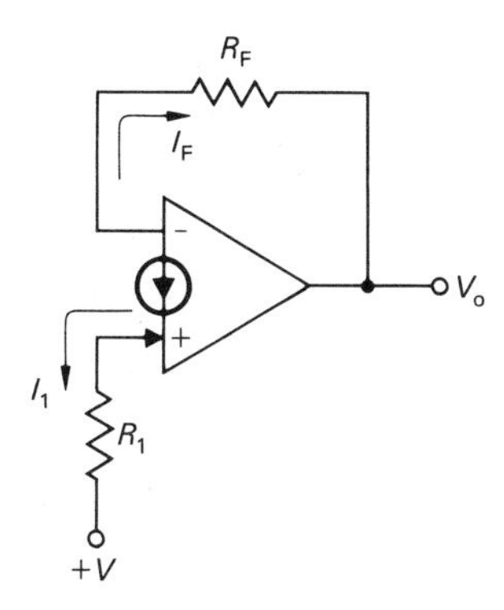

Figure 2.51 Current Mirror Biasing the CDA

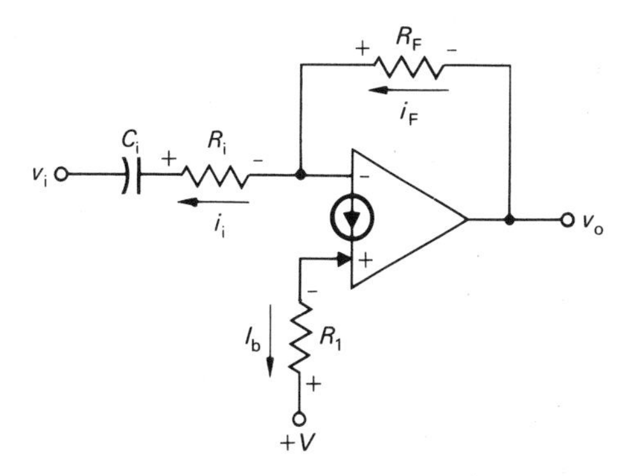

Figure 2.52 CDA Inverting Amp

Since $I_1 = I_F$ (from the mirror principle), we have the following equation:

$$V_o = \left(\frac{+V}{R_1}\right) R_F \tag{2.39}$$

The output DC voltage in quiescent conditions is determined by the bias supply ($+V$), R_F, and R_1. For centered operation (where $V_o = +V/2$), $R_1 = 2R_F$. If $2R_F$ is substituted into Equation 2.32, then $V_o = +V/2$ (centered operation).

CDA Inverting Amplifier

The CDA inverting amplifier is similar in many ways to the comparable op amp circuit. The gain equation is identical. Figure 2.52 shows the basic CDA inverting amplifier. It is generally used in AC applications because the DC bias may be affected by the previous stage if there is no coupling capacitor.

Let us suppose that a positive-going input signal is applied to this circuit. Current i_i (the current due to the AC voltage applied to the input) flows as illustrated. The current mirror, which tries to keep the same current flowing out of the inverting terminal, causes the total current in R_F to decrease by an amount equal to i_i. So the change in R_F current, denoted as i_i, is then equal but opposite to i_F, which is the AC component of the total current in R_F. The gain equation is as follows:

$$A_V = \frac{v_o}{v_i} \tag{2.40}$$

where v_i is the AC input voltage and v_o is the AC output voltage.

Thus, we can state v_o and v_i in terms of currents and resistances:

$$\frac{v_o}{v_i} = \frac{-i_F R_F}{i_i R_i} \tag{2.41}$$

Since $i_i = i_F$, this equation reduces the gain equation to the following:

$$A_V = \frac{v_o}{v_i} = \frac{-R_F}{R_i} \tag{2.42}$$

The output is now reduced from the $+V/2$ level by the amount A_V times v_i.

Figure 2.53A illustrates a basic inverting amplifier using a CDA. Current through R_1 (which is designated I_1) is calculated from Equation 2.37 as follows:

$$I_1 = \frac{+V - 0.7\text{ V}}{R_1} = \frac{+10\text{ V} - 0.7\text{ V}}{2\text{ M}\Omega}$$
$$= 4.6\ \mu\text{A} \tag{2.43}$$

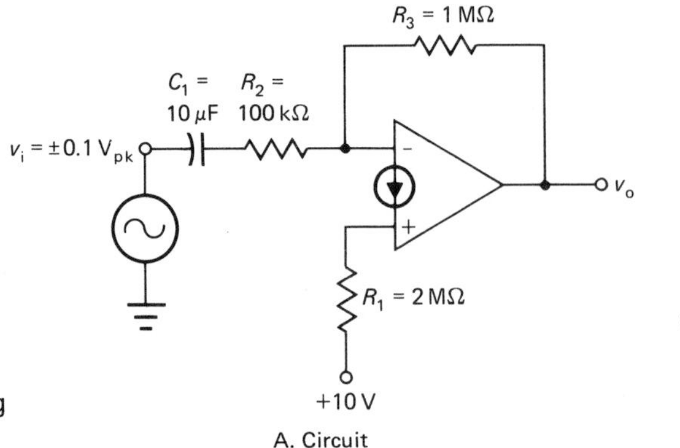

Figure 2.53 CDA Inverting Amp with AC Input

A. Circuit

B. Input and Output Waveforms

From our previous discussion, we know that I_3 equals I_1, where I_3 is the current through R_3. The output voltage, then, is as follows:

$$V_o = (I_F R_F) + 0.7 \text{ V} = 4.6 \text{ V} + 0.7 \text{ V} = 5.3 \text{ V}$$
$$\approx \frac{+V}{2} = 5 \text{ V} \qquad \textbf{(2.44)}$$

If we inject a 0.1 V peak signal into this amplifier, the output voltage change, where $A_V = -R_F/R_1 = -10$, is as follows:

$$v_o = A_V v_i = 10 \times 0.1 \text{ V}$$
$$= 1 \text{ V}_p \qquad \textbf{(2.45)}$$

Thus, a 0.1 V peak input produces a 1 V peak output riding on a +5 V DC level. This relationship is diagramed in Figure 2.53B.

CDA Noninverting Amplifier

The circuit in Figure 2.54 connects a CDA so that it operates as a noninverting amplifier. The gain equation for this amp is the same as that of the inverting amp (except for the sign):

$$A_V = \frac{R_3}{R_2} \qquad \textbf{(2.46)}$$

As in the inverting amplifier, the output voltage is +5 V, with about 4.6 μA flowing through R_1 and

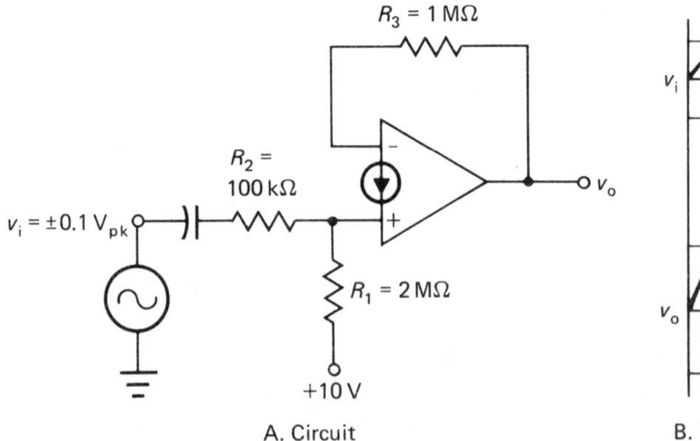

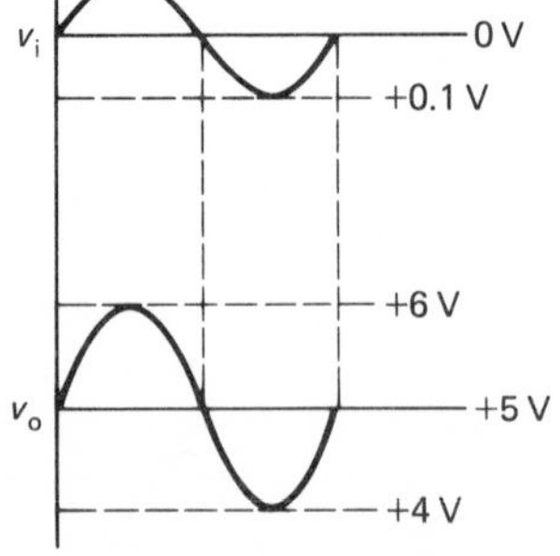

Figure 2.54 Noninverting CDA Amp with AC Input

A. Circuit

B. Input and Output Waveforms

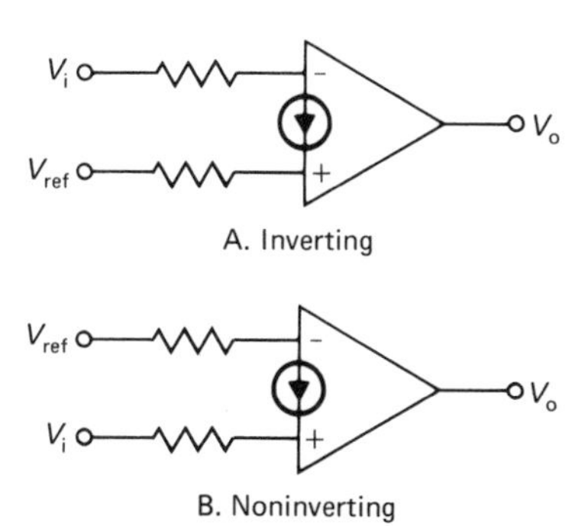

Figure 2.55 CDA Comparator

R_3 in the no-input-signal condition. The gain is also identical ($A_V = 10$). A 0.1 V peak input produces a 1 V peak output with no phase difference.

CDA Comparator

The CDA comparator is illustrated in Figure 2.55. The output of the device in Figure 2.55A is 0 V as long as the input voltage V_i is below the reference voltage V_{ref}. When the input voltage rises above V_{ref}, the output goes to saturation voltage and stays there as long as the input remains above the reference. Comparators such as these find extensive application in digital systems.

We have seen from this discussion that the CDA can substitute for the op amp in many applications. In fact, most of the op amp configurations can be duplicated by the CDA.

Operational Transconductance Amplifier (OTA)

A *transconductance amplifier* is a circuit or component that changes an input voltage into an output current. An *operational transconductance amplifier* (*OTA*) is an IC transconductance amplifier. In many ways, the OTA resembles the op amp device discussed earlier. It possesses a differential voltage input, has a high input impedance, and high open-loop gain. The most striking difference in the two devices can be seen in the schematic diagram in Figure 2.56. Note the addition of a terminal labeled I_{ABC} in the OTA. This terminal is a control input that is called the *amplifier bias current control.* Many of the OTA's parameters can be controlled by varying the current in this terminal.

Table 2.3 shows the changes in OTA parameters with an increase in I_{ABC}. Notice that the parameter *transconductance* is given in Table 2.3 instead of voltage gain. Recall that transconductance is the change in output current divided by the change in input voltage.

The OTA, as its name implies, differs in regard to its output characteristics. Recall that the op

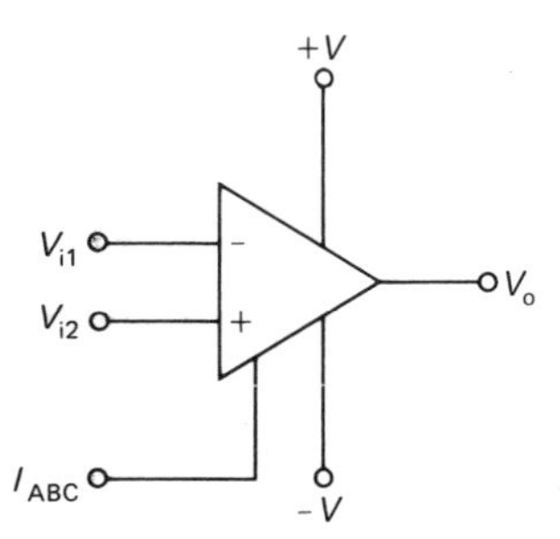

Figure 2.56 Schematic Diagram of the OTA

Table 2.3 *Changes in OTA Parameters with Increased I_{ABC}*

Parameters	*Changes*
Input resistance	Decrease
Output resistance	Decrease
Input capacitance	Decrease
Output capacitance	Decrease
Transconductance	Increase
Input bias current	Increase
Input offset current	Increase
Input offset voltage	No change
Slew rate	Increase

amp's output was defined in terms of voltage. In contrast, the OTA's output is defined in terms of current. In this regard, the OTA resembles an FET (field effect transistor) or vacuum tube. Therefore, transconductance is the suitable parameter to express this relationship, being the change in output current divided by the change in input voltage.

Like FETs and vacuum tubes, the OTA can be used as a voltage amplifier with the addition of load resistance or impedance. The voltage gain in this case equals the (mutual) transconductance (g_m) times the load resistance ($g_m R_L$). As expected, the OTA, because of its transconductance properties, has an output impedance much higher than that of the standard op amp.

Figure 2.57 shows the input stage of the OTA. Notice that it is a differential amplifier, just like the op amp. Transistor Q_3 and diode D_1 form the current source for the differential amplifier. This circuit differs from the op amp circuit in that the collector current of Q_3 can be changed by varying the current in the I_{ABC} terminal. Changing this current varies the transconductance of the differential amplifier.

Numerous applications exist for OTAs. It is probably best suited to telemetry and communications use by virtue of the unique control provided by the I_{ABC} terminal. The OTA is also used in measurement circuits in a sample-and-hold configuration. Other applications include nonlinear mixing, gain control (including automatic gain control), waveform generation, and multiplication.

An example of an OTA application is the circuit shown in Figure 2.58, which has an amplitude modulator. Here, the output current is equal to the transconductance times the input voltage. In this amplifier, the level of the unmodulated carrier signal (V_c) is controlled by the average current into the I_{ABC} terminal. The carrier frequency is amplitude-modulated when the modulating signal V_m causes changes in I_{ABC}. As the input-modulating

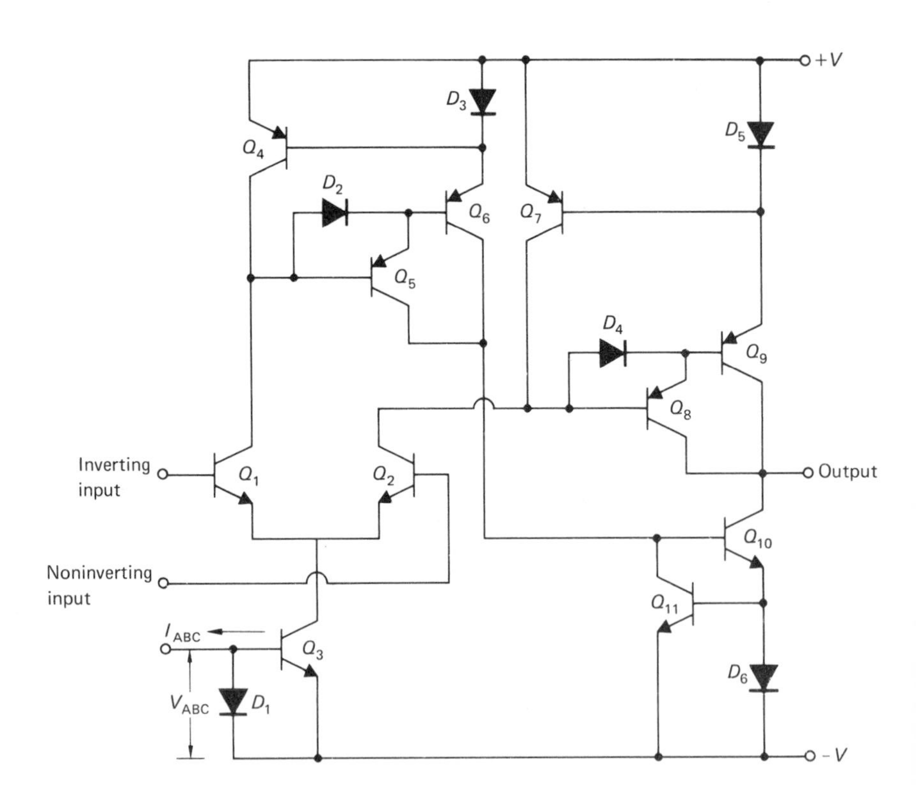

Figure 2.57 Internal Construction of the OTA

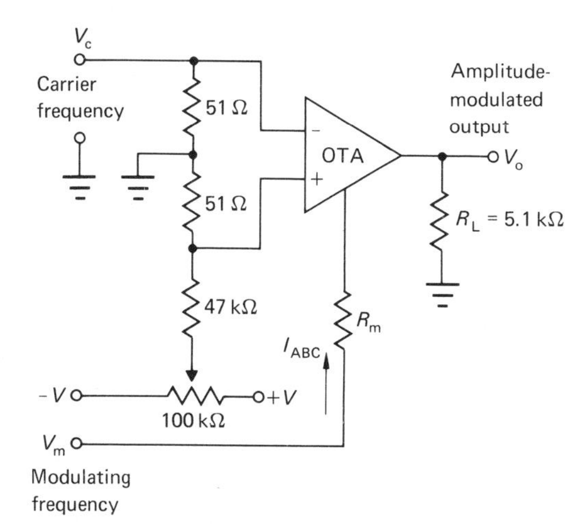

Figure 2.58 Amplitude Modulation Using an OTA

signal (V_m) goes positive, the current in I_{ABC} increases, thus increasing the transconductance of the OTA. As V_m goes negative, the current in I_{ABC} decreases, decreasing the OTA's g_m. The output, developed across the 5.1 kΩ resistor, is the familiar amplitude-modulated waveform.

Programmable Op Amps

Most op amps have characteristics that the user has little or no control over. For instance, the input bias current for an op amp is generally fixed for each device according to its internal construction. *Programmable op amps*, however, allow the user to change some of these characteristics. Often, circuit designers are forced to make trade-offs when choosing a particular device for a job. The programmable op amp gives the engineer more flexibility in fitting an op amp to a particular application.

The programmable op amp has a special terminal similar to the I_{ABC} terminal on the transconductance amp. The control terminal on the programmable op amp is called the *set terminal*. Current flow from this terminal can change an op amp's slew rate, gain-bandwidth product, input offset current and voltage, and noise.

Troubleshooting Op Amps

Once a problem has been isolated to an IC op amp, the op amp itself must be checked. Fault analysis of any kind can be broken down into three categories: no output, low output, or distorted output. A simple check of the op amp's output determines in which of these general areas a malfunction is occurring.

No Output

A no-output condition may be caused by any of the following faults:

- Lack of power supply—check the op amp terminals for proper supply voltages.
- Lack of input voltage—check for the proper input signal or voltage.
- Saturated op amp—check for undesired DC inputs that may be causing saturation, check for a shorted input resistor or an open feedback resistor, and check for proper grounding of components and terminals.
- Malfunctioning op amp—take out the op amp and check it in an inverting amp configuration or replace it. (The inverting mode of operation ensures that the op amp is working in order to give an inverted output. A short or open in the noninverting mode may cause it to appear to be working poorly.)

Low Output

A low-output condition may be caused by any of the following faults:

- Low power supply—check the op amp terminals for proper supply voltages.
- Low input voltage—check the op amp input for proper signal or voltage levels.
- Change in component values—check the input and feedback resistances for changes in resistance.
- Malfunctioning op amp—replace it with a good unit.

Distorted Output

A distorted-output condition may be caused by any of the following faults:

- Distortion in the power supply or low power supply—check the op amp terminals for proper supply voltages.
- Distorted input—check the input to the op amp for distortion.
- Malfunctioning op amp—replace the op amp with a good unit.
- Operation outside of published specifications—check the slew rate, gain bandwidth, and so on.

Troubleshooting Breadboarded Circuits

As a technician, you may either breadboard circuits of your own or be required to check breadboarded circuits. In addition to observing the troubleshooting hints mentioned above, you should closely check your circuit's wiring. Both beginner and expert often do not wire a circuit correctly. Make sure that the IC pinout from the data sheet is observed.

Conclusion

We have only scratched the surface of op amp applications in this chapter. But we have seen that the IC op amp has applications perhaps undreamed of by those who developed it in the 1960s. Since this device is a major analog circuit building block, it is important that you have a firm grasp on its operation.

Questions

1. The voltage follower is a special application of the ______ amplifier. Its input impedance is considered to be ______.
2. The difference amplifier is a combination of the inverting and non-inverting amplifiers. As such, it has ______ inputs.
3. The instrumentation amplifier is a ______ amplifier with each input connected to a voltage follower. It amplifies difference-mode signals while ______ common-mode noise.
4. The output of the integrator is a voltage that is produced by ______ the input signal over time. The output of a differentiator is proportional to the ______ of the input.
5. In the log amp, the output is proportional to the ______ of the input. The mathematical operation of ______ can be accomplished with log amps.
6. A filter that passes everything above a specified cutoff frequency is called a ______ filter. The roll-off of a first-order filter is ______ per decade.
7. The CDA is compatible with ______ logic, since it can operate on one 5 V supply.
8. In the OTA, the output ______ is proportional to the input voltage.
9. Describe the gain and input impedance of the voltage follower and some applications of this circuit.
10. Describe the advantages of the difference amplifier when compared with single-input circuits.
11. Contrast the integrator and differentiator in terms of function and input and output waveforms.
12. What is the major advantage of a comparator with hysteresis over one without it?
13. List the four different types of pass filters, and give an example of each in an active filter.
14. What are the advantages of active filters over passive ones?
15. Describe the function of a gyrator.
16. Describe the need for and construction of a CDA.
17. What is the phase relationship between input and output in the circuit shown in Figure 2.54?
18. Compare the characteristics and applications of op amps, CDAs, and OTAs.

Problems

1. Suppose the difference amp in Figure 2.3 has the following resistances: $R_F = 20\ k\Omega$, $R_i = 10\ k\Omega$, $R_A = 5\ k\Omega$, and $R_B = 5\ k\Omega$.
   a. Calculate V_o with $V_{i1} = +1$ V and $V_{i2} = -2$ V.
   b. Draw the resultant output waveform caused by

applying a 2 V peak input sine wave at each input (assume both signals are in phase).

2. Suppose the instrumentation amp in Figure 2.4 has $R_1 = 45\ k\Omega$, $R_2 = 10\ k\Omega$, and $R_3 = 100\ k\Omega$, with R_A adjusted to 10 kΩ. With a differential input of 1 V, what is the output voltage?
3. The integrator in Figure 2.8 has a 1 kHz, 1 V peak input sine wave. What is the output waveform amplitude? How does the output waveform compare with the input in phase and waveshape?
4. The integrator in Figure 2.8 has a 2 kHz, 0.5 V peak, square wave input. Draw the output waveform in the correct phase with the input, and indicate the output amplitude.
5. Draw the output waveform, indicating proper phase and amplitude, with the inputs to Figure 2.11 as follows:
   a. 2.5 kHz, 1 V peak sine wave.
   b. 1.0 kHz, 0.5 V peak triangle wave.
6. Draw the output waveform of the comparator circuit shown in Figure 2.19A with a 5 V peak, sine wave input and a V_{ref} of -1 V (assume $\pm V_{sat} = \pm 10$ V).
7. Calculate upper and lower trip points (V_{ref}) for the circuit shown in Figure 2.21 (assume $\pm V_{sat} = \pm 15$ V, $R_A = 4$ kΩ, and $R_B = 1$ kΩ).
8. Assume the following resistance and voltage values in Figure 2.52: $R_i = 100$ kΩ, $R_F = 470$ kΩ, $R_1 = 1$ MΩ, and $+V = +20$ V. Calculate (a) A_V, (b) $V_{o(DC)}$, (c) $V_{o(AC)}$ with a 2 V peak input sine wave, and (d) I_b.
9. With the supply voltage shown in Figure 2.53, a DC voltage of 7.5 V is measured at the output of the CDA.
   a. What is the ratio of R_3 to R_1 that produces this voltage?
   b. With this bias point, what is the maximum peak voltage that can be applied at the input without distorting the output. *Hint:* How far can the output voltage go in a positive direction?
10. Assume the following resistance and voltage values in Figure 2.54: $R_1 = 470$ kΩ, $R_2 = 50\ k\Omega$, $R_3 = 220$ kΩ, $+V = 5$ V, and the input sine wave is $+0.1$ V peak. Calculate (a) A_V, (b) $V_{o(DC)}$, (c) $V_{o(AC)}$, and (d) I_b.

DC Motors and Generators

Objectives

On completion of this chapter, you should be able to:

- Describe the characteristics of ideal, series, shunt, separately excited, and compound DC motors and generators;
- Identify the schematic diagrams of the series, shunt, and separately excited DC motors and generators;
- Predict the performance of DC motors and generators;
- Give examples of each of the configurations of series, shunt, separately excited, and compound DC motors and generators.

Introduction

In these days of rapid electronic advancement, technicians, especially digital electronic technicians, frequently overlook the importance of the electric motor as a key system element. In fact, in some educational systems, digital electronics students are not required to take industrially related courses, such as courses related to motors and generators. In this chapter on DC motors and generators, we have attempted to correct that situation by presenting a condensed version of the topic. While, of course, some topics are omitted here, the important principles are presented in some detail.

CHAPTER

3

Many applications in industry and in process control (see Chapter 8) require electric motors as critical functional elements. Because such motors are electromechanical rather than purely electrical, their response sets the entire system's performance limits. So, technicians who wish to be completely functional in the industrial electronics environment must be familiar with the varieties of motors and generators. In addition, technicians must understand basic operating principles if they are to perform intelligent troubleshooting (and system design).

In discussions involving both DC motors and DC generators, it is convenient to refer to a generalized electric machine, because much of what is said about electric motors is equally applicable to generators. The generalized electric machine is referred to as a *dynamo*, a machine that converts either mechanical energy to electric energy or electric energy to mechanical energy. When a dynamo is driven mechanically by a power source such as a gas or diesel engine, or a steam or water turbine, and provides electric energy to a load, it is called a *generator*. If electric energy is supplied to

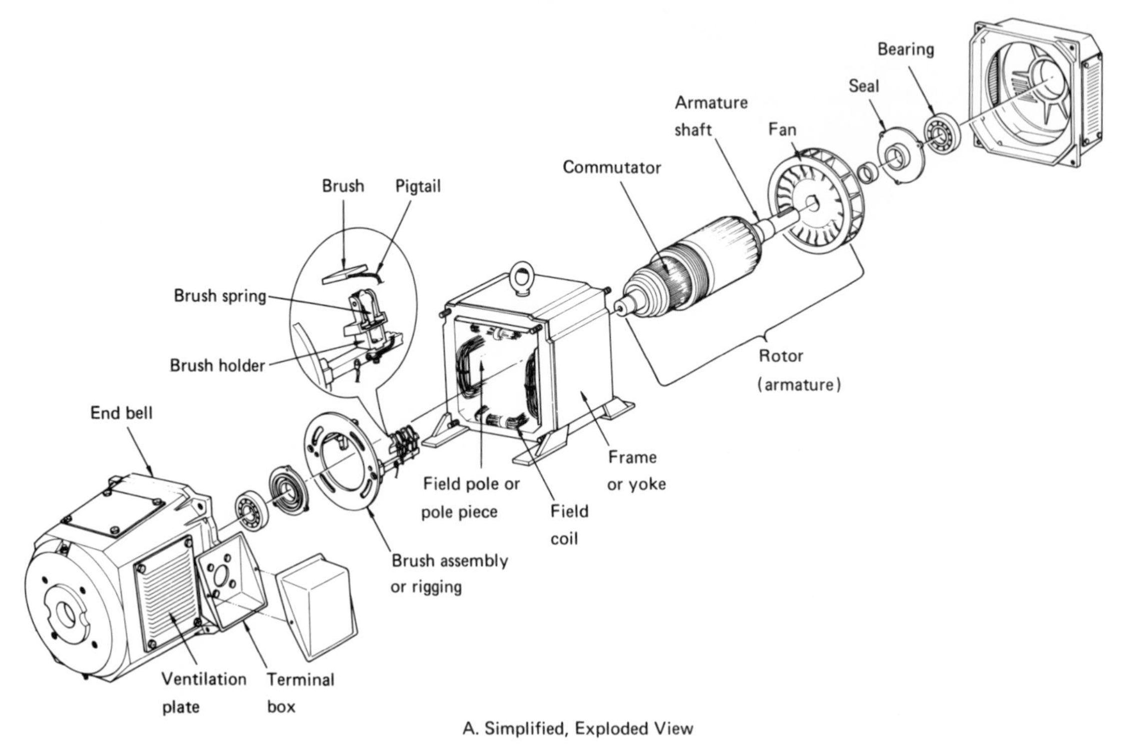

A. Simplified, Exploded View

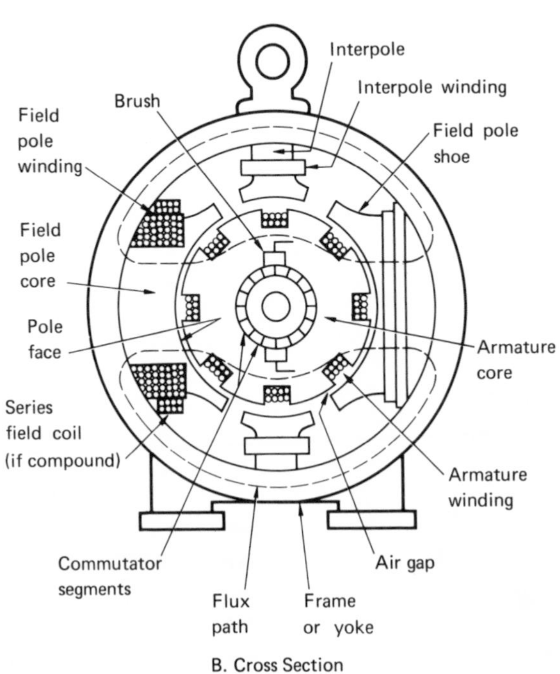

B. Cross Section

Figure 3.1 DC Dynamo

the dynamo, and its output is used to provide mechanical motion or *torque* (a force acting through a distance), it is called a *motor*. Generators are rated at the kilowatts they can deliver without overheating at a rated voltage and speed. Motors are rated at the horsepower they can deliver without overheating at their rated voltage and speed.

Dynamo Construction

Figure 3.1A shows a simplified, exploded view of a DC dynamo. Figure 3.1B shows the dynamo in cross section.

The dynamo construction consists of two major subdivisions, the *rotor* (*armature*), or rotating part, and the *stator*, or stationary part. We will discuss these two parts in detail in the following sections.

Rotor

The rotor consists of armature shaft, armature core, armature winding, and commutator.

The *armature shaft* is the cylinder on which the rotor components are attached. The armature core, armature winding, and commutator (mechanical switch) are attached to the armature shaft. This entire assembly rotates. Sometimes, fins are attached to the shaft to provide cooling of the dynamo.

The *armature core* is constructed of laminated (thin-sheet) layers of sheet steel. These layers provide a low-reluctance (magnetic resistance) path between the magnetic field poles. The laminations are insulated from each other and attached together securely. The core is laminated to reduce *eddy currents*, which are circular-moving currents. The grade of sheet steel used is selected to produce low loss due to hysteresis (lagging magnetization). The outer surface of the core is slotted to provide a means of securing the armature coils.

Two basic armature, or field pole, forms are available in the construction of the dynamo: salient (standing out from the surrounding material) and nonsalient (minimal projection) poles. Figure 3.2 illustrates these forms for the armature of the dynamo.

The *armature winding* consists of insulated coils, which are insulated from each other and from the armature core. The winding is embedded in the slots in the armature core face, as shown in Figure 3.1B.

The *commutator* consists of a number of wedge-shaped copper segments that are assembled into a cylinder and secured to the armature shaft. The segments are insulated from each other and from the armature shaft, generally with mica. The commutator segments are soldered to the ends of the armature coils. Because of the armature rotation, the commutator provides the necessary switching of armature current to or from the circuit external to the armature.

The armature of the dynamo performs the following four major functions:

1. It permits rotation for mechanical generator action or motor action.
2. Because of rotation, it provides the switching action necessary for commutation.
3. It provides housing for the armature conductors, into which a voltage is induced or which provide a torque.

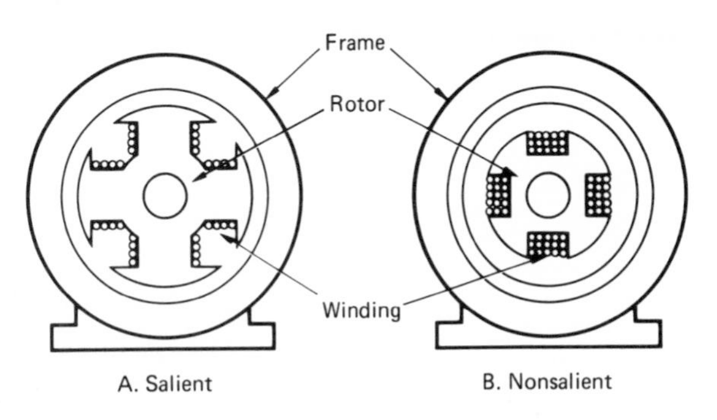

Figure 3.2 Pole Construction

4. It provides a low-reluctance flux (magnetic lines) path between the field poles. Recall that *flux* or *flux lines* are invisible magnetic lines of force.

Stator

The stator consists of eight major components: field yoke, field windings, field poles, air gap, interpoles, compensating windings, brushes, and end bells. See Figure 3.1.

The *field yoke*, or *frame*, is made of cast or rolled steel. The yoke supports all the parts of the stator. It also provides a return path for the magnetic flux produced in the field poles.

The *field windings* are wound around the pole cores and may be connected either in series or in parallel (shunt) with the armature circuit. The windings consist of a few turns of heavy-gauge wire for a series field, or many turns of fine wire for a shunt field, or both for a compound (both series and shunt field) dynamo.

The *field poles* are constructed of laminated sheet steel and are bolted or welded to the yoke after the field windings have been placed on them. The *pole shoe* is the end of the pole core next to the rotor. It is curved and is wider than the pole core. This widening of the pole core spreads the flux more uniformly, reduces the reluctance of the air gap, and provides a means of support for the field coils.

The *air gap* is the space between the armature surface and the pole face. The air gap varies with the size of the machine, but generally, it is between 0.6 and 0.16 cm.

The *interpoles*, or *commutating poles*, are small poles placed midway between the main poles. The interpole winding consists of a few turns of heavy wire, since interpoles are connected in series with the armature circuit. The interpole magnetic flux is, therefore, proportional to the armature current.

The *compensating windings*, shown in Figure 3.3, are used only in large DC machines. They are connected in the same manner as the interpole windings but are physically located in slots on the pole shoe.

Brushes rest on the commutator segments and are the sliding electrical connection between the armature coils and the external circuit. Brushes are made of carbon of varying degrees of hardness; in some cases, they are made of a mixture of carbon and metallic copper. The brushes are held in place by springs in brush holders, the entire assembly being called the *brush rigging*. Electrical connection between the brush and brush holder is made with a flexible copper braid called a *pigtail*.

End bells support the brush rigging as well as the armature shaft. The end bells generally determine the type of protection the dynamo receives; that is, the construction can be open, semiguarded, dripproof (as in Figure 3.1), or totally enclosed. Semiguarded types have screens or grills to protect rotating or electrically live parts and to provide ventilation. A dripproof housing protects the equipment from dripping liquids. The totally enclosed construction is completely sealed and must dissipate the heat buildup by radiation from the enclosing case.

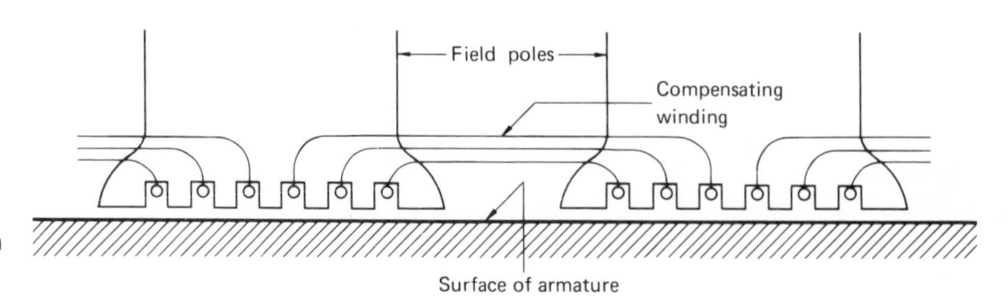

Figure 3.3 Location of Compensating Winding in the Face of Field Poles

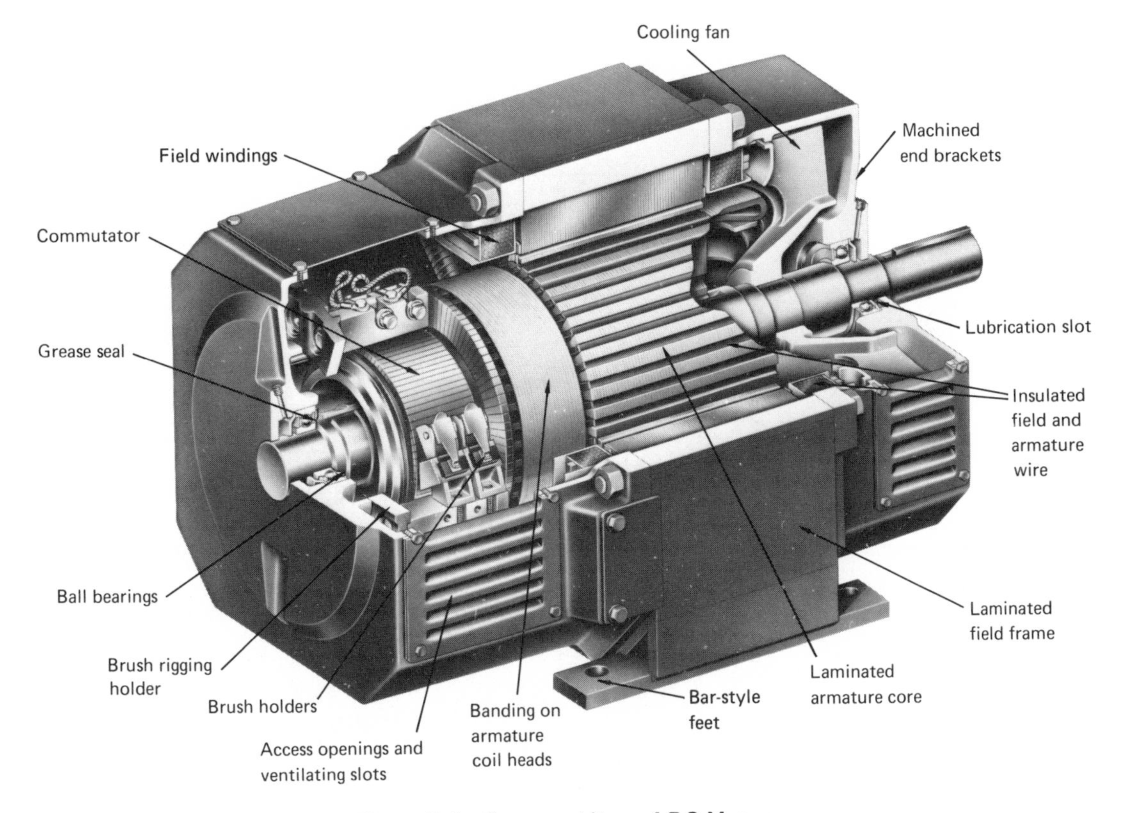

Figure 3.4 Cutaway View of DC Motor

Figure 3.4 shows a cutaway view of a completely assembled DC machine, in this case, a DC motor. Notice the dual brushes in each brush rigging, provided for high currents. Also, the side access covers are of dripproof construction.

Dynamo Classification

The construction shown in Figures 3.1 and 3.4 is known as a wound-field dynamo. There are many other classifications of dynamos. Figure 3.5 charts selected DC motor classifications and shows a rich variety from which to choose. Each class of DC motors has its advantages, depending on the desired applications. The types of DC motors shown in Figure 3.5 will be briefly discussed later; a table of their characteristics is given at the end of the chapter. However, the conventional permanent-magnet (PM) and wound-field DC motors are the subject of most of the discussion in this chapter.

Unlike DC motors, DC generators are primarily of the conventional permanent-magnet and wound-field types. Consequently, they will be the only types discussed.

Basic Principles of the DC Generator

The simplest generator that can be built is an AC generator, shown in Figure 3.6. The basic generator

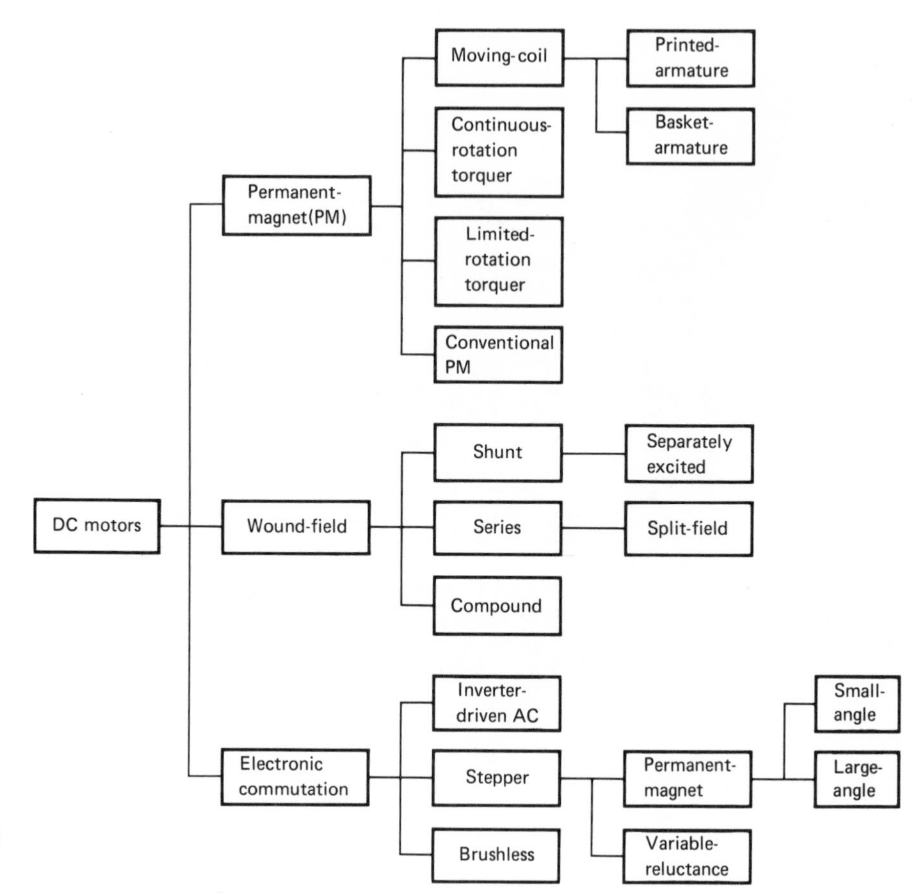

Figure 3.5 Selected DC Motor Types

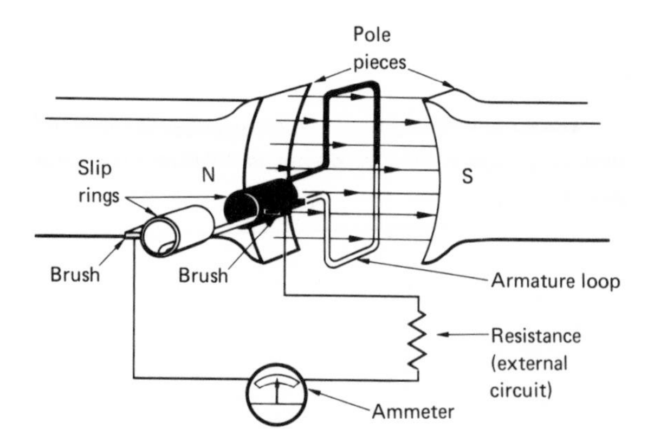

Figure 3.6 Elementary AC Generator

principles are most easily explained through the use of the elementary AC generator. For that reason, the AC generator will be discussed first. The DC generator will be developed next.

Elementary AC Generator

The elementary generator of Figure 3.6 consists of a single wire loop (the armature) placed so that it can be rotated in a stationary magnetic field. The magnetic field is shown in Figure 3.6 with arrows going from the N (north) pole to the S (south) pole.

The pole pieces are shaped and positioned as shown to concentrate the magnetic field as close as possible to the wire loop. The rotation of the loop will produce an induced voltage, or emf (electromotive force), in the loop. Brushes sliding on the slip rings connect the loop to an external circuit resistance, allowing current to flow because of the induced emf. *Slip rings* are metal rings around the rotor shaft. Each ring attaches to the separate ends of the rotor coil. If the brushes were not contacting the slip rings but the loop was rotating, an emf would still appear at the slip rings but no current would flow.

The elementary generator produces a voltage in the following manner: Assuming the armature loop is rotating in a clockwise direction, its initial or starting position is as shown in Figure 3.7A. This position is called the 0° position, or the *neutral plane.* At the 0° position, the armature loop is perpendicular to the magnetic field, and the black and white portions of the loop are moving parallel to the field. At the instant the conductors are moving parallel to the magnetic field, they do not cut any lines of force, and no emf is induced in the loop. The meter in Figure 3.7A indicates zero current at this time.

As the armature loop rotates from position *A* to position *B* (see Figure 3.7F), the sides of the loop cut through more and more lines of force, at a continually increasing angle. At 90° (position *B*), the sides of the loop are cutting through a maximum number of lines of force and at a maximum angle. The result is that between 0° and 90°, the induced emf in the conductors builds up from zero to a maximum value. Notice that from 0° to 90° rotation, the black side of the loop cuts down through the field while the white side cuts up through the field. The induced emf's in the two sides of the rotating loop are series adding. Thus, the resultant voltage across the brushes (the armature voltage, V_A) is the sum of the two induced voltages, and the meter in Figure 3.7B indicates a maximum value.

As the armature loop continues rotating from position *B* to position *C*, the sides of the loop that were cutting through a maximum number of lines of force at position *B* now cut through fewer lines. At position *C* (180°), they are again moving parallel

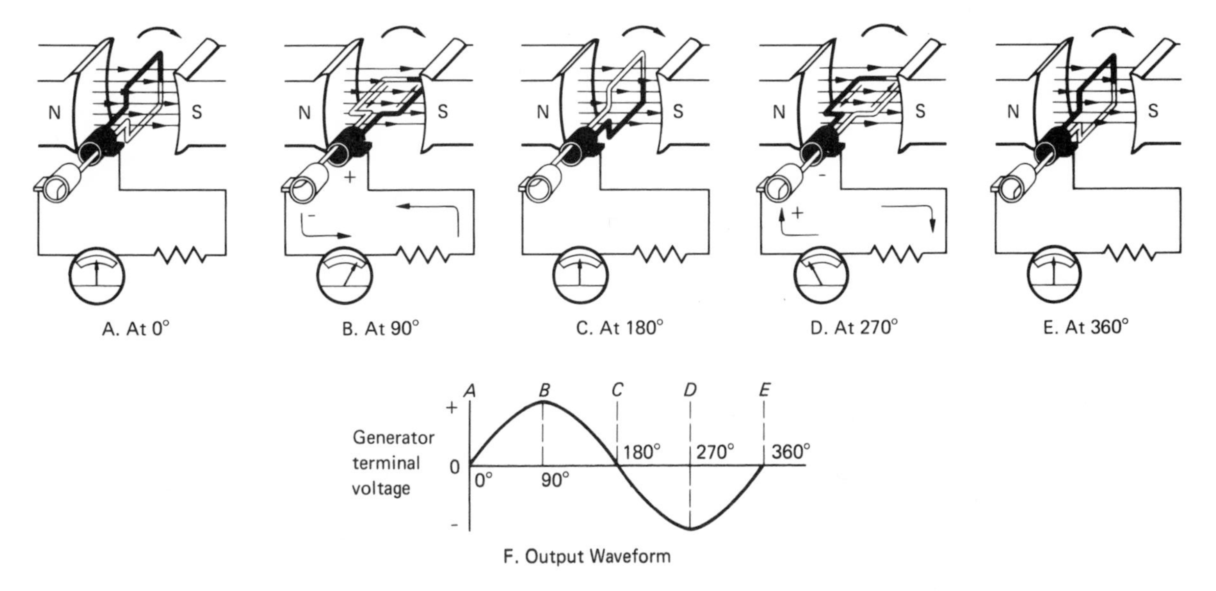

Figure 3.7 Output Voltage of Elementary AC Generator during One Revolution

to the magnetic field. As the armature rotates from 90° to 180°, the induced voltage will decrease to zero, just as it had increased from 0° to 90°. The meter again indicates zero.

From 0° to 180°, the sides of the loop were moving in the same direction through the magnetic field. Therefore, the polarity of the induced voltage remained the same. However, as the loop starts rotating beyond 180°, from position *C* through *D* to position *E*, the direction of the cutting action of the sides of the coil through the magnetic field reverses. As a result, the polarity of the induced voltage reverses, as shown in the graph in Figure 3.7F.

Elementary DC Generator

A single-loop generator with each end of the loop connected to a segment of a two-segment metal ring is shown in Figure 3.8. The two segments of the split metal ring are insulated from each other. The ring is called the commutator. The commutator in a DC generator replaces the slip rings of the AC generator and is the main difference in their construction. The commutator mechanically reverses the armature loop connections to the external circuit at the same instant that the polarity of the voltage in the loop reverses. Through this process, the commutator changes the generated AC voltage to a pulsating DC voltage, as shown in the graph of Figure 3.8F. This action is known as *commutation.*

We can follow the rotation of the loop in Figure 3.8 as we did in Figure 3.7. If we do so, we will see that as the segments of the commutator rotate with the loop, they contact opposite brushes. Also, the direction of current flow through the brushes and the meter is the same. The voltage developed across the brushes is pulsating and unidirectional, peaking twice during each revolution between zero and maximum. This variation is called *ripple.*

The pulsating voltage of a single-loop DC generator is unsuitable for most applications. Therefore, in practical generators, more armature loops and more commutator segments are used to produce an output voltage with less ripple.

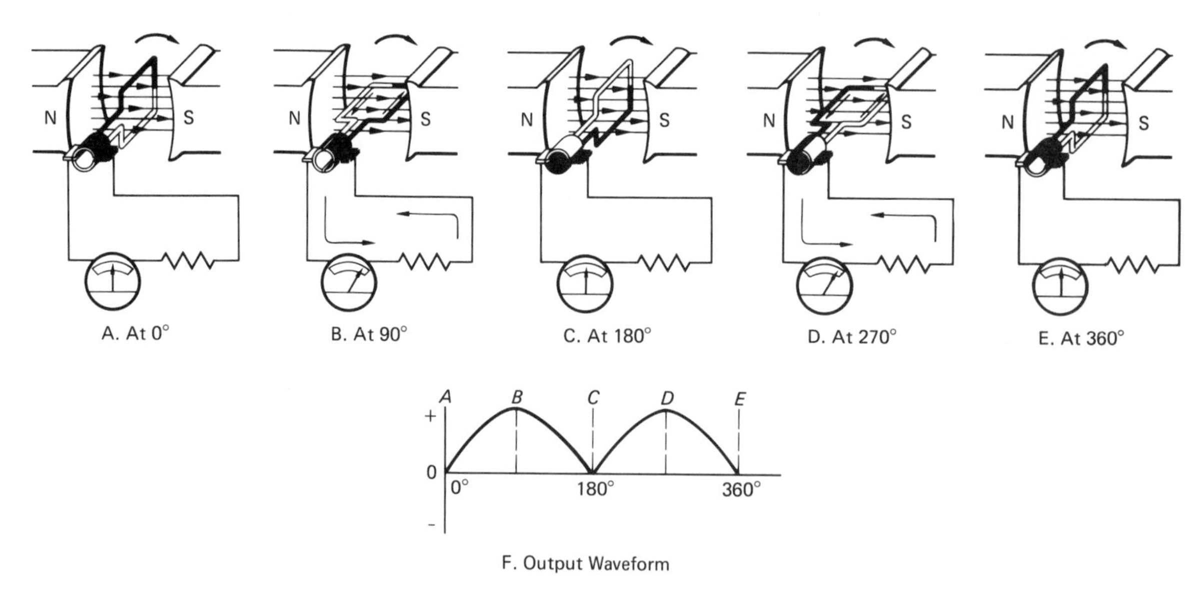

Figure 3.8 Elementary DC Generator and Effects of Commutation

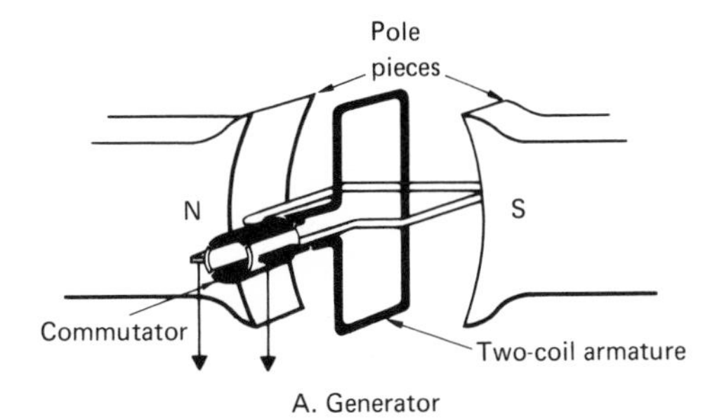

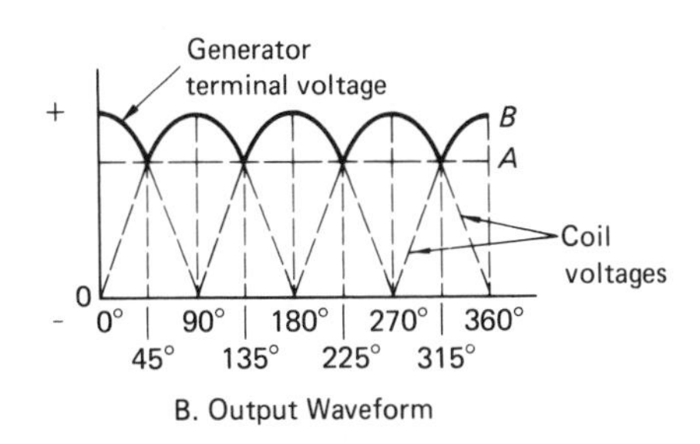

Figure 3.9 Effects of Additional Coils

Additional Coils and Poles

The effects of additional coils may be illustrated by the addition of a second coil to the armature. The commutator must now be divided into four parts since there are four coil ends; see Figure 3.9A. Since there are four segments in the commutator, a new segment passes each brush every 90° instead of every 180°. This action allows the brush to switch from the white coil to the black coil at the instant the voltages in the two coils are equal.

The graph in Figure 3.9B shows the ripple effect of the voltage when two armature coils are used. As the figure shows, the ripple is limited to the rise and fall between points *A* and *B* on the graph. With the addition of more armature coils, the ripple effect is further reduced. Decreasing ripple in this way increases the effective voltage of the generator output.

Practical generators use many armature coils. They also use more than one pair of magnetic poles. The additional magnetic poles have the same effect on ripple as the additional armature coils have. In addition, the increased number of poles provides a stronger magnetic field. This effect, in turn, allows an increase in output voltage, because the coils cut more lines of flux per revolution.

Electromagnetic Poles

Nearly all practical generators use electromagnetic poles instead of the permanent magnets found in the elementary generator. The electromagnetic field poles consist of coils of insulated copper wire wound on soft iron cores, as shown in Figure 3.1. The main advantages for using electromagnetic poles are (1) increased field strength and (2) a means of controlling the strength of the field. By variation of the input voltage to the field windings, the field strength is varied. By variation of the field strength, the output voltage of the generator is controlled.

Generator Voltage Equation

The average generated emf, in volts, of a generator may be calculated from the following equation:

$$\text{emf} = \frac{pN_{\text{cond}}\Phi n_{\text{A}}}{10^8 \times 60b_{\text{A}}} \quad \textbf{(3.1)}$$

where p = number of poles

N_{cond} = total number of conductors on armature

Φ = flux per pole

n_{A} = speed of armature, in revolutions per minute

b_{A} = number of parallel paths through armature, depending on type of armature winding

For any given generator, all the factors in Equation 3.1 are fixed values except the flux per pole Φ and the speed n_{A}. Therefore, Equation 3.1 can be simplified as follows:

$$\text{emf} = C_{\text{emf}}\Phi n_{\text{A}} \quad \textbf{(3.2)}$$

where C_{emf} = all fixed values or constants for a given generator

DC Motor

Stated very simply, a *DC motor* rotates as a result of two magnetic fields interacting with each other. These two interacting fields are the field due to the current flowing in the field windings and the field produced by the armature as a result of the current flowing through it. As shown in Figure 3.10A, the loop (armature) field is both attracted to and repelled by the field from the field poles. This action causes the armature to turn in a clockwise direction, as shown in Figure 3.10B.

After the loop has turned far enough so that its north pole is exactly opposite the south field pole, the brushes advance to the next commutator segment. This action changes the direction of current flow through the armature loop and, therefore, changes the polarity of the armature field, as shown in Figure 3.10C.

Torque

Torque is a twisting action on a body, tending to make it rotate. Torque (T) is the product of the force (F) times the perpendicular distance (d) between the axis of rotation and the point of application of the force:

$$T = F \times d \tag{3.3}$$

The torque is in newton-meters, the force is in newtons, and the distance is in meters. The unit of measure of torque in the British system of units is the pound-foot.

Recall from physics that work is also defined as a force acting through a distance. However, work requires that the motion be in the direction of the force; torque does not. A distinction between work and torque in the British system of measurement is made by the units: Torque is in pound-feet, while work is in foot-pounds. For the metric system, torque is measured in dyne-centimeters or newton-meters, while work is measured in ergs (which are the same units as dyne-centimeters) or joules (the same as newton-meters).

The metric measurement for torque with units of newton-meters will be used in this text. However, conversion to the British system with units of pound-feet can easily be made with the following conversion factor:

$$0.73756 \text{ lb-ft} = 1 \text{ Nm} \tag{3.4}$$

Motor Torque Equation

The torque for the DC motor can be calculated from the following equation:

$$T = \frac{pN_{\text{cond}}\Phi I_A}{2\pi b_W} \tag{3.5}$$

where I_A = current in external armature circuit

b_W = number of parallel paths through winding

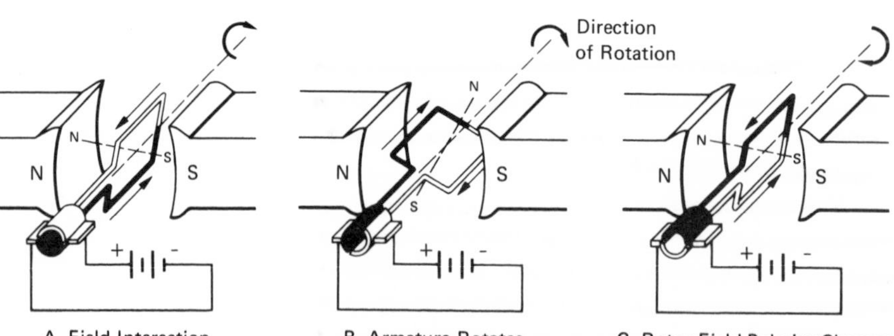

Figure 3.10 Armature Rotation of Elementary DC Motor

For any given motor, all the factors in Equation 3.5 are fixed values except the flux per pole Φ and the armature current I_A. Therefore, Equation 3.5 can be simplified as follows:

$$T = C_T \Phi I_A \tag{3.6}$$

where C_T = all fixed values or constants for a given motor

Ideal DC Machine

In many problems involving the behavior of a DC dynamo as a system component, the machine can be described with satisfactory accuracy in terms of an idealized model having the following properties:

1. The stator has salient poles, and the air gap flux distribution is symmetrical.
2. The armature can be considered as a finely distributed (spread evenly) winding.
3. The brushes are narrow, and commutation occurs when the coil sides are in the neutral zone.

If these conditions exist, then the air gap flux Φ is linearly proportional to field current I_{field}. Then, Equations 3.2 and 3.6 can be simplified further, as follows:

$$\text{emf} = K_{\text{emf}} I_{\text{field}} n_A \tag{3.7}$$

$$T = K_T I_{\text{field}} I_A \tag{3.8}$$

where K_{emf} = constants C_{emf} and Φ/I_{field}

K_T = constants C_T and Φ/I_{field}

Equations 3.7 and 3.8 are very useful and easily measured. The procedure for evaluating K_T will be shown now. The evaluation of K_{emf} will be given later in this chapter in the section titled "Evaluating Variables."

To determine K_T, we use the following procedure: Attach a lever to the armature shaft, and secure a scale to the end of the lever. Apply and measure the armature current and the field current, and note the reading on the force scale. The arrangement is shown in Figure 3.11. In Figure 3.11, the scale should be perpendicular to a line through the motor shaft and the lever on the shaft. The distance d is from the center of the shaft to the point of scale attachment.

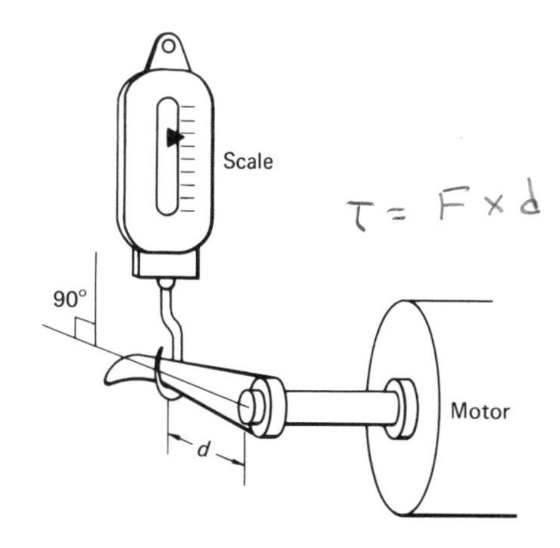

Figure 3.11 Arrangement for Measuring Torque

The torque is the scale reading times the distance in Figure 3.11. The value of K_T is then determined from the following equation:

$$K_T = \frac{T}{I_{\text{field}} I_A} \tag{3.9}$$

This value of K_T should remain relatively constant for this motor. Factors that could affect the value of K_T are those factors that alter the motor in any way, such as heating of the wire in the motor. When the wires in the armature and field heat because of the current flow, the resistance increases and, therefore, reduces the current (and field strength) for a constant voltage. Also, as the armature turns faster, the air friction increases, causing K_T to change. The value of K_T should not change greatly, however.

After K_T has been determined, any value of torque can be approximated by measuring the values of armature and field currents and solving Equation 3.8 for torque T. Equation 3.8 holds whether the motor armature is rotating or stationary. For purposes of calculation, we will use a value of 0.5 Nm/A^2 in this chapter for K_T, unless indicated otherwise. This value of K_T is for a representative 1/8-horsepower (hp) motor.

Counter emf in the Motor

In the development of the torque and voltage equations, recall that the motor had armature current and field current as inputs and torque as an output. Look at Equation 3.7 for emf. Observe that when the armature speed (n_A) and field current (I_{field}) interact, a voltage is generated. This action occurs in the motor as well as the generator. As armature and field currents are applied to the motor to produce torque and armature speed, the speed and the field current produce a voltage in the armature of the motor. This self-generated voltage opposes the external armature voltage that was applied to turn the motor. This self-generated and opposing voltage is termed *counter electromotive force* and is abbreviated cemf. The equation for determining cemf in the motor is exactly the same as the equation for determining emf in the generator:

$$\text{cemf} = K_{emf} I_{field} n_A \qquad (3.10)$$

Conveniently, the cemf generally does not equal or exceed the value of armature voltage (V_A) applied to the motor. If the cemf did exceed the applied armature voltage, the motor would be acting like a generator and would drive current back into the line. If a voltmeter were placed across the armature, it would indicate the larger of the two voltages, applied armature voltage or cemf.

Counter Torque in the Generator

Likewise, in the generator, the rotation of the generator causes a current (I_A) to flow in the armature. Then, torque is produced in the generator because of armature current (I_A) and field current (I_{field}) interaction. This torque is in opposition to the external torque turning the armature and is termed *counter torque* (cT). The counter torque equation for the generator is the same as the torque equation for the motor:

$$cT = K_T I_{field} I_A \qquad (3.11)$$

Counter torque produced by the generator is generally less than the torque being supplied by the device that is turning the generator. From the previous development, we see that both the torque and voltage equations are in operation in both the motor and the generator.

Let us now apply these equations to the basic motor and generator configurations and observe the results.

Dynamo Configurations

A cross-sectional view of a dynamo is shown in Figure 3.12A. Schematic diagrams of the various ways to connect the field and armature in the dynamo are shown in Figures 3.12B through 3.12F. The armature is drawn as a circle representing the end view of the armature, with two squares representing the brushes, while the field is represented by the standard inductor symbol.

The general configurations of DC dynamos take their names from the type of field excitation used. When the dynamo supplies its own excitation, it is called a *self-excited dynamo* (Figures 3.12C through 3.12F). If the field of a self-excited dynamo is connected in parallel with the armature circuit, it is called a *shunt-connected dynamo* (Figure 3.12C). If the field is in series with the armature, it is called a *series-connected dynamo* (Figure 3.12D). If both series and shunt fields are used, it is called a *compound dynamo*. Compound dynamos may be connected as a *short shunt*, with the shunt field in parallel with the armature only (Figure 3.12E), or as a *long shunt*, with the shunt field in parallel with both the armature and series

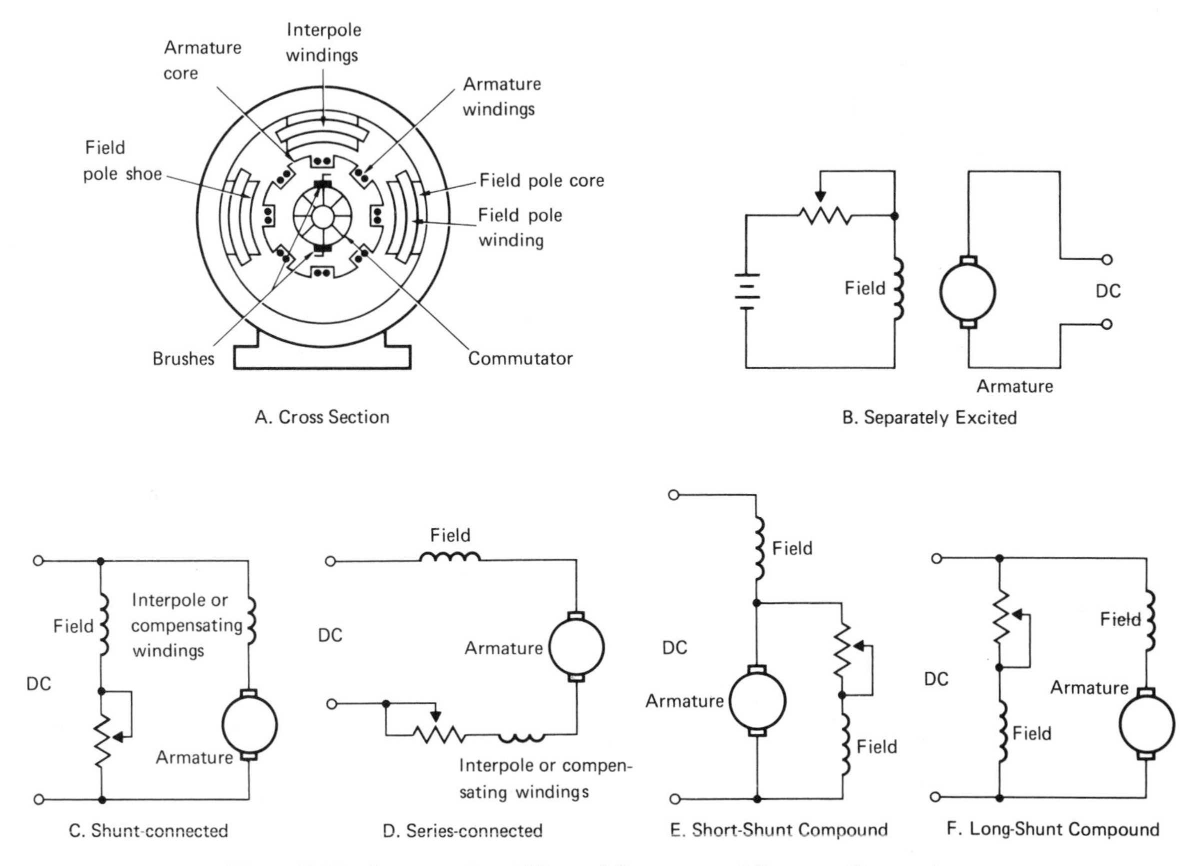

Figure 3.12 Cross-sectional View of Dynamo and Dynamo Connections

fields (Figure 3.12F). The series and shunt machines show the electrical placement of the interpole or compensating winding; their physical placement was discussed earlier.

Field potentiometers, as shown in this figure, are adjustable resistors placed in the field circuits to provide a means of varying the field flux, which controls the amount of emf generated or motor torque produced. These resistors are not placed in the armature circuit if that can be avoided, because of the current being higher in the armature than in the field. In keeping with basic control system principles, it is more efficient and economical to control the low-power circuits, which, in turn, control the higher-power circuits. The current control of the lower-power field circuit has a profound effect on the output of the higher-power armature circuit, whether it is generating emf or producing torque.

Constant Line Voltage

The following analyses are somewhat long and involved, but we feel this part of the chapter on DC motors and generators is the most important. In the following pages, we will produce the DC motor *characteristic curves,* which are graphs of the motor's electrical and mechanical responses. These curves are derived from applications of Equations 3.7 and 3.8, Ohm's law, and Kirchhoff's law. To truly understand DC dynamo principles, you must be very proficient with the characteristic curves and the equations mentioned above.

The first topic we deal with concerns the series motor with constant line voltage. This condition can occur when the torque load of a motor varies while the line voltage remains fixed. The characteristic curves developed help illustrate what happens in the motor.

Series Motor

The series-connected motor is the first machine considered for analysis. A simplified series motor is shown in Figure 3.13.

Schematic

The field may, in actuality, be split into two windings, one on each side of the armature. But in the schematic, it is generally drawn as one inductor. The symbol V_T in this figure and others in this text represents the line voltage. It symbolizes the connection made to the motor or generator external to its case. The field and the armature together are the schematic representation of the series motor.

Measuring Field and Armature Resistance

The DC resistances of the field and armature are important factors and must be measured. If the field and armature leads do not come out of the motor, some disassembly of the machine may be necessary to obtain these measurements. In this text, a 1/8 hp machine will be used for illustration purposes.

As shown in Figure 3.13, the series motor has all the armature current flowing through the field. As a consequence, the field windings should be made of heavy conductors in order to carry the armature current without heating or dropping an excessive amount of line voltage. A typical value of field resistance in this motor might be 50 Ω, as measured with an ohmmeter.

The resistance of the armature may be measured with an ohmmeter or with a method called the locked-rotor–voltage-current method. If the ohmmeter method is used, the armature should be *slowly* rotated while measuring resistance. The ohmmeter reading will fluctuate owing to the slight amount of voltage and current being generated in the armature and to the fluctuating resistance because of the brushes sliding from one commutator segment to another. The armature resistance may be on the order of 10 Ω.

In the *locked-rotor–voltage-current method*, a voltage is applied to the motor while the armature is held stationary (locked rotor). The armature voltage and current are measured, and the armature resistance is determined by applying Ohm's law to these measurements. The test voltage should not be applied for a long time, because of the heating of the conductors and possible damage to the insulation.

For the motor just described, we now have the circuit shown in Figure 3.14, where R_{field} is the field resistance and R_A is the armature resistance.

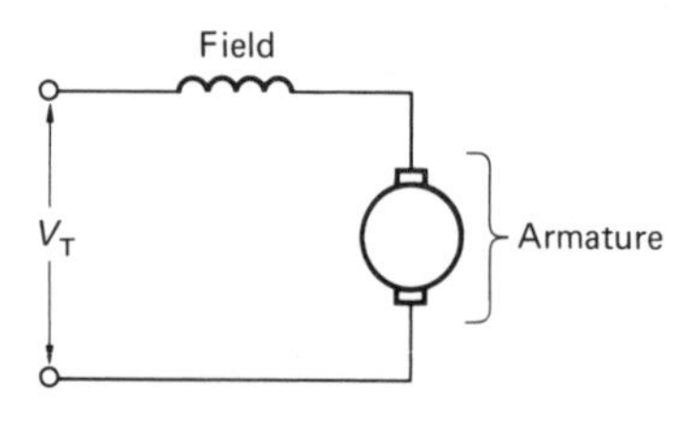

Figure 3.13 Basic Series-connected Motor

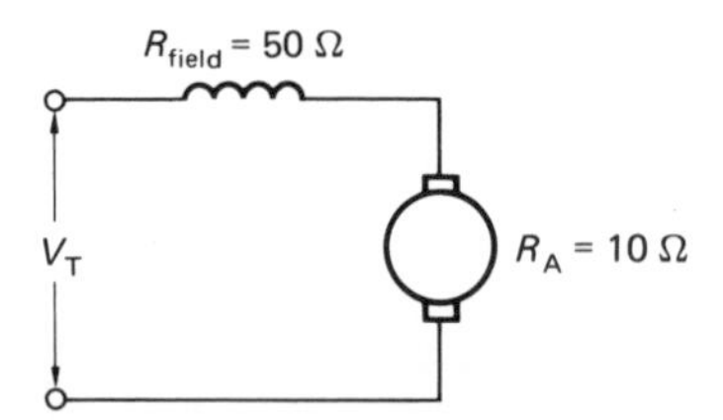

Figure 3.14 Series-connected Motor with Measured Resistances

Evaluating Variables

The next procedure is to apply a line or terminal voltage (V_T) and evaluate the value of K_{emf} in the emf equation. Some provision should be made for applying an armature load or external counter torque to the motor since n_A is one of the variables we want to control. This load can be provided by having the motor drive a hydraulic pump or some other rotational device such as a generator.

The hydraulic pump arrangement is illustrated in Figure 3.15. Here, the pump circulates hydraulic fluid in the pipe. If more load is required on the motor, the valve is closed further to restrict the fluid flow. This restricted flow produces an increased back pressure on the pump, which, in turn, produces an increased load on the drive motor. This loading method would work well for motors of 1/4 hp or less.

When considering the torque and emf equations and the schematic of the series motor, we see that there are at least four variables: V_T, I_A or I_{field} (since the currents are the same in a series circuit), torque T, and speed n_A. In most system analyses, it is instructive to hold one of the variables constant while allowing a second variable to change and observe the results on the remaining variables. For example, in Ohm's law, if V_T is held constant and resistance R is increased, then current I must decrease in order to keep the equation balanced. This technique will be applied to the series motor system.

For the first case, we will hold V_T constant at 100 V while the torque is varied and observe the effects on I_A and n_A. The value of K_T was determined previously to be 0.5 Nm/A^2. We will arbitrarily pick an n_A of 1000 r/min and an I_A of 0.5 A for illustration purposes. These values would, of course, be measured in actual practice.

For the circuit of Figure 3.14, then, where I_A and I_{field} are equal (because it is a series circuit), the voltage drop across the field (V_{field}) is as follows:

$$V_{field} = I_{field}R_{field} = (0.5\text{ A})(50\ \Omega) = 25\text{ V} \quad \textbf{(3.12)}$$

The voltage drop across the field winding is due to the resistance of the wire; there is no reactive component of impedance in the field.

Figure 3.16 is a convenient way of representing the armature of the motor as an equivalent circuit. The cemf in the armature in Figure 3.16 is a function of field current (I_{field}) and armature speed (n_A) (see Equation 3.10). In the motor, current flows from the line into the armature against the cemf. Applying Kirchhoff's law to the series motor yields the following equation:

$$V_T = \text{cemf} + I_AR_A + I_{field}R_{field} \quad \textbf{(3.13)}$$

The line voltage must be balanced by the IR drops in the armature and the field and the cemf at all times. If the cemf were ever to become larger than line voltage, then current would flow back into the line against the line voltage.

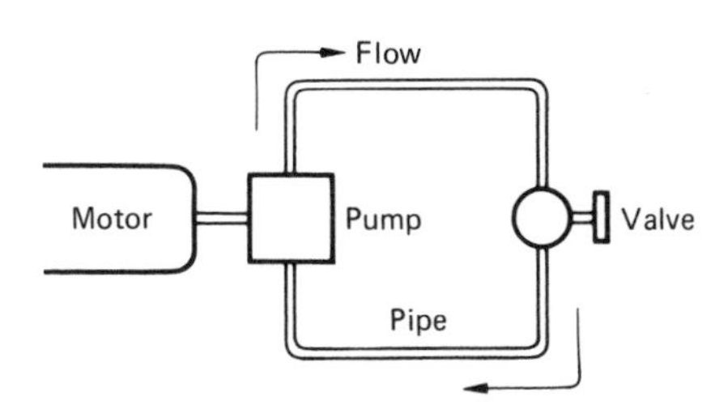

Figure 3.15 Hydraulic Pump Arrangement for Providing a Load to Motor

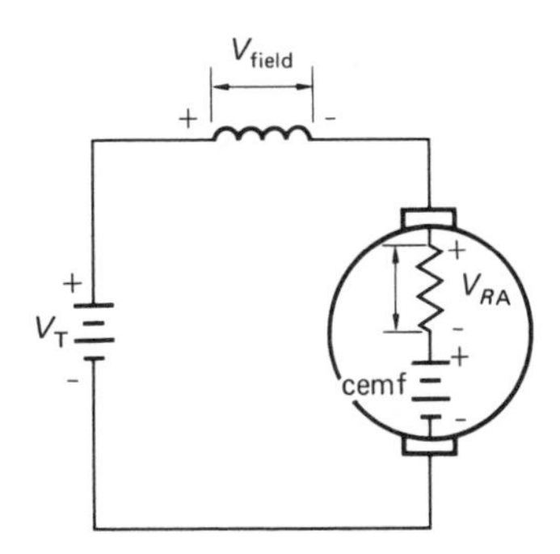

Figure 3.16 Voltage Distribution around Series Motor Circuit

Table 3.1 *Series Motor, V_T Constant*

V_T (V)	Torque (Nm)	$I_A(I_{field})$ (A)	V_{field} (V)	V_A (V)	V_{RA} (V)	cemf (V)	n_A (r/min)
100	—	—	—	—	—	—	decreasing to 0
100	0.926	1.36	68	32	13.6	18.4	96.6
100	0.500	1.00	50	50	10.0	40.0	286.0
100	0.463	0.962	48	52	9.6	42.0	314.0
100	0.245	0.70	35	65	7.0	58.0	592.0
100	0.125	0.50	25	75	5.0	70.0	1000.0
100	0.080	0.40	20	80	4.0	76.0	1357.0
100	0.045	0.30	15	85	3.0	82.0	1952.0
100	0.020	0.20	10	90	2.0	88.0	3143.0
100	0.005	0.10	5	95	1.0	94.0	6714.3
100	0	0	0	V_T	0	V_T	∞

The voltages V_{RA} (equal to $I_A R_A$, the drop across R_A) and V_{field} are voltage drops, while cemf is a voltage rise. It should be noted that a voltmeter placed across the armature would measure V_A (75 V), not the cemf generated.

At this point, the constant K_{emf} in the cemf equation (3.10) can be calculated. Rearranging the cemf equation yields the following:

$$K_{emf} = \frac{\text{cemf}}{I_{field} n_A} = \frac{70 \text{ V}}{(0.5 \text{ A})(1000 \text{ r/min})} = 0.14 \text{ V min/A r} \tag{3.14}$$

This value of K_{emf} should remain relatively constant as long as the motor is not taken apart and physically altered.

Torque can also be calculated, as follows:

$$T = K_T I_{field} I_A = (0.5 \text{ Nm/A}^2)(0.5 \text{ A})(0.5 \text{ A}) = 0.125 \text{ Nm} \tag{3.15}$$

Now, we have completed the calculations for one set of data. Additional sets of data need to be calculated and graphed so that we can visualize what is occurring when the torque of the motor changes but the line voltage remains fixed. This procedure can be repeated for different values of torque. These calculations have been done and are tabulated in Table 3.1. The table also shows that as n_A decreases to zero, torque and armature current increase to some maximum value.

Locked Rotor

The condition when n_A is zero is called a *locked rotor*. When n_A is zero, voltage is applied to the motor, and it is as if the rotor or armature were locked. In this condition, maximum armature current and torque are produced because the opposing cemf is zero. This condition also occurs momentarily each time the motor is started from a dead stop.

When power is first applied to the motor, the armature is stationary; maximum current and torque are produced. As the motor starts to turn, the armature speed produces the opposing cemf, which, in turn, reduces current and, therefore, torque. As armature speed continues to increase, the armature current and torque are further reduced, until a stable operating condition is reached in which the torque produced by the motor just balances the torque required by the load to turn it.

Another name for the locked-rotor torque is *stall torque*. The motor is stalled until it produces enough torque to overcome the torque demanded and begins to rotate. The maximum rate of change of speed occurs when power is first applied. The

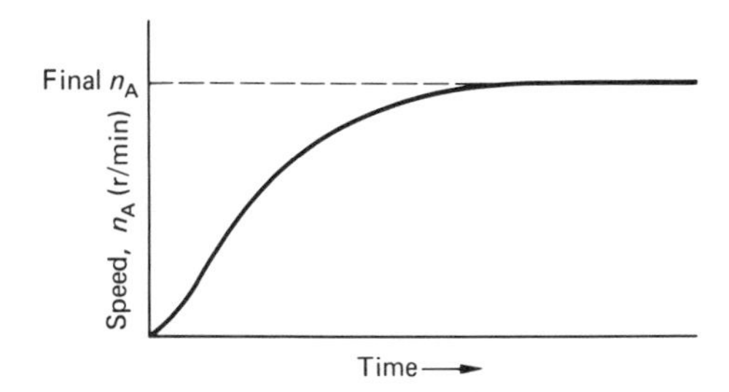

Figure 3.17 Time Response of Motor Starting from Standstill

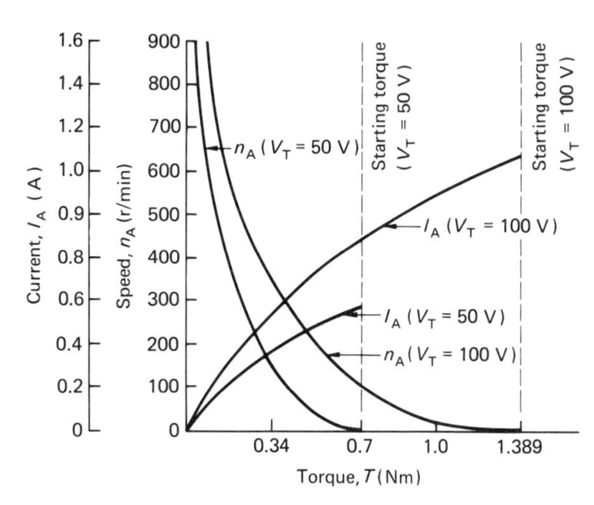

Figure 3.18 Characteristic Curves for Series Motor at Constant Voltage

rate of change in speed decreases until a stable speed is reached, at which time the rate of change is zero. This curve of armature speed versus time, shown in Figure 3.17, looks very much like the capacitance charge curve. In a typical motor, this action occurs within seconds.

Zero Torque

The bottom of Table 3.1 shows the results of the torque demand going to zero. The cemf then becomes equal to the line voltage (V_T), and no current flows in the armature or field. Of course, these conditions can only exist in the ideal machine. The real machine has friction and other losses that limit the maximum armature speed.

Characteristic Curve

A graph of torque, current, and armature speed (from Table 3.1) is shown in Figure 3.18. These curves are the characteristic curves, or responses, for the series motor.

As shown in Figure 3.18, armature speed is by no means a linear function of torque. In fact, as T (demand) decreases, n_A increases dramatically. A dangerous condition can occur when the torque demand becomes very low and the speed very high. This condition is generally referred to as *run away.*

Two sets of graphs are shown in Figure 3.18: a set for $V_T = 100$ V and a set for $V_T = 50$ V. The resulting curves show that the characteristic curves are similar in shape. These curves are also referred to as *constant–line voltage curves.* A different set of curves results if the line voltage is allowed to vary and the torque is held constant (they are shown later).

Shunt Motor

In the shunt-connected motor, shown in Figure 3.19, the field is connected in parallel, or shunt, with the armature. Field current and armature current are not the same.

Since the field winding no longer carries armature current, it can be made of fine wire and many turns. Thus, the shunt motor field has higher resistance than the series motor field. A typical value of shunt field resistance for a 1/8 hp motor may be 1000 Ω.

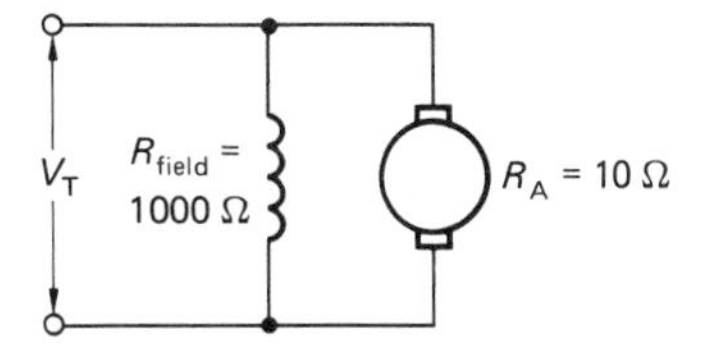

Figure 3.19 Basic Shunt-connected Motor

Evaluating Variables

Again, line voltage is held constant at 100 V, the torque required is varied, and armature current (I_A) and speed (n_A) are observed. The value of K_T is evaluated, as done previously; it is found to be 0.2 Nm/A^2. An initial set of values, $n_A = 1000$ r/min and $I_A = 6.25$ A, is measured.

The field voltage drop (V_{field}) is the same as the line voltage (V_T) and armature voltage (V_A) since they are all in parallel. To determine field current (I_{field}), we apply Ohm's law:

$$I_{field} = \frac{V_{field}}{R_{field}} = \frac{100\ \text{V}}{1000\ \Omega} = 0.1\ \text{A} \tag{3.16}$$

Then:

$$T = K_T I_{field} I_A = (0.2\ \text{Nm/A}^2)(0.1\ \text{A})(6.25\ \text{A}) = 0.125\ \text{Nm} \tag{3.17}$$

To determine cemf, we first evaluate the internal voltage drop due to R_A:

$$V_{RA} = I_A R_A = (6.25\ \text{A})(10\ \Omega) = 62.5\ \text{V} \tag{3.18}$$

The cemf, then, is as follows:

$$\text{cemf} = V_T - V_{RA} = 100\ \text{V} - 62.5\ \text{V} = 37.5\ \text{V} \tag{3.19}$$

At this point, K_{emf} for the shunt motor can be evaluated:

$$K_{emf} = \frac{\text{cemf}}{I_{field} n_A} = \frac{37.5\ \text{V}}{(0.1\ \text{A})(1000\ \text{r/min})} = 0.375\ \text{V min/A r} \tag{3.20}$$

Knowing the value of K_{emf} now allows us to determine any value of n_A for the shunt motor. First, we select a value of torque required of the motor—for example, 0.1 Nm—and solve for armature current (I_A):

$$I_A = \frac{T}{K_T I_{field}} = \frac{0.1\ \text{Nm}}{(0.2\ \text{Nm/A}^2)(0.1\ \text{A})} = 5\ \text{A} \tag{3.21}$$

Then:

$$V_{RA} = I_A R_A = (5\ \text{A})(10\ \Omega) = 50\ \text{V} \tag{3.22}$$

$$\text{cemf} = V_T - V_{RA} = 100\ \text{V} - 50\ \text{V} = 50\ \text{V} \tag{3.23}$$

$$n_A = \frac{\text{cemf}}{K_{emf} I_{field}} = \frac{50\ \text{V}}{(0.375\ \text{V min/A r})(0.1\ \text{A})} = 1333\ \text{r/min} \tag{3.24}$$

Table 3.2 shows the results of this procedure repeated for various values of torque. Similar to the

Table 3.2 *Shunt Motor, V_T Constant*

V_T (V)	Torque (Nm)	I_{field} (A)	I_A (A)	V_{field} (V)	V_A (V)	V_{RA} (V)	cemf (V)	n_A (r/min)
100	0.2	.1	—	—	—	—	—	decreasing to 0
100	0.15	.1	7.50	100	100	75.0	25.0	667
100	0.133	.1	6.65	100	100	66.5	33.5	893
100	0.125	.1	6.65	100	100	62.5	37.5	1000
100	0.10	.1	5.0	100	100	50.0	50.0	1333
100	0.75	.1	3.75	100	100	37.5	62.5	1667
100	0.05	.1	3.33	100	100	33.3	66.7	1779
100	0.025	.1	1.25	100	100	12.5	87.5	2333
100	0	.1	0	100	100	0	V_T	approaches maximum

results for the series motor, Table 3.2 shows that as armature speed decreases to zero, torque and armature current increase to some maximum value.

Maximum Armature Speed

How is maximum armature speed determined in the shunt motor? From the cemf equation, we have the following:

$$n_{A(max)} = \frac{\text{cemf}_{max}}{K_{emf} I_{field}} = \frac{100\ \text{V}}{(0.375\ \text{V min/A r})(0.1\ \text{A})} = 2667\ \text{r/min} \quad (3.25)$$

Maximum speed for the shunt motor can be determined in another way without knowing line voltage. Look at the previous equation:

$$n_{A(max)} = \frac{\text{cemf}_{max}}{K_{emf} I_{field}} \quad (3.26)$$

where $\text{cemf}_{max} = V_T - I_A R_A$ (3.27)

But $I_A = 0$ since cemf is maximum. Therefore, $\text{cemf} = V_T$.

Solving for I_{field} in terms of voltage and resistance gives the following:

$$I_{field} = \frac{V_T}{R_{field}} \quad (3.28)$$

Substituting yields the following:

$$n_{A(max)} = \frac{\text{cemf}_{max}}{K_{emf} I_{field}} = \frac{V_T}{K_{emf}(V_T/R_{field})} = \frac{V_T R_{field}}{K_{emf} V_T} \quad (3.29)$$

V_T cancels. Thus:

$$n_{A(max)} = \frac{R_{field}}{K_{emf}} = \frac{1000\ \Omega}{0.375\ \text{V min/A r}} = 2667\ \text{r/min} \quad (3.30)$$

This equation shows that maximum armature speed of the shunt motor is not a function of line voltage but of circuit constants. In other words, theoretically, no matter what line voltage is applied, as torque demand goes to zero, speed will always go to the same value. This result is shown in Figure 3.20 for line voltage of 100 V and 50 V.

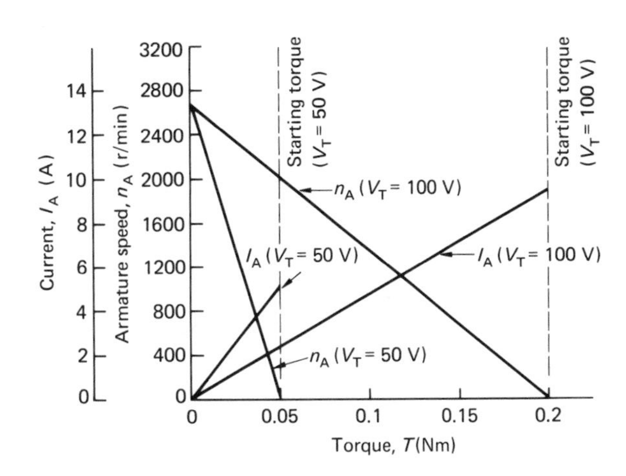

Figure 3.20 Characteristic Curves for Shunt Motor at Constant Voltage

Doubling Line Voltage

Also observe in Figure 3.20 what occurs when line voltage is doubled: Maximum torque produced is quadrupled because both I_A and I_{field} are doubled. Thus, some high-voltage motors produce disproportionately more available torque.

Comparing Series and Shunt Motors

As shown in Figure 3.20, armature current in the shunt motor is directly proportional to torque:

$$I_A = \frac{T}{K_T I_{field}} \quad (3.31)$$

(I_{field} stays constant as long as V_T is constant.)

But in the series motor (Figure 3.18), armature current is a second-order function of torque:

$$I_A = I_{field} \quad (3.32)$$

Therefore:

$$I_A^2 = \frac{T}{K_T} \qquad (3.33)$$

This equation results in the parabolic-shaped curve of Figure 3.18.

Constant Torque

When we examined the series and shunt motors in the previous sections, we observed that there were at least four variables: V_T, I_A or I_{field} (since they are the same in a series motor), torque T, and n_A. It is obvious that there are more variables than can be easily observed on a simple, two-dimensional graph. Our approach, then, is to hold one variable constant, vary another, and observe the effect on the remaining variables. In the previous sections, we examined the effects on I_A and n_A as torque was varied while V_T was held constant. In the next investigation, we will maintain torque at a constant value, vary the line voltage V_T, and again observe the effects on armature current I_A and speed n_A.

Series Motor

Since the same series motor will be used in this section as in the previous one, K_T (0.5 Nm/A^2) and K_{emf} (0.14 V min/A r) will also be the same because we have not physically altered the motor. Therefore, from Table 3.1, when $V_T = 100$ V, torque T is 0.125 Nm and n_A is 1000 r/min.

The next step is to vary V_T while torque is held at 0.125 Nm. Armature and field currents will be constant at 0.5 A:

$$T = K_T I_{field} I_A \qquad (3.34)$$

$$I_A = \frac{T}{K_T} = \frac{0.125 \text{ Nm}}{0.5 \text{ Nm/A}^2} = 0.5 \text{ A} \qquad (3.35)$$

Therefore:

$$V_{field} = I_A R_A = (0.5 \text{ A})(50\ \Omega) = 25 \text{ V} \qquad (3.36)$$

Also:

$$V_{RA} = I_A R_A = (0.5 \text{ A})(10\ \Omega) = 5 \text{ V} \qquad (3.37)$$

Setting V_T to 50 V produces the following results:

$$V_A = V_T - V_{field} = 50 \text{ V} - 20 \text{ V} = 30 \text{ V} \qquad (3.38)$$

$$\text{cemf} = V_A - V_{RA} = 30 \text{ V} - 5 \text{ V} = 25 \text{ V} \qquad (3.39)$$

Therefore:

$$n_A = \frac{\text{cemf}}{K_{emf} I_{field}} = \frac{25 \text{ V}}{(0.14 \text{ V min/A r})(0.5 \text{ A})} = 357 \text{ r/min} \qquad (3.40)$$

Table 3.3 *Series Motor, Torque Constant*

V_T (V)	Torque (Nm)	$I_A(I_{field})$ (A)	V_{field} (V)	V_A (V)	V_{RA} (V)	cemf (V)	n_A (r/min)
stall voltage	.125	.5	25	decreasing to minimum	5	0	0
50	.125	.5	25	30	5	25	357
100	.125	.5	25	75	5	70	1000
150	.125	.5	25	130	5	125	1786
200	.125	.5	25	180	5	175	2500
∞	.125	.5	25	∞	5	∞	∞

Table 3.3 shows the above procedure repeated for different values of V_T. From Table 3.3, we see that as the line voltage continues to increase, so does armature speed; no theoretical limit is reached. In practice, maximum armature speed is determined by how long the motor can hold together.

Minimum line voltage is reached when n_A goes to zero, or stalls. At zero armature speed, cemf is also zero:

$$n_A = \frac{\text{cemf}}{K_{\text{emf}} I_{\text{field}}} = \frac{0}{(0.14 \text{ V min/A r})(0.5 \text{ A})} = 0 \quad \textbf{(3.41)}$$

Also:

$$V_A = V_T - V_{\text{field}} \quad \textbf{(3.42)}$$

Or after rearrangement:

$$\begin{aligned} V_T &= V_A + V_{\text{field}} = (\text{cemf} + V_{RA}) + V_{\text{field}} \\ &= (\text{cemf} + I_A R_A) + V_{\text{field}} \\ &= 0 + (0.5 \text{ A})(10\ \Omega) + 25 \text{ V} \\ &= 30 \text{ V} \end{aligned} \quad \textbf{(3.43)}$$

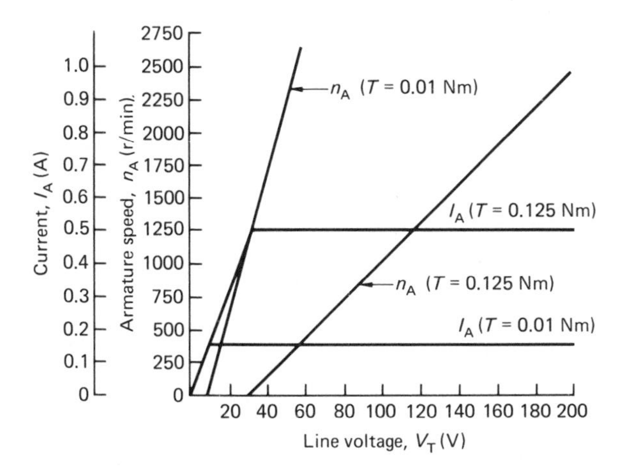

Figure 3.21 Characteristic Curves for Series Motor at Constant Torque

This equation shows that when the line voltage decreases to 30 V, the motor will stall if it still is required to produce 0.125 Nm of torque.

The graph in Figure 3.21 is the characteristic curve for the series motor with constant torque. Two armature speed curves have been drawn to show what happens at a higher, fixed torque. Notice that these lines are not parallel. Also, the higher-torque example does not start rotating until a line voltage of 60 V is exceeded.

Shunt Motor

In the shunt motor under varying line voltage but constant torque, there are five possible variables: V_T, I_A, I_{field}, T, and n_A. Field current in the shunt motor varies directly as the line voltage. If the torque is held constant and field current changes, then armature current must vary inversely.

The calculations required to produce the characteristic curves are as follows: Increase the line voltage from 100 to 150 V. Then, we have the following calculations:

$$I_{\text{field}} = \frac{V_{\text{field}}}{R_{\text{field}}} = \frac{150 \text{ V}}{1000\ \Omega} = 0.15 \text{ A} \quad \textbf{(3.44)}$$

$$I_A = \frac{T}{K_T I_{\text{field}}} = \frac{0.125 \text{ Nm}}{(0.2 \text{ Nm/A}^2)(0.15 \text{ A})} = 4.17 \text{ A} \quad \textbf{(3.45)}$$

$$V_{RA} = I_A R_A = (4.17 \text{ A})(10\ \Omega) = 41.7 \text{ V} \quad \textbf{(3.46)}$$

$$\text{cemf} = V_A - V_{RA} = 150 \text{ V} - 41.7 \text{ V} = 108.3 \text{ V} \quad \textbf{(3.47)}$$

$$n_A = \frac{\text{cemf}}{K_{\text{emf}} I_{\text{field}}} = \frac{108.3 \text{ V}}{(0.375 \text{ V min/A r})(0.15 \text{ A})} = 1925 \text{ r/min} \quad \textbf{(3.48)}$$

Table 3.4, which shows the above procedure repeated for various values of line voltage, is another verification that line voltage does not

Table 3.4 *Shunt Motor, Torque Constant*

V_T (V)	Torque (Nm)	I_{field} (A)	I_A (A)	V_{field} (V)	V_A (V)	V_{RA} (V)	cemf (V)	n_A (r/min)
stall voltage	.125	decreasing to 0	increasing to maximum	stall voltage	stall voltage	increasing to maximum	decreasing to 0	decreasing to 0
90	.125	6.09	6.94	90	90	69.4	20.6	610
100	.125	0.1	6.25	100	100	62.5	37.5	1000
120	.125	0.12	5.21	120	120	52.1	67.9	1509
150	.125	0.15	4.17	150	150	41.7	108.3	2250
160	.125	0.16	3.91	160	160	39.1	120.9	2015
200	.125	0.2	3.125	200	200	31.3	168.8	2250
300	.125	0.3	2.08	300	300	20.8	279.2	2482
∞	.125	0	∞	—	—	0	—	maximum

determine maximum possible speed for the shunt motor. Maximum armature speed is determined as follows:

$$n_{A(max)} = \frac{R_{field}}{K_{emf}} = \frac{1000\ \Omega}{0.375\ \text{V min/A r}} = 2667\ \text{r/min} \tag{3.49}$$

We also need to determine *stall voltage*, the voltage at which the motor stops turning. At stall, we have the following:

$$\text{cemf} = 0 \tag{3.50}$$

$$V_T = V_{field} = V_{RA} = I_A R_A \tag{3.51}$$

or

$$I_{field} R_{field} = I_A R_A \tag{3.52}$$

Solving for I_A yields the following:

$$I_A = \frac{I_{field} R_{field}}{R_A} \tag{3.53}$$

Also:

$$T = K_T I_{field} I_A \tag{3.54}$$

Substitution gives us the following:

$$T = K_T I_{field} \left(\frac{I_{field} R_{field}}{R_A} \right) \tag{3.55}$$

Solving for I_{field}, we have the following result:

$$T = \frac{K_T I_{field}^2 R_{field}}{R_A} \tag{3.56}$$

$$I_{field} = \sqrt{\frac{T R_A}{K_T R_{field}}} = \sqrt{\frac{(0.125\ \text{Nm})(10\ \Omega)}{(0.2\ \text{Nm/A}^2)(1000\ \Omega)}} = 0.07906\ \text{A} \tag{3.57}$$

Also:

$$I_A = \frac{T}{K_T I_{field}} = \frac{0.125\ \text{Nm}}{(0.2\ \text{Nm/A}^2)(0.07906\ \text{A})} = 7.905\ \text{A} \tag{3.58}$$

$$V_T = I_A R_A = (7.905\ \text{A})(10\ \Omega) = 79.05\ \text{V} \tag{3.59}$$

The data in Table 3.4, along with lower-constant-torque data, are used to obtain the curves in Figure 3.22. Notice that as torque decreases, the curve becomes more angular. Thus, if the torque demand on the shunt motor is low, maximum speed is reached quickly with the application of a relatively small line voltage. In the ideal or

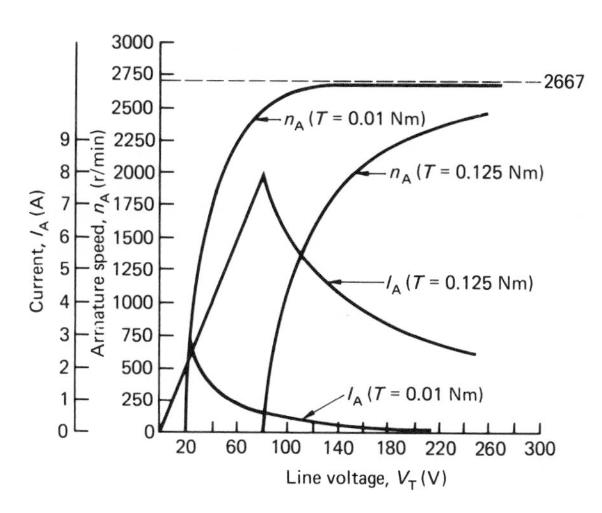

Figure 3.22 Characteristic Curves for Shunt Motor at Constant Torque

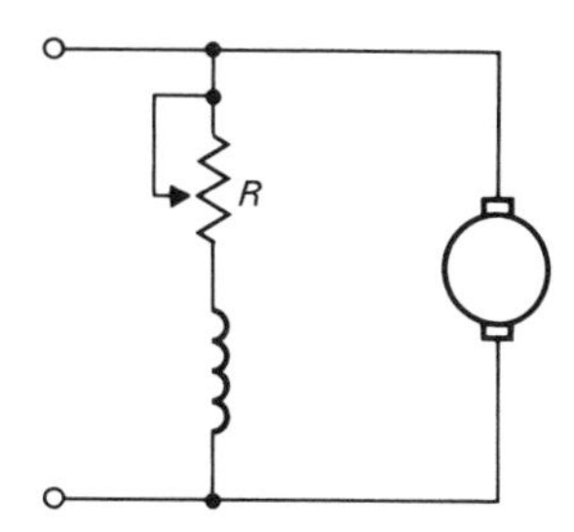

Figure 3.23 Shunt Motor with Field Current Control

theoretical motor, with a torque demand of zero, maximum speed will be reached any time the line voltage exceeds 0 V. Figure 3.22 also shows that armature current is linear up to the point where the motor starts rotating. After that point, armature current decreases owing to increased cemf.

Motor Control

Now that we have completed the basic characteristic curves, we must discuss how to control the motor. Methods and circuitry for electronic motor control will be presented in Chapter 6 but a discussion of what parts of a motor lend themselves to control can be undertaken now.

A general rule for control systems is to place the control mechanism in the low-power part of the system so that the circuit can then control the higher-power part of the system. For example, a signal in a transistor amplifier is generally applied to the base because it is the low-power circuit that controls the higher-power output circuit. However, just as the transistor is not always base-driven, neither are all control systems controlled in the low-power circuitry. Much depends on the application and the component(s) used for the control.

In the shunt motor, it is relatively easy to control the speed by controlling field current. Field current is generally much smaller than armature current and, as we will see, has a great effect on armature current.

For example, consider the shunt motor in Figure 3.23. A variable resistor R has been placed in the field circuit to control field current. Let us see what effect a changing field current has on armature speed.

For convenience, we consider this motor to be the same as the one in the previous section pertaining to shunt motors with constant line voltage. With $V_T = 100$ V, Table 3.5 shows the measured and calculated values. We will assume that the motor load needs the same amount of torque to turn it, no matter what speed it has.

Table 3.5 *Measured and Calculated Values for Shunt Motor*

V_T (V)	Torque (Nm)	I_{field} (A)	I_A (A)	V_{field} (V)	V_A (V)	V_{RA} (V)	cemf (V)	n_A (r/min)
100	.01	.1	.5	100	100	5	95	2533

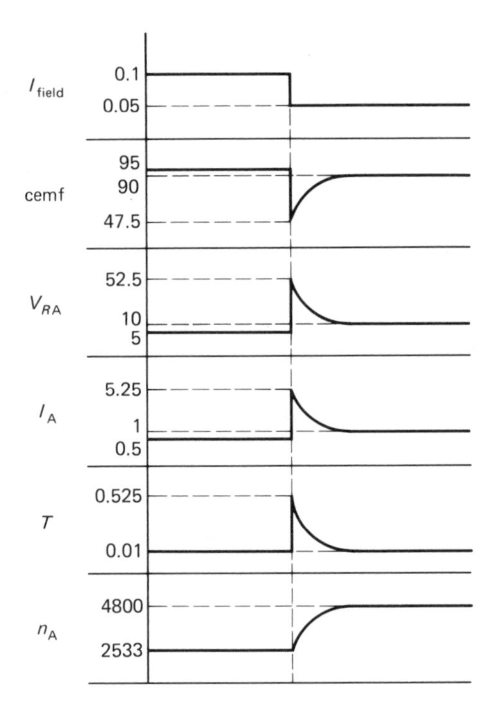

Figure 3.24 Synchrogram of Changes That Occur within Shunt Motor

Now, suppose resistor R is increased suddenly so that field current is cut in half. A synchrogram of this result is shown in Figure 3.24. A *synchrogram* is a vertical alignment of related graphs to show a time relationship.

We will assume that the speed of the motor does not change instantly whenever a change in torque is applied because of inertia of the armature. If at the instant field current changes, n_A is still 2533, then we have the following results:

$$\begin{aligned}\text{cemf} &= K_{\text{emf}} I_{\text{field}} n_A \\ &= (0.375 \text{ V min/A r})(0.05 \text{ A})(2533 \text{ r/min}) \\ &= 47.5 \text{ V}\end{aligned} \tag{3.60}$$

$$\begin{aligned}V_{RA} &= V_T - \text{cemf} = 100 \text{ V} - 47.5 \text{ V} \\ &= 52.5 \text{ V}\end{aligned} \tag{3.61}$$

$$\begin{aligned}I_A &= \frac{V_{RA}}{R_A} = \frac{52.5 \text{ V}}{10\ \Omega} \\ &= 5.25 \text{ A}\end{aligned} \tag{3.62}$$

$$\begin{aligned}T &= K_T I_{\text{field}} I_A = (0.2 \text{ Nm/A}^2)(0.05 \text{ A})(5.25 \text{ A}) \\ &= 0.525 \text{ Nm}\end{aligned} \tag{3.63}$$

With this increased torque produced, the motor will increase in speed. As the motor increases in speed, more cemf is produced, which, in turn, decreases armature current, causing reduced torque. A stable speed is reached when torque returns to its original value, as the following calculations show:

$$\begin{aligned}I_{A(\text{final})} &= \frac{T}{K_T I_{\text{field}}} = \frac{0.01 \text{ Nm}}{(0.2 \text{ Nm/A}^2)(0.5 \text{ A})} \\ &= 1 \text{ A}\end{aligned} \tag{3.64}$$

$$\begin{aligned}V_{RA(\text{final})} &= I_A R_A = (1 \text{ A})(10\ \Omega) \\ &= 10 \text{ V}\end{aligned} \tag{3.65}$$

$$\begin{aligned}\text{cemf}_{\text{final}} &= V_T - V_{RA} = 100 \text{ V} - 10 \text{ V} \\ &= 90 \text{ V}\end{aligned} \tag{3.66}$$

$$\begin{aligned}n_{A(\text{stable})} &= \frac{\text{cemf}}{K_{\text{emf}} I_{\text{field}}} = \frac{90 \text{ V}}{(0.375 \text{ V min/A r})(0.05 \text{ A})} \\ &= 4800 \text{ r/min}\end{aligned} \tag{3.67}$$

These calculations show that a 0.05 A change in field current can cause a 0.5 A change in armature current. The surprising aspect is that when field current is reduced, armature speed is *increased*. Thus, if the shunt motor field is accidentally disconnected, the motor could go into runaway.

Observation of the speed curve in Figure 3.24 shows that the previous shunt motor maximum speed of 2667 r/min has been greatly exceeded (4800 r/min). This result occurs because the field resistance must now include the control resistor's resistance:

$$\begin{aligned}R_{\text{field}} &= \frac{V_T}{I_{\text{field}}} = \frac{100 \text{ V}}{0.05 \text{ A}} \\ &= 2000\ \Omega\end{aligned} \tag{3.68}$$

$$\begin{aligned}n_{A(\max)} &= \frac{R_{\text{field}}}{K_{\text{emf}}} = \frac{2000\ \Omega}{0.375 \text{ V min/A r}} \\ &= 5333 \text{ r/min}\end{aligned} \tag{3.69}$$

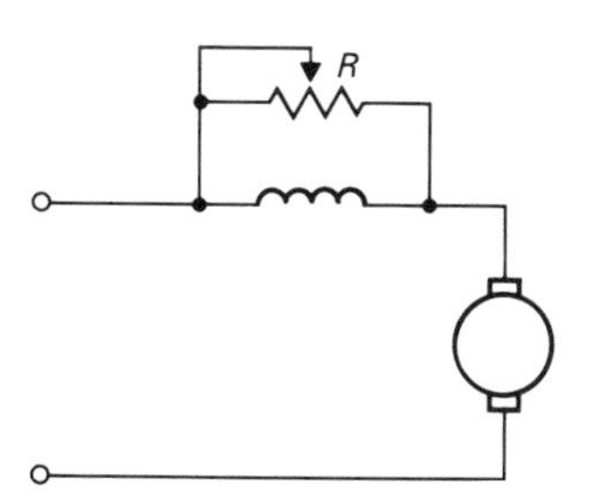

Figure 3.25 Series Motor with Control Resistor

Note: Equations 3.72 and 3.73 were used to calculate *maximum* possible speed. Figure 3.24 is not representing maximum speeds.

The series motor, shown in Figure 3.25, has a control resistor in parallel with the field. In the series motor, as resistor R is decreased, the actual field current is reduced, thus reducing field strength. However, armature current is increased because the total field resistance is decreasing. The net effect is that torque is increased, just as it was in the shunt motor previously discussed.

Characteristic Curves

The characteristic curves mathematically derived in the previous sections show that the emf and torque equations do represent the motor's physical operation. It is not our intent, though, to have you, as a technician, become proficient in manipulating numbers. Rather, we expect you to perceive basic motor operation through the equations and characteristic curves. As examples, consider the following.

1. In the series motor with constant line voltage, what would happen to the armature speed if torque were decreased? With the curve of Figure 3.18, we see that speed would increase. You would probably not need to know how much the speed would increase, but you should know if it increased nonlinearly.

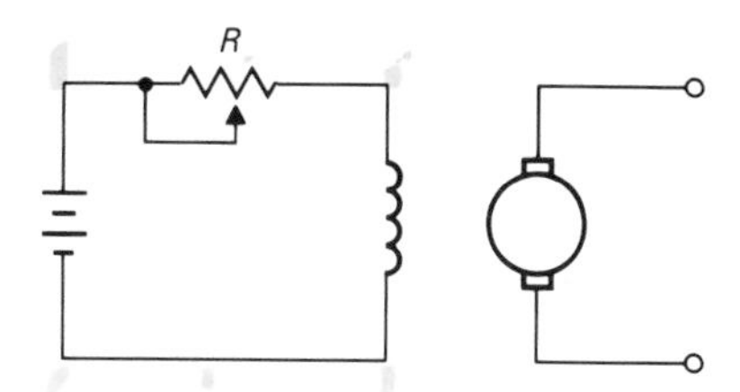

Figure 3.26 Separately Excited Motor

2. If the field current in a shunt motor of constant line voltage is decreased, what happens to the motor armature speed? As discussed in the previous section, speed generally increases because cemf is typically larger than V_{RA}. This action is not easily observable in either Equation 3.8 or 3.10, since cemf is related to speed n_A as follows:

$$n_A = \frac{\text{cemf}}{K_{\text{emf}} I_{\text{field}}} \tag{3.70}$$

And since cemf and I_{field} both decrease, they both have counteracting effects on n_A or the characteristic curves.

Separately Excited Motor

The schematic for the separately excited motor is shown in Figure 3.26. The equations and characteristic curves for this motor are nearly identical to those for the shunt motor. The derivations are left as exercises.

Compound Motors

The compound motor has both a shunt and a series field. As a review, Figure 3.27 shows the two possible ways of connecting the fields, long shunt and short shunt. Of these two possibilities, the long shunt is most often used since the shunt field has a relatively constant current—and, therefore, field strength—as long as the line voltage remains

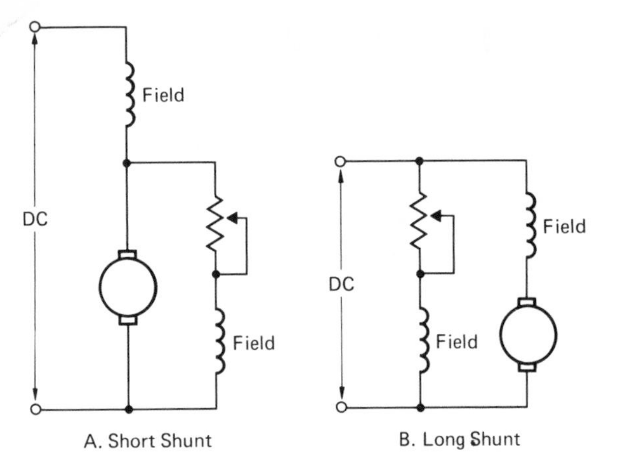

Figure 3.27 Compound Motor Configurations

Figure 3.29 Characteristic Curves for Compound, Series, and Shunt Motors

constant. The long-shunt arrangement also produces a series field strength that is proportional to armature current.

In most cases, the series winding is connected so that its field aids that of the shunt winding, as shown in Figure 3.28A. Motors of this type are called *cumulative compound motors*. If the series winding is connected so that its field opposes that of the shunt winding, the motor is called a *differential compound motor*, as shown in Figure 3.28B. In these figures, Φ represents the flux per pole, and the arrows indicate relative direction of the flux.

The characteristics of a cumulative compound motor are between those of a series motor and those of a shunt motor. If the field strength of the series field is stronger than that of the shunt field, the characteristics are more like those of a series motor, and vice versa. Figure 3.29 shows the relationship.

As the figure indicates, the cumulative compound motor has higher starting torque than the shunt motor and better speed regulation than the series motor. Unlike the series motor, however, it does have a definite no-load speed.

In some operations, it is desirable to use the cumulative series winding to obtain good starting torque. When the motor comes up to speed, the series winding is shorted out. The motor then has the improved speed regulation of a shunt motor.

The differential compound motor is seldom used because of two basic problems. One problem

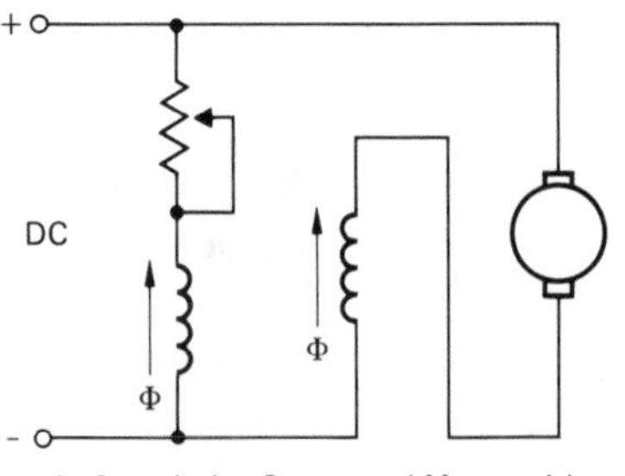

A. Cumulative Compound Motor with Fields Aiding

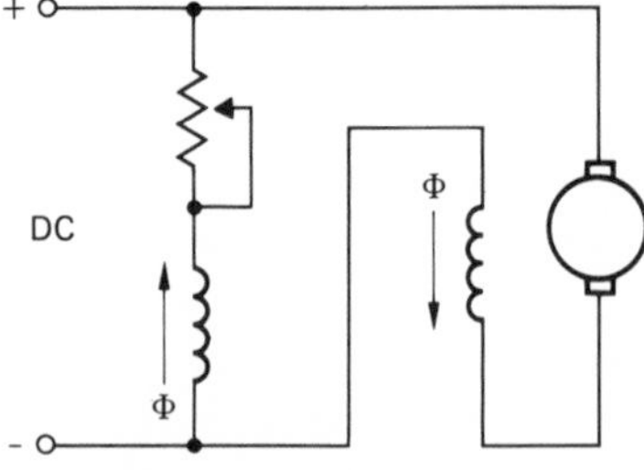

B. Differential Compound Motor with Fields Opposing

Figure 3.28 Compound Motors

is speed stability under heavy load. The speed regulation is very good under light loads. But as the load increases, so does the speed of the motor. As the speed increases, the armature current does, also. The increased current causes the field strength of the series' opposing field to eventually exceed the shunt's field strength, and the motor starts to run backwards. As shown in Figure 3.29, the other problem is that starting torque is low for a relatively high armature current. That is, it uses energy in opposing magnetic fields when armature current is high.

General Considerations

Reversing Direction

Reversing the direction of any DC motor requires that the current through the armature with respect to the field be reversed. If the currents in the armature and field are both reversed, there will be no change in rotational direction. This situation is shown in Figure 3.30, where the direction of F (force) remains unchanged when both conductor and field currents are reversed.

To change the direction of a motor, we must reverse the armature rather than the field, for the following four reasons:

1. The field is more inductive than the armature, and frequent reversals produce switch contact arcing and erosion.
2. In a compound motor, both fields must be reversed. Otherwise, the motor changes from a cumulative compound motor to a differential compound motor.
3. The armature circuit connections are usually opened for various types of braking, as will be discussed later.
4. If the field-reversing circuit fails to close, runaway could result.

Motor Starters and Controllers

In large motors, a motor starter is required in order to prevent excessive starting currents. A *motor starter* is a switching device that is intended to start and accelerate the motor. A *motor controller*, on the other hand, is a device that controls the power flow to a motor, usually resulting in some form of speed control. Controllers will be discussed later in Chapter 6.

A motor starter generally consists of a resistive bank inserted in the motor circuit to prevent full line voltage from being applied to the motor. Some method is used to reduce this resistance as the motor speeds up, thus keeping motor currents from reaching destructive levels. This reduction may be done manually or automatically.

Figure 3.31 shows a typical manual starter circuit. A spring on the starter switch holds it in the off position. But once in the run position, the switch will be held there by an electromagnet. It is

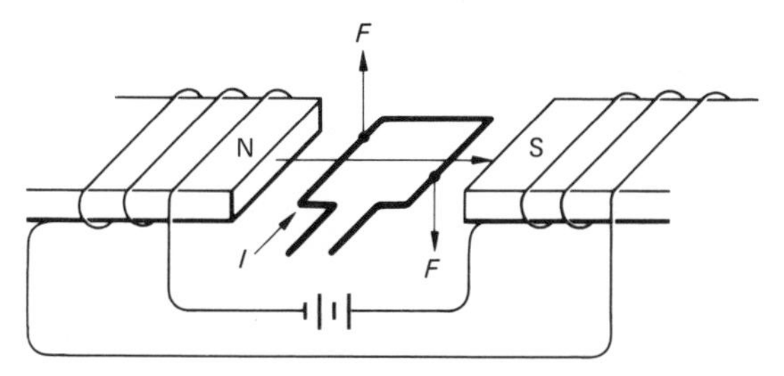

A. Original Direction

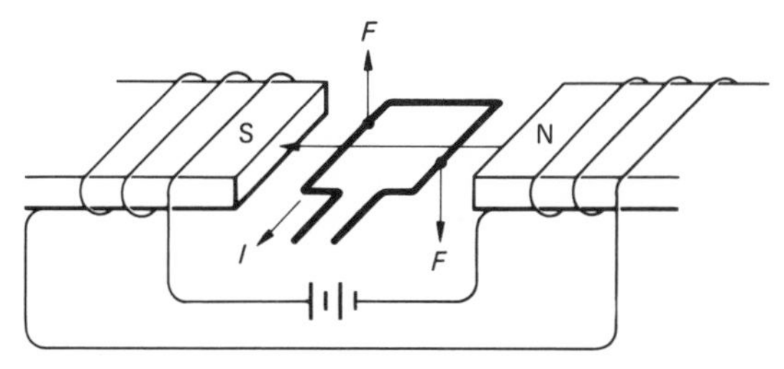

B. Direction Reversed

Figure 3.30 Magnetic Field and Armature Current Reversed

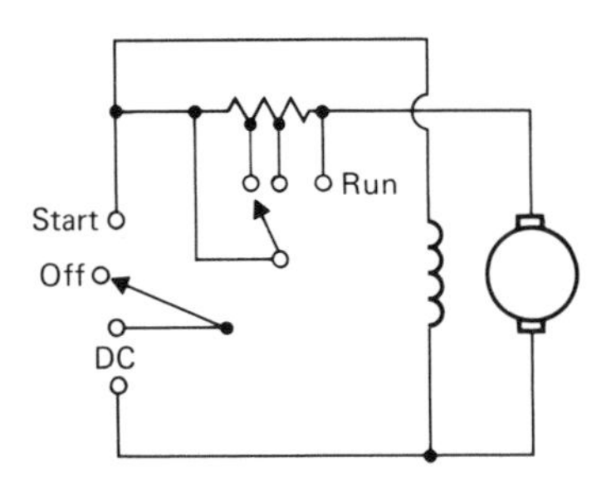

Figure 3.31 Manual Motor Starter Circuit

up to the operator to provide the necessary time delay between switch positions to allow the speed to stabilize and prevent high currents. An automatic starter typically uses the same resistive configuration, but the switching is done automatically with time-delay relays.

Most starters and controllers include one or more devices known as *overload relays*. These devices protect the motor from overheating caused by loads above the rated value. There are four types of overload protection systems in use: (1) thermal relays, (2) magnetic relays, (3) electronic relays, and (4) thermistors.

Thermal overload relays have a small heating element connected in series with the motor. Some consists of a bimetallic strip with a heating coil around it. As the current heats the strip, it opens and stops the motor. Other devices resemble fuses, which melt to break the circuit. Some method of resetting the circuit, either manually or automatically, is also provided.

Figure 3.32 shows the schematic symbol for the thermal relay. The question mark shapes represent the heating elements. At the right in Figure 3.32 is the normally closed (NC) contact that would open.

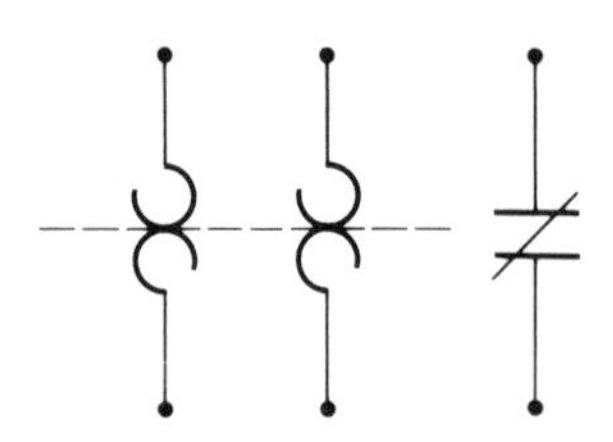

Figure 3.32 Schematic Symbol for Thermal Overload Relay

Magnetic overload relays are actuated by an electromagnetic coil, much as the typical circuit breaker is. This relay would be ineffective against heat buildup due to a large number of successive starts and stops.

Electronic overload relays sense both the voltage applied and the current flow through the motor. They electronically simulate the iron loss and copper loss within the motor.

A *thermistor* is a resistor whose resistance changes greatly as the temperature increases. The temperature characteristic of most thermistors is negative. That is, as temperature increases, the thermistor's resistance decreases. However, in heat-sensing applications of motors, the thermistor with a positive temperature coefficient (discussed in Chapter 7) is generally used.

The thermistor is embedded in the windings of the motor and is monitored electronically to detect heat. When the windings get too hot, the power is shut off. Thermistors are generally not placed in the motor rotor because of the connection problem. Thus, it is unlikely that a thermistor would be used in a DC machine.

Stopping a Motor

Large motors with heavy loads may take a very long time to coast to a stop. The motor itself, however, can be used to slow the rotation. Two methods used for electromechanical braking are dynamic braking and plugging.

Dynamic braking is shown in Figure 3.33. Here, the armature of a shunt motor is disconnected from the line and connected to a resistor. The motor armature is still rotating within a magnetic field and therefore producing an armature voltage due to generator action. If the armature circuit is completed through a resistor, current will flow and a counter torque will be produced. This counter torque will slow the motor. As the motor slows down, the counter torque will decrease because the current is decreasing. Therefore, most of the

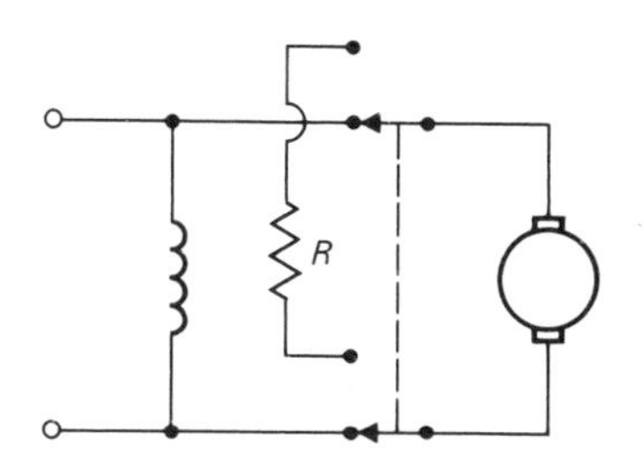

Figure 3.33 Dynamic-Braking Circuit

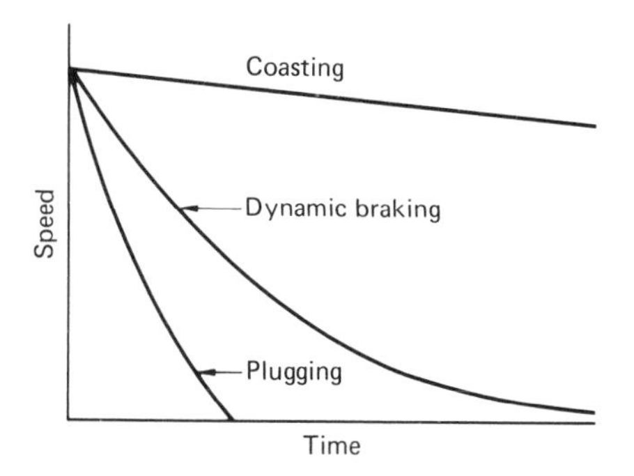

Figure 3.34 Time Relationship for Various Braking Methods

slowing effect occurs at higher speeds, as indicated in Figure 3.34.

Plugging occurs when the armature is disconnected from the line and then is reconnected to the line in the opposite direction. This technique causes the motor to slow down very rapidly. When the motor comes to rest, it is automatically disconnected from the power line. Figure 3.35 is a simplified diagram illustrating plugging.

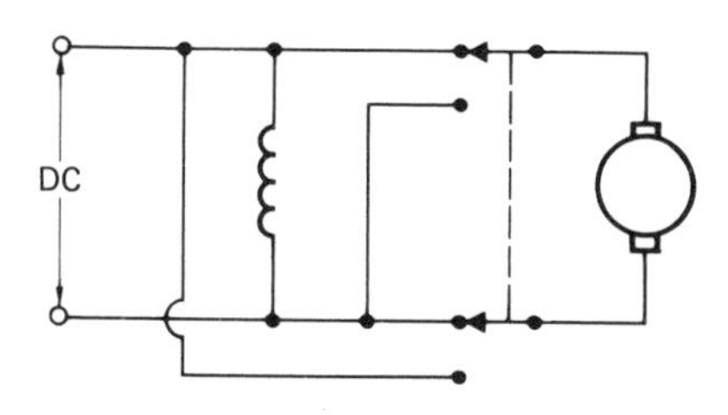

Figure 3.35 Plugging Circuit

DC Generator

Classification

DC generators are classified by their method of supplying excitation current to the field coils. The two major classifications are separately excited and self-excited generators. Self-excited generators are further classified by the method in which the field coils are connected. Like the DC motor, the generator can be connected as series, shunt, or compound. The schematic diagrams for these generator configurations are identical to the corresponding motor schematics. These configurations will be discussed in more detail next.

Separately Excited Generator

A DC generator that has its field supplied by another generator, batteries, or some other outside source is referred to as a *separately excited generator*. The circuit is shown in Figure 3.36.

The field excitation in this generator is the same as the excitation in the separately excited motor. However, the armature circuit here is different. Whereas the motor has an external voltage applied to the armature so that a torque is produced, the generator has an external torque supplied to rotate the armature and a voltage (emf) is produced in the armature.

If the armature circuit is completed, current flows through the external load resistor R_L. The voltage distribution is shown in Figure 3.37, which shows that emf is the only voltage rise in the armature circuit. Thus, armature voltage must be

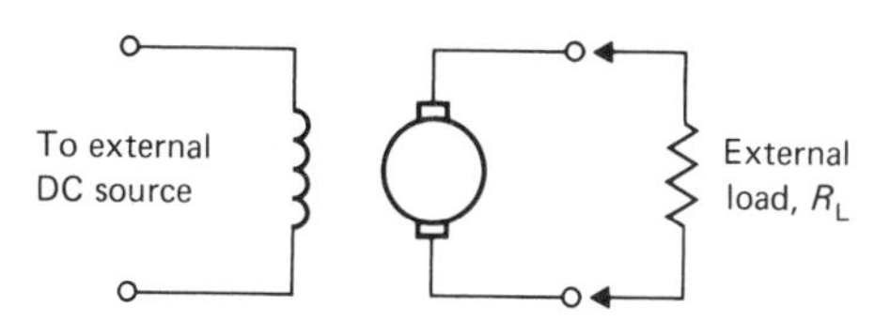

Figure 3.36 Separately Excited Generator Connection

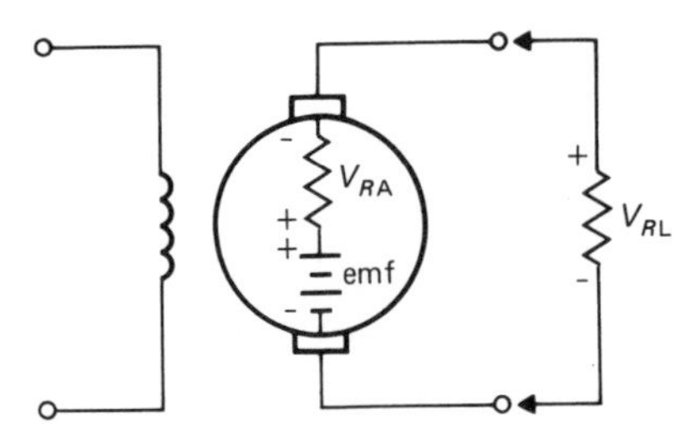

Figure 3.37 Voltage Distribution in Armature Circuit of Generator

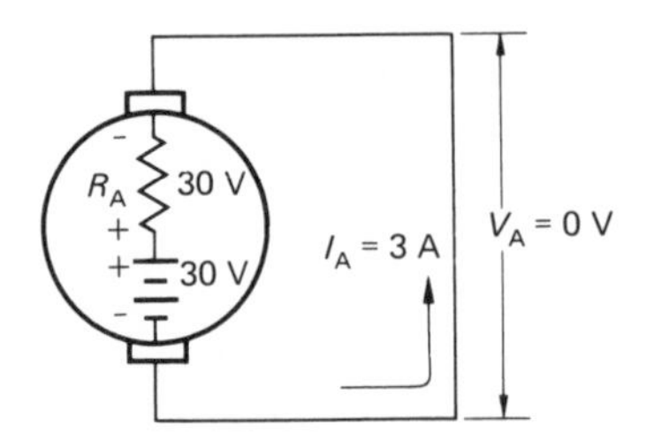

Figure 3.38 Voltage Distribution in Armature Circuit of Shorted Generator

less than—or, at best, equal to—the emf generated. Therefore, the generator armature voltage (V_{RL}) will also be lower than the emf generated. This result is an obvious departure from the motor characteristics, where the motor armature voltage is always higher than the cemf generated.

Now, let us look at an example that illustrates the development of the separately excited characteristic curve. Assume a constant-armature-speed drive is turning the generator at 1000 r/min. Armature resistance (R_A) is measured at 10 Ω, load resistor (R_L) is 50 Ω, K_T is 0.5 Nm/A^2, armature current (I_A) is 0.5 A, and field current (I_{field}) is 0.5 A. With these values determined, the armature terminal voltage, or load resistor voltage, is easily found:

$$V_{RL} = I_A R_L = (0.5\text{ A})(50\ \Omega) = 25\text{ V} \tag{3.71}$$

Generated voltage (emf) can also be found, as follows:

$$\text{emf} = I_A R_A + I_A R_L = (0.5\text{ A})(10\ \Omega) + 25\text{ V} = 30\text{ V} \tag{3.72}$$

Also:

$$K_{emf} = \frac{\text{emf}}{I_{field} n_A} = \frac{30\text{ V}}{(0.5\text{ A})(1000\text{ r/min})} = 0.06\text{ V min/A r} \tag{3.73}$$

These calculations show that the emf of 30 V is greater than the armature terminal voltage of 25 V due to the internal drop across R_A. As the load resistor decreases, the armature current increases, and more voltage is dropped across R_A. If the load resistor goes to zero (or shorts), maximum armature current flows:

$$I_A = \frac{\text{emf}}{R_A} = \frac{30\text{ V}}{10\ \Omega} = 3\text{ A} \tag{3.74}$$

This maximum current causes all the generated emf to be dropped internally, and the armature terminal voltage is zero. Figure 3.38 shows this result.

This same situation is encountered when a battery's terminals are shorted with a wire. A high current flows in the shorting wire, causing it to become hot. All the battery's open-circuit terminal voltage is dropped across its own internal resistance.

Maximum counter torque (cT) is also produced in the generator when the output is shorted since the following relation holds:

$$cT = K_T I_{field} I_A \tag{3.75}$$

Field current is constant and armature current is maximum. The characteristic curve is shown in Figure 3.39.

Some applications require the armature speed to vary but the output voltage to remain constant, such as in the automobile charging system. This result can be achieved by changing the field current as an inverse function of speed.

As an example, suppose the output voltage of 25 V of the previous generator were to be

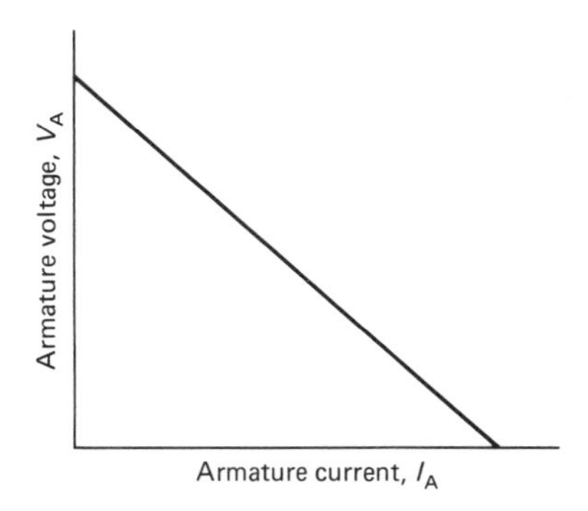

Figure 3.39 Characteristic Curve for Separately Excited Generator at Constant Armature Speed

maintained but the speed increased to 2000. Let us find the required field current. As long as the load resistance remains constant, so must armature current and emf. Therefore, the emf equation can be used to solve for I_{field}, as follows:

$$I_{field} = \frac{emf}{K_{emf}n_A} = \frac{30 \text{ V}}{(0.06 \text{ V min/A r})(2000 \text{ r/min})} = 0.25 \text{ A} \quad (3.76)$$

The automotive charging system is one that has a constantly changing speed drive for the generator but needs a constant output voltage for charging the battery. The automotive system can perform its task by using some method of sensing output voltage and adjusting field current to keep the voltage constant.

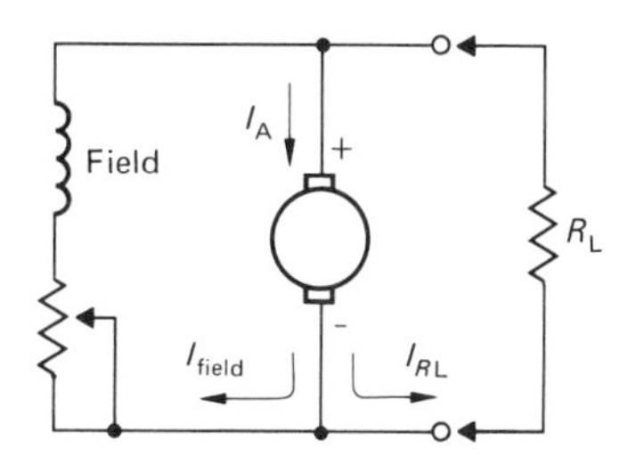

Figure 3.40 Shunt-connected Generator

Self-Excited Generators

Shunt Generator

When the field windings are connected in parallel with the armature, as in Figure 3.40, the generator is shunt-connected.

Self-excited generators, of which the shunt generator is one, do not produce an output voltage when their armatures are rotated, for several reasons. Consider what happens when the generator in Figure 3.40 starts from 0 r/min. At 0 r/min, armature, field, and load currents are all zero. As the armature starts turning, is an output voltage produced? Ideally, no. Since no field current is flowing, there is no magnetic field across the armature. Remember that voltage is only produced when a conductor moves within a magnetic field.

An output voltage generally is produced, though, because of a small residual magnetism remaining in the soft iron of the field core. Figure 3.41 shows how this residual magnetism appears. Consider the circuit of Figure 3.41A, where field

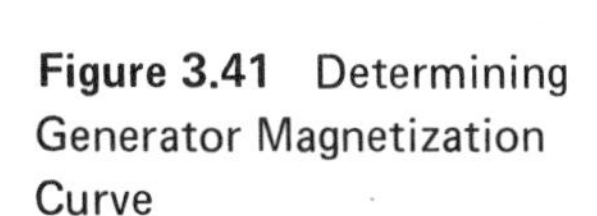

Figure 3.41 Determining Generator Magnetization Curve

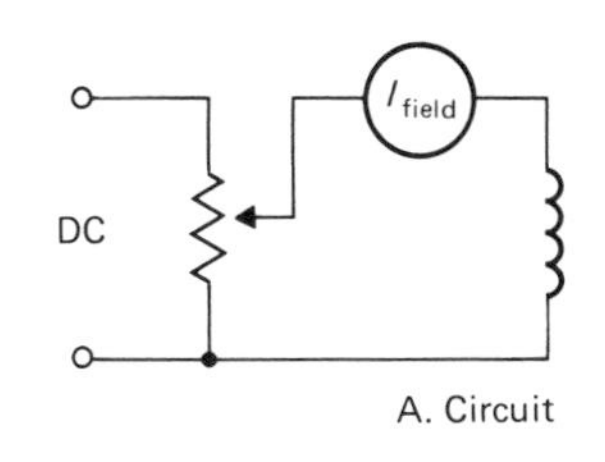

A. Circuit

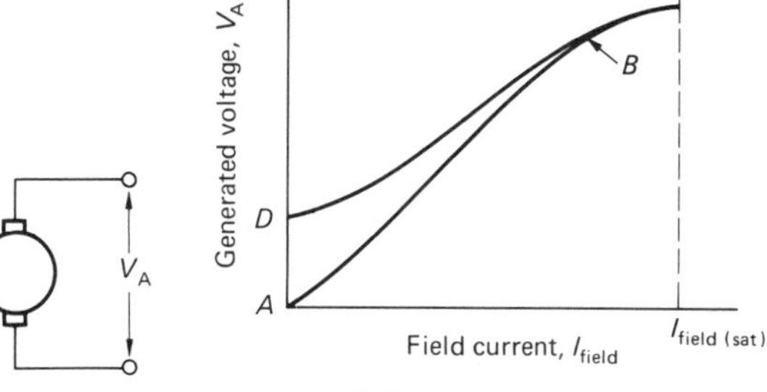

B. Magnetization Curve

current starts from zero and increases to $I_{field(sat)}$ (field current saturation). As the armature speed is held constant, the generated voltage increases from 0 V to some voltage at point C on the curve in Figure 3.41B. Notice that the curve is approximately linear from point A to point B but then curves to point C. This entire curve is generally called the *magnetization curve*; it shows field strength as a function of field current.

So, field strength increases linearly with field current to point B. After that point, called the knee of the curve, it takes proportionately much more field current to produce the same increase in field strength. From point B to point C and beyond is called the saturation part of the curve. Realize, though, that there is no point reached where an increase in field current does not produce an increase in field strength.

As field current is decreased to zero, some magnetism is retained because some of the molecular domains (groups of molecules that act together) remain aligned in the direction they were when field current was present. This small magnetism is called *residual magnetism* and is a property of the material called hysteresis.

Another problem arises in a self-excited generator when the field is connected in the wrong direction. Figure 3.42 shows a shunt-connected generator with residual magnetism. The solid lines represent the field due to the residual magnetism, and the dashed lines represent the field due to the generated field current. As the generator starts to rotate, current increases in the direction that opposes the residual magnetism. At some point, the field current will completely cancel out the field magnetism, and the output voltage will be zero.

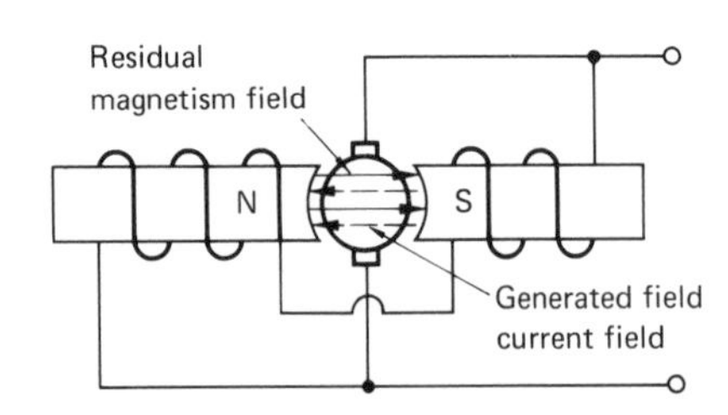

Figure 3.42 Field Connected in Incorrect Direction

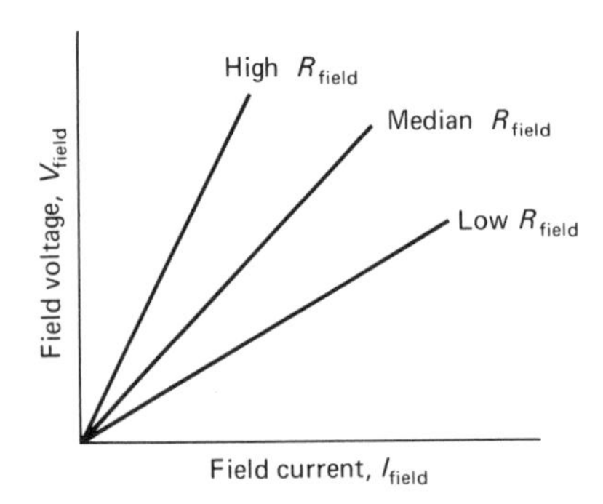

Figure 3.43 Voltage-Current Characteristics for Various Field Resistances

A field resistance that is too large can also prevent voltage buildup in the generator. Figure 3.43 shows a graph of field current and voltage for various field resistances R_{field}. For the same field voltage, as field resistance decreases, field current increases.

If this graph is superimposed on the magnetization curve (Figure 3.41), Figure 3.44 results. This graph shows that for a field resistance (R_{field}) and constant speed, a small initial voltage (V_{A1}) will be produced. As current starts to flow in the field, more voltage will be produced, generating more

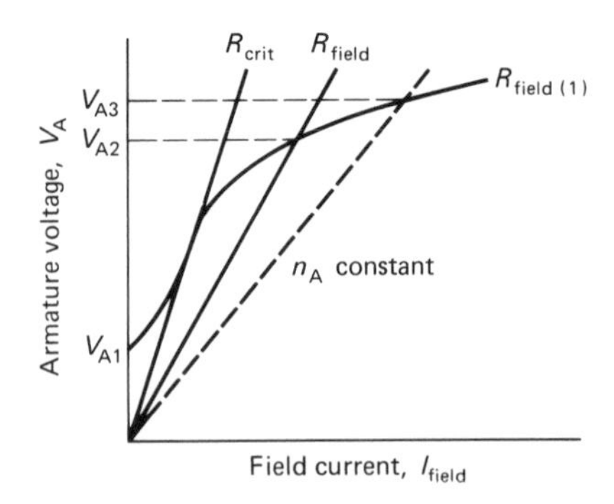

Figure 3.44 Magnetization and Field Resistance Curves Superimposed

current, and so on. A stable armature voltage (V_{A2}) will be reached when the voltage and current produced are all dropped across the field, assuming no other losses. If the field resistance is decreased to a new value $R_{field(1)}$ (corresponding to the low R_{field} curve in Figure 3.43), armature voltage stabilizes at V_{A3}. If, however, the field resistance is increased above R_{field} to R_{crit} (the critical resistance), the output voltage drops to a value below V_{A2}. If the field resistance is increased above R_{crit}, the generator field will *unbuild* (or collapse). In other words, the output voltage will drop back down to V_{A1}.

In summary, the following four conditions may cause a self-excited generator not to build an output voltage.

1. There may be lack of residual magnetism. Residual magnetism may be lost by physical shock (such as being dropped on a floor), heat, vibration, or lack of use for a period of time. This condition can be corrected by *flashing the field*, that is, by applying a DC voltage to the field in the proper direction to reestablish the residual magnetism.
2. The field circuit may be reversed with respect to the armature. A simple test shows whether the field is reversed. Connect a voltmeter to the generator output. If the output increases slightly when the field is disconnected, then the field is reversed.
3. The field circuit resistance may be higher than the critical value. An open or a high-resistance field circuit may be the problem. Check the circuit with an ohmmeter.
4. An open or a high resistance may exist in the armature circuit. Again, check the circuit with an ohmmeter.

The characteristic curve for a shunt generator with varying speed is shown in Figure 3.45. This graph shows that the generator cannot build voltage until it reaches a speed that overcomes the resistance of the field and load. This point is similar to the critical resistance point in Figure 3.44.

Figure 3.46 shows the characteristic curve at a constant speed. As the load current increases, the armature voltage drops. At a point called the breakdown point, the current drain of the load and the field exceeds what the generator can supply, and so the generator unbuilds. Over most of the load current range, however, the armature voltage is relatively constant.

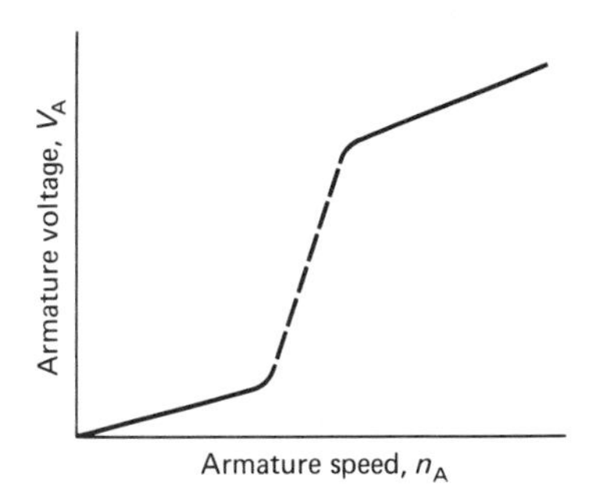

Figure 3.45 Characteristic Curve for Shunt Generator with Varying Armature Speed

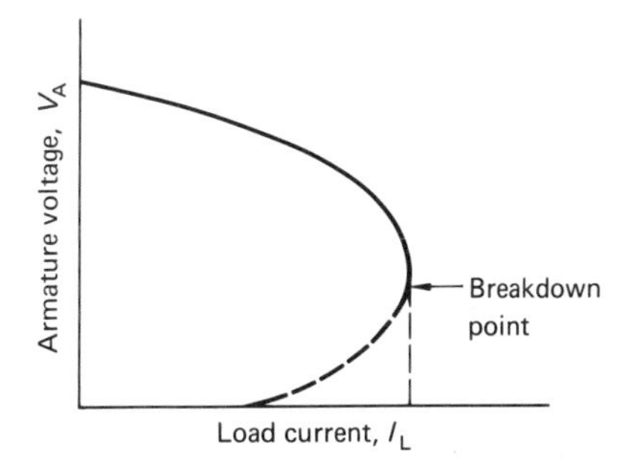

Figure 3.46 Characteristic Curve for Shunt Generator at Constant Armature Speed

The nonlinearity of the magnetization curve prevents easy use of the emf and torque equations with self-excited generator circuits.

Series Generator

When the field is in series with the armature, as in Figure 3.47, the generator is series-connected.

The characteristic curve for the shunt generator at constant armature speed is shown in Figure 3.48. The curve shows that at zero load

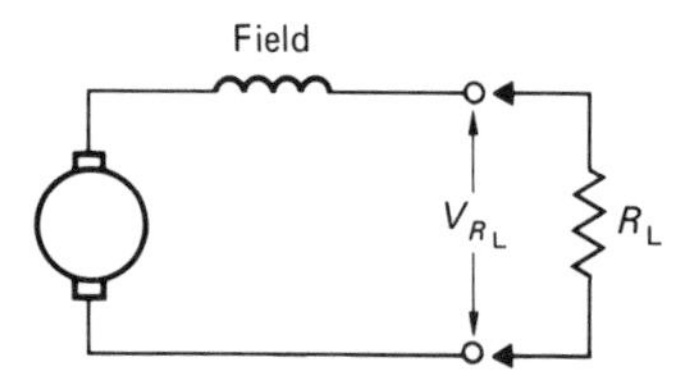

Figure 3.47 Series-connected Generator

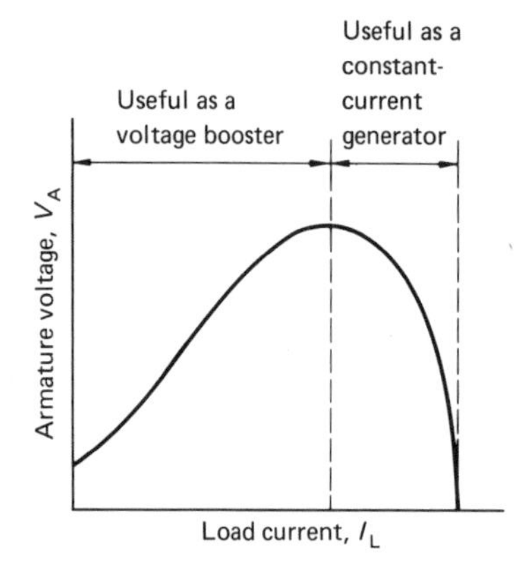

Figure 3.48 Characteristic Curve for Series Generator at Constant Armature Speed

current (output open-circuited), no armature current can flow to build up field strength. Therefore, the output voltage (V_{A1}) will be a result of residual magnetism only. As the load resistance decreases, more field current can flow and more output voltage is generated. At the peak of the curve in Figure 3.48, an increase in field current does not increase output voltage because the field has become saturated.

Past the peak of the curve, no additional voltage is produced, but more voltage continues to be dropped across R_A and R_{field}. Thus, the output voltage eventually goes to zero. This dropping part of the curve can be useful in welding generators. In such generators, the constantly changing arc length causes voltage to fluctuate, but the current must be constant in order to produce a consistent heat.

Compound Generator

As we have seen previously, terminal voltages associated with series-connected and shunt-connected generators vary in opposite directions with load current. If both a series and a shunt field were included in the same unit, it would be possible to obtain a generator with characteristics somewhere between the characteristics of these two types. The resulting device is a compound generator. The compound configurations and characteristic curves are shown in Figures 3.49 and 3.50.

If the number of turns in the series field is changed, three distinct types of compound generators are obtained: overcompounded, flat-compounded, and undercompounded. See Figure 3.49A.

When the number of turns of the series field is more than necessary to give approximately the same voltage at all loads, the generator is *overcompounded.* Thus, the terminal voltage at full load will be higher than the no-load voltage. This feature is desirable when the power must be transmitted

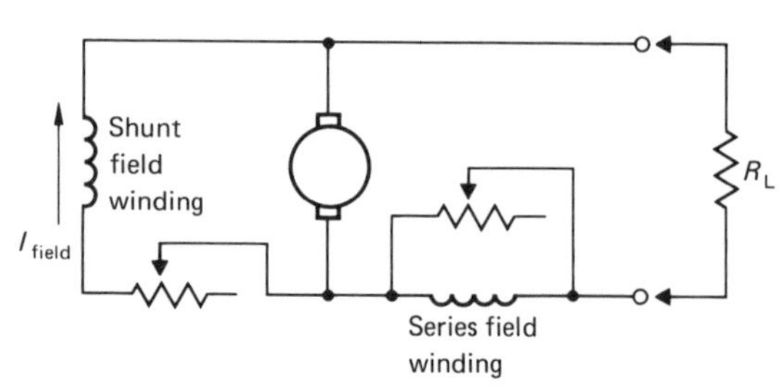

A. Overcompounded, Flat-compounded, and Undercompounded

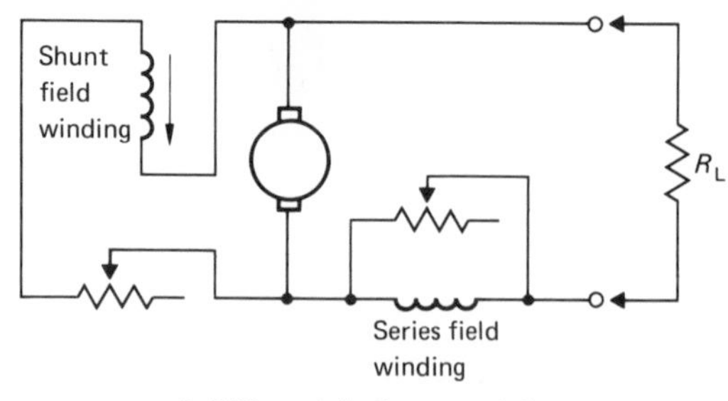

B. Differentially Compounded

Figure 3.49 Compound Generators

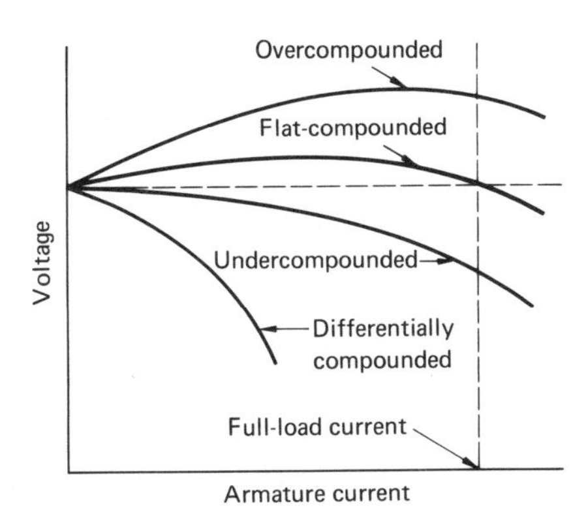

Figure 3.50 Characteristic Curves for Compound Generators

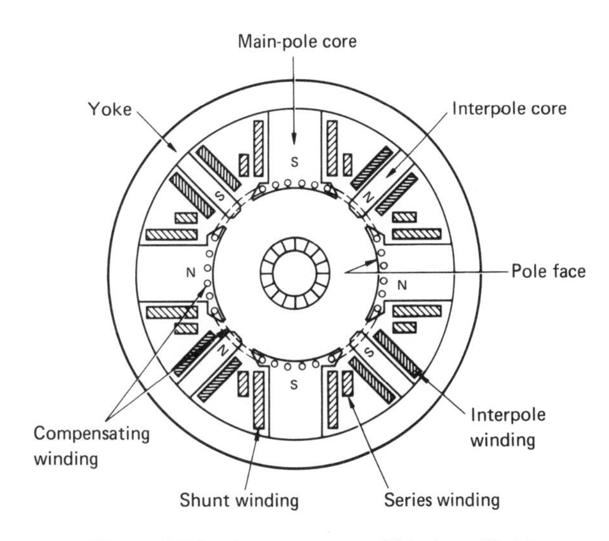

Figure 3.51 Arrangement of Various Field Windings on DC Machine

some distance. The rise in generated voltage compensates for the drop in the transmission line.

When the relationship between the turns of the series and the shunt fields is such that the terminal voltage is approximately the same over the entire load range, the unit is *flat-compounded.*

When the series field is wound with so few turns that it does not compensate entirely for the voltage drop associated with the shunt field, the generator is *undercompounded.* In this type of generator, the voltage at full load is less than the no-load voltage.

An undercompounded generator in which the series and shunt fields are connected so as to oppose rather than aid one another is referred to as a *differentially compounded generator.* See Figure 3.49B. With this type of generator, the terminal voltage decreases rapidly as the load increases. Undercompounded generators are sometimes used in welding machines.

Armature Reaction

As we noted earlier in the chapter, interpoles, sometimes called commutating poles, are small auxiliary poles placed midway between the main poles, as shown in Figure 3.51. They have a winding in series with the armature. Their function is to improve commutation and to reduce sparking at the brushes to a minimum.

To describe interpole function, we first must discuss *armature reaction,* which is the effect that the armature magnetic field has on the field distribution. Figure 3.52 illustrates how armature reaction is produced. Figure 3.52A shows the flux lines produced by the pair of poles when no current is flowing through the armature coils. A vertical line through the field indicates the zero axis (neutral plane) of the field. Figure 3.52B shows the flux lines produced by current flowing through the armature coils alone. Figure 3.52C shows the resultant field of the two fluxes superimposed. Note that the zero axis of the resultant field is displaced, as indicated by the line designated "new neutral plane." This displacement results in a shift in the position of the old neutral plane. Shifting of the neutral plane results in sparking, burning, and pitting of the commutator. As the load current and the resulting armature reaction increase, this effect becomes more pronounced.

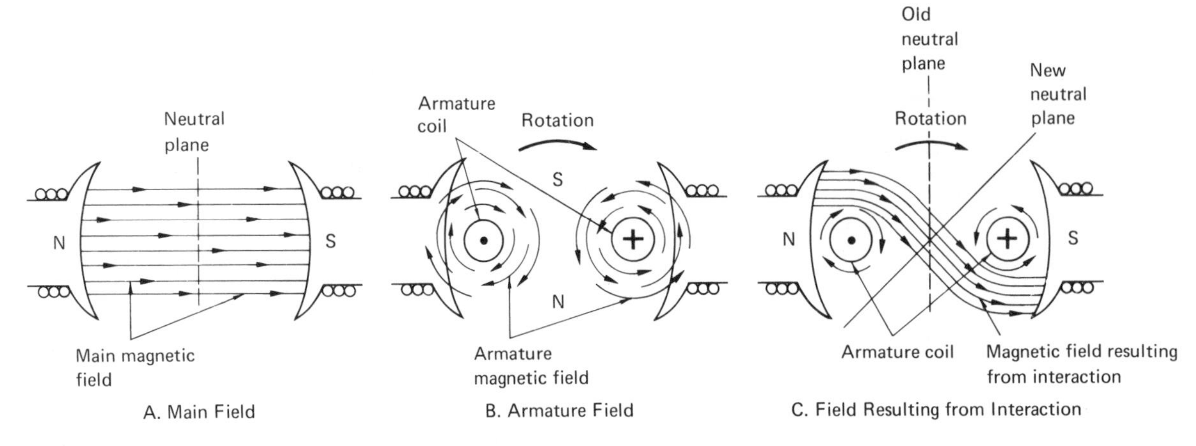

Figure 3.52 Armature Reaction

The short-circuiting effect can be counteracted by adding interpoles and compensating windings to the generator. Figure 3.53 shows the schematic diagram for a compound generator with interpoles (R_{pole}) and compensating windings (R_{comp}) added. The neutral plane, or load neutral, in the motor is shifted in the opposite direction with the same direction of armature rotation as shown in Figure 3.52. This example shows that motors and generators cannot be interchanged for best results.

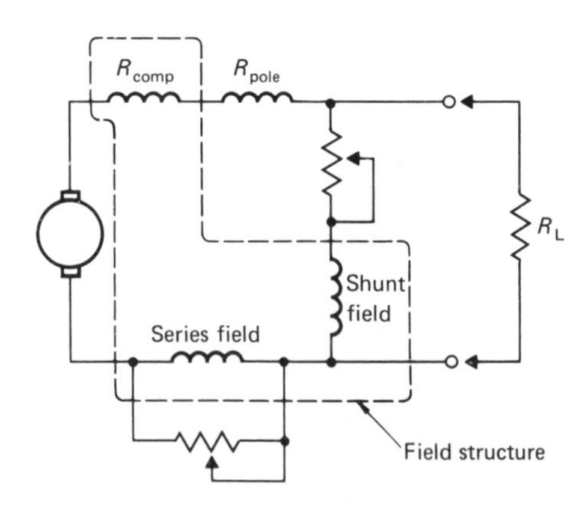

Figure 3.53 Compound Generator with Interpoles and Compensation Windings Added

Power and Efficiency

Power, P, is defined as the rate of doing work. The original unit of power was developed from James Watt's determination of horsepower. Power is such an important and useful concept that the basic electrical unit of power, the watt, was named after him, even though he did not work with electricity. Watt defined 1 hp as 550 ft-lb/s. Conversion to metric or SI units uses the following equation:

$$550 \text{ ft-lb/s} = 1 \text{ hp} = 745.7 \text{ Nm/s} = 745.7 \text{ W} \quad (3.77)$$

Torque and speed can also be related to power:

$$1 \text{ hp} = \frac{T \times n_A}{5252.1} \quad (3.78)$$

where torque is in pound-feet and speed is in revolutions per minute. In SI units, the conversion is as follows:

$$1 \text{ W} = 1 \text{ Nm/s} \quad (3.79)$$

The *efficiency*, η, of any device, such as a motor or generator, is its output power (P_o) divided by its input power (P_i), when they are in the same units:

$$\eta\% = \frac{P_o}{P_i} \times 100 \quad \textbf{(3.80)}$$

(η is the lowercase Greek letter eta.)

For example, a 1/8 hp motor operating at 110 V DC requires 1.02 A when operating at rated conditions. To determine the efficiency, we must find input and output power. Input power (P_i) is as follows:

$$\begin{aligned} P_i &= IV = (1.02 \text{ A})(110 \text{ V}) \\ &= 112.2 \text{ W} \end{aligned} \quad \textbf{(3.81)}$$

At rated conditions, the motor should be producing 1/8 the horsepower:

$$\begin{aligned} \frac{1}{8}\text{ hp} &= \frac{1}{8}(745.7) \\ &= 93.2 \text{ W} \end{aligned} \quad \textbf{(3.82)}$$

Therefore:

$$\begin{aligned} \eta &= \frac{P_o}{P_i} \times 100 = \frac{93.2 \text{ W}}{112.2 \text{ W}} \times 100 \\ &= 83.1\% \end{aligned} \quad \textbf{(3.83)}$$

Nameplate Specifications

The nameplate of a dynamo indicates the power, voltage, speed, and so forth, of the machine. These specifications, or nominal characteristics, are the values guaranteed by the manufacturer. The following information punched on the nameplate of a 100 kW generator is an example:

Power	100 kW	Speed	1200 r/min
Voltage	250 V	Type	Compound
Exciting current	20 A	Class	B
Temperature rise	50°C		

These specifications show that the machine can deliver, continuously, a power of 100 kW at a voltage of 250 V, without exceeding a temperature rise of 50°C. It can then supply a load current of 100,000/250 = 400 A. It is a compound motor with both series and shunt windings. The shunt field current is 20 A. The class B designation refers to the class of insulation used in the machine. These ratings are designed by the manufacturer and should not be exceeded. Characteristics of selected DC motor types are given in Table 3.6.

Conclusion

The theory of operation of DC motors and generators is quite old and well established, compared with the theory of some electronic components such as the transistor and microprocessor. However, the electric motor is still the major dynamo used in industry. Because of its widespread use, all electronics technicians need to have a thorough understanding of the basic principles of DC motor and generator operation. With the development of new electronic motor control systems, new motor applications will continue to be created.

Questions

Fill in the missing words with increase(s) (I), decreases(s) (D), or stay(s) the same (S).

1. The torque on a series motor is held constant. The armature speed ______ and the armature current ______ as the applied voltage is decreased.
2. In a shunt-connected motor, if you were to hold line voltage constant and increase torque, then armature speed would ______, armature current would ______, and cemf would ______.
3. A separately excited motor has a fixed armature voltage. If the torque demanded of the motor remains constant and the field current is increased, then cemf would ______, armature current would ______, and armature speed would ______.
4. A shunt generator is operating at a constant speed and has unlimited power available from the prime

Table 3.6 *Characteristics of Selected DC Motor Types*

Motor Type and Basic Characteristics	*Performance Ranges*	*Application Areas*
CONVENTIONAL PERMANENT-MAGNET* A simple alternative to wound-field shunt. PM field plus wound armature. Linear torque-speed relationships in small units. Life limited by brushes in high-speed or severe applications. Readily controlled by transistors or SCRs.	Outputfrom 1 W to a fraction of a horsepower. Time constants (to 63.6% of no-load speed) to less than 10 ms. Efficiencies from 60% to 70% in 10 W sizes. With new magnet materials, can deliver high peak powers (horsepower range).	For full range of inexpensive, good performance drive and control applications. With appropriate environmental precautions, suitable for military and aerospace use. Preferred as a high-performance, general-purpose servo motor.
LIMITED-ROTATION DC TORQUER No commutator wear or friction. Unlimited life. Infinite resolution. Smooth, cog-free rotation. No electromagnetic interference generation. Available as motor elements or fully housed.	Travel range typically to 120°. Torque from a few ounce-inches to greater than 40 lb-ft. Mechanical time constants from 10 to 50 ms.	Very high accuracy positioning or velocity control over a limited angle.
CONTINUOUS-ROTATION DC TORQUER Slow speed, high torque. Relatively low power output. Available as pancake-shaped components. Wide dynamic range. Large number of coils give smooth operation.	From tens of ounce-inches to hundreds of foot-pounds. Moderate mechanical time constants. Control to seconds of arc. Relatively expensive.	For direct coupling to load. For very precise control. Alternative to geared types.
MOVING-COIL, PRINTED-ARMATURE (IRONLESS ROTOR)† Similar to permanent-magnet DC units. Linear torque-speed characteristics. Smooth, noncogging rotation. Handles very high, short-duration peak loads. Fast response (less than 10 ms).	Outputs from less than 1 W to fractional horsepower. High efficiencies. Very low mechanical and electrical time constants.	Computer peripherals where smooth control and fast response are needed. Control applications needing high-response bandwidth, fast starting and stopping.
VARIABLE-RELUCTANCE STEPPER Brushless and rugged. High stepping rate dependent on driver circuitry. No locking torque at zero energization. Poor inherent damping. Low power efficiency. Can exhibit resonance. Operates open loop. Wide dynamic range. Easily controlled. Very reliable and low in cost in popular frame sizes.	Several hundred to thousands of pounds per second. Power output up to a few hundred watts.	Alternative to synchronous motor. Used in control applications where fast response rather than high power is the principal requirement. Interfaces well to digital computers.
SMALL-ANGLE, PERMANENT-MAGNET STEPPER Uses vernier principle to give very small stepping angles. High stepping rates. High cogging torques with zero input power. Efficiency usually very low.	Stepping rate from less than 100 lb/s to many thousands. Dependent on driver electronics. Power up to a few hundred watts. Single step takes a few milliseconds.	Useful in numerical control and actuator application where control is digital. Provides fast slewing and high-resolution tracking.
INVERTER-DRIVEN AC Operates from DC line using a switching inverter. Somewhat less efficient than AC induction motors; otherwise similar in performance. Single-phase (capacitor) or two-phase versions most common.	Outputs from less than 1 W to fractional horsepower. Efficiencies from 20% to 80% in larger models. Speeds up to 30,000 r/min and higher.	Use where DC is only power available. For universal applications where AC supplies vary widely, as in foreign applications. Use where brushes might not be sufficiently reliable, as in very high speeds or in severe environments. Variable-frequency versions used in accelerating high inertial loads.
BRUSHLESS DC PM units using electronic commutation of stator armature. Exhibits conventional DC motor characteristics, but torque modulation with rotation is higher. Lack of brushes gives reliability in difficult applications.	From less than 1 W to 1–2 hp. Relatively high time constants. Speeds to 30,000 r/min. Efficiencies to 80%. Voltages to hundreds of volts DC.	For brushless, long-life applications requiring superior efficiency and control. May be operated at very high altitudes or may be totally submerged.

Comparisons with Other Motors	*Selection and Application Factors*
gher efficiency, damping, lower electrical time constants than comparable AC control motors, except in very low power applications. ır more efficient than stepper drives. ore easily controlled than other motor types.	Select for safe operation with acceptable temperature rises. Check operating conditions for abrupt starts or reversals, which can demagnetize PM fields. Check for altitude or environmental effects on brushes, especially over 10,000 r/min. High stall currents are drawn in efficient or high-power motors. Low–output impedance electronic control required to utilize inherent motor damping.
uch simpler than continuous-rotation torquers with or without commutators.	Suitable for direct-drive, wide-band, high-accuracy mechanical control. Similar wide-angle brushless tachometers available. Requires high-power driving amps.
pplies most precise control, smooth and accurate tracking for continuous-rotation applications. equires higher-powered amps compared with geared units.	Stiff direct coupling to load preferred. Pulse-width modulation amps preferred for high control power.
ıster response than iron-rotor motors. xcellent brush life. ɔwer starting voltage, limited only by brush friction. uch more efficient than stepper motors.	Recommended for low-cogging, low–starting voltage, fast-response applications. Rotor heats up quickly. Thermal transients and heat removal can be important factors. Larger, high-performance units can be expensive. Low armature inductance permits commutation of very high current surges.
ɔwer output and efficiency generally very low compared with DC control motors.	Care required in application. Performance dependent on electronic driver circuitry. Heat dissipation a possible problem. At certain pulse rates, resonances can occur, which reduce load-handling capability. Load inertia reduces performance. Friction can improve damping. Damping, gearing, and mechanical couplings require special attention.
fficiency of shaft power generation is low. ore flexible than comparable means. mple and inexpensive alternative to synchronous or wide-speed-range drives. andles higher load inertia than variable-reluctance stepper and has better damping.	Choose where special control characteristics are preferred over efficiency. Check for resonances at all pulse rates. Gearing can require extra safety margins because of impacts inherent in stepper operation. Coupling compliances can help in accelerating load inertia, but additional resonances can be introduced. Driver circuit design is critical. Standard drivers available.
ess efficient than true brushless motors using electronic commutation. ore complex, expensive, and noisier than brush-type DC motors. ess suitable for control than other DC types. ery long life with properly designed inverter.	Inverter can be separate or packaged with motor. High line circuit spikes. Electromagnetic interference generation, with bulky filter capacitors required for suppression. SCR inverters preferred in higher-power uses, but transistor inverters are easier to switch and more reliable. Power supply capacitors can be required and must withstand supply transients.
ore efficient, easier to control, generates less electromagnetic interference than inverter-type motors. ommutating transistors can be used for speed control, reversing current, and torque limiting without a separate controller, unlike other types. elivers highest sustained output in a given package size.	Electronics can be packaged externally or within motor housing. High peak line currents. Bulky line filter required if electromagnetic interference is a problem. Power supply capacitors could be required. With properly designed electronics, life is limited only by bearings. Temperatures can set limits to some commutation sensors.

mover. If the load current on the generator is increased, the output voltage of the generator will ______.

5. If the output of any operating DC generator were suddenly shorted, the torque required to keep the speed constant would ______.
6. Define these devices: (a) dynamo, (b) generator, and (c) motor.
7. Name the parts of a dynamo's rotor and stator.
8. Define torque. Name its units (British and SI).
9. When a generator supplies load current, the terminal voltage is not the same as the generated emf. Is it higher or lower? Why?
10. How may the direction of rotation of a DC motor be reversed?
11. Explain why the generator magnetization curve is not a straight line.
12. When a motor is in operation, why is the armature current not equal to the line voltage divided by the armature resistance?
13. What is an interpole? What is its purpose? How is its winding connected?
14. What is the effect of armature reaction in a motor?
15. What four factors may prevent the buildup of voltage for a self-excited shunt generator?
16. Why is it dangerous to open the field of a shunt motor running at no load?
17. When is a compound generator said to be (a) flat-compounded and (b) overcompounded?
18. Why is the speed regulation of a series motor poorer than that of a shunt motor?
19. Define efficiency.
20. Why should a series motor never be operated without load?
21. Describe what is meant by dynamic braking.
22. Draw the schematic configurations for the series, shunt, separately excited, and long-shunt and short-shunt compound dynamos.
23. Draw the constant-torque and constant-line-voltage characteristic curves for the series and shunt motors and generators.

Problems

1. Find the armature speed of a series motor with 10 Ω field resistance, 5 Ω armature resistance, $K_{emf} = 0.311$ V min/A r, $K_T = 1$ Nm/A^2, an applied voltage of 200 V, and $T = 0.09$ Nm.
2. Given a shunt motor with a field resistance of 500 Ω, an armature resistance of 10 Ω, $K_{emf} = 0.311$ V min/A r, and $K_T = 0.5$ Nm/A^2. If the torque demanded is 0.2 Nm and the applied voltage is 100 V, what is the motor speed?
3. In the motor of Problem 2, suppose torque and armature current remain constant. What would the field resistance need to be in order to change the armature speed to 2500 r/min?
4. Given a separately excited motor with $R_{field} = 500$ Ω, $I_{field} = 0.2$ A, $I_A = 0.45$ A, $R_A = 20$ Ω, $V_A = 100$ V, and $K_T = 0.5$ Nm/A^2. The torque demand is constant for any n_A, and only the field voltage is reduced by half (V_A remains fixed). Find the new I_A, V_{RA}, cemf, and n_A.
5. In the motor of Problem 4, V_A and V_{field} are returned to what they were originally, and what the motor was turning suddenly locks up (stops turning). Find the armature current and torque produced by the motor. What is the power being dropped across R_A?
6. A separately excited generator has a field current of 0.3 A, with $R_A = 5$ Ω, load resistance (R_L) of 150 Ω, output voltage (V_{RL}) of 100 V, and an armature speed of 1500 r/min. Find the no-load voltage at 1000 r/min (R_L is infinite).
7. With the generator of Problem 6, what armature speed would be needed to give an output voltage of 150 V with a load resistance of 100 Ω?
8. With the generator of Problem 6, if you wished to keep the voltage across the 100 Ω load resistor at 150 V while changing the speed of the armature to 2000 r/min, the field current would have to be changed. What field current would be needed?
9. Use the generator of Problem 6. What is its output voltage when the load resistance goes to 0 Ω? What is the armature current?
10. Derive the characteristic curves for the separately excited motor of Figure 3.26. Use the appropriate motor values, similar to those used in the shunt motor characteristic curves of Figures 3.20 and 3.22.

AC Motors

CHAPTER 4

Objectives

On completion of this chapter, you should be able to:

- Classify AC motors by horsepower and internal construction;
- Explain the concept of the rotating field and calculate its speed;
- Explain how torque is produced in an induction motor;
- Calculate the slip of an induction motor;
- List and describe the different methods of starting single-phase motors;
- Justify the need for special starting methods for synchronous motors.

Introduction

Alternating current (AC) has one property that allows it to be transported long distances by wire more efficiently than direct current. Alternating current circuits can be stepped up (transformed) in voltage and at the same time stepped down in current. AC transmission lines can thus carry high voltages at low currents and keep I^2R (power) losses much lower than the equivalent DC power lines can. Because of this property, most power-generating systems today produce AC power. Consequently, the majority of motors used throughout industry are designed to use AC power.

There are other advantages to AC motors besides the wide availability of AC power. In general, AC motors of the same horsepower rating are smaller and, therefore, less expensive than DC motors. AC induction motors do not use brushes and commutators. They are thus less prone to mechanical wear and sparking. This feature decreases maintenance requirements and the possibility of igniting explosive gases. AC motors are well suited to constant-speed applications. Several AC motors can be made to run in synchronization at the same speed.

Industry uses AC motors in a wide variety of shapes, sizes, and power ratings. One type of AC motor is shown in Figure 4.1.

In this chapter, we will examine the characteristics of AC motors, and what makes them different from DC motors. We will briefly consider the universal motor, which is basically a series DC motor. As its name implies, the universal motor can be operated with either DC or AC voltage. The remainder of the chapter will deal with the two other classifications of AC motors, the induction motor and the synchronous motor.

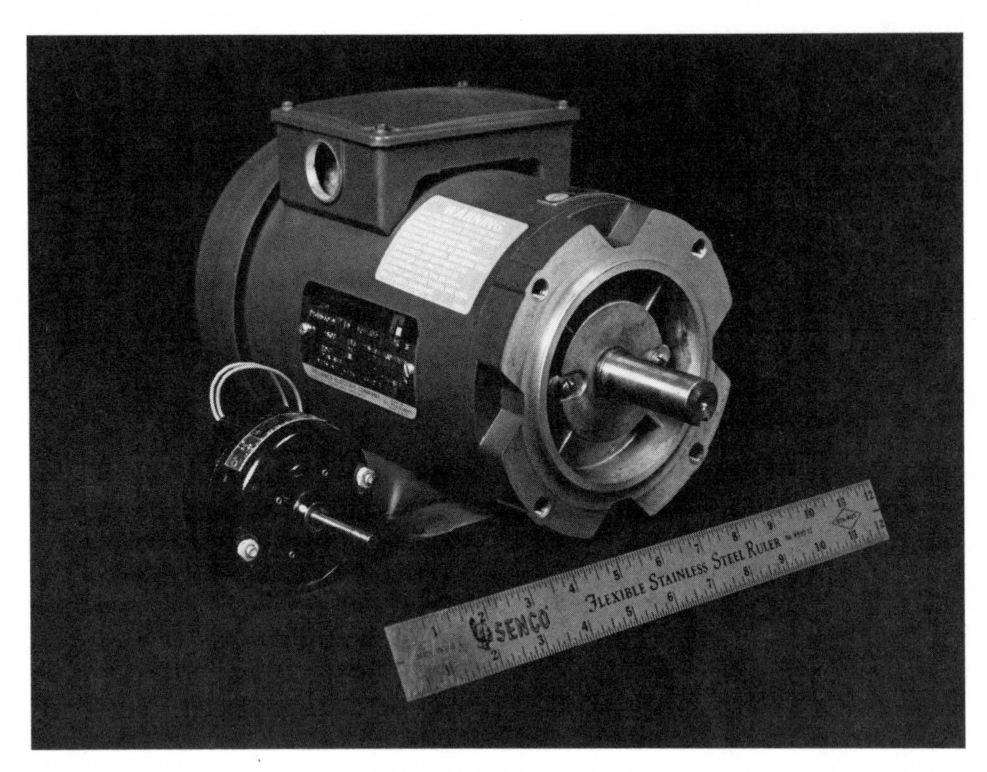

Figure 4.1 AC Motor

Classification of AC Motors

AC motors may be classified in several ways. First, they may be classified in terms of their horsepower (hp) ratings. *Fractional hp motors* are motors rated under 1 hp. *Integral hp motors* are motors rated at 1 hp or more. Integral hp motors may range from 1 to 10,000 hp.

The number of phases (usually one or three) applied to the stator windings is another basis for classification. Single-phase motors are found in domestic, business, farm, and small-industry applications. Three-phase power is generally used in heavy industry.

In this chapter, we will divide motors into three areas: (1) universal, (2) induction, and (3) synchronous. This classification is based on the internal construction of the motor. We cannot, in one chapter, cover all aspects of AC motors. Consequently, we will deal with the most important operating principles of the most common types of AC motors. For more information on AC motors, consult the bibliography for this chapter at the end of the book.

Universal Motor

The construction of the universal motor is shown in Figure 4.2. Note that the *universal motor* is electrically the same as the series DC motor. Current flows from the supply through the field, through the armature windings, and back to the supply.

If you use the left-hand rule for coils, you can determine the directions of the field shown in the figure. The polarities of the armature and field flux

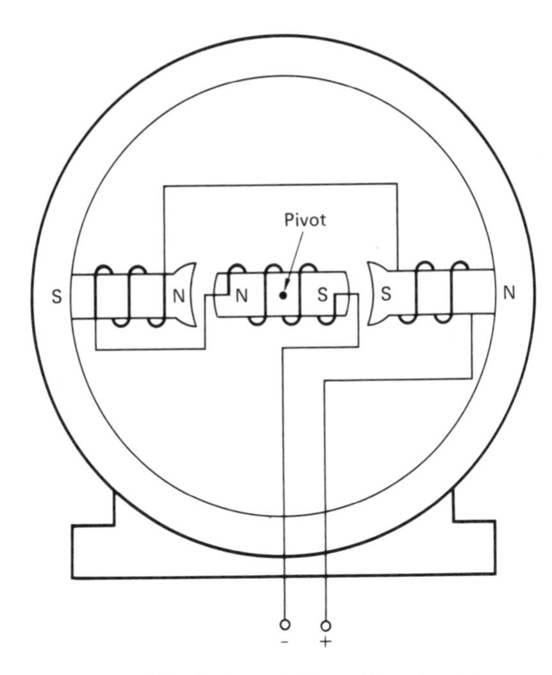

Figure 4.2 Universal Motor Construction

oppose each other. If current is reversed, both fields reverse. The magnetic flux of the armature and the field still oppose one another. The motor, therefore, tends to turn in one direction no matter which way current flows. When 60 Hz AC is applied to the input terminals, the current (and the magnetic fields) changes direction 60 times per second. These motors are called universal motors because they operate equally well on AC and on DC power.

Universal motors differ somewhat from DC motors in their physical construction. Recall that transformers have eddy current and hysteresis losses. AC motors tend to have these same losses. So that these losses are reduced, AC motors are constructed with special metals, laminations, and windings. We can say, therefore, that a universal motor works equally well on AC or DC. A series DC motor, however, does not work well on an AC supply because of the losses mentioned.

The operating characteristics of the universal motor are similar to those of the DC motor. For example, starting torque is high in the universal motor. As in the DC motor, speed can become dangerously high if the load is taken off. Universal motors are normally fractional hp devices powered by single-phase AC.

Rotating Field

Synchronous and induction motor construction can be divided into two basic parts, the rotor and the stator. Stators for both types of motors are similar in construction. So that eddy currents and hysteresis losses are reduced, the stator is made of many laminated steel discs. Coil slots are punched around the inside bore. The stator windings are wound around these slots. A typical single-phase stator winding is shown in Figure 4.3.

Principle of the Rotating Field

Since stators differ very little in induction and synchronous motors, the operating principle of rotating fields is basically similar for both. The principle of the *rotating field* is the key to the operation

Figure 4.3 Single-Phase Stator Winding

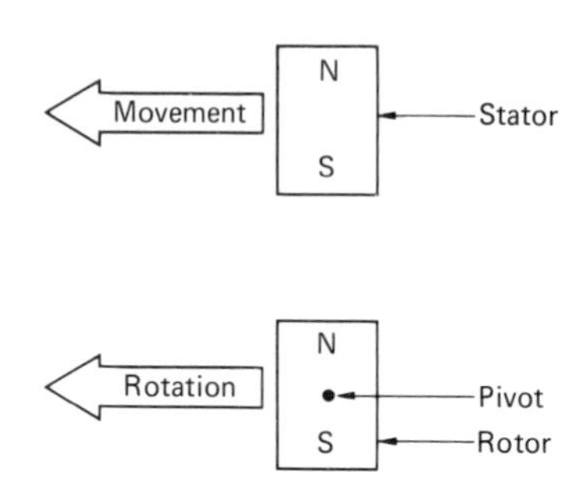

Figure 4.4 Rotating Rotor by Physically Moving Stator

of synchronous and induction AC motors. The idea is relatively simple. Let us, for the moment, assume that the rotor is a permanent magnet. One way we can get the rotor to rotate is to place a magnet near it and move the magnet; see Figure 4.4.

However, this technique is obviously unsatisfactory. One of these two parts must be stationary. Normally, the magnet that pulls the rotor around is a series of fixed windings called the stator windings. If these windings are fixed and cannot move physically, how can we get the rotor to rotate? The answer is to use a magnetic field that can be made to rotate electrically.

Let us see how this result can be obtained. Refer to Figure 4.5A. The principle of the rotating field is most easily seen in a two-phase system, as shown in the figure. But note that this principle is most frequently used in three-phase power systems.

Let us begin with a motor with two stator windings each displaced from the other by 90°. Stator 1 pole pairs are vertical, and stator 2 pole pairs are horizontal. If the voltages applied to stator windings are 90° out of phase, then the currents that flow in those windings are 90° out of phase also. If the currents are out of phase by 90°, the magnetic fields produced by those currents are 90° out of phase.

To see how the fields rotate, refer to Figure 4.5B. Remember that phase 1 is applied to stator winding 1 and phase 2 is applied to stator winding 2. At time 1, note that the voltage at stator 1 is maximum while the voltage across stator 2 is zero. The current in stator 1 is then maximum, producing the vertical magnetic field shown. Note that it is north on top and south on the bottom. Also note that there is no horizontal field. The total resultant field is shown by the arrow.

At time 2, the current has decreased in stator 1 and increased in stator 2. Thus, there are two fields present, one in stator 1 and one in stator 2. The total resultant field is again shown by the arrow.

At time 3, the current in stator 1 has decreased to zero, while stator 2 current has reached a maximum. There is a strong horizontal field but no vertical field. The resultant field is again indicated by the arrow.

Looking at the total resultant field direction, we see a counterclockwise movement of the field. It has moved a total of 90°. The field starts out in the six o'clock position and ends in the nine o'clock position. In the figure, this rotation continues for one full cycle. When the two-phase voltages have finished one complete cycle, the resultant field has moved one complete 360° rotation.

We have, then, created a rotating field by doing two things. First, we have placed two field windings (called the stator) at right angles to each other. Second, we have excited those windings with voltages 90° out of phase.

Two-phase motors are rarely used except in specialized equipment. We have discussed them here only as an aid to understanding. This same principle of the rotating field is used, however, in three-phase and single-phase systems. In the three-phase system, the phases are 120° out of phase, not 90°. In the single-phase system, the required phase difference is most commonly produced by adding inductance or capacitance to one of the pairs of windings.

Calculating Field Speed

In Figure 4.5B, note that if the supplied current completes 60 cycles each second, the resultant field rotates at 60 r/s, or 3600 r/min. If we double the number of stator coils, the stator field rotates only half as fast. In a motor with a rotating stator field,

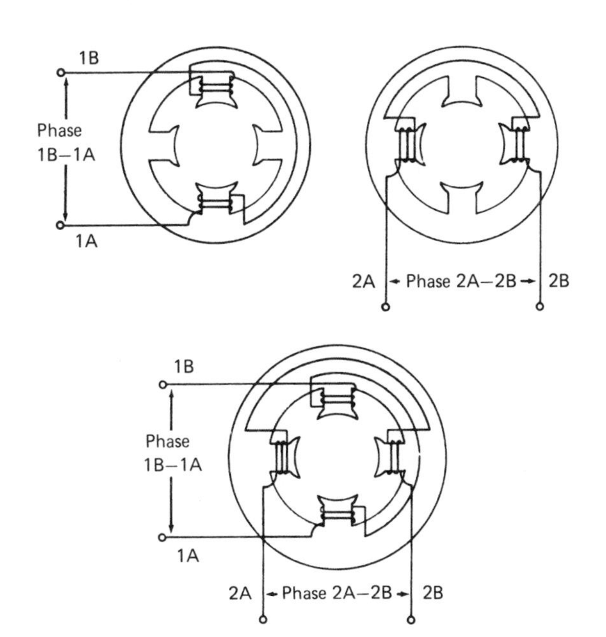

A. Two Phases Applied to Opposite Poles

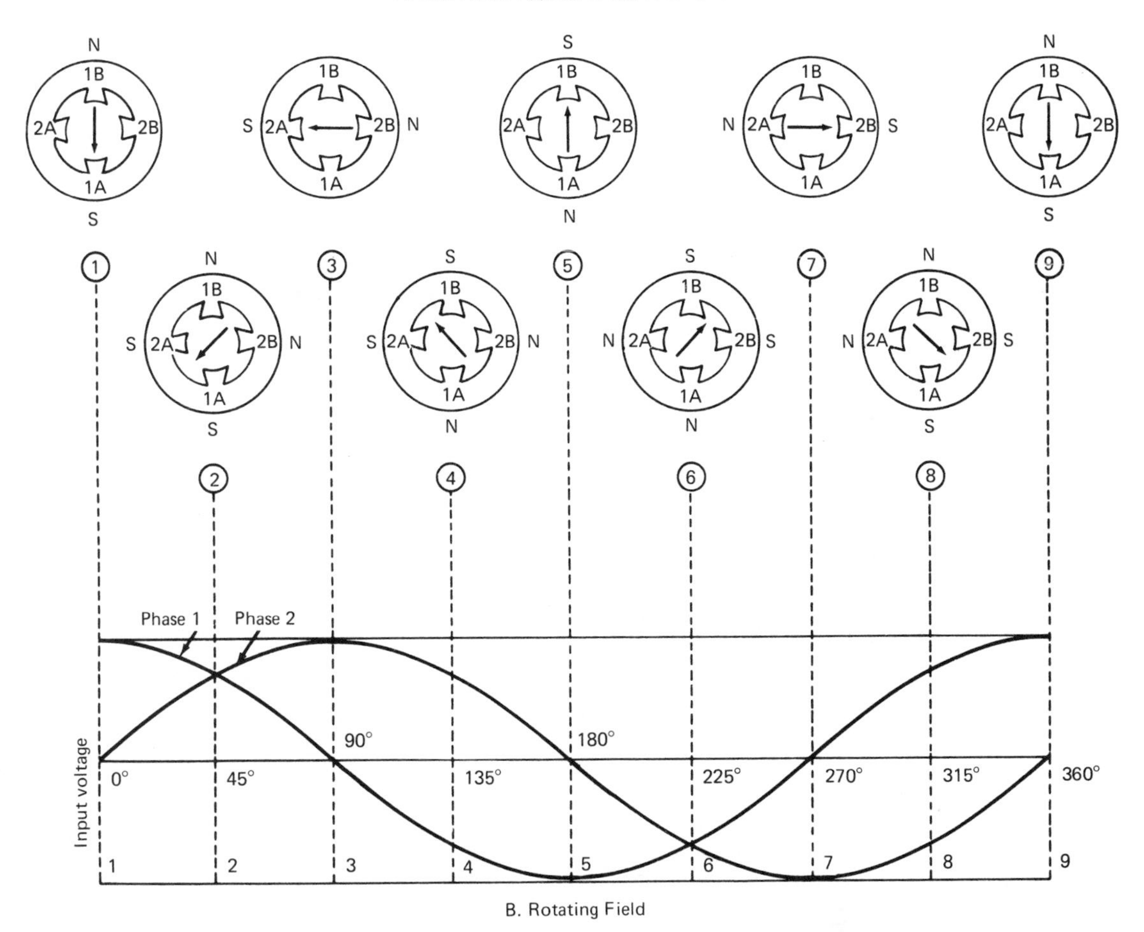

B. Rotating Field

Figure 4.5 Rotating Field in a Two-Phase System

the field travels one rotation per pole pair, per winding, for each cycle of applied voltage. If we double the number of poles for each winding, the magnetic field has more pole pairs to travel. It, therefore, takes twice as long to complete one rotation of all the pole pairs in one winding. This relationship can be stated in an equation:

$$n_{st} = \frac{120f}{p} \tag{4.1}$$

where N_{st} = stator speed, in revolutions per minute
f = frequency of voltage applied to stator windings
p = number of magnetic poles per phase

For example, let us say we have a three-phase motor with four magnetic poles. If the frequency of the voltage applied to the stator is 60 Hz, what is the stator speed? The stator speed is as follows:

$$n_{st} = \frac{(120)(60 \text{ Hz})}{4} = 1800 \text{ r/min} \tag{4.2}$$

The speed at which the stator field moves is called the *synchronous speed.* This name is used because the field is synchronized to the frequency of the supply voltage at all times. Note that as we increase the number of poles, the stator speed decreases. Thus, we see that the speed of this type of motor can be changed by switching in different numbers of poles. However, this method of speed control is limited to a small number of discrete speed changes.

Induction Motors

The driving torque of both DC and AC motors comes from the interaction of current-carrying conductors in a magnetic field. Recall that in the DC motor, the armature moves, while the field is stationary. Currents in both field and armature windings create the magnetic fields that produce the torque. The rotor current comes from the supply through the brushes and commutator. In the *induction motor*, electromagnetic induction produces the rotor currents.

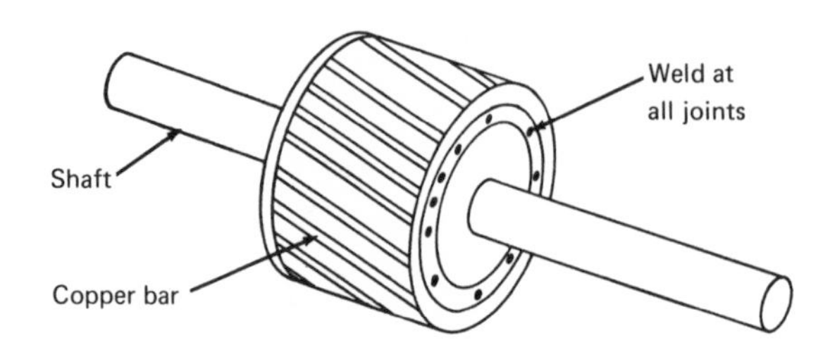

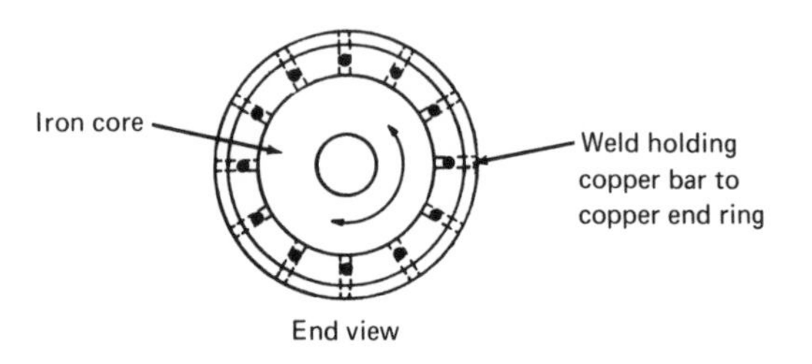

Figure 4.6 Squirrel-Cage Rotor Construction

Rotor Construction

Before we turn to the operation of the induction motor, we will examine the rotor construction in Figure 4.6. A squirrel-cage rotor is shown in Figure 4.7.

Figure 4.7 Squirrel-Cage Rotor

The rotor conductors are usually made of heavy-gauge copper. All these bars are connected at either end by copper or brass end rings. The end rings short-circuit the copper bar at both ends. No insulation is needed between the bars because the voltages generated are low and the bars represent the lowest resistance path for the current to flow. The core of the rotor (not shown in the diagram) is made of iron. This construction technique reduces air gap reluctance. It also concentrates the flux lines between the rotor conductors. Because of its unique appearance, this rotor is often called the *squirrel-cage rotor.*

Induction Motor Operation

Let us proceed now to the operation of the induction motor. Assume that we have a motor with a two-phase rotating field. If this field rotates over the stationary rotor windings, current is induced in the copper bars. One of these bars is shown in Figure 4.8.

The current generation follows the left-hand rule for generators. Remember that although the stator field (shown as a magnet) is moving right, the relative motion of the conductor is to the left. From the dot convention, current is generated, flowing into the page. This current flow generates a magnetic field around the conductor.

The interaction of the stator field and the rotor field produces a torque. The torque generated tends to move the conductor toward the right. We see, then, that the rotor tends to move in the direction of the stator rotation. Note that there is no physical contact between the rotor and stator. The only thing that links them is the electromagnetic field.

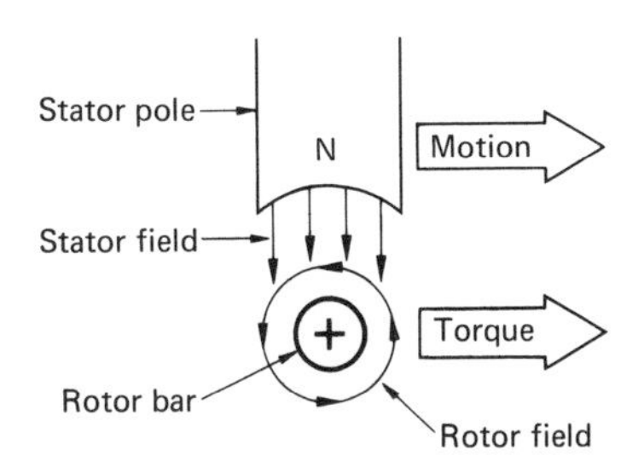

Figure 4.8 Interaction of Stator Field and Rotor Bars

Since this motor has no physical connection between rotor and stator, it is a relatively maintenance-free device. There are no brushes to replace. In addition, since no brushes exist, there is no sparking to ignite explosive gases and vapors.

Slip

Note that it is impossible for the rotor of an induction motor to turn at the same speed as the stator. If it could, no current would be produced in the rotor. No rotor current means no torque, so the motor will not run on its own. There must be relative motion between the rotor and the stator field. The rotor, therefore, runs at a slower speed than the stator.

The difference between the rotor speed and the stator field speed is called *slip.* Slip (s) is normally expressed in equation form, as follows:

$$s = \frac{n_{st} - n_{rt}}{n_{st}} \tag{4.3}$$

where n_{st} = stator speed, in revolutions per minute

n_{rt} = rotor speed, in revolutions per minute

In this equation, slip will range between 0 and 1. Sometimes, slip is expressed as a percentage. In that case, multiply the calculated value by 100. Then, slip ranges between 0% and 100%. In actual loaded AC motors, slip ranges between 1% and 10%.

When the motor starts, the stator immediately goes to synchronous speed. At the first instant of starting, the rotor is stationary. The slip at starting is, therefore, 1, or 100%. Gradually, the rotor picks up speed, rotor currents decrease, and the rotor reaches a point where the torque produced equals the torque demanded by the load. The rotor speed ceases to increase but remains constant as long as this condition exists.

Rotor Reactance

As the motor picks up speed, the rate at which the rotating field cuts the rotor conductors decreases. The amount of rotor voltage and current then decreases. We observe that the frequency of the rotor current varies directly with slip. As slip increases, so does rotor current frequency.

Recall the equation for inductive reactance (X_L) from basic electronics:

$$X_L = 2\pi f L \tag{4.4}$$

where L = inductance
f = frequency of current flow through that inductance

In our example, if L is the inductance of the rotor and f is the stator frequency, we can add slip (s) to the equation because the reactance decreases as rotor frequency decreases. Our new equation for *rotor reactance* is, then, as follows:

$$X_{\text{rt}} = 2\pi f L s \tag{4.5}$$

where X_{rt} = rotor reactance at slip s

Torque

Note that when the two speeds, rotor and stator, are close to each other, slip is nearly zero. As slip approaches zero, so does rotor reactance. When slip is near zero, the only impedance to rotor current flow is resistive. Thus, rotor current and voltage are in phase. At starting, however, slip is one, and the impedance is mostly reactive. In this case, rotor current lags rotor voltage by about 90°.

With this value of slip (100%), the high rotor reactance produces a low power factor. Recall that the *power factor* of an AC circuit is the power delivered to or absorbed by the circuit divided by the apparent power of the circuit. A resistor, because it does not have any reactance, has a power factor of one. A perfect coil or capacitor has a power factor of zero because it consumes no DC power.

In the near-zero-slip motor application, a low power factor means low torque due to the small rotor current. At the other end of the scale, when slip approaches one, torque is low because the power factor angle is large. The torque produced by an induction motor, then, varies approximately as the power factor of the rotor current. The torque equation is as follows:

$$T = K I_{\text{st}} I_{\text{rt}} \cos\theta \tag{4.6}$$

where K = a constant (as in the DC motor)
I_{st} = stator current
I_{rt} = rotor current
$\cos\theta$ = power factor of the rotor current

With no load connected, the rotor of an induction motor rotates at a value close to synchronous speed. If the motor is loaded, slip increases, causing more rotor current to flow. Remember that rotor resistance is small. Under the loaded condition, more rotor current produces more torque output. This increased torque speeds up the motor to a value close to the value where it started. Induction motors seldom change speed more than 5% from no-load to full-load conditions. They are, therefore, classified as constant-speed devices.

There is a limit to the amount of torque the motor produces. If we keep increasing the torque demand, the motor produces more torque, up to a point. When the motor produces as much torque as it can (at a slip of about 25%), further torque demands simply decelerate the motor to a stop. The point at which this happens is called the *pull-out*, or *breakdown, torque*. This point is illustrated in the curve in Figure 4.9.

Figure 4.9 shows the torque and current curves for a three-phase induction motor with a squirrel-cage rotor. As load increases, slip and rotor reactance increase. The pull-out torque occurs at about 25% slip. Note that the torque at this point is about three times the normal, full-load value.

Another important point to notice is the change in rotor current when the motor stalls. It increases by a factor of five. As the rotor current increases, so does the stator current because of the induction or transformer action of the rotor and

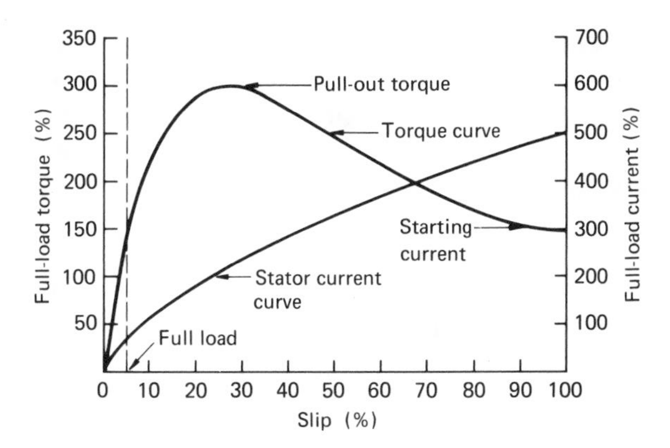

Figure 4.9 Torque and Current Curves for Three-Phase Induction Motor

stator. The rotor can be considered the secondary and the stator the primary of a transformer. As the current in the secondary of a transformer increases, so does the primary current in order to keep the power in the two windings approximately the same. Thus, in the induction motor at stall, the stator current increases by approximately a factor of five over its full-load current. Current this large damages most motor windings. Designers normally protect motors from this kind of damage by installing circuit breakers of appropriate size.

Some tests require that the rotor of a motor be locked. In this case, the stator voltage should be lowered to about 50% of its normal value. Lower stator voltage decreases locked-rotor current, thus protecting the motor.

Classifications of Squirrel-Cage Induction Motors

Squirrel-cage induction motors have several classifications, all based on differing types of rotor construction. Changes in the rotor construction produce differences in torque, speed, slip, and current characteristics. These motor types have been standardized by the National Electrical Manufacturers Association (NEMA). Examples of this classification scheme are NEMA motor types B, C, and D.

The NEMA type B motor is one with normal starting torque and current. A low rotor resistance produces low slip and high running efficiency. Generally, applications include constant-speed and constant-load requirements, where the motor does not need to be turned on and off frequently. Centrifugal pumps, blowers, and fans are examples of such applications.

The NEMA type C motor has the same slip and starting currents but a higher starting torque and poorer speed regulation than the type B. This motor is used with loads that are hard to start, such as compressors, conveyors, and crushers.

The NEMA type D motor has a high-resistance rotor. Slip increases and efficiency decreases in this motor. It is used in applications that require frequent starting, stopping, and reversing. Since type D has the highest starting torque, it is used to drive loads with very high torque demand, such as cranes, hoists, and punch presses. Because of its specialized rotor construction, the type D motor is the most expensive of the three. The type B is the most commonly used.

Double Squirrel-Cage Construction

Another special type of rotor construction is illustrated in Figure 4.10. This rotor construction is known as the *double squirrel-cage rotor*. In this

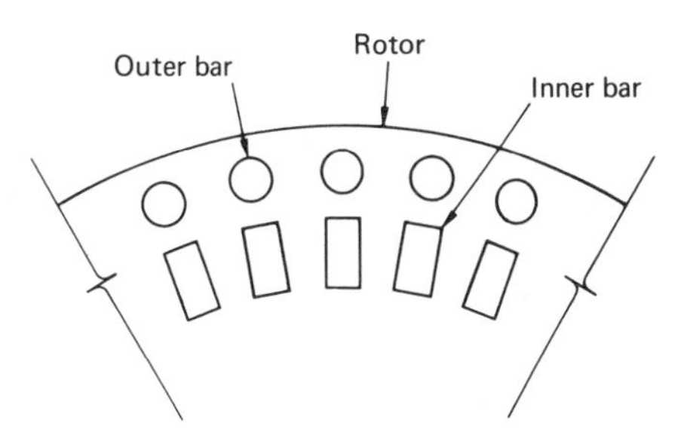

Figure 4.10 Double Squirrel-Cage Rotor Construction

rotor, the set of bars closest to the surface have a small cross-sectional area and a resistance of a few tenths of an ohm. The bottom bars have a much lower resistance, on the order of several thousandths of an ohm.

When the motor starts, the frequency induced into the bars is high. Since the lower bars have more inductance, the high frequency causes them to have a higher reactance than the top bars have. The current flow at starting is, therefore, higher in the top bars. Since these bars have a high resistance, they cause a reduction in the phase angle between the stator flux field and the rotor current. This reduction tends to increase starting torque. As speed picks up, the reactance of the bottom bars decreases, causing more current to flow in them. The effect of adding these extra bars is more torque to the motor under varying load conditions. The NEMA type C induction motor is an example of the double squirrel-cage construction.

Single-Phase Induction Motors

As their name implies, *single-phase motors* run when supplied with only one phase of input AC power. They are used in fractional hp sizes in small commercial, domestic, and farm applications. Single-phase AC motors are generally less expensive to manufacture in fractional hp sizes than three-phase motors. Of course, they eliminate the need to run three-phase power lines, which can be expensive and may not be necessary. Single-phase motors are used in blowers, fans, compressors in refrigeration equipment, portable drills, grinders, and so on.

The single-phase induction motor does not produce a rotating field and, therefore, has no starting torque. But if the rotor is accelerated, it behaves as if it were driven by a rotating field. With a single-phase supplied to the stator, the stator field reverses at each alternation of the supply voltage. Thus, we can visualize the stator field as rotating in half-turn steps, in either direction.

Because the single-phase induction motor does not produce any starting torque, it must depend on another system to start it. So, single-phase induction motors are classified by the method used to start them. These classifications are discussed next.

Shaded-Pole Motor

One of the lowest-cost starting methods is the *shaded-pole motor*. Shaded-pole motor construction is illustrated in Figure 4.11. Notice that a portion of each pole is surrounded by a copper strap, or shading coil. When current is increasing through the windings, the flux increases also. But as the flux travels through the portion of the pole piece shaded by the copper bar, the current induced into the copper strap builds up a field of its own. This field opposes the field buildup caused by the stator winding. The total flux is concentrated on the left side of the pole, as shown in Figure 4.12A.

As the applied voltage reaches a peak, the rate of change of flux is zero. At this time, the flux is evenly spread over the pole piece, as shown in Figure 4.12B.

When the applied voltage decreases, the flux decreases also. At this time, however, the flux around the copper bar reverses direction, opposing

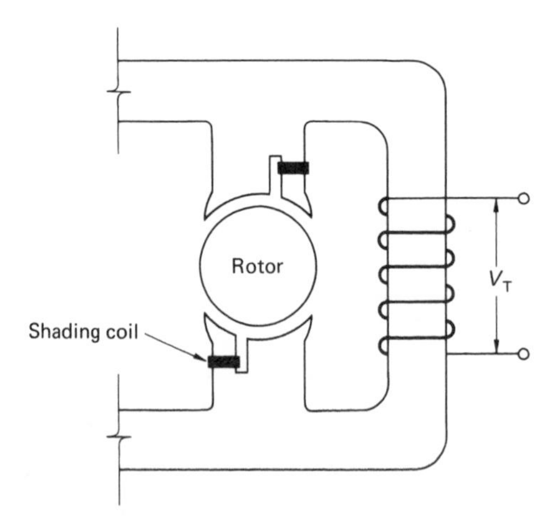

Figure 4.11 Shaded-Pole Motor Construction

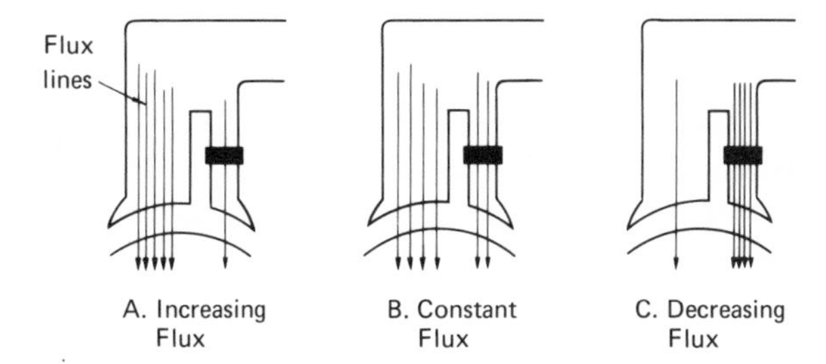

A. Increasing Flux B. Constant Flux C. Decreasing Flux

Figure 4.12 Movement of Flux in Shaded-Pole Motor

the decrease in flux around the bar. This reversal produces a total flux concentration on the right side of the pole piece (Figure 4.12C).

Note that the total flux density shifts from left to right with respect to time. In other words, the flux moves toward the shaded pole. The rotor then follows this shift as if it were a rotating field.

The shaded-pole motor produces starting torques in the range of 50% of full-load torque. It is manufactured in fractional hp sizes to about 1/20 hp, and it is not very efficient. A typical application of this motor is in a small fan or inexpensive phonograph.

Split-Phase Motor

The *split-phase motor* has more starting torque than the shaded-pole motor. From Figure 4.13A, we see that the split-phase motor has a stator with two sets of windings. One set, called the start winding, has a few turns of small wire. It has a high resistance and a low reactance compared with the other windings, the run windings. The axes of these windings are displaced by 90°.

The current in the start winding (I_{start}) lags the line voltage by about 30°, as shown in Figure 4.13B. It is also smaller than the current in the run winding because the impedance of the start winding is larger. The current in the run winding (I_{run}) lags the applied voltage by about 45°. This phase shift produces a rotating field effect such as we discussed earlier in the chapter. The interaction of this field and the one produced by induction in the rotor causes a starting torque to be generated. This motor is sometimes called the *resistance-start motor*.

When the motor comes up to about 75% of its running speed, a centrifugal switch opens. A *centrifugal switch* is a device that disconnects the start winding in an AC motor at about 3/4 of the full-load rotor speed. The centrifugal switch is mounted on the rotor and, as its name implies, is operated by centrifugal force. This switch disconnects the start winding. The motor continues to turn as long as a single phase is applied to the stator. Thus, the split-phase motor has good speed regulation. Starting torques range from 150% to 200% of full-load torque.

More starting torque is produced by adding an inductor in series with the run winding, as shown in

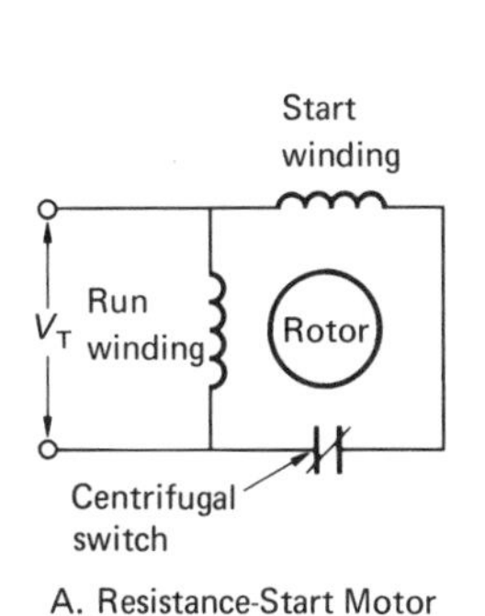

A. Resistance-Start Motor

B. Vector Graph of Phase Relations

C. Reactor-Start Motor

Figure 4.13 Split-Phase Motor

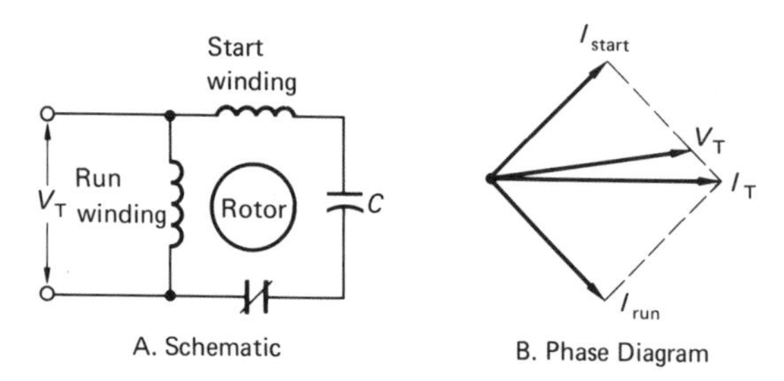

Figure 4.14 Capacitor-Start Motor

Figure 4.13C. When the centrifugal switch opens, it not only disconnects the start winding but also shorts out the inductor. This motor is sometimes called the *reactor-start motor*. Direction of rotation is reversed by reversing either the start or run windings.

Applications of the split-phase motor include washers, ventilating fans, and oil burners.

Another type of split-phase motor is the *capacitor-start motor*. This motor has more starting torque than the reactor-start motor because a capacitor is added in series with the start winding. This motor is illustrated in Figure 4.14A.

The capacitor produces a larger phase difference between currents in the start and run windings than the resistance-start motor does. The current in the start winding is about 90° out of phase with the current in the run winding, as shown in Figure 4.14B. This phase difference produces a larger starting torque than is available in the resistance-start motor. Starting torque can reach 350% of full-load torque.

If the capacitor is disconnected by a centrifugal switch, the motor is called a *capacitor-start motor*. If the capacitor is left in the circuit, the motor is called a *capacitor-run motor*.

The capacitor in the capacitor-start motor must be a nonpolarized one. Sizes range from 80 μF for a 1/8 hp motor to 400 μF in a 1 hp motor. The capacitor-start motor is normally used in applications that demand high starting torque where DC motors cannot be used.

Wound-Rotor Induction Motor

Another type of induction motor, called the *wound-rotor motor*, is sometimes seen in industry, although not as often as the squirrel-cage. Its rotor construction is closer to the DC motor armature than to the squirrel-cage rotor construction. The windings are also similar to the stator windings. They are usually wye-connected, with the free ends of the windings connected to three slip rings mounted on the rotor shaft, as indicated in Figure 4.15. An externally variable wye-connected resistance is connected to the rotor circuit through the slip rings. In Figure 4.15, ϕ_1, ϕ_2, and ϕ_3 indicate the various phases in a three-phase system.

Rotor resistance is high at starting, producing high starting torque. The higher rotor resistance increases the rotor power factor, which accompanies the higher starting torque. As the motor

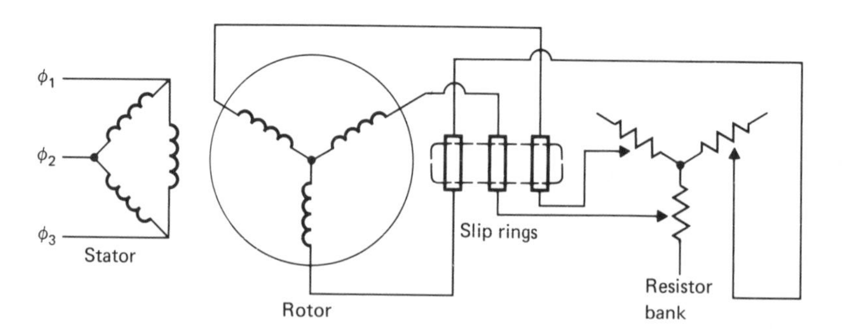

Figure 4.15 Wound-Rotor Induction Motor

accelerates, the variable resistance is reduced. When the motor reaches full-load speed, the resistance is shorted out. The motor then acts like a squirrel-cage motor.

The advantages of the wound-rotor motor over the squirrel-cage motor are (1) smoother acceleration under heavy loads, (2) no excessive heating when starting, (3) good running characteristics, (4) adjustable speed, and (5) high starting torque. There are two major disadvantages of the wound-rotor motor: Its initial cost is high, and it has higher maintenance costs.

The wound-rotor motor is not used as much today as it was in the past. Its decline is due to improvements in speed control by varying stator frequency and to the development of the double squirrel-cage rotor.

Synchronous Motors

The *synchronous motor* differs from the induction motor in several ways. A large polyphase synchronous motor needs a separate source of DC for its field. It also requires special starting methods, since it does not produce enough starting torque to start on its own. The rotor of the synchronous motor is essentially the same as an alternator rotor. Recall, from the previous chapter, that a DC motor may be run as a DC generator. Likewise, an AC generator (called an *alternator*) may be run as an AC motor.

Unlike the induction motor, the synchronous motor runs at synchronous speed. Also, the synchronous motor can be used to correct the low power factors caused by large inductive loads. As mentioned previously, synchronous motors range from fractional sizes to thousands of horsepower. Synchronous motors are found in a wider range of sizes than any other type of motor.

Synchronous Motor Operation

The basic operating principle of the synchronous motor may be described by visualizing the rotor as a magnet. Indeed, in fractional hp motors, the rotor may be a permanent magnet. If the stator field is sweeping around at synchronous speed, the magnet is literally yanked around in step with the revolving field.

Starting the Motor

The synchronous motor suffers from the disadvantage of not being able to start on its own. Refer to the simplified diagram in Figure 4.16. Current flowing in the direction shown causes interaction of the two fields. Before the rotor has a chance to turn, the current reverses, also reversing the field direction. Thus, total net torque produced over several cycles is zero. Special measures must be used to overcome this disadvantage.

One way to solve the problem is to add squirrel-cage winding slots to the rotor. In this way, the motor can accelerate as an induction

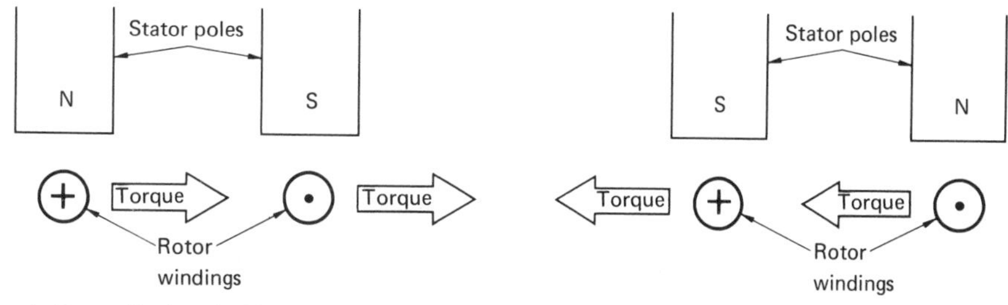

Figure 4.16 Synchronous Motor Starting

motor. When it approaches synchronous speed, the rotor is excited with DC. If the rotor is excited at the proper time, it locks into step with the stator field. The torque generated when the rotor locks into step with the stator field is called the *pull-in torque.*

Special sensing circuits are used to detect the precise instant at which excitation needs to be applied. If excitation is applied at the wrong time, the poles repel instead of attracting. The motor quickly comes to a stop. The motor also comes to a stop if too much torque is demanded by the motor. This torque, as mentioned previously, is called the pull-out torque.

Another method used to start the synchronous motor is to drive it to nearly synchronous speed with another motor, either mechanical or electrical. Such a motor is called a *pony motor.*

Loaded Operation

When the synchronous motor has no load connected, the rotor is positioned almost directly under the stator field, as shown in Figure 4.17A. If a load is connected, the rotor is pulled behind the stator, as shown in Figure 4.17B. At this time, the motor is drawing lagging current. However, if we increase the amount of excitation, the strength of the rotor field increases. This increased strength produces a stronger attraction for the stator pole. If the excitation is strong enough, the rotor comes back under the stator pole. From this point, if we increase the excitation, the rotor moves ahead of the stator field. The motor then draws leading current. That is, the current and induced voltage lead the applied voltage. The motor then looks like a capacitor to the line.

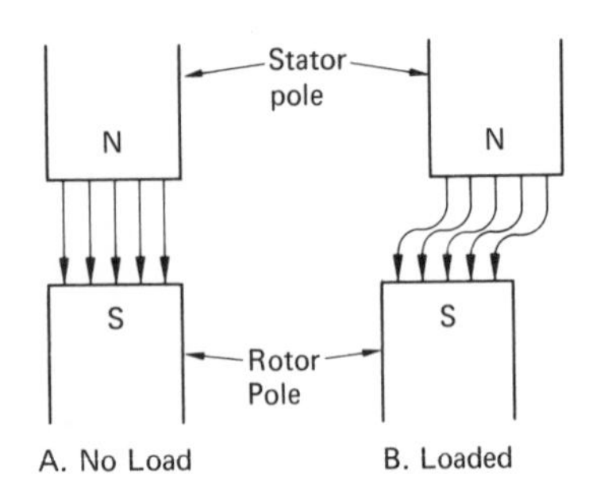

Figure 4.17 Synchronous Motor's Pole Positions

Many industrial loads draw considerable inductive current, which degrades the power factor and increases utility rates. The synchronous motor can be adjusted to compensate for these inductive loads, thus improving power factors and efficiency.

Fractional hp Synchronous Motors

Up to this time, we have been discussing integral hp synchronous motors. But there are many fractional hp synchronous motors in use today. Most, however, do not use a separately excited rotor. It is too expensive and usually not necessary. Motors without a separately excited rotor are called nonexcited synchronous motors. There are two major types in use: reluctance motors and hysteresis motors.

Reluctance Motor

The *reluctance motor* uses differences in reluctance between areas of the rotor to produce torque. The reluctance principle utilizes a force that acts on magnetically permeable material. When such a material is placed in a magnetic field, it is forced in the direction of the greatest flux density. A diagram of the special rotor construction is shown in Figure 4.18.

The rotor has slots cut into it at regular intervals, as shown in the figure. These slots form salient poles (see Chapter 3). Proper motor operation demands that the number of salient poles be equal to the number of stator poles. The rotor also has bars, which cause it to accelerate as an induction motor. When the motor nears synchronous speed, the field magnetizes the rotor and pulls it into step. The torque to drive the motor is generated because the reluctance is greater in the areas that do not have any bars. The areas of the rotor that have bars have a lower reluctance.

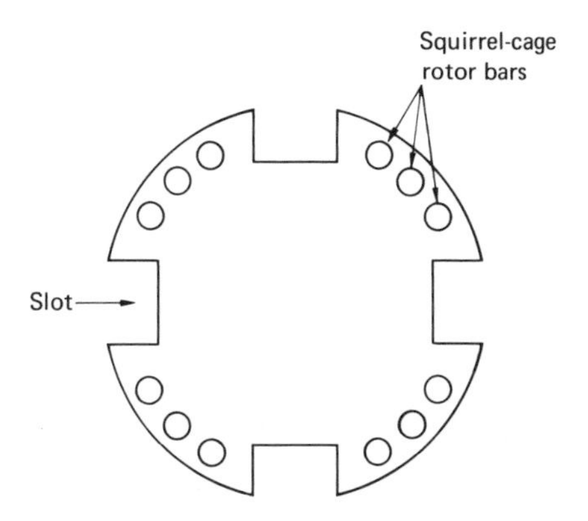

Figure 4.18 Rotor Construction of Reluctance Motor

This motor has very low starting torque as well as low pull-out torque. Although the reluctance motor is normally found in fractional hp sizes, it is also encountered in integral sizes up to 100 hp.

Hysteresis Motor

Another interesting synchronous motor uses the hysteresis effect as a driving principle. The rotor of the *hysteresis motor* is made of a cobalt steel alloy. This alloy retains a magnetic field well and has high permeability.

Recall, from basic electricity, that when a magnetically permeable material is placed in a magnetic field, the domains in the metal align with the field. If the fields rotate, the domains still align themselves with the field, as shown in Figure 4.19. Power is consumed when the domains maintain their alignment against the rotation of the rotor. This power loss is called hysteresis loss. During the time these domains are changing, starting torque is produced by another method, such as squirrel-cage bars. As the rotor approaches synchronous speed, current in the squirrel-cage bars decreases, producing less torque. At this time, the rotor domains lock into alignment, resulting in a fixed magnetic polarity in the rotor. The retained magnetic field is now pulled into synchronization with the rotating field.

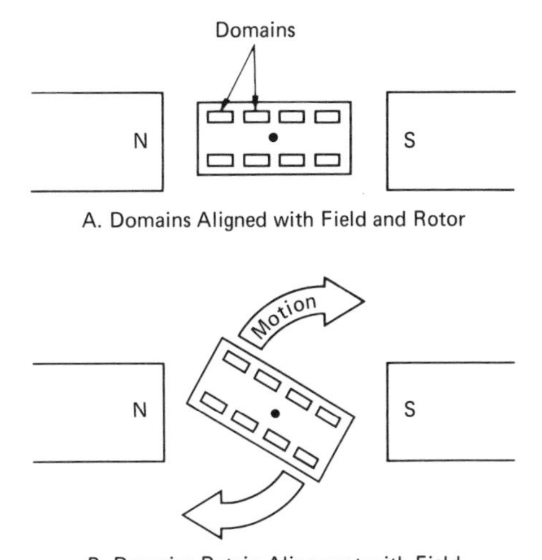

Figure 4.19 Rotor Behavior of Hysteresis Motor

One interesting feature of the hysteresis motor is that the torque produced remains constant, regardless of load or speed changes. Consequently, it is used in clock and timer motors as well as audio turntables and tape drives.

Conclusion

The AC motor has special characteristics that differ from those of the DC motor. The induction motor, for example, is one of the most common motors in industry owing to its need for little maintenance. Since many AC motors are used in industry, technicians must have a general understanding of how they operate.

Questions

1. The universal motor is similar in construction to the ______-connected DC motor. The universal motor, then, can run on ______ as well as AC power.

2. Both induction and synchronous motors use the principle of the ______ stator field to generate torque. This principle requires that separate pairs of windings have applied voltages that differ in ______.
3. In the induction motor, energy is coupled from the stator to the rotor by ______ ______. Therefore, there is no ______ connection between rotor and stator.
4. Because only one phase is present, single-phase motors must have special circuitry to ______ them.
5. The rotors of synchronous motors run at ______ speed. Synchronous motors develop very little torque at ______. This problem is sometimes solved by placing squirrel-cage bars in the synchronous motor's ______.
6. Explain how the universal motor can be used in AC and DC circuits. Explain the difference in construction in a series DC motor and a universal motor.
7. Describe how a rotating magnetic field is created.
8. Is it possible to change the direction of an AC motor? If so, explain how.
9. What factors influence the rotor speed of an AC motor?
10. Describe how an induction motor produces torque.
11. Define slip in an induction motor.
12. Explain the NEMA classification system for induction motors.
13. Describe the construction of a double squirrel-cage motor. What effect does this construction have on motor operation?
14. Rank the types of single-phase induction motors in terms of the torque produced as a percentage of full-load torque.
15. Explain how starting torque is produced in the following engines: (a) split-phase, (b) capacitor-start, and (c) shaded-pole.
16. Describe the operation of the centrifugal switch.
17. Describe the method of rotor excitation in the wound-rotor motor.
18. List the advantages and disadvantages of the wound-rotor motor compared with the squirrel-cage motor.
19. At what speed does the synchronous motor run?
20. Why doesn't the synchronous motor develop enough torque to start on its own? How is the synchronous motor started?
21. What purpose does the synchronous motor serve other than the obvious one of turning mechanical loads?
22. Explain the method by which torque is produced in the hysteresis and reluctance motors.

Problems

1. Calculate the synchronous speed of a 12-pole motor with three-phase, 60 Hz power applied to its stator.
2. How many poles would a three-phase motor need to turn at a synchronous speed of 900 r/min?
3. If an induction motor's rotor is turning at 1750 r/min, what is the most likely stator synchronous speed? *Hint*: Examine Equation 4.2. On the basis of this speed, what is the slip?

Control Devices

CHAPTER 5

Objectives

On completion of this chapter, you should be able to:

- Categorize individual control devices into one of four areas;
- Summarize the advantages and disadvantages of electromagnetic relays and reed relays;
- Summarize the advantages and disadvantages of solid-state relays;
- Identify the parts of the electromagnetic relay;
- Identify the schematic symbols of thyristors in common industrial use;
- Calculate the firing angle of an AC thyristor circuit;
- Calculate the junction temperature and power dissipation of a thyristor;
- Give examples of manually operated switches and their applications;
- Give examples of mechanically operated switches and their applications;
- List four methods by which thyristors are triggered.

Introduction

Control devices lie at the heart of modern industrial systems. Very simply, a *control device* is a component that governs the power delivered to an electrical load. The load may be a motor or generator load, an electronic circuit, or even another control device. Indeed, every industrial system uses a control device. Since control devices are so common, a good background in the different types of control devices—how to recognize them and how they work—is essential.

Basically, control devices can be broken down into four categories: manually operated switches, mechanically operated switches, electromagnetic switches (relays), and electronic switches (thyristors). A *switch* may be described as a device used in an electrical circuit for making, breaking, or changing electrical connections. In this chapter, we discuss each of these basic categories in terms of the control devices in each category and the operating principles of these devices.

Manually Operated Switches

A *manually operated switch* is one that is controlled by hand. Many types and classifications of switches have been developed. Frequently, switches are classified by the number of poles and throws they have.

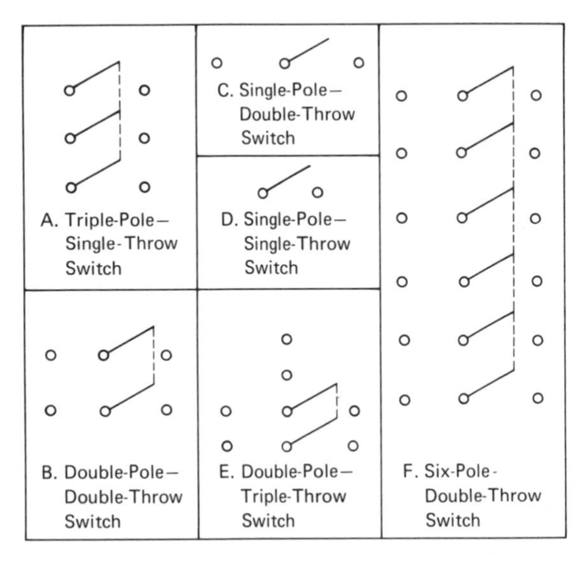

Figure 5.1 Switch Schematic Diagrams

The *pole* is the movable part of the switch. The *throw* of a switch signifies the number of different positions in which the switch can be set. The number of poles can be determined by counting the number of movable contacts. The number of throws can be found by counting how many circuits can be connected to each pole. Figure 5.1 shows several schematic diagrams of different combinations of poles and throws.

Toggle Switches

A *toggle switch* is an example of a manually operated switch. One type of toggle switch is shown in Figure 5.2. Toggle switches have many uses, most of them centered around energizing low-power lighting and other low-power electronic equipment. Electrical ratings are in terms of volts and amperes. These ratings should not be exceeded; otherwise, damage to the switch may result. If a switch needs to be replaced but an exact replacement cannot be found, substitute a switch with higher voltage and current ratings.

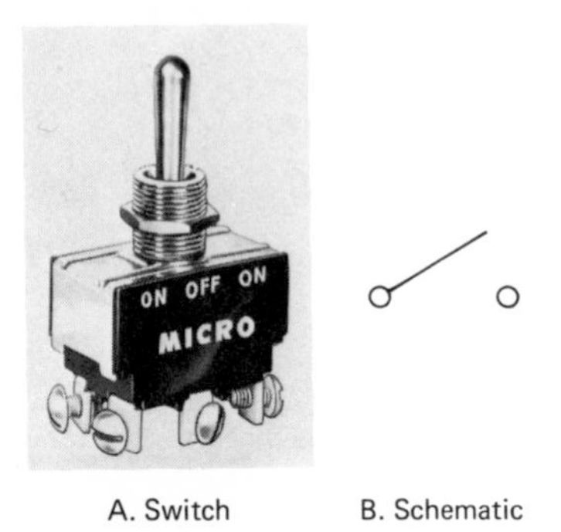

Figure 5.2 Toggle Switch

Push-Button Switches

Another common manually operated switch in industrial use is the *push-button switch.* This switch and its schematic diagram are shown in Figure 5.3.

Push-button switches can be broken down into two categories: momentary-contact and maintained-contact. In the *momentary-contact switch,* actuation occurs only when the switch is pressed down. This push-button switch resets itself when released. (Resetting is the process of returning the switch back to its original state.) In the *maintained-contact switch,* as its name suggests, the switch is actuated even when the button is released. It must be pressed again for it to change states. This switch is most commonly used in turning electrical equipment on and off. A good example of its application is the dimmer switch in a car.

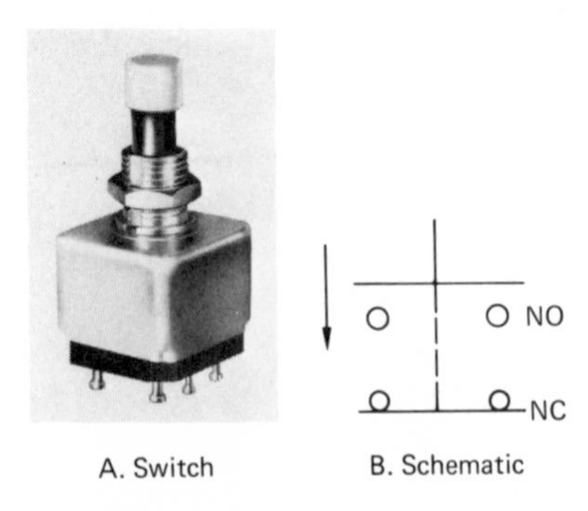

Figure 5.3 Push-Button Switch

Note the abbreviations NO and NC in Figure 5.3B. These letters show the electrical state of the switch contacts when the switch is not actuated. The abbreviation NC means that the switch contacts are normally closed; NO means that they are normally open.

Knife Switches

The *knife switch* is a manually operated switch used extensively for controlling main power circuits. The knife switch is shown in Figure 5.4. The knife switch is typically a bar of copper with an attached handle. The switch is actuated by pushing the copper bar down into a set of spring-loaded contacts. Normally, for safety reasons, the knife switch is enclosed in a box.

Rotary-Selector Switches

The *rotary-selector switch* is another common manually operated switch. This switch is shown in Figure 5.5. As the knob of the switch is rotated, circuits can be opened and closed according to the construction of the switch. Some rotary switches have several layers, or wafers, as shown in Figure 5.5B. The addition of wafers makes it possible to control more circuits with a single change of position. The rotary switch is often used on test equipment and other multifunction, low-power equipment. The range selector switch of a voltmeter is an example of the use of a rotary switch.

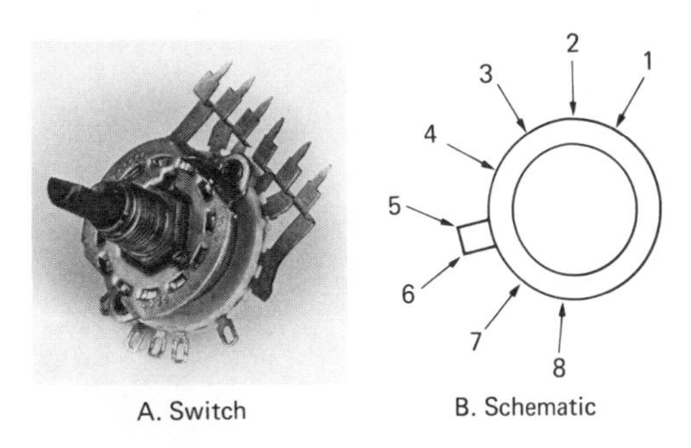

A. Switch B. Schematic

Figure 5.5 Rotary Switch

A. Switch B. Schematic

Figure 5.4 Knife Switch

Mechanically Operated Switches

Mechanically operated switches find many uses in industry. A *mechanically operated switch* (sometimes called an automatic switch) actuates automatically in the presence of some specific environmental factor, such as pressure, position, or temperature.

Limit Switches

The *limit switch*, shown in Figure 5.6, is a very common industrial device. It is usually actuated by horizontal or vertical contact with an object, such as a cam. Note in the figure that the actuation is accomplished by moving a lever. Often, the lever is the roller type, as shown in the figure.

Limit switches have two basic uses. First, they serve as safety devices to keep objects between certain physical boundaries. For instance, in a machine that uses a moving table, a limit switch can keep the table from moving past its safe physical limits by shutting off the table's drive motor. Second, limit switches are used to control industrial processes that use conveyors or elevators.

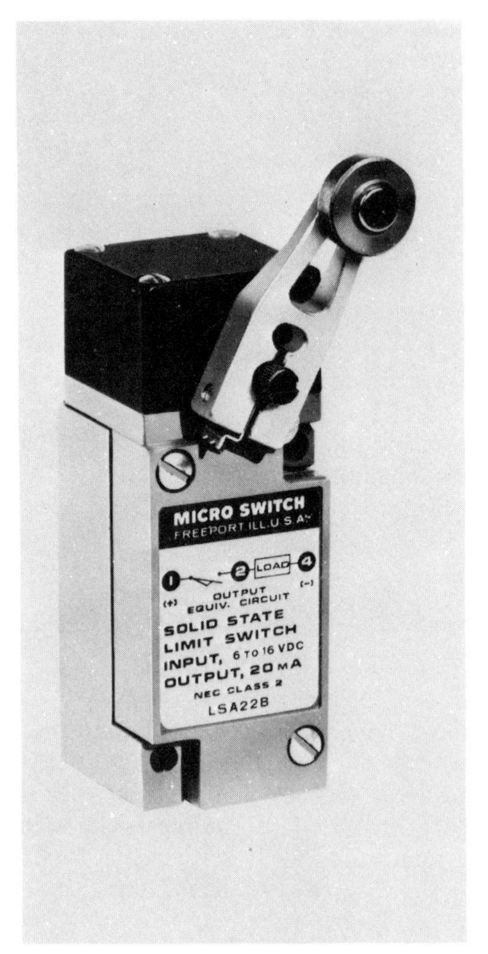

Figure 5.6 Limit Switches

A. Lever Type

B. Wand Type

Mercury Switches

The *mercury switch* is nothing more than a glass enclosure containing mercury and several switch contacts placed so that the mercury will flow over and around them. Several examples of mercury switches are shown in Figure 5.7.

Because of the simplicity of the mercury switch, it is very versatile. It can be used as a position sensor; it can sense the amount of centrifugal or centripetal force and changes in speed, inertia, and momentum. In addition, it has little contact bounce (a major problem in relays and mechanical switches), is hermetically sealed against environmental impact, and has a relatively long life.

Most mercury switches are used with currents up to 4 A and voltages up to 115 V. When actuated, the contact resistance can be less than a tenth of an ohm. As shown in Figure 5.7, the mercury switch comes in many different configurations. The glass envelope is usually filled with hydrogen gas to prevent arcing.

An application of a mercury switch is shown in Figure 5.8. When the mercury switch closes, trigger voltage is provided to the silicon-controlled rectifier (SCR), which turns on. The SCR is an

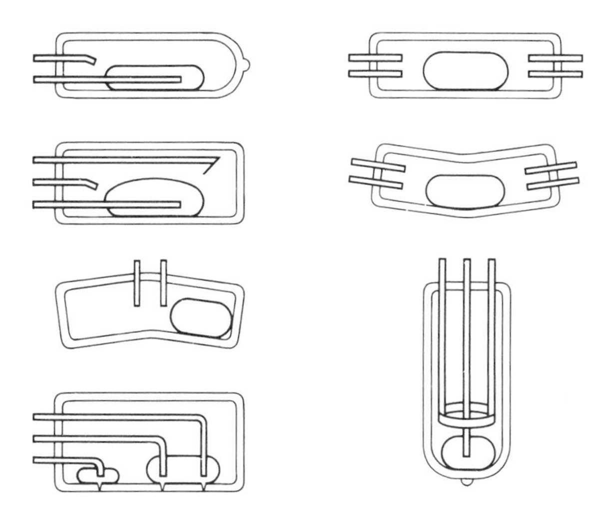

Figure 5.7 Mercury Switch Configurations

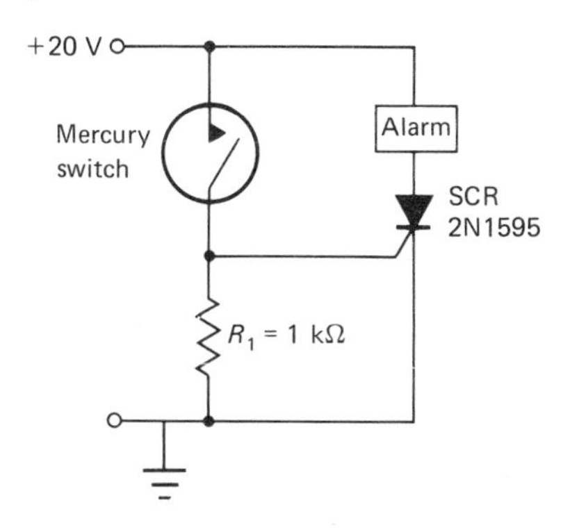

Figure 5.8 Mercury Switch Controlling SCR

electronic switch; its operation is discussed later in this chapter.

Snap-acting Switches

A *snap-acting switch* is a switch in which the movement of the switch mechanism is relatively independent of the activating mechanism movement. In a snap-acting switch, like toggle switch, no matter how fast or slow the toggle is moved, the actual switching of the circuit takes place at a fixed speed. In other switches, like the knife switch, the speed with which the switch closes is determined by the mechanism driving it. In the snap-acting switch, the switching mechanism is a leaf spring, which snaps between positions.

A *precision snap-acting switch* is one in which the operating point is preset and very accurately known. The operating point is the point at which the plunger or lever causes the switch to change position. The precision snap-acting switch is commonly called a microswitch, which is actually a trade name of the Micro Switch Division of Honeywell Corporation. An example of a microswitch is shown in Figure 5.9. Frequently, limit switches, like those in Figure 5.6, contain snap-acting switches as their contact assembly. Much less arcing results from the rapid movement of the contacts.

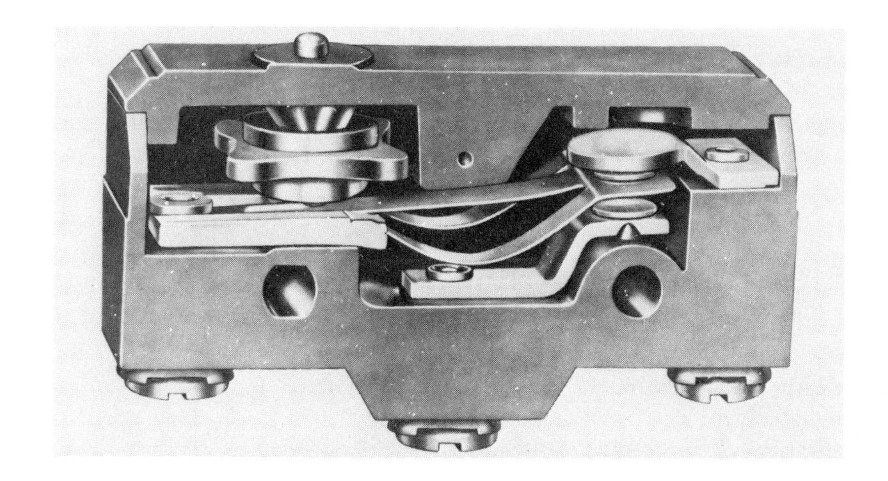

Figure 5.9 Cutaway Showing Construction of Precision Snap-Action Switch

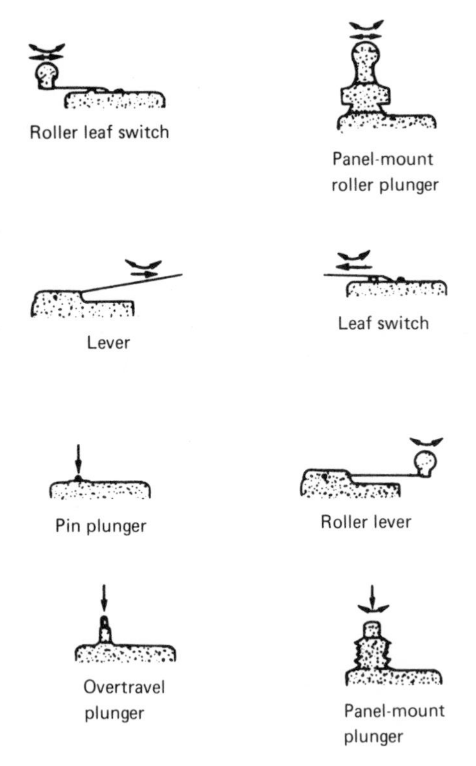

Figure 5.10 Common Actuators Used in Precision Snap-Acting Switches

The basic precision snap-acting switch is used in many applications as an automatic switch. It is most often used as a safety interlock to remove power from a machine when there is danger to the operator or technician.

Several different methods are used to actuate this type of switch. Some of the more common actuators and their uses are shown in Figure 5.10.

Electromagnetic Relays (EMRs)

Relays are undoubtedly one of the most widely used control devices in industry. A *relay* is an electrically operated switch. Relays may be divided into two basic categories: control relays and power relays.

Control relays, as their name implies, are most often used to control low-power circuits or other relays. Control relays find frequent use in automatic relay circuits, where a small electrical signal sets off a chain reaction of successively acting relays performing various functions.

Power relays are sometimes called contactors. The power relay is the workhorse of large electrical systems. The *power relay* controls large amounts of power but is actuated by a small, safe power level. In addition to increasing safety, power relays reduce cost, since only lightweight control wires are connected from the control switch to the coil of the power contactor.

The components of both power and control relays are the same; see Figure 5.11. Each has a coil wound around an iron core, a set of movable and stationary relay contacts (the switch), and the mounting. A switch is usually used to start or stop current through the coil. When current flows through the coil, a strong magnetic field is created. This electromagnet pulls the armature, which, in turn, moves the relay contact down to make electrical connection with the stationary contact. The physical movement of the armature occurs only when current flows through the coil. Any number

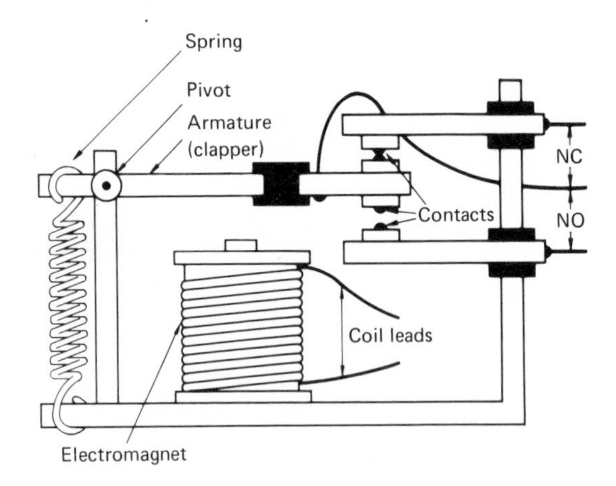

Figure 5.11 Basic Parts of Electromagnetic Relay

of sets of contacts may be built onto a relay. Thus, it is possible to control many different circuits at the same time.

Like switches, the relay contacts have maximum voltage and current ratings. If these ratings are exceeded, the life of the contact may be decreased greatly. Relays also have pull-in voltage and current ratings. *Pull-in current* is the amount of current through the coil needed to operate the relay. *Pull-in voltage* is the voltage needed to produce that current. Usually, the actual steady-state (continuous) voltage applied to the relay coil is somewhat higher than the pull-in voltage. This condition guarantees enough current so that the relay actuates when needed and holds under vibration.

Not all relays have the same construction as the armature relay of Figure 5.11. Another type of relay, called the *reed relay*, is also widely used in industry. The reed relay, like the armature relay, uses an electromagnetic coil, but its contacts consist of thin reeds made of magnetically permeable material. These contacts, called the reed switch, are usually placed inside the coil. A cutaway view of a reed relay is shown in Figure 5.12A.

When current flows through the relay coils, a magnetic field is produced. The field is conducted through the reeds, magnetizing them. As shown in Figure 5.12B, the magnetized reeds are attracted to one another, and the relay contacts are closed.

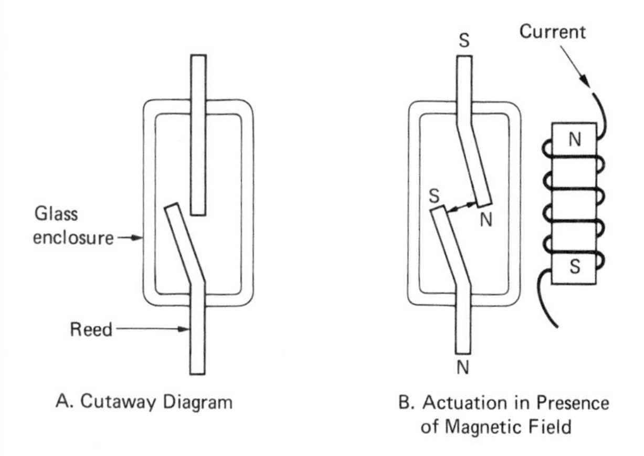

Figure 5.12 Reed Relay

Reed relays are superior to armature relays in many ways. Reed relays are faster, more reliable, and produce less arcing than the armature relay. However, the current-handling capabilities of the reed relay are limited. In addition, because of the glass case that houses the reed relay contacts, the reed relay is more susceptible to damage by shock than the armature relay is.

Arcing and pitting can be further reduced in the reed relay by wetting the relay contacts with mercury. Since mercury is a liquid metal, it is drawn up the bottom relay contact by capillary action. The mercury completely covers the relay contact, thus protecting it from arcing and extending contact life.

Semiconductor Electronic Switches: Thyristors

While the electromagnetic relay has advantages in numbers of circuits controlled, the *electronic switch* is faster, cheaper, and more energy efficient, and it lasts longer. The electronic switch does not exhibit contact bounce, a major problem in digital circuits. Also, the electronic switch is not acceleration-sensitive or vibration-sensitive, works better in hostile environments, and does not arc or spark. It is, therefore, explosionproof. Because of these advantages, electronic switches, or thyristors, have replaced many electromagnetic relays and thyratron tubes in industry. In addition, thyristors are used in rectifiers, alarms, motor controls, heating controls, and lighting controls. Thyristors are used wherever large amounts of power are being controlled.

A *thyristor* is a four-layer, PNPN device. This PN sandwich, then, has a total of three junctions. As we will see later, this device has two stable switching states: the on state (conducting) and the off state (not conducting). There is no linear area in between these two states, as there is in the transistor. That is, a switch is either on or off, never (hopefully) in between.

Silicon-controlled Rectifier (SCR)

The most common thyristor you will encounter as a technician is the *silicon-controlled rectifier* (*SCR*). It is widely used because it can handle higher values of current and voltage than any other type of thyristor. Presently, SCRs can control currents greater than 1500 A and voltages greater than 2000 V.

The schematic symbol for the SCR is shown in Figure 5.13A. Notice that the schematic symbol is very much like that of the diode. In fact, the SCR resembles the diode electrically since it only conducts in one direction. In other words, the SCR must be forward-biased from anode to cathode for current conduction. It is unlike the diode because of the presence of a gate (G) lead, which is used to turn the device on.

SCR switching operation is best understood by visualizing its PNPN construction, as shown in Figure 5.13B. If we slice the middle PN junction diagonally, as shown in Figure 5.14, we end up with an NPN transistor connected back to back with a PNP transistor. A forward-bias voltage between the gate-to-cathode leads turns on transistor Q_1. Transistor Q_1 conducts, turning on Q_2. Transistor Q_2 drives Q_1 further into conduction. This regenerative process, shown in Figure 5.15, continues until both transistors are driven into saturation. Only microseconds are needed to complete the process. This regenerative action is called *latching*

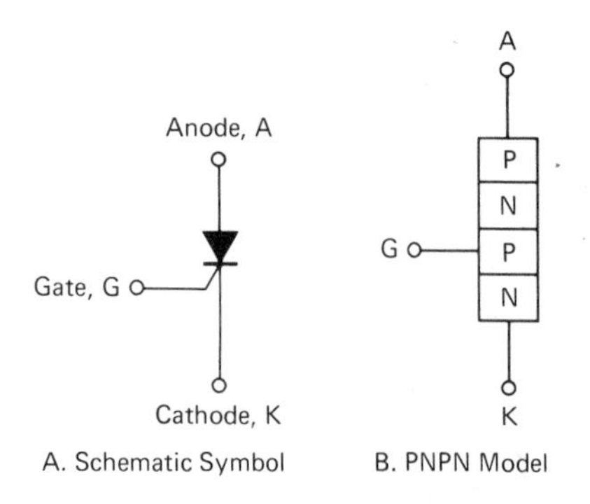

A. Schematic Symbol B. PNPN Model

Figure 5.13 SCR

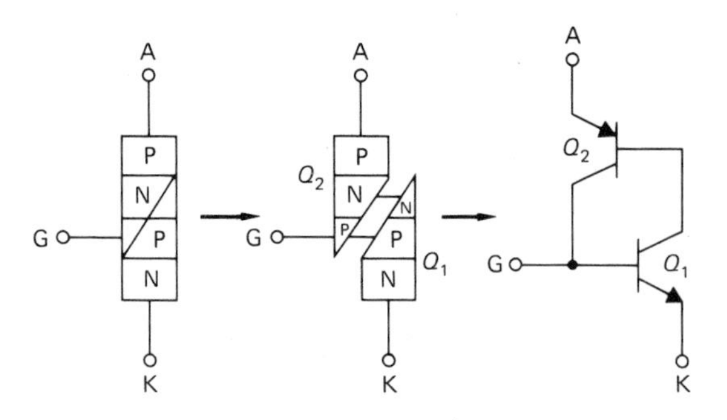

Figure 5.14 SCR Two-Transistor Analogy

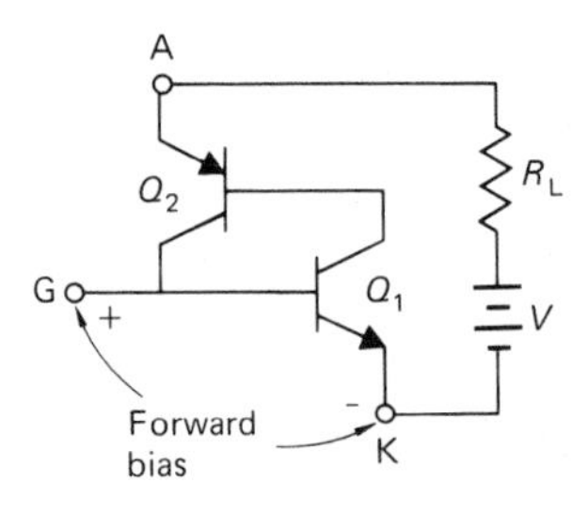

Figure 5.15 SCR Two-Transistor Analogy with Regenerative Feedback

because the transistors continue to conduct even if the gate-to-cathode, forward-bias voltage is removed.

Why does the SCR handle power efficiently? When it is in the off state, it does not draw much current. When it is on, the voltage across it is around 1 V, regardless of the current through the device. Recall the power equation:

$$P = IV \tag{5.1}$$

where P = power dissipated by device
I = current through device
V = voltage across device

If the voltage remains at about 1 V, the power is approximately equal to the size of the current flow. For instance, if the voltage drop across the device stays at 1 V with 1000 A flowing through it, the power dissipated is only 1000 W. All the rest of the power is transferred to the load.

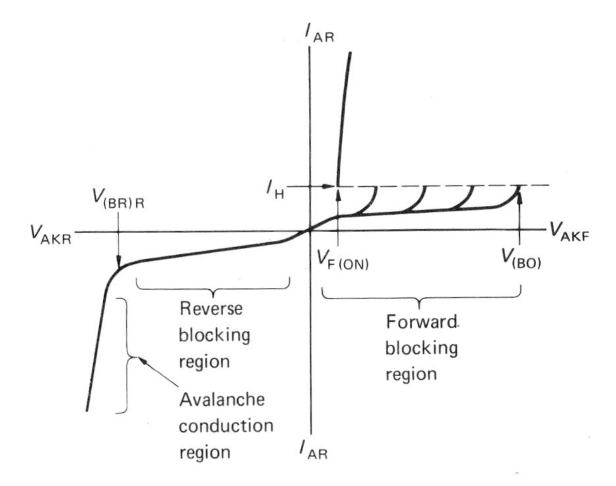

Figure 5.16 SCR Current-Voltage Characteristic Curve

Turning on the SCR

The gate-to-cathode voltage required to turn on the SCR is referred to as the *gate turnon voltage* (V_{GT}). The amount of gate-to-cathode current needed to turn the device on is called *gate turnon current* (I_{GT}). Voltage V_{GT} usually ranges from 1 to 3 V, while I_{GT} ranges from 1 to 150 mA. The SCR can be turned on by methods other than the gate voltage. However, with the exception of the radiation method of triggering, little use is made of other triggering methods.

Refer to Figure 5.16, which shows the current-voltage behavior of an SCR with no gate current flowing and the gate grounded. When the SCR is forward-biased (positive voltage on the anode with respect to the cathode), very little current flows. This region is called the *forward blocking region.* However, if the *forward voltage* (V_{AKF}) is high enough, minority carriers will be accelerated to high velocities and create carriers in the gate-cathode junction. When enough carriers are generated, the device breaks into conduction, just as if a gate current were drawn. The voltage at which the device breaks over is represented by $V_{(BO)}$ and is called the *breakover voltage.*

Regardless of how the current flow starts, once it begins, it keeps on flowing. However, the diagram in Figure 5.16 also gives a clue as to how to shut off the SCR. Note the part of the curve labeled I_H. If anode current falls below this current, called the *holding current,* the device goes back into the off state.

Note that in the reverse direction, the SCR curve looks like a normal PN diode. Very little reverse current flows when the SCR is reverse-biased. This region is called the *reverse blocking region.* When the reverse breakdown voltage $V_{(BR)R}$ is reached, the device will quickly go into avalanche conduction, which usually destroys the device.

Although the most common method of turning on an SCR is gate voltage control, there are other ways to turn it on. SCRs are often turned on by radiation, particularly light. When radiation falls on the middle junction, it creates electron-hole pairs. If enough pairs are created, the device breaks into conduction. Such a device is called a *light-activated SCR* (*LASCR*).

The LASCR is similar in construction to the SCR, except that it has a window to let light in. The construction of the LASCR is shown in Figure 5.17A. The schematic symbol for the LASCR is shown in Figure 5.17B. Light is directed through the lens to fall on the silicon pellet. When light of great enough intensity falls on the middle PN

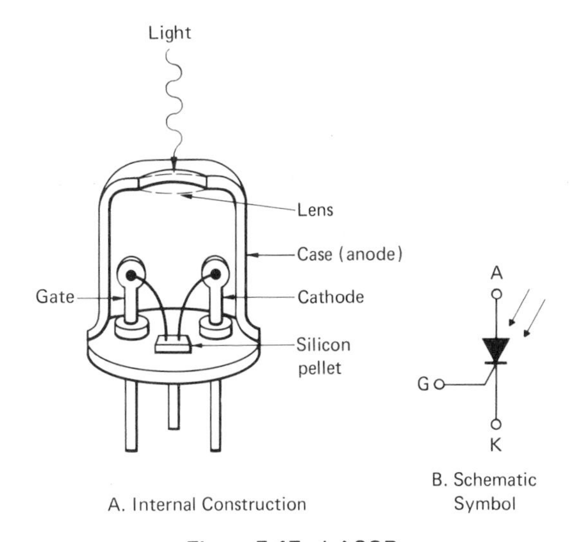

Figure 5.17 LASCR

junction, the device conducts. The LASCR is often used in industry to detect the presence of opaque objects; it is also used in cases where electrical isolation between circuits is needed.

There are two remaining methods by which SCRs are turned on. The first is called the *rate effect turnon method*, which operates as follows: Recall that every transistor has PN junctions, which act as capacitors. If a quickly rising voltage is applied between the anode and gate of an SCR, charge current flows through the device. If this charge current is high enough, it can trigger the SCR into conduction. Most specification sheets specify this parameter in terms of volts per microsecond. It is called the *critical rate of rise*, or dV/dt. For example, the 2N1595 SCR has a critical rate of rise of 20 V/μs. In other words, if the anode voltage rises any faster than 20 V in 1 μs, the device turns on. The critical rate of rise also limits the maximum line frequency that can be connected to the device.

Engineers usually try to prevent rate effect turnon because the device turns on when they do not want it to. Firing by exceeding the critical rate of voltage rise is prevented by the addition of an *RC snubber circuit*, shown in Figure 5.18. The *RC* network delays the rate of the anode voltage rise.

Finally, the SCR can be turned on by temperature. As junction temperature increases, minority carrier current flow (leakage) increases. In fact, it doubles every 14°F (8°C). If the temperature rises high enough, this increased current is enough to turn on the SCR. This thermal effect is usually prevented by using a heat sink.

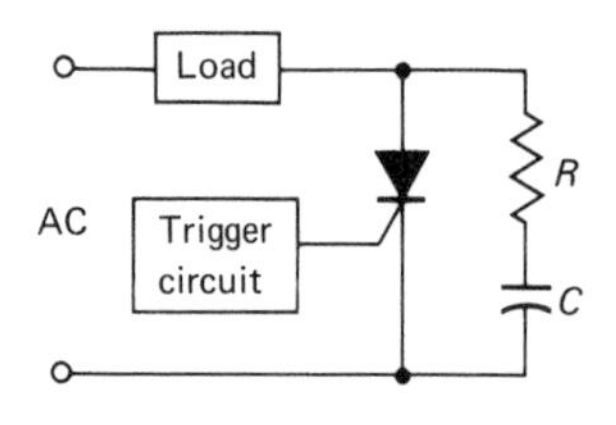

Figure 5.18 SCR Circuit with Snubber to Decrease dV/dt

Another problem with SCRs and temperature is the temperature dependence of some of the SCR parameters. Specifically, V_{GT}, I_{GT}, and $V_{(BO)}$ all decrease with increasing temperature. This decrease will cause a change in firing angle with temperature fluctuations in circuits where the firing angle depends on these parameters.

Heat Sink for the SCR

Most SCRs are devices that handle large amounts of power. Consequently, the case of the SCR is often not large enough to dissipate the heat generated by the thyristor. The maximum power that the thyristor can dissipate safely is a function of (1) the temperature at its junction, (2) the ambient (air) temperature, and (3) the size of its thermal resistance R_θ.

Thermal resistance R_θ is a measure of how well the device will conduct heat away from the junction. Thermal resistance determines the ratio between the change in junction temperature and the amount of junction power dissipation. If the thermal resistance is small, there is less heat buildup at the junction.

Thermal resistances are normally given in the manufacturer's data sheets. For small devices that are operated with or without a heat sink, two thermal resistances are given: (1) case to ambient ($R_{\theta CA}$), the condition without a heat sink, and (2) junction to case ($R_{\theta JC}$), the condition with a heat sink. In the case where a heat sink is used, the sink has its own thermal resistance from case to ambient ($R_{\theta CA}$). To determine the total thermal resistance ($R_{\theta JA}$) for a device using a heat sink, we add the thermal resistances ($R_{\theta JC} + R_{\theta CA} = R_{\theta JA}$).

Thermal resistance $R_{\theta CA}$ can be divided further into $R_{\theta CS}$ (case to sink) and $R_{\theta SA}$ (sink to ambient). The case-to-sink thermal resistance $R_{\theta CS}$ is a function of the thermal connection between the thyristor case and the heat sink. With a good-quality thermal grease between the case and the sink, we can assume that $R_{\theta CS} = 0$°C/W. In this situation, then, we can equate $R_{\theta CA}$ and $R_{\theta SA}$.

The *maximum junction temperature* $T_{J(max)}$ (or T_{JM}) is normally given in the data sheets. This temperature ranges between 150°C and 200°C. The maximum power that the junction can safely dissipate is calculated by the following equation:

$$P_{J(max)} = \frac{T_{J(max)} - T_A}{R_{\theta JA}} \quad \textbf{(5.2)}$$

where $P_{J(max)}$ = maximum junction power dissipation, in watts

T_A = ambient temperature, in degrees Celsius

For example, a 2N3005 operated at a maximum ambient temperature of 50°C has a thermal resistance, junction to ambient, of 275°C/W and a maximum junction temperature of 200°C. Maximum junction power dissipation is 545 mW without a heat sink. Adding a heat sink with a thermal resistance, case to ambient, of 5°C/W to the 2N3005 with a junction-to-case thermal resistance of 75°C/W increases the maximum junction power dissipation to 2 W.

We have assumed that the case-to-sink thermal resistance is negligible. With a good-quality thermal grease to improve the case-to-sink heat flow, this assumption is not unreasonable. Most manufacturers assume case-to-sink thermal resistance to be less than 0.5°C/W.

If we look at this problem from another angle, we can also calculate the desired thermal resistance needed for a particular power dissipation application.

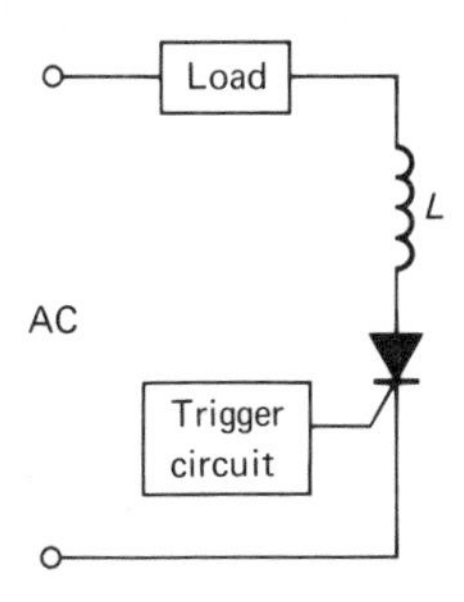

Figure 5.19 SCR Circuit with Inductor to Limit Current Rise

Critical Rate of Current Rise

Some data sheets specify a parameter called *critical rate of current rise.* Critical rate of current rise is expressed in amperes per microsecond. For example, the critical rate of current rise of the 2N1595 is about A/μs. If this parameter is exceeded, hot spots develop within the device, which may destroy it.

One method used to prevent current damage is shown in Figure 5.19. The inductor L in the anode lead opposes any change in current, thus slowing down the rise of anode current.

Gate Triggering for the SCR

Figure 5.20 shows a simple SCR circuit with a DC supply. Note that the SCR is forward-biased from anode to cathode.

Assume that gate voltage V_G is 0 V. If we close S_1 no current flows because of the absence of gate current I_G. What value of V_G will fire the SCR? Assume that gate turnon voltage V_{GT} is 0.6 V, and I_{GT} is 20 mA. Notice that the current I_G flows through R_G. If I_G flows through the 100 Ω resistor, it should drop about 2 V. Therefore, by Kirchhoff's law, V_G is as follows:

$$V_G = V_{GK} + V_{RG} = 0.6 \text{ V} + 2.0 \text{ V} = 2.6 \text{ V} \quad \textbf{(5.3)}$$

where V_{GK} = DC voltage from gate to cathode

V_{RG} = DC voltage drop across R_G

By this calculation, we find the gate voltage necessary to fire this device.

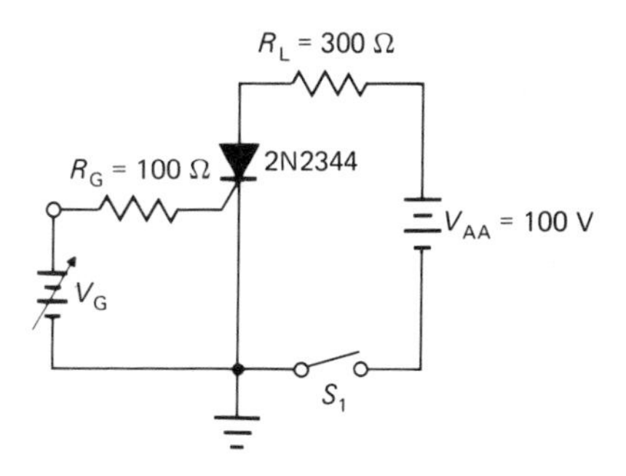

Figure 5.20 SCR in DC Circuit

The next question we might ask is, "Will the SCR stay on after the gate voltage is removed?" Anode current must rise to a value called the latching current for the SCR to remain on. The *latching current* is defined as that amount of anode current necessary to keep the device on after switching from the off state and after the trigger has been removed. The latching value is normally two to three times the holding current value. If we assume that the latching current is 300 mA, then the anode current must be more than this value if the SCR is to remain on.

Anode current (I_A) is calculated by assuming that the voltage drop across the SCR is zero (actually about 1 V). Thus, we assume that the entire supply voltage (V_{AA}) is dropped across the 300 Ω load (R_L). By Ohm's law, then, we have the following:

$$I_A = \frac{V_{AA}}{R_L} = \frac{100\ \text{V}}{300\ \Omega} = 333\ \text{mA} \quad \textbf{(5.4)}$$

So, when the SCR fires, about 333 mA of current flows. Since this value exceeds the latching current value (300 mA), the device stays latched when the gate voltage is reduced to zero.

The gate trigger current and voltage in this example are assumed to be DC values. They may, however, be pulses with a certain pulse width. This pulse width must be long enough to allow the anode current to build up to the latching current value. The time it takes the anode current to build up to the latching current level is called *turnon time*.

In this example, we have also assumed that the load is resistive. Many times, the load is inductive, such as in a relay or a motor. This inductance may keep the SCR from firing. As we know, an inductor opposes any change in current. With pulse triggering, the anode current may be kept below the latching current for the duration of the pulse. One way to overcome this problem is to make the pulse width longer, which is not always possible. Another solution is to bypass the inductive load with a resistor. This solution allows the anode current to rise quickly to the latching value.

Suppose the load in Figure 5.20 is inductive instead of resistive. How large does the resistor need to be? We can assume that 100 V is applied across the load when the SCR fires. If we divide this voltage by the latching current value, we have the minimum value of resistance needed to keep the SCR on. In this case, the resistance bypassing the inductive loads need to be at least 333 Ω.

Turning off the SCR

In most DC circuits, the problem is not how to turn the device on; the problem is how to turn it off. The only way to turn an SCR off is to reduce anode current below the holding current value. In DC circuits, this reduction is usually accomplished by adding a manual reset switch (S_1 in Figure 5.20). When the reset switch is opened, the blocking junction reestablishes itself.

Another common turnoff method for DC circuits is the commutation capacitor illustrated in Figure 5.21A. To simplify the schematic diagrams

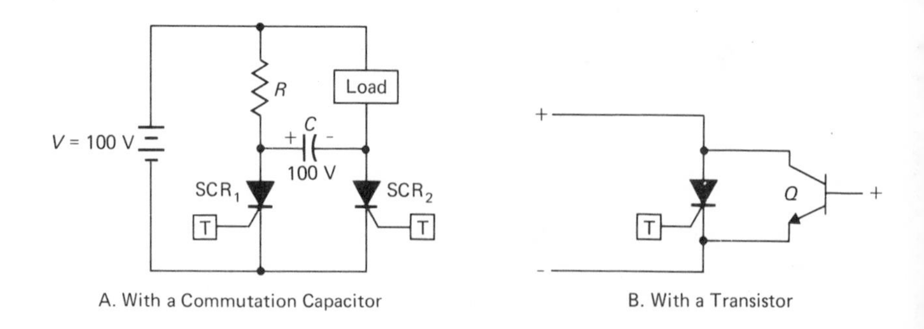

Figure 5.21 Turning Off SCR

in this text, we represented the thyristor trigger circuits with a T at the gate. When SCR_2 is turned on by a gate pulse, anode current flows through the load and also charges C through R (SCR_1 is off). Because SCR_2 drops so little voltage in its on state, the capacitor C charges to about 100 V. SCR_2 may be turned off by triggering SCR_1 into conduction. When SCR_1 turns on, it reverse-biases SCR_2 by placing -100 V across it from anode to cathode.

Alternatively, the SCR may be turned off by turning transistor Q on, as shown in Figure 5.21B. When Q is forward-biased to saturation, it conducts current around the SCR. If the circuit is properly designed, the anode current falls below the holding current, and the thyristor turns off. This method depends on the use of a power transistor with a low collector-to-emitter saturation voltage.

The problem of SCR turnoff does not occur in AC circuits. The SCR is automatically shut off during each cycle when the AC voltage across the SCR approaches zero. As zero voltage is approached, anode current falls below the holding current value. The SCR stays off throughout the entire negative AC cycle since the SCR is reverse-biased.

A simple AC application of the SCR is shown in Figure 5.22. Here, the positive AC half cycle is shown. As the supply goes positive, an increasing amount of gate current is drawn. The SCR blocks (between points A and B) until V_T rises to V_{GT} (point B). At point B, the SCR quickly turns on and load current flows. The current through the SCR rapidly rises to a peak and decreases sinusoidally as the AC supply voltage decreases toward zero. Finally, at point C, the anode current falls below the holding current value. The SCR turns off.

Note that, as shown in Figure 5.22, the SCR blocks current flow during the negative half cycle. No current flows through the load at this time. Reverse-biased diode D_1 prevents potentially damaging current from flowing from the gate during the negative half cycle. Resistor R_1 is placed in this circuit so that the total resistance cannot be reduced to a point where damaging gate current flows.

The distance from point A to B in Figure 5.22 represents the time in the positive half cycle when the SCR is off. The distance between A and B, in degrees, is called the *firing angle* (θ_{fire}). Note that the SCR is conducting between points B and C. This angle is called the *conduction angle* (θ_{cond}). As this circuit is drawn, the firing angle can range between 10° and 90°.

Decreasing the resistance of R_2 in the circuit allows more gate current to flow, thus firing the SCR earlier in the cycle. This technique applies a higher anode current to the load by increasing the average DC current and voltage to the load. In many cases, an ammeter is used to monitor the current through the load. Recall that most DC ammeters are calibrated to read average current. The following equation allows calculation of the

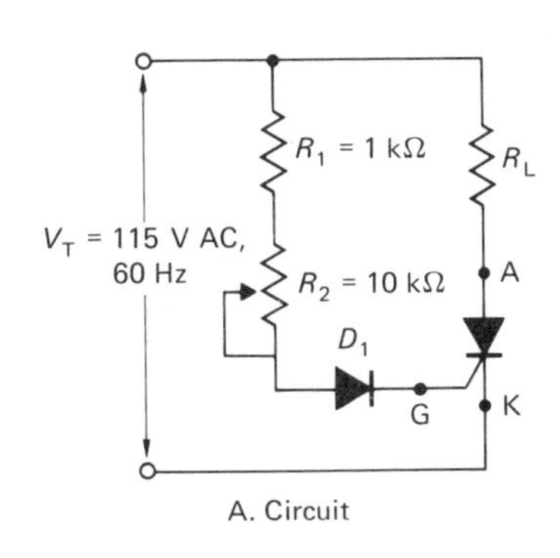

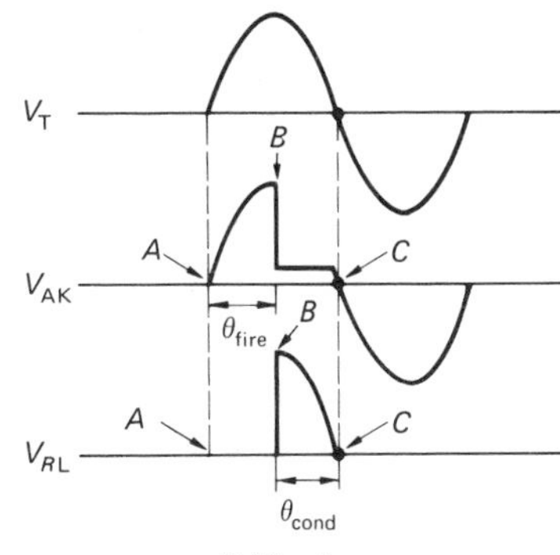

Figure 5.22 SCR in AC Circuit

average load current:

$$I_{L(av)} = \frac{V_{pk}}{2\pi R_L}(1 + \cos\theta_{fire}) \quad \textbf{(5.5)}$$

where $I_{L(av)}$ = average load current, in amperes
V_{pk} = peak supply voltage, in volts
R_L = resistance of the load, in ohms
θ_{fire} = firing angle, in degrees

If load current is measured with an ammeter, the firing angle may be calculated by solving this equation.

Alternatively, we can calculate the firing angle if we know R_1, R_2, I_{GT}, and V_{GT}. For $I_{GT} = 10$ mA, $V_{GT} = 1$ V, the forward-biased diode drop of D_1 at 0.6 V, and $R_2 = 5$ kΩ, we can calculate the supply voltage at which the SCR fires. At the firing point, I_{GT} is flowing through the gate circuit. This current drops 10 V across R_1, 50 V across R_2, 1 V across V_{GT}, and 0.6 V across D_1. These voltages total 61.6 V. In other words, the SCR will trigger when the supply reaches 61.6 V. Recall that any instantaneous voltage on a sine wave can be calculated from the following equation:

$$v = V_{pk}\sin(2\pi ft) \quad \textbf{(5.6)}$$

where v = instantaneous voltage
V_{pk} = peak sine wave voltage
$2\pi ft$ = firing angle, in degrees

Solving for the firing and using 162.6 V for V_{pk}, we get a firing angle of 22.3°.

An improvement to this circuit is shown in Figure 5.23. Here, a capacitor has been added between R_2 and the cathode of the SCR. The delayed gate voltage rise across the capacitor allows the SCR to fire past the 90° point. This circuit can control the SCR firing angle between about 15° and 170°. This type of AC control is called *phase control.*

Triac

After the SCR, the most commonly used thyristor is the triac. Like the SCR, the *triac* acts as a switch, with a gate that controls the switching state. Unlike the SCR, the triac can conduct in both directions. Thus, this device is a *bilateral,* or *bidirectional, device.* Recall that the SCR triggers only on a positive gate voltage (with respect to the cathode). The triac, on the other hand, fires on either positive or negative polarity gate voltages. The triac is superior to the SCR in that it can be used in full-wave AC control applications, such as control of AC motors and AC heating systems. As yet, triac current and voltage ratings do not approach those of the SCR, though.

The triac schematic symbol is shown in Figure 5.24A. Note that none of the terminals are labeled

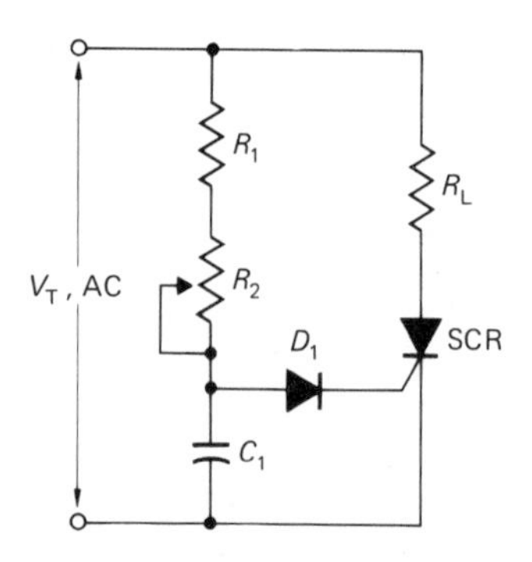

Figure 5.23 SCR Phase Control in AC Circuit

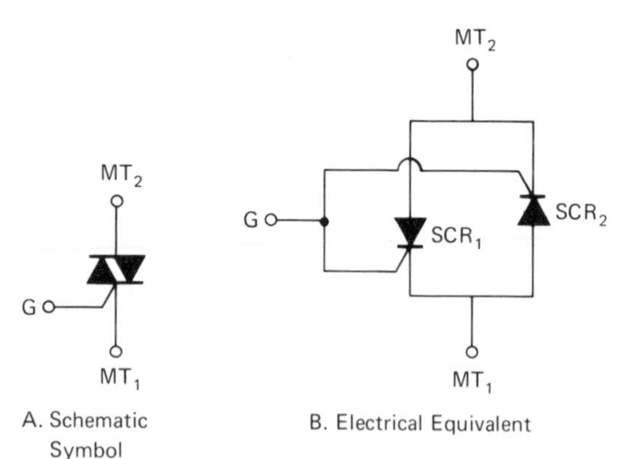

Figure 5.24 Triac

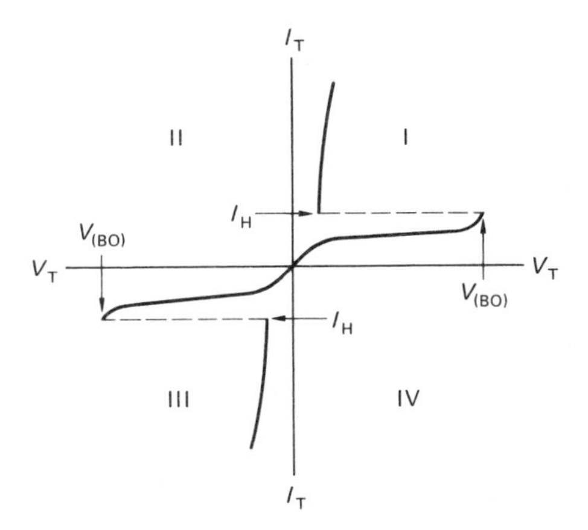

Figure 5.25 Triac Voltage-Current Characteristic Curves

anode or cathode. The bidirectional nature of this device makes that labeling impossible. Instead, the two comparable terminals are called *main terminal* 1 (MT_1) and main terminal 2 (MT_2). The gate lead retains the same name and function as in the SCR.

The electrical behavior of the triac is depicted in Figure 5.24B. Quite simply, the triac can be thought of as two SCRs in inverse, parallel connection. When both high-current and full-wave operation are required, two inverse, parallel SCRs are used. Notice that the gate leads of both SCRs are connected together. When SCR_1 fires, current flows from MT_1 to MT_2. When SCR_2 fires, current flow is in the opposite direction. Thus, we have a bidirectional device. Its bidirectional nature is further shown in the current-voltage graph in Figure 5.25. Notice that the curve in quadrant III of the graph has the same shape as the one in quadrant I.

The triac is ideal for use in AC control circuits. Since the SCR is a unidirectional device, it has no control over half of each input cycle. The triac, however, can control current flow through a load driving both halves of the input cycle. A simple triac-controlled AC circuit is shown in Figure 5.26. The capacitor C_1 charges in either direction until $V_{(BO)}$ is reached. The triac then fires and stays on throughout the remainder of that half cycle. The triac turns off when the AC input voltage nears zero. The main terminal current falls below the holding current.

Phase control is achieved by varying the resistance of R_2. Increasing R_2 resistance causes the capacitor to charge up more slowly, thus firing the triac later in the cycle. This increase thus delivers less energy to the load. Decreasing R_2 resistance causes the capacitor to charge up quickly, firing the thyristor earlier in the cycle. This decrease thus delivers more energy to the load.

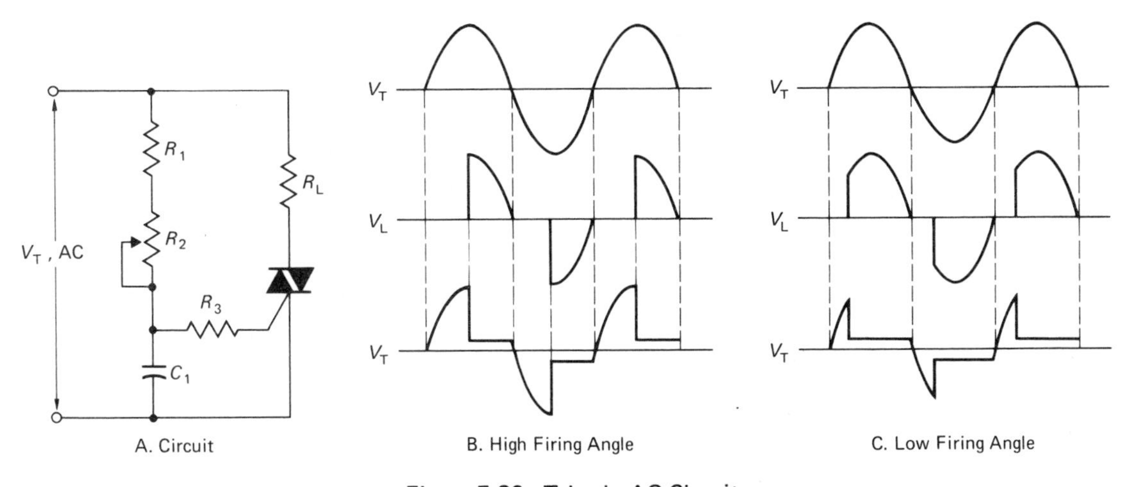

Figure 5.26 Triac in AC Circuit

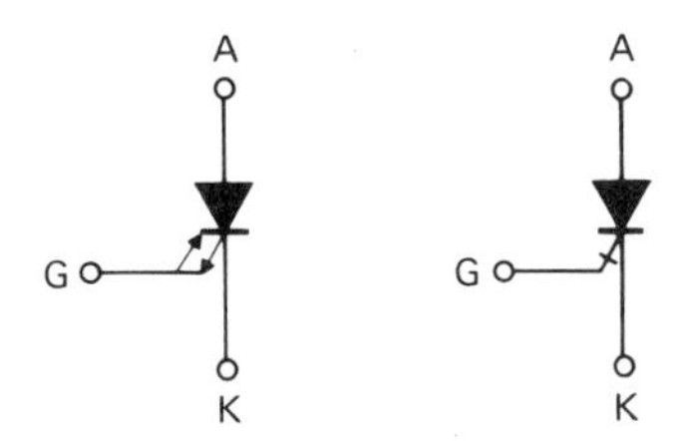

Figure 5.27 Schematic Diagrams for Gate-controlled Switch

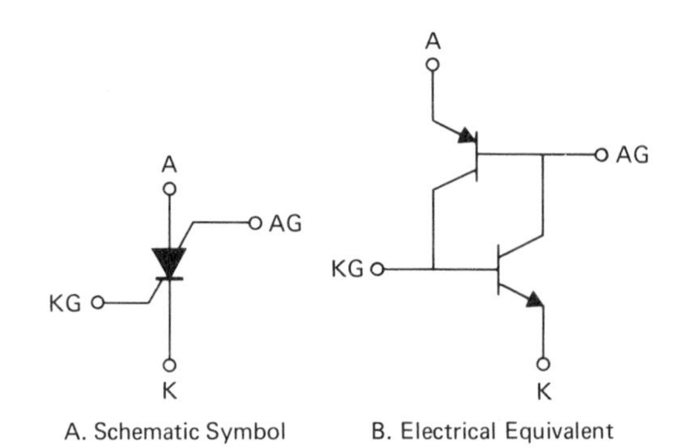

Figure 5.28 Silicon-controlled Switch

Gate-controlled Switch

As we have mentioned previously, the SCR and the triac can only be turned off by low-current drop-out. In other words, anode (or main terminal) current must fall below the holding current. The *gate-controlled switch* (*GCS*), or *gate-turnoff SCR* (*GTO*), as it is sometimes called, turns on like a normal SCR, that is, with a gate signal. But unlike the SCR, it may be turned off by a negative trigger voltage. Positive and negative trigger spikes can be provided by differentiating a square wave and applying the signal to the gate. (Recall from Chapter 2 that a differentiated square wave is a series of alternating positive and negative voltage spikes.) The GCS schematic symbols are shown in Figure 5.27.

One problem with the GCS is that it takes from 10 to 20 times more gate current to turn off the GCS than to turn it on. However, this disadvantage is more than outweighed when using the GCS in DC circuits. At best, an additional SCR and commutation capacitor are required when SCRs are used in DC circuits (see the discussion on commutation in Chapter 3). But the GCS may be used in DC circuits without these extra components.

The GCS finds applications in automobile ignition systems, hammer drivers in computer printers, and television horizontal deflection circuits.

Silicon-controlled Switch

The final gate-controlled thyristor we will discuss is the *silicon-controlled switch* (*SCS*). Figure 5.28 gives a schematic diagram of the SCS. The SCS has two gates: an anode gate (AG) and a cathode gate (KG).

Figure 5.28B reveals an interesting fact. The electrical equivalent for Figure 5.28A is just the basic SCR circuit with both transistor bases accessible to trigger pulses. The device may be turned on with a positive gate pulse (with respect to the cathode), just as in the SCR. The anode gate exercises control over the thyristor when a negative gate voltage (with respect to the anode) is received. The SCS is used in logic applications and in counters and lamp drivers.

As a technician, you do not need to know how to design thyristors. However, you should know, in a general sense, how much current a thyristor is designed to carry. As an aid to understanding thyristors, we have included drawings, approximate case sizes, and approximate current-handling capability of the most popular case configurations in Figure 5.29.

Thyristor Triggering

The thyristors discussed so far are used primarily as power control devices, high power in the case of the SCR and the triac and somewhat lower power in the others. We have also seen that these devices are triggered by gate signals. The next thyristor devices to be discussed are breakover devices. In *breakover devices*, there are usually no gate structures, in contrast with the SCR. Breakover devices

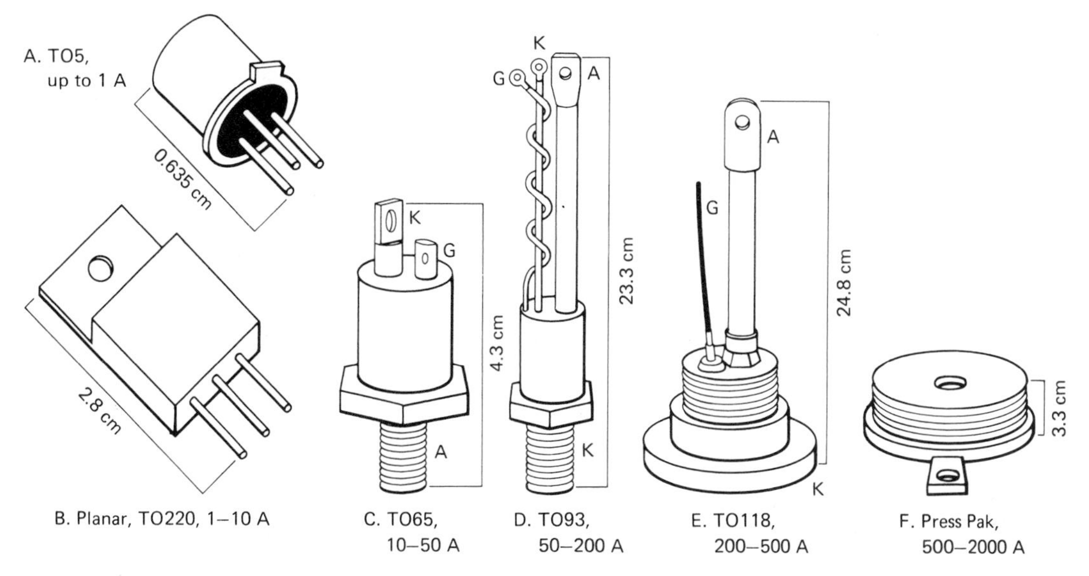

Figure 5.29 Thyristor Case Configurations

are triggered by placing a voltage across the device high enough to break over the middle junction. Once the breakover occurs, the device quickly turns on by regeneration, just as in the SCR.

Shockley Diode

The *Shockley diode* (or *four-layer diode*) is a unidirectional thyristor used as a breakover device. Its structure is similar to the SCR without the gate, as shown in Figure 5.30A. The commonly used schematic symbol is shown in Figure 5.30B.

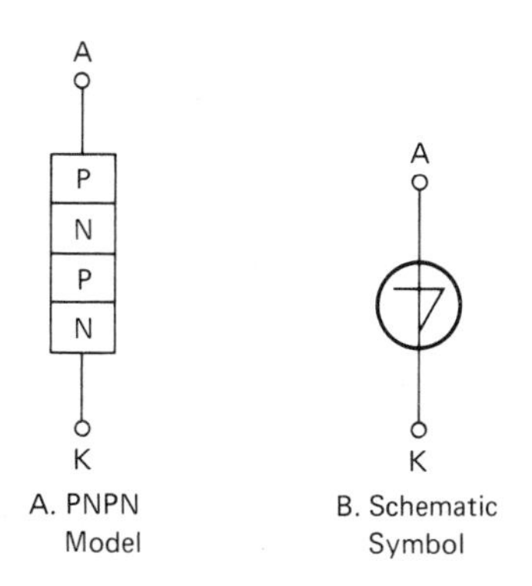

Figure 5.30 Shockley Diode

A circuit using the Shockley diode is shown in Figure 5.31. This circuit represents a relaxation oscillator. When the +15 V supply is connected, capacitor C_1 charges exponentially through the resistor R_1. When the breakover voltage is reached (5 V in this case), the capacitor discharges very quickly through the thyristor. At this time, the voltage falls quickly toward zero. Near 0 V, the current through the device will fall below the holding current value. The output waveform approaches that of the sawtooth oscillator (see Figure 5.31B).

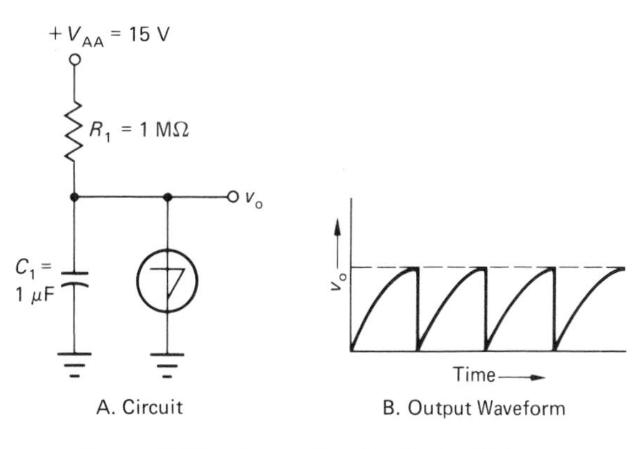

Figure 5.31 Relaxation Oscillator Using a Shockley Diode

The oscillator period is calculated by using the following equation:

$$T = R_1 C_1 \ln \frac{1}{1 - (V_{(BO)}/V_{AA})} \quad \textbf{(5.7)}$$

where
T = time period, in seconds
V_{AA} = voltage applied to circuit
$V_{(BO)}$ = device's breakover voltage
R_1 = resistance, in megohms
C_1 = capacitance, in microfarads

If we use the circuit in Figure 5.31 as an example, we have the following calculation:

$$\begin{aligned} T &= R_1 C_1 \ln \frac{1}{1 - (V_{(BO)}/V_{AA})} \\ &= (1\ \text{M}\Omega)(1\ \mu\text{F}) \ln \frac{1}{1 - (5\ \text{V}/15\ \text{V})} \\ &= 0.405\ \text{s} \end{aligned} \quad \textbf{(5.8)}$$

Since the frequency f of oscillation is the reciprocal of the period, we also have the following:

$$\begin{aligned} f &= \frac{1}{T} = \frac{1}{0.405\ \text{s}} \\ &= 2.47\ \text{Hz} \end{aligned} \quad \textbf{(5.9)}$$

This relaxation oscillator, then, is oscillating at 2.47 Hz.

Diac

We have seen that the thyristor family contains a unilateral breakover device called the Shockley diode. The *diac* is the part of the family that is bidirectional. Like the Shockley diode, the diac does not have a gate. The only way to get it to conduct is by exceeding the breakover voltage. The diac schematic symbol is shown in Figure 5.32. Notice in Figure 5.32B that, electrically, the diac is similar to the triac, with no gate leads available. The diac characteristic curve is shown in Figure 5.33.

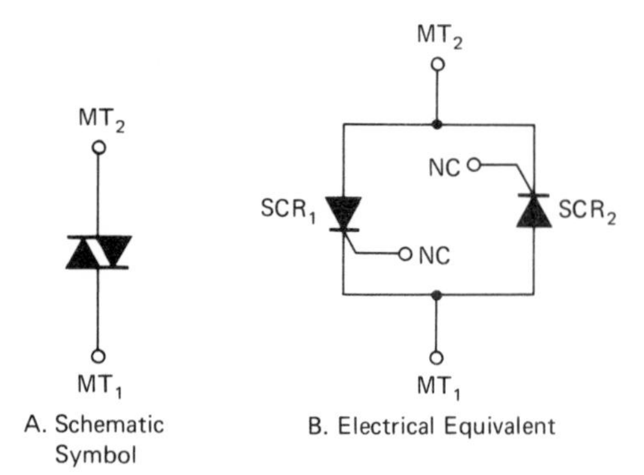

Figure 5.32 Diac

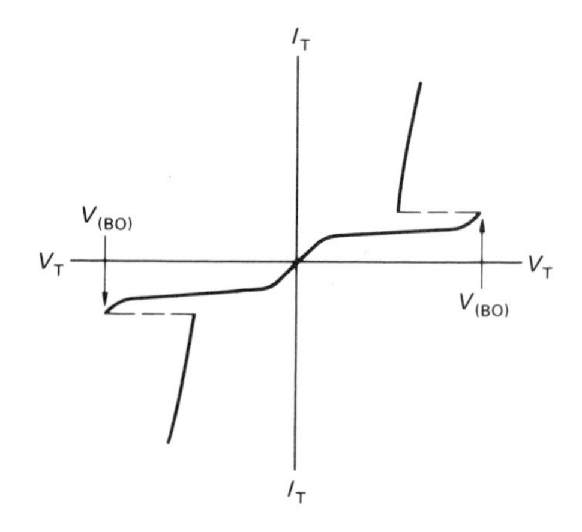

Figure 5.33 Diac Voltage-Current Characteristic Curve

The diac is most often used to trigger a triac into conduction, an application discussed in Chapter 6. However, it may also be used in the same relaxation oscillator circuit as the Shockley diode.

Unijunction Transistor

Although the *unijunction transistor* (*UJT*) is not strictly a thyristor, we consider it here because it is often used to trigger an SCR or a triac. The UJT schematic symbol and construction are shown in Figure 5.34. The UJT's physical construction consists of an evenly doped block of N material with a portion of P material grown into its side.

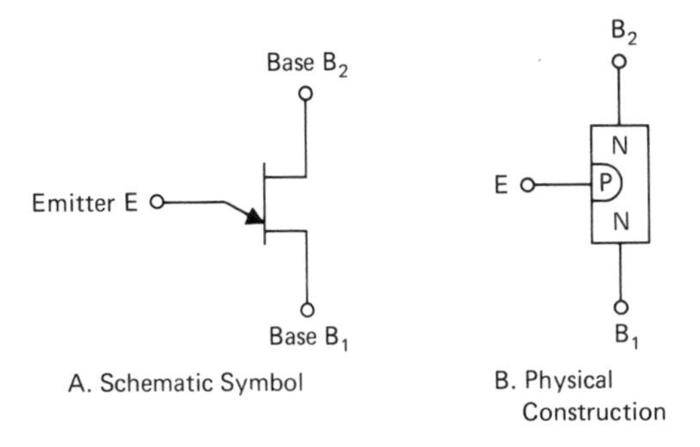

Figure 5.34 Unijunction Transistor

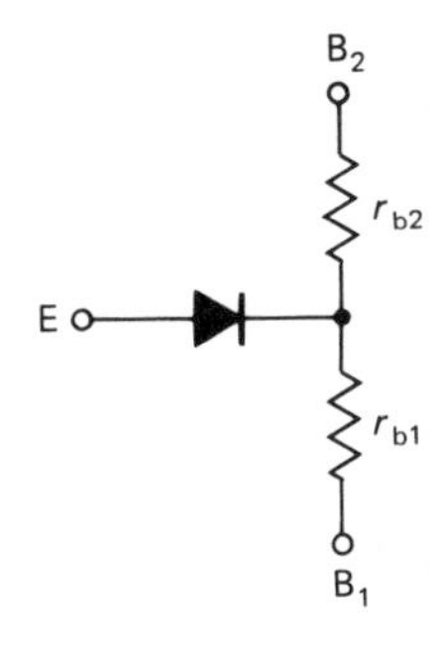

Figure 5.35 UJT Electrical Equivalent

Electrically, the UJT is represented by the circuit in Figure 5.35. The resistance between the PN junction and base B_1 is designated by r_{b1}; the resistance between the PN junction and base B_2 is called r_{b2}. The PN junction is represented by a diode symbol.

Since the UJT is used often as a relaxation oscillator, we will discuss that circuit, as shown in Figure 5.36. When supply voltage V_{BB} is applied to this circuit, the capacitor C_1 starts to charge at a rate determined by the time constant R_1C_1. Let us say that the UJT's P section is midway between base B_1 and base B_2. Since the block of N material is evenly doped, a voltage of about +10 V (with respect to ground) exists at the PN junction. The PN junction is reverse-biased until the capacitor voltage rises to about 0.6 V above the +10.6 V potential, or about +10.6 V. When this +10.6 V potential is reached across the capacitor, the PN junction becomes forward-biased. The capacitor discharges quickly through the forward-biased junction and R_3. When the capacitor voltage goes down near zero, the current flow goes below the holding current value, and the diode becomes reverse-biased again. The waveforms at the two outputs are shown in Figure 5.36B. Notice that the output taken across R_3 is a spiked waveform, created by the rapid discharge of C_1 through R_3.

The period T of the oscillations (in seconds) is calculated from the following equation:

$$T = R_1C_1 \ln \frac{1}{1 - \eta} \tag{5.10}$$

where η = intrinsic standoff ratio

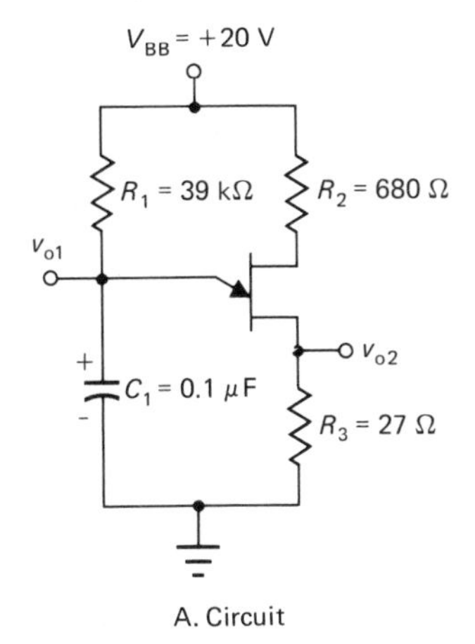

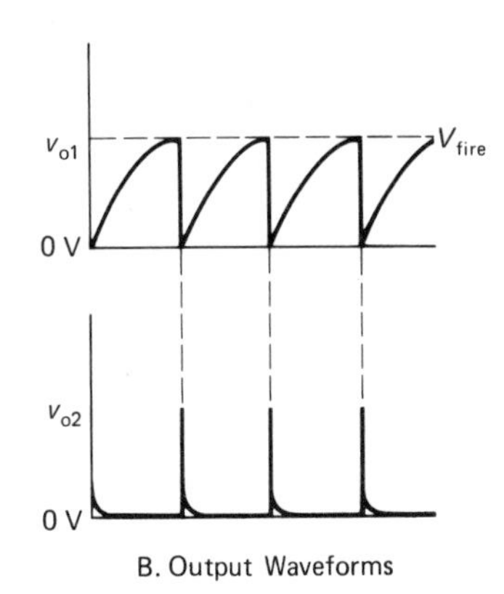

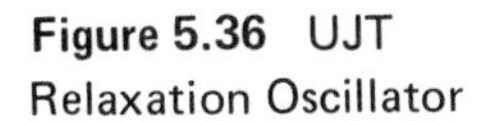
Figure 5.36 UJT Relaxation Oscillator

The *intrinsic standoff ratio*, electrically, is as follows:

$$\eta = \frac{r_{b1}}{r_{b1} + r_{b2}} \tag{5.11}$$

This parameter tells us basically how far the PN junction is from B_1. For instance, if the length of the N material is 10 mils, then an intrinsic standoff ratio of 0.6 means that the PN junction is 6 mils from B_1. UJTs have intrinsic standoff ratios that vary from 0.5 to 0.8.

The waveform's period (using $\eta = 0.5$) is, then, as follows:

$$\begin{aligned} T &= R_1C_1 \ln \frac{1}{1-\eta} \\ &= (39\ \text{k}\Omega)(0.1\ \mu\text{F}) \ln \frac{1}{1-0.5} \\ &= 2.7\ \text{ms} \end{aligned} \tag{5.12}$$

Taking the reciprocal of the period gives a frequency of 370 Hz.

The firing voltage may also be calculated. The intrinsic standoff ratio η times the supply voltage V_{BB} gives the voltage potential at the PN junction. Adding 0.6 V to this value yields the firing voltage, that is, the voltage at which the PN junction is forward-biased. The equation for the firing voltage V_{fire} is as follows:

$$V_{fire} = \eta V_{BB} + 0.6\ \text{V} \tag{5.13}$$

In the example shown in Figure 5.36, where η is 0.5 and V_{BB} is 20 V, the PN junction is forward-biased when the capacitor charges to 10.6 V. Note also that the oscillation frequency of this circuit can be changed by varying the resistance of R_1.

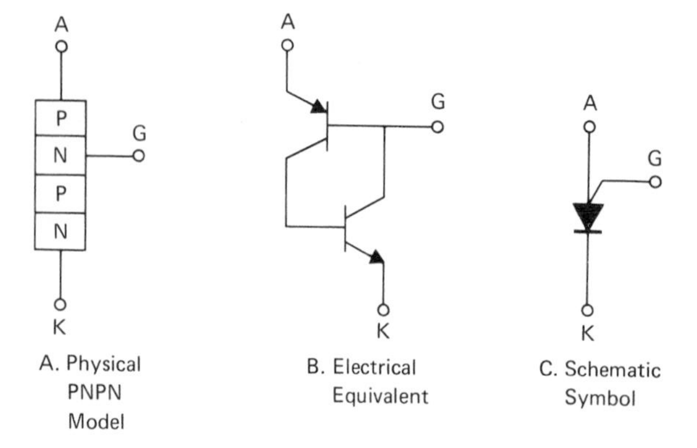

Figure 5.37 Programmable Unijunction Transistor

Programmable Unijunction Transistor (PUT)

The *programmable unijunction transistor* (*PUT*) is a thyristor that has a function similar to that of the UJT. It is commonly used in relaxation oscillators. Unlike the UJT, the PUT is a thyristor with a gate connected to the anode PN junction. The schematic symbol and electrical equivalent are shown in Figure 5.37. If you conclude from looking at the figure that the PUT is more like an SCR than an UJT you are right. The PUT has an anode gate instead of a cathode gate. Thus, to trigger the PUT, we need a negative voltage on the gate with respect to the anode.

A relaxation oscillator using a PUT is shown in Figure 5.38. When supply voltage V_{BB} is applied to the circuit, capacitor C_1 charges through R_1 at a rate determined by the time constant. The R_2–R_3 voltage divider applies 10 V to the gate. When the voltage at the anode reaches 10.6 V, the PN junction is forward-biased and the PUT fires. The capacitor discharges quickly toward zero, eventually turning off the thyristor when anode current falls below holding current.

The functional difference between the PUT and UJT is that the PUT has an intrinsic standoff ratio that is variable or programmable, whereas the UJT does not. In Figure 5.38, the intrinsic standoff ratio is as follows:

$$\eta = \frac{R_3}{R_2 + R_3} \tag{5.14}$$

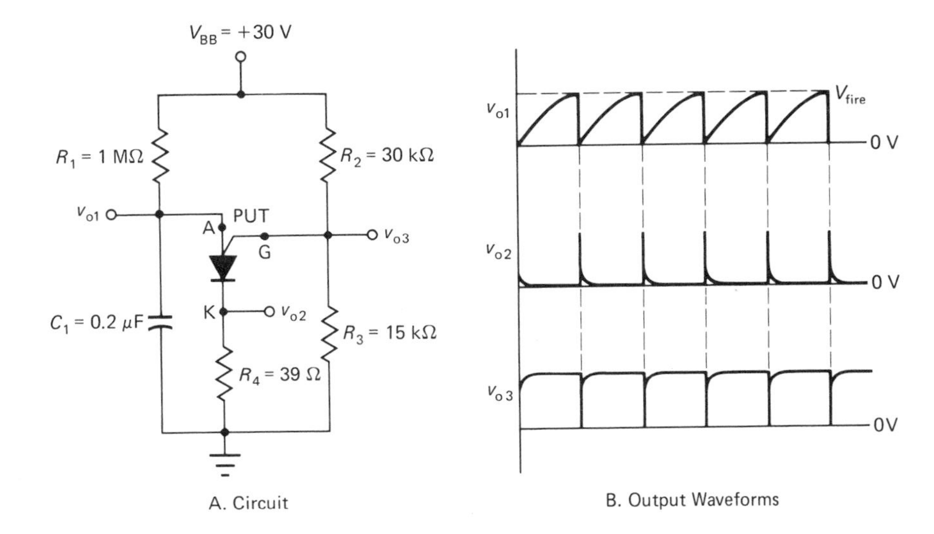

Figure 5.38 PUT Relaxation Oscillator

Calculations of peak firing voltage and frequency are identical to those for the UJT relaxation oscillator circuit discussed previously.

Solid-State Relays

After performing switching tasks for several decades, the electromagnetic relay (EMR) is being replaced in some applications by a new type of relay, the solid-state relay (SSR). A *solid-state relay* differs from an EMR in that the SSR uses a thyristor to do the switching. An SSR has no moving parts.

A block diagram of an SSR is illustrated in Figure 5.39. A DC voltage turns on the light-emitting diode. The diode turns on the phototransistor, which triggers the triac. The degree of isolation between the control circuit and the power circuit depends on two factors: the dielectric characteristics and the distance between the source and sensor.

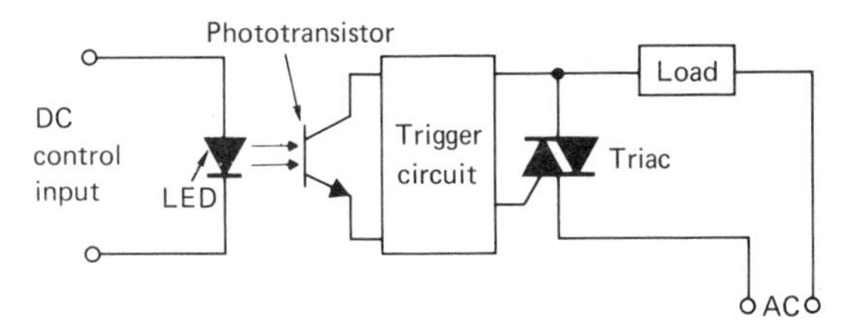

Figure 5.39 Functional Diagram of Optically Coupled SSR

The SSR has many advantages over the EMR. The SSR is more reliable and has a longer life because it has no moving parts. It is compatible with IC circuitry and does not generate as much electromagnetic interference. The SSR is more resistant to shock and vibration, has a much faster response time, and does not exhibit contact bounce.

In most applications, SSRs are used to interface between a low-voltage control circuit and a higher AC line voltage. These devices are used to switch lamps, motors, solenoids, and heating elements in all kinds of industrial equipment and systems.

Troubleshooting Thyristors

The most appropriate in-circuit test for thyristors is one performed with the oscilloscope. You should become familiar with the waveforms shown with each of the thyristors in this chapter. Bear in mind that most thyristors are open circuits when off and have an anode-to-cathode drop of about 1 V when on.

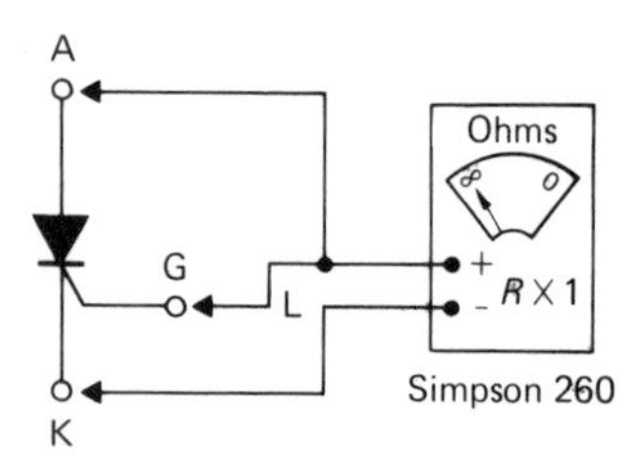

Figure 5.40 Checking SCR with Ohmmeter

A simple static test for SCRs and triacs may be made with an ohmmeter connected as in Figure 5.40. If the anode to cathode is forward-biased with the ohmmeter's internal battery, the middle junction should block current flow. The ohmmeter should read infinite ohms. When lead L is connected to the gate, the middle junction will be forward-biased and the thyristor should turn on. The resistance should decrease and should stay low if the gate lead is removed.

Care is required with this method of checking SCRs and triacs. If the voltage applied by the ohmmeter to the gate-cathode junction is too high, the thyristor could be destroyed. Also, the resistance in the ohmmeter may be too high. The thyristor may turn on but turn right off if the anode current is lower than the holding current value. The ohmmeter scale should be changed to lower the resistance.

Troubleshooting relaxation oscillators is made more difficult by the fact that circuit resistances can determine if the oscillator will work. For example, in Figure 5.36, the value of resistor R_3 is critical to circuit operation. If the UJT appears to be good, try checking the value of each of the resistors around the circuit.

The entire thyristor family is summarized in tabular form in Figure 5.41.

Conclusion

In recent years, more has been heard about the new integrated circuit technology than about power semiconductor devices, such as the thyristor. However, the thyristor is still the workhorse of power control applications. Recent developments in power semiconductors have increased the voltage and current ratings to levels where vacuum tubes are virtually unnecessary. While the thyristor is far from an ideal switch, its usefulness can only increase in industrial applications.

Questions

1. The toggle switch, normally operated by hand, is an example of a ______ switch. Switches that are actuated automatically are called ______ ______ switches. A switch often used to start and stop motors by the switch's being contacted by a moving part of the system is called a ______ switch.
2. The relay, quite commonly used in industry, is an example of an ______ switch. In many applications, the use of relays has given way to the electronic switch, sometimes referred to as a ______.
3. When the thyristor is triggered, it turns on quickly, or latches, by means of ______ feedback. The SCR, the most popular thyristor, is usually triggered by a voltage on its ______.
4. The SCR is a thyristor that conducts in only one direction. Such a device is called a ______ device. The triac can conduct in both directions, so it is a ______ device.
5. The diac and Shockley diode are examples of ______ devices; they are used to turn on other thyristors.
6. The UJT has a ______ intrinsic standoff ratio that determines relaxation oscillator behavior. The PUT has an intrinsic standoff ratio that is variable or ______.
7. List the four classifications of control devices.
8. Give examples of each of the four switch classifications and their applications.
9. Describe each of the main parts of the armature type of electromagnetic relay.
10. List the advantages of thyristors over electromagnetic relays.
11. List the advantages of reed relays over the electromagnetic relay.
12. Describe the regenerative switching action that makes a thyristor latch.

Type	Number of Leads	IEC Official Name	Common Name	Schematic Symbol: Usage	Schematic Symbol: USASI	Equivalent Cross Section	Main Trigger Means	Maximum Ratings Available	Major Applications
Unidirectional (reverse blocking)	2 (diode)	Reverse blocking diode thyristor	Four-layer (Shockley) diode			Anode p n p n Cathode	Exceeding anode breakover voltage	1200 V 300 A peak pulse	Triggers for SCRs, overvoltage protection, timing circuits, pulse generators
		Reverse blocking diode thyristor	Light-activated switch (LAS)	✲		Anode p n p n Cathode	Infrared and visible radiation	200 V 0.5 A	Static switches, triggers for high-voltage SCR applications, photoelectric controls
	3 (triode)	Reverse blocking triode thyristor	Silicon controlled rectifier (SCR)	† ✲	†	† Anode p n p n Gate Cathode	Gate signal	1800 V 550 A avg.	Phase controls, inverters, choppers, pulse modulators, static switches
		Reverse blocking triode thyristor	Light-activated SCR (LASCR)	✲		Anode p n p n Gate Cathode	Gate signal or radiation	200 V 1 A avg.	Position monitors, static switches, limit switches, trigger circuits, photoelectric controls
		Turnoff thyristor	Gate-controlled switch (GCS, GTO)			Anode p n p n Gate Cathode	Gate signal turns GCS off as well as on	500 V 10 A	Inverters, pulse generators, choppers, dc switches
		—	Silicon unilateral switch (SUS)		—	Anode Cathode p n p n+ p Gate	Exceeding breakover voltage or gate signal	10 V 0.2 A	Timer circuits, trigger circuits, threshold detector
		—	Complementary unijunction transistor (CUJT)		—	Base 1 p n p n Emitter Base 2	When B1-emitter voltage reaches predetermined fraction of B1-B2 voltage	30 V 2 A peak pulse	Interval timing, trigger circuits, level detector, oscillator
	4 (tetrode)	Reverse blocking tetrode thyristor	‡ Silicon controlled switch (SCS)	‡ ✲	‡	Cathode gate Anode Cathode p p n n Anode gate	‡ Gate signal on either gate lead	200 V 1 A avg.	Lamp drivers, logic circuits, counters, alarm and control circuits
Bidirectional	2 (diode)	Bidirectional diode thyristor	Biswitch, diac, SSS		—	n p n p n	Exceeding breakover voltage in either direction	400 V 60 A rms	Overvoltage protection, ac phase control, triac trigger
	3 (triode)	Bidirectional triode thyristor	Triac		—	Anode 2 n n p n n n p Gate Anode 1	Gate signal or exceeding breakover voltage	500 V 20 A rms	Switching and phase control of ac power
		—	Silicon bilateral switch (SBS)		—	Two SUS structures inverse-parallel on same chip	Exceeding breakover voltage in either direction or gate signal	10 V 0.2 A	Threshold detector, trigger circuits, overvoltage protection

Figure 5.41 Summary of Thyristor Family

13. Draw the schematic symbols for the thyristors listed, and describe the usual method of triggering each: (a) SCR, (b) triac, (c) diac, (d) UJT, (e) PUT.
14. Describe how an LASCR is triggered.
15. Why is it necessary to have a snubber across a thyristor? How does a snubber work?
16. Distinguish between the PUT and UJT in terms of function and construction.
17. Define the intrinsic standoff ratio for a UJT.
18. List four advantages of a solid-state relay over the electromagnetic version.
19. Define critical rate of voltage and current rise. How do these parameters affect the operation of thyristors?
20. Describe how temperature affects the following parameters: (a) I_{GT}, (b) V_{GT}, and (c) $V_{(BO)}$.
21. Draw the schematic symbol of the following devices and describe how each one is triggered: (a) GCS, (b) SCS, and (c) Shockley diode.

Problems

1. Assume you have the following voltages and resistances for the circuit shown in Figure 5.20: $R_L = 2\ k\Omega$, $R_G = 40\ \Omega$, $V_{AA} = 50$ V, and $V_{F(ON)} = 1$ V. The SCR used is a 2N881 with $I_{GT} = 200$ mA, $V_{GT} = 0.8$ V, and $I_H = 5$ mA.
   a. Will a V_G of 10 V fire the SCR?
   b. Calculate the power dissipated by the SCR, assuming that it turns on.
2. Given $R_{\theta JA} = 345°C/W$, $R_{\theta JC} = 124°C/W$, an ambient temperature of 40°C, and $T_{J(max)} = 200°C$.
   a. Calculate the junction temperature when the thyristor is dissipating 300 mW without a heat sink.
   b. Calculate $R_{\theta CA}$ of the heat sink needed to dissipate 1.2 W at the junction. Assume $R_{\theta CS} = 0°C/W$.
   c. Calculate the power dissipated at the junction with a sink having an $R_{\theta CA}$ of 30°C/W and a junction temperature of 100°C.
3. Suppose the load in Figure 5.20 is resistive. What size bypass resistor is necessary to ensure that the SCR will turn on? Assume $I_L = 15$ mA, $V_{F(ON)} = 1$ V, and $V_{AA} = 50$ V.
4. If R_L in Figure 5.22A is 10 Ω, what is the firing angle with R_2 set at 3 kΩ? Assume $I_{GT} = 200\ \mu A$ and $I_H = 10$ mA.
5. At the firing angle calculated in Problem 4, what is the average current flowing through the load?
6. For the following values for the circuit shown in Figure 5.31A, calculate the frequency and the period of the relaxation oscillator: $R_1 = 25\ k\Omega$, $C_1 = 0.1\ \mu F$, $V_{(BO)} = 30$ V, and $+V_{AA} = 50$ V.
7. Calculate the peak firing voltage and the oscillator frequency for the circuit shown in Figure 5.38A. Use the following values: $R_1 = 220\ k\Omega$, $C_1 = 0.01\ \mu F$, $R_2 = 16\ k\Omega$, $R_3 = 27\ k\Omega$, $R_4 = 22\ \Omega$, and $V_{BB} = 25$ V.

Power Control Circuits

CHAPTER 6

Objectives

On completion of this chapter, you should be able to:

- Describe how hysteresis produces the snap-on effect in phase control;
- Describe the operation and advantages of the ramp-and-pedestal circuit, zero-voltage switching, and chopper motor control;
- Describe open-loop and closed-loop motor control;
- Explain how pulse-width modulation can accomplish motor speed control;
- List and describe the methods by which an AC motor's speed may be varied;
- Describe how different levels of power are applied to a load with phase control circuitry;
- Define a converter, and describe the conversion that takes place;
- Describe how the phase-locked loop is used to control the speed of a DC motor;
- List the advantages of transistors and power MOSFETs as power control devices.

Introduction

In the preceding chapter, we discussed the basic building blocks of semiconductor power control circuits, the thyristors. Now that we have provided a background in the operation of these devices, we are ready to discuss applications. Applications will fall into two basic categories: (1) general power control circuit principles, such as phase control and zero-voltage switching, and (2) motor control. Motor control will be broken down further into AC and DC motor control. Since motor control is one of the major applications of thyristors in industry, we will use most of this chapter to discuss this important subject.

As a technician, you will need to be able to analyze power control circuits. Consequently, in our presentation, we not only show simplified schematic diagrams, but we also give an idea of what the voltage waveforms should look like around the circuit.

Phase Control

In the previous chapter, we discussed AC phase control circuits using SCRs and triacs as power control devices. Examples of these circuits are shown in Figures 5.23 and 5.26. These circuits work

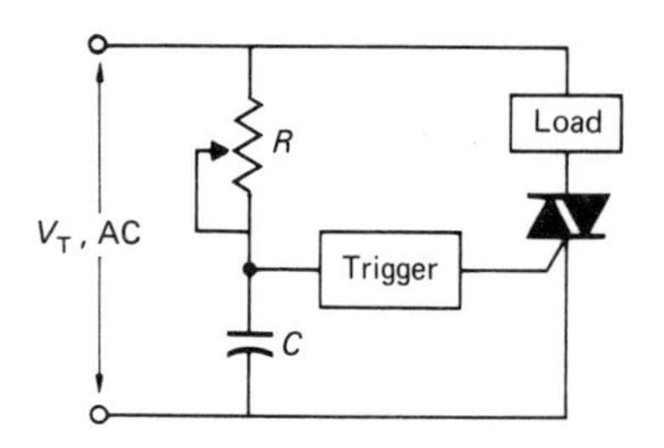

Figure 6.1 Triac Phase Control Circuit

well but do have some problems. In this section, we will show circuits that overcome some of the problems inherent in the circuits presented in Chapter 5.

Hysteresis in Phase Control

One major problem is one that designers call the *hysteresis effect.* The hysteresis effect can be explained through the use of Figure 6.1. From examining this circuit, we would expect that the load (say a lamp) would have AC power applied to it gradually. Such is not the case. In reality, the lamp will turn on suddenly with moderate brightness. After the lamp comes on, its intensity can be varied from low to high levels. This *snap-on effect* is due to the hysteresis inherent in this design.

During the time when the triac is off, the capacitor charges through the variable resistor R. Since the capacitor is a reactive component, the capacitor voltage lags the line voltage by about 90°. When the triac fires, the capacitor discharges through the gate–main terminal 1 junction of the triac and the trigger device. During the next half cycle, the capacitor voltage will now exceed the trigger device's breakover voltage sooner, because it will start to charge from a lower voltage.

This undesirable hysteresis effect can be reduced or eliminated in several ways. Two ways of accomplishing the reduction are shown in Figure 6.2. The addition of the extra RC network in Figure 6.2A allows C_1 to partially recharge C_2 after the diac has fired, thus reducing hysteresis. This addition also adds a greater range of phase shift across C_2.

The second method of reducing hysteresis, Figure 6.2B, uses a device called an asymmetrical trigger diode. This diode has a greater breakover voltage in one direction than in the other. It is so made that when it triggers for the first time, the capacitor discharges into the triac gate. During the next half cycle, however, the triggering voltage is equal to the original breakover voltage plus the capacitor's voltage decrease. This result allows the capacitor voltage to have the same time relationship to the applied voltage. Thus, hysteresis is reduced.

The circuits shown in Figure 6.2 can be used in virtually any AC low-power control circuits, such as universal and induction motors, lamps, and heaters. In fact, one designer calls this circuit the universal power controller.

UJT Phase Control

As discussed in Chapter 5, when thyristors are used in power control circuits, the thyristor is fired at some variable phase angle. But if we want maxi-

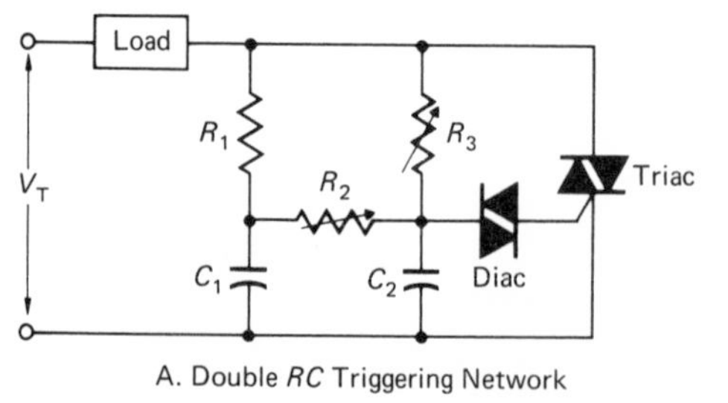

A. Double RC Triggering Network

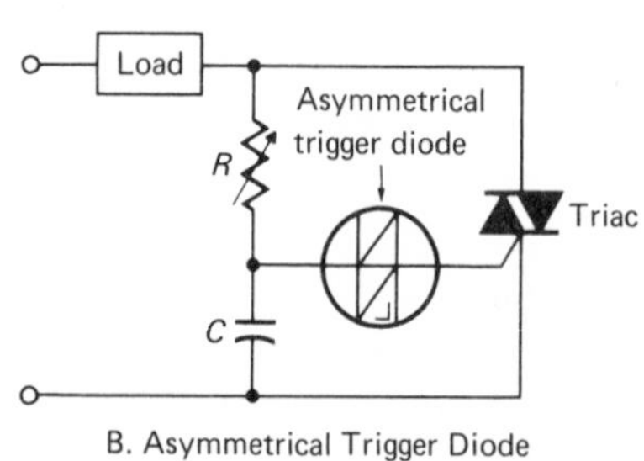

B. Asymmetrical Trigger Diode

Figure 6.2 Reducing Hysteresis

mum power delivered to a load, the thyristor must fire as soon as the AC voltage across it goes positive (and/or negative in the triac). The thyristor must conduct for 180° after firing for maximum power. If less than full power is needed, the thyristor firing must be delayed past 0° or 180°. A control circuit is needed, therefore, to control the firing angle between 0° and 180° for the SCR and also between 180° and 360° for the triac.

We have seen that the *RC* lag network can accomplish this control. These circuits, however, are affected by loading when connected to thyristor gate leads. They are also affected by supply voltage variations. The UJT oscillator, on the other hand, has none of the disadvantages of the *RC* networks. Thus, the UJT can be used as a phase control device in control circuits using either triacs or SCRs.

Figures 6.3A and 6.3B show a UJT oscillator controlling both an SCR circuit and a triac circuit. Both circuits function in essentially the same way. The thyristor is in the off state until the UJT fires. In Figure 6.3A, when the UJT fires, the capacitor will discharge quickly through the resistor R_4. This positive pulse will turn on the SCR. The SCR will remain on until the line voltage approaches zero. At this point, the SCR will turn off and remain off throughout the entire negative half cycle.

The triac circuit, Figure 6.3B, works in the same way. The positive spike is coupled to the triac's gate by the pulse transformer. When the triac turns on, the voltage to the control section of the circuit is taken away. No more pulses can occur during that half cycle of the input voltage. Since the line voltage is rectified by the bridge, the operation is the same for both positive and negative half cycles. This circuit can control up to 97% of the power available to the load.

Ramp-and-Pedestal Phase Control

Most of the preceding phase control circuits work well in manually controlled applications. But signals in automatic control situations are generally not large enough to produce adequate control. We might say that these circuits have relatively low gain. That is, in the *RC* networks used in phase control, a relatively large change in resistance is needed.

An alternative to the *RC* network is the *ramp-and-pedestal control circuit*, which provides much higher gain. Such a control circuit is illustrated in Figure 6.4A. When the line voltage goes positive, the zener diode keeps point *A* at a positive 20 V. The capacitor C_1 charges quickly through diode

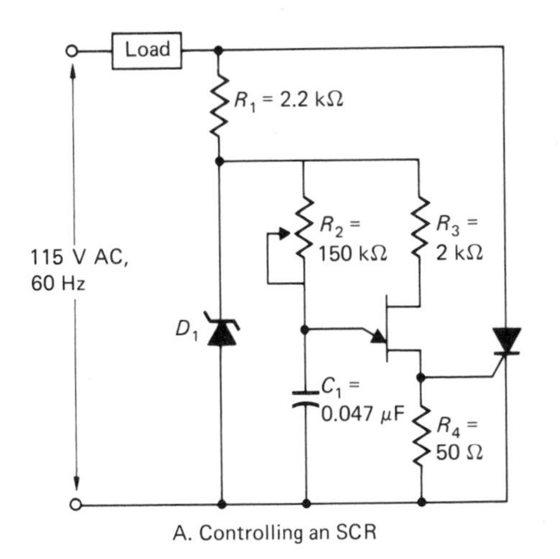

A. Controlling an SCR

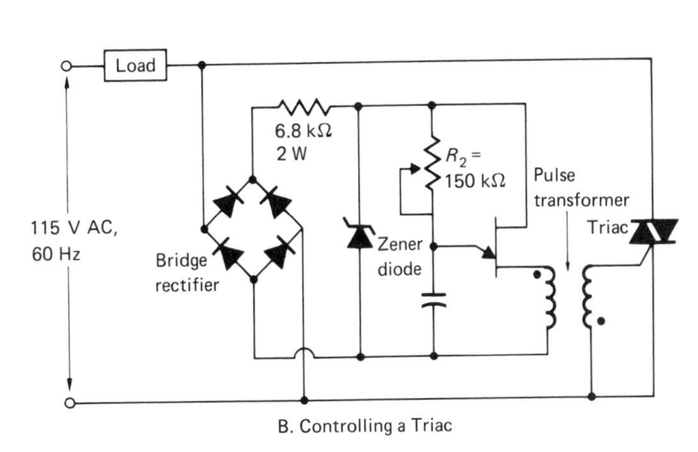

B. Controlling a Triac

Figure 6.3 UJT Oscillator Control Circuits

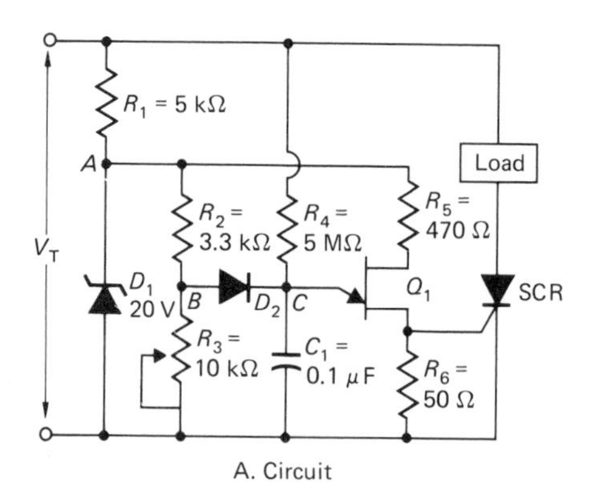

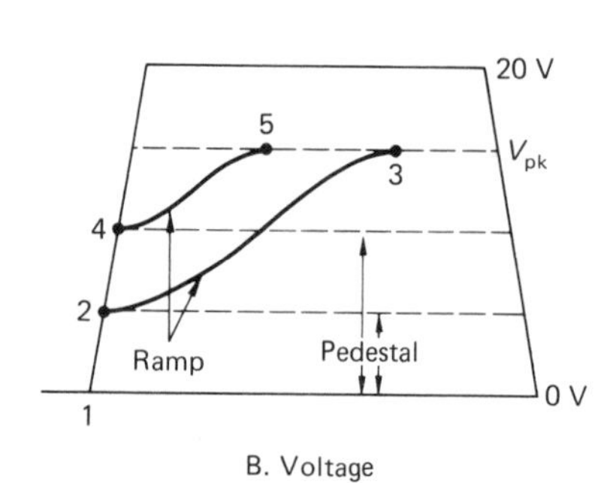

Figure 6.4 Discrete-Component, Ramp-and-Pedestal Control

D_2, R_2, and R_1. Let's say that resistor R_3 is adjusted to 3.3 kΩ. With this resistance, R_3 will drop about 10 V. The capacitor will continue to charge until the voltage at point *C* reverse-biases the diode D_1, which will occur at about 10 V. The capacitor, which was charging through about 8.3 kΩ, will now charge much more slowly through the 5 MΩ resistance of R_4. Eventually, the voltage at point *C* will reach the firing potential of Q_1. The UJT will then turn on and fire the SCR.

The voltage rise at point *C* is diagramed in Figure 6.4B. Note that the voltage rises very quickly between points 1 and 2 (the pedestal). At point 2, diode D_1 becomes reverse-biased and charges more slowly from point 2 to 3 (the ramp). At point 3, the UJT will fire, turning on the SCR. If we increase the resistance of R_3, the capacitor will charge quickly to a higher voltage before D_1 becomes reverse-biased. This higher voltage is represented by point 4. Note that since the capacitor started charging slowly from a higher voltage, it will reach the firing potential sooner in the input cycle. Thus, more power will be delivered to the load.

IC Ramp-and-Pedestal Phase Control

The simple ramp-and-pedestal circuit we have just discussed can be improved by adding more components. Of course, these extra components add more cost as well as improved performance. Since one of the advantages of integrated circuits is their ability to replace large and sometimes costly collections of discrete devices, ICs are well suited to this application.

Several manufacturers make an IC that does the phase control task we discussed above. One such IC is the PA436 phase controller manufactured by GE. Figure 6.5A shows the ramp-and-pedestal waveform used by the PA436. Note the difference between this waveform and the UJT waveform (Figure 6.4B). The UJT ramp-and-pedestal has a positive-going ramp, while the PA436 has a negative-going cosine ramp and a positive pedestal and reference. (A *cosine ramp* is a waveform whose average DC level is changing at an inverse sine rate.) In the UJT circuit, the reference, or firing, voltage remains constant and the pedestal changes. In the PA436 system, the pedestal remains constant and the reference changes.

The block diagram of the PA436 is illustrated in Figure 6.5B. The input signal is a DC voltage that establishes the pedestal level. The cosine ramp is developed from the supply voltage and is adjustable externally by the gain adjust. The ramp-and-pedestal waveform is then compared to a reference waveform by the differential comparator. The differential comparator produces an output waveform only when the ramp is below the reference level. The lockout gate keeps the differential comparator signal from reaching the trigger circuit until the AC supply voltage passes through zero.

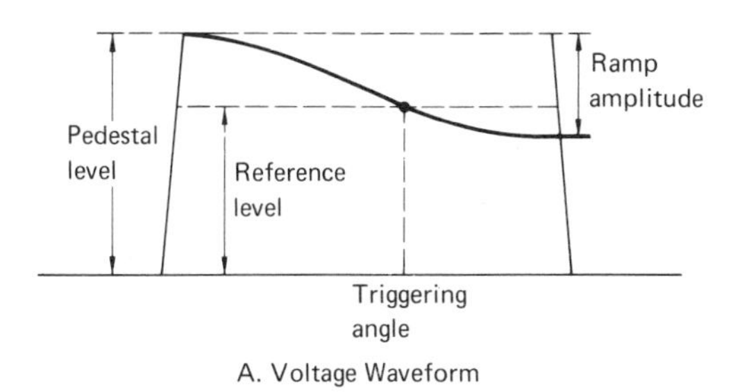

A. Voltage Waveform

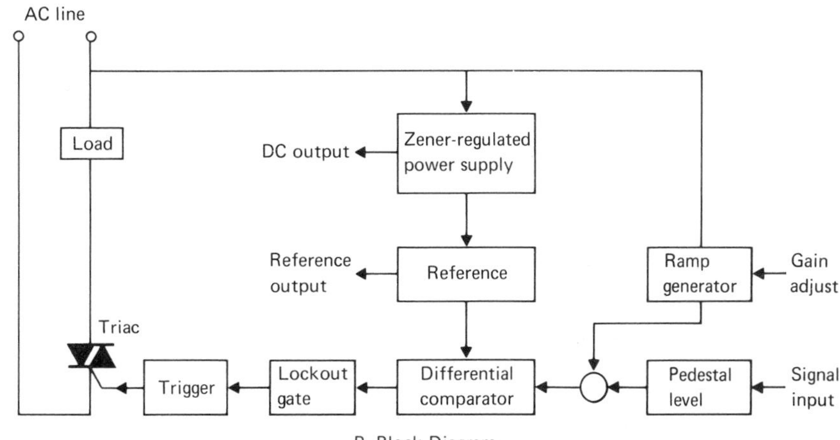

B. Block Diagram

Figure 6.5 IC Ramp-and-Pedestal Control

Zero-Voltage Switching

Several problems arise in the design of thyristor power control circuits such as the ones we have described previously. Opening and closing switches with applied voltages cause large variations in line current during short periods of time. These rapid current variations produce unwanted radio frequency interference and potentially damaging inductive-kick effects. (Recall that an inductor with current flowing through it will produce a large magnetic field around it. The field will collapse when current stops, generating a very high voltage.)

One solution to this problem is to make sure that the thyristor fires only when the supply is at or near 0 V. With no voltage across the switch, no current should flow. Several manufacturers provide ICs that accomplish this type of switching, called *zero-voltage switching*.

One such circuit is the CA3058 manufactured by RCA. This device is a monolithic IC designed to act as a triggering circuit for a thyristor. Internally, the CA3058 contains a threshold detector, a diode limiter, and a Darlington output driver. Together, these circuits provide the basic switching action. The DC operating voltages for the device are provided by an internal power supply. This power supply is large enough to supply external components, such as transistors and other ICs.

An example of a circuit using this IC zero-voltage switch (ZVS) is shown in Figure 6.6. Note that the ZVS output is connected directly to the gate of the thyristor. Whenever the DC logic level is high, the output to the gate of the triac is disabled. A low-level input enables the gate pulses, thus turning on the triac and providing current to the load.

Since the ZVS only turns on at or near the zero-voltage point of the applied voltage, only

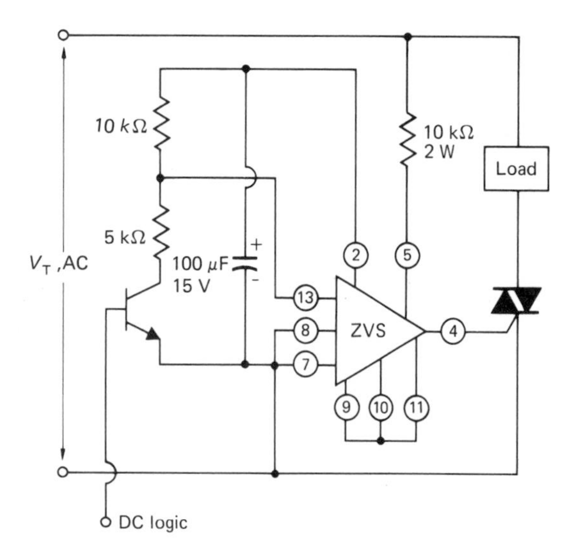

Figure 6.6 IC Zero-Voltage Switch Controlling a Triac

complete half or full cycles are applied to the load. This point is illustrated in Figure 6.7. The amount of power delivered to the load depends on how many full or half cycles are applied during a particular time. The average power delivered to the load, then, depends on the ratio of how long the thyristor was on to how long the thyristor was off.

Zero-voltage switching, while suitable for controlling heating elements, is not very useful in controlling the speed of motors. Motors have a tendency to slow down during the time that power is applied to the circuit, especially under heavy loads. This factor is not bothersome in heating elements because of their long time constant.

The IC zero-voltage switch is a very versatile device. It is used in industry in relay and valve control applications as well as heating and lighting controls.

DC Motor Control

DC motors are widely used in industry today for many reasons. Speed is relatively easily controlled with precision in DC motors. Furthermore, speed can be controlled continuously over a wide range, both above and below the motor's rated speed. Many industrial applications call for a machine that has very high starting torque, such as those found in traction systems. The DC motor satisfies most of these requirements adequately. Compared with AC drives, DC drives are generally less expensive, an important advantage in keeping initial equipment costs down.

As we have seen in this chapter, the thyristor is the major power control element in power con-

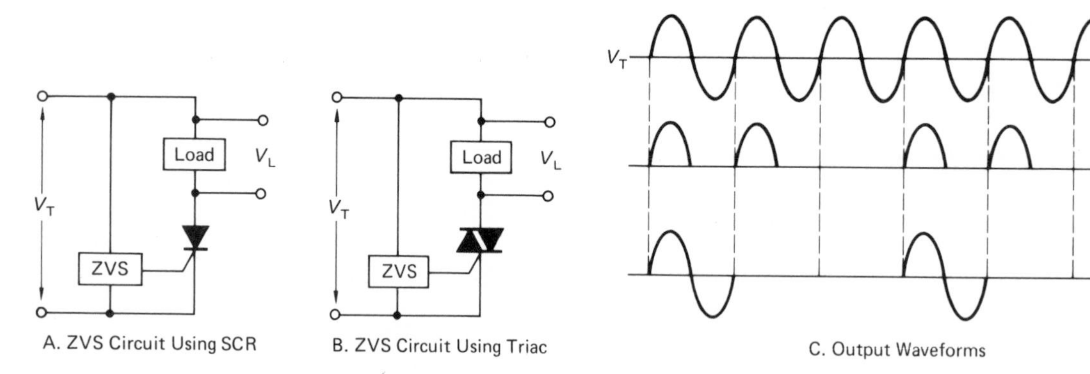

A. ZVS Circuit Using SCR B. ZVS Circuit Using Triac C. Output Waveforms

Figure 6.7 ZVS Applying Voltage to Load

trollers. And the SCR is the thyristor most often used in DC motor control applications. SCRs are used in two types of DC motor control systems: phase control circuits and chopper circuits. The phase control circuits, as we have seen above, control power by changing the firing angle (α). When the SCR fires early in the cycle, corresponding to a small α, more average power is delivered to the load. As firing angle increases, the SCR fires later in the cycle, delivering less power to the load.

Some DC motors have power supplies that are also DC. A good example of this application is the traction motor found in industrial delivery vehicles and forklifts. In the past, variable-resistance control was used in these situations. But such control is very inefficient. Today, the electronic chopper is used to control machines with DC power sources. A *chopper* is a thyristor switching system in which an SCR is turned on and off at variable intervals, producing a pulsating DC. The longer the SCR is on, or the higher the switching frequency, the more power is delivered to the load.

Phase Control

Many small industrial plants, farms, and homes are supplied with only single-phase power. DC motor speed control systems in these cases depend on the rectifying capabilities of the SCR to change the AC into the DC needed to develop torque in the motor. Because the reverse-blocking thyristor converts AC to DC, this circuit is often referred to as a *converter*.

Since in these cases, the thyristors are supplied with AC voltage, commutation of the thyristor is not a problem. The SCR naturally commutates when the line voltage passes near 0 V. In addition to natural commutation with AC line voltages, reverse-blocking, thyristor converter circuits are very efficient. Because of the low voltage drop across the device, as much as 95% of available power can be delivered to the load.

Basic Converter

The simplest of the converters, the *half-wave converter*, is illustrated in Figure 6.8A. During the time

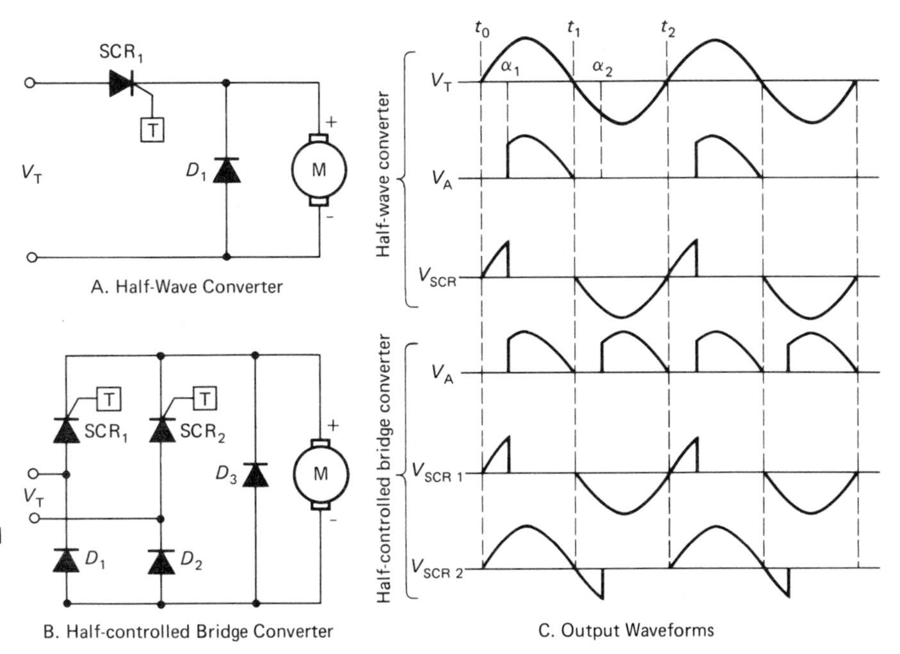

Figure 6.8 Half-Wave and Half-controlled Bridge Converters for Driving DC Motors

that the supply goes negative, SCR_1 blocks current flow. As a result, no power is applied to the motor (M). When the input goes positive, SCR_1 is forward-biased. It will conduct when a positive trigger pulse is applied to the gate (T represents the trigger circuit in the figure). The SCR will remain in conduction until the supply goes negative.

When SCR_1 conducts, it applies power to the motor. Diode D_1 is a flywheel, or freewheeling, diode. After current flows through the armature and the SCR turns off, the magnetic field built up around the armature will collapse. The field reverses polarity and conducts through the diode. The collapsing field supplies energy to the load during the time the SCR is off.

As shown in the armature voltage (V_A) waveform (Figure 6.8C), the half-wave converter has a large amount of ripple. As a result, this form of control is not used very much in DC motor speed control.

Half-controlled Bridge Converter

Figure 6.8B shows a more useful converter, the *half-controlled bridge converter.* When the supply goes positive, SCR_1 and D_2 are forward-biased. Current will flow (when SCR_1 is triggered) through D_2, the armature, and SCR_1. Near 0 V, SCR_1 commutates (at time t_1 in Figure 6.8C) and the armature field collapses, causing current to flow in the flywheel diode. At α_2, SCR_2 is triggered, causing current to flow through D_1 and the motor armature. Current will continue to flow until time t_2, when SCR_2 commutates. From that point, the cycle starts over again.

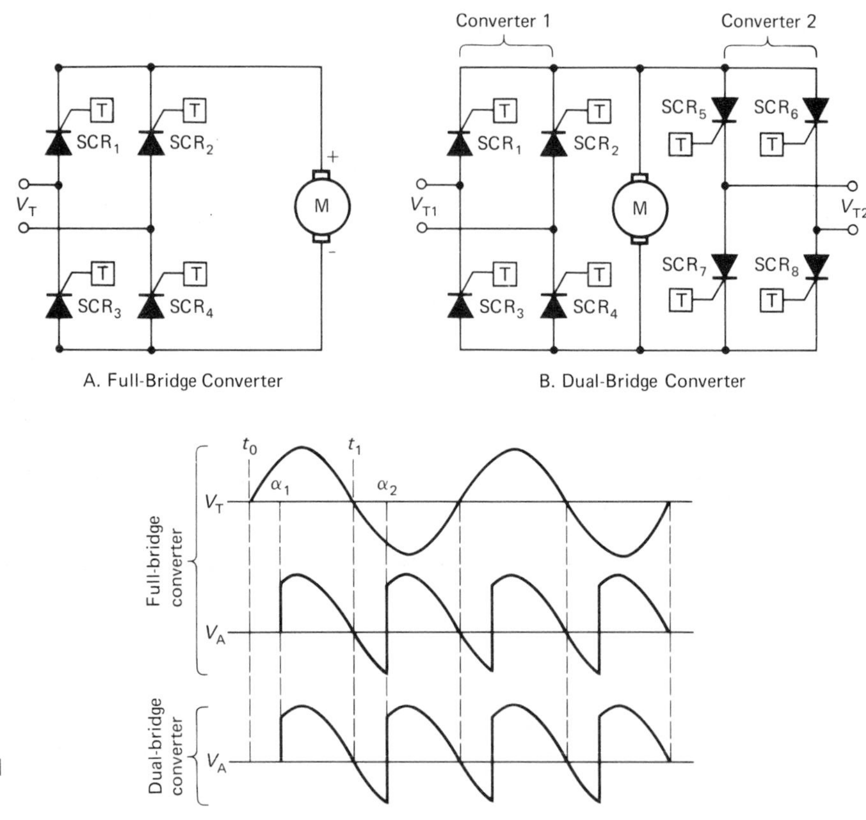

Figure 6.9 Full-Bridge and Dual-Bridge Converters Driving DC Motors

If you compare the armature waveforms of these two systems, you will note that the half-controlled bridge has far less ripple than the half-wave system. As in any phase control system, decreasing the firing angle will cause more power to be delivered to the load, increasing motor speed.

Full-Bridge Converter

A *full-bridge converter* is illustrated in Figure 6.9A. When the line voltage goes positive, SCR_1 and SCR_2 are forward-biased. They will conduct when triggered. Note in Figure 6.9C that the armature voltage goes negative between time t_1 and α_2. At time t_1, the SCRs have commutated. Beyond this point, the motor armature field collapses and reverses polarity, supplying power back to the supply.

If the firing angle goes past 90°, the average armature voltage will be negative. That is, the motor will be returning more power than it is using. Any rotating system that returns more power than it uses will decrease rotational speed. In fact, some motors use this system as a brake. Such a system is called a *regenerative braking system.*

Recall from Chapter 3 that a DC motor can reverse direction by reversing the direction of current in the armature. One way to accomplish this reversal is to have a switch that, when energized, will reverse the armature connections. In a shunt or a separately excited motor, we could reverse the direction of the current in the field winding by the same method. But mechanical or electromechanical switching is often too slow for some applications. However, almost instantaneous reversals of current flow can be achieved by using the circuit shown in Figure 6.9B, the *dual-bridge converter*. In this circuit, changing from converter 1 to converter 2 will reverse the direction of current flow in the motor.

Three-Phase Systems

Large DC motors in industry use a three-phase power source. One reason three-phase power is used to supply DC drive systems is ease of filtering. The three-phase system has significantly less ripple voltage and therefore takes less filtering to achieve the same result.

We will not discuss three-phase DC motor drive systems, because they do not differ significantly in theory of operation from single-phase systems. Of course, there will be two additional phases to contend with, each displaced 120° from each other. Because of the additional number of phases, circuit complexity increases. Also, waveform analysis of such a system becomes very complex, but little added understanding is gained through such analysis.

Chopper Control

In those applications where the power source is DC and the load is a motor, the DC chopper motor speed control is one of the most efficient methods of control. Chopper controls are used to control the speed of electric vehicles such as forklifts, delivery cars, and electric trains and trolleys. Chopper control is used not only because it is the most efficient but also because it provides smooth acceleration characteristics.

Basic Chopper Control

The basic chopper control circuit is illustrated in Figure 6.10A. Note that the SCR provides DC power to the motor by switching on and off. The average DC voltage presented to the motor is controlled by keeping the frequency constant and increasing and decreasing the amount of time that the SCR is on. When the SCR is on 50% of the time and off by the same amount (Figure 6.10B), an average of 50% of the voltage is delivered to the load. With a duty cycle of 25% (the SCR on 25% of the time), only a quarter of the applied voltage is delivered to the load (Figure 6.10C). Changing from a duty cycle of 50% to one of 25% will be accompanied by a decrease in the motor's speed. You may recognize this technique as pulse-width modulation.

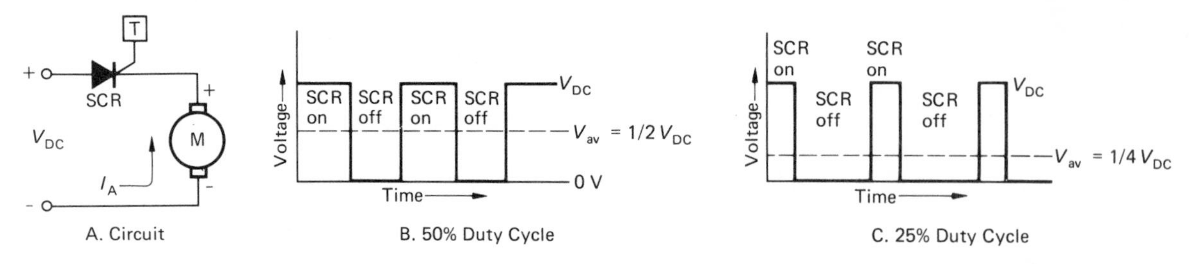

Figure 6.10 Basic Chopper Control Circuit

Another method of chopper control keeps the pulses the same width and increases or decreases the frequency. This method is not often used in controlling DC motors because it requires too great a frequency change to be practical.

Jones Circuit

Figure 6.11 shows a circuit often used in chopper control of DC motors, the *Jones circuit.* In this circuit, when SCR_1 is fired, current flows through the motor, L_1, and SCR_1. Because of the mutual inductance between L_1 and L_2, current flow in L_1 induces current flow in L_2. Current flows down through D_1 and C_1, resulting in the charge shown. As long as SCR_1 is conducting, power is applied to the motor.

SCR_1 is commutated as follows: SCR_2 is triggered into conduction, causing C_1 to discharge through SCR_1. This discharge path is opposite the path of the load current flowing through SCR_1. Total anode current is reduced in SCR_1 to the point where it falls below the holding current level. SCR_1 then turns off.

External Commutation

Chopper circuits may be commutated externally. *External commutation* is shown in Figure 6.12. When SCR_1 is fired, current flows through the motor from the supply, developing torque to turn the motor. The thyristor is commutated by turning on the transistor Q_1. The transistor applies a reverse-biased voltage V_{com} across the thyristor, turning it off. Power to the motor is regulated by proper choice of the transistor turnon point.

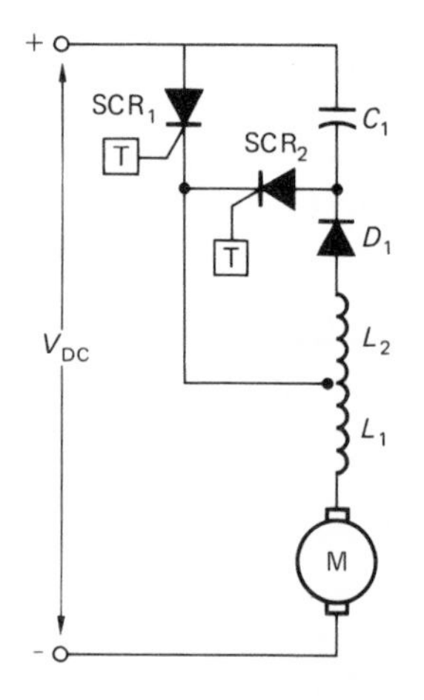

Figure 6.11 Jones Circuit

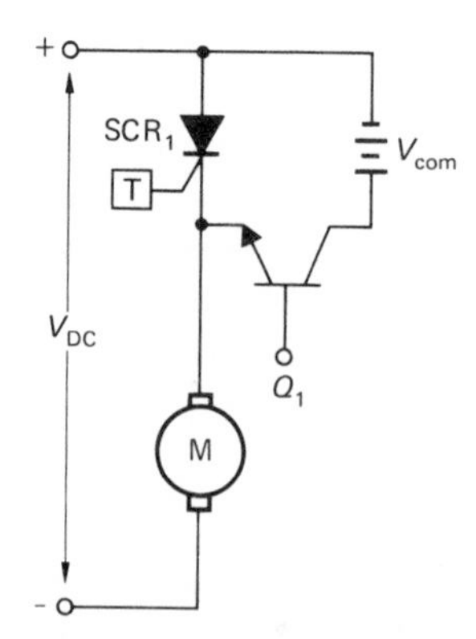

Figure 6.12 External Commutation Circuit

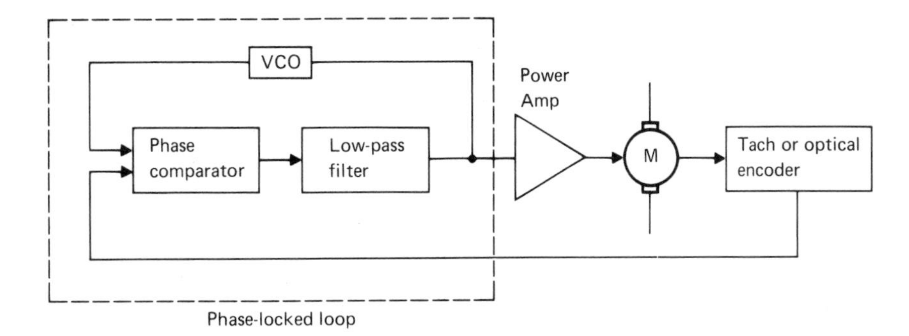

Figure 6.13 Phase-locked Loop Motor Control

Closed-Loop Speed Control

Up to this point, we have assumed that the control of the motor's speed was adjusted in an open-loop situation. *Open-loop control*, as we will see in a later chapter, is control without feedback. *Feedback control* of DC motors is an excellent method to regulate the motor's speed precisely. There are many methods used to control a motor by using feedback. The one we will discuss uses a device called a phase-locked loop.

Phase-locked Loop (PLL)

The *phase-locked loop (PLL)*, discussed in detail in Chapter 9, is an electronic feedback control system having a phase detector, a low-pass filter, and a voltage-controlled oscillator (VCO). A phase-locked loop can also be considered a frequency-to-voltage converter.

Every system using negative feedback has the same characteristics. Negative (180° out of phase) feedback is fed from the output back to the input. The feedback signal or voltage is then compared with a reference. If the feedback is not equal to the reference, an error voltage or signal is generated. This error component is proportional to how different the two values are. This error component is then used to control the system and reduce the error to zero.

In the PLL system shown in Figure 6.13, the reference is a *voltage-controlled oscillator (VCO)*. The VCO is nothing more than a free-running multivibrator whose frequency of oscillation is controlled by an *RC* network. This frequency is compared with a frequency fed back from the motor. These two frequencies are compared in the phase comparator. If they are different in frequency, the phase comparator will produce an output signal proportional to the difference. This signal is then rectified by the low-pass filter and used to control a DC amplifier. The DC amplifier adjusts the motor's speed until there is little or no difference between the reference signal and the feedback signal.

This type of control can regulate motor speed to within 0.001% of the desired speed. Other methods of analog feedback are used to control DC motor speed, but none is as efficient as the PLL.

Pulse-Width Modulation

Figure 6.14 illustrates *pulse-width modulation* for DC motor control. Here, pulse-width modulation is generated by the LM3524 integrated circuit. The

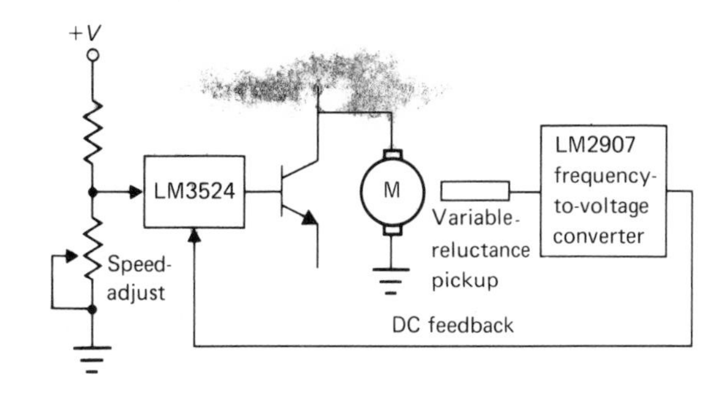

Figure 6.14 Pulse-Width Modulation

pulse-modulated output of the LM3524 drives a transistor, which, in turn, controls the speed of the DC motor. A variable-reluctance sensor acts as a tachometer, converting the motor's mechanical motion to a variable frequency. This frequency is converted to a voltage by the LM2907 frequency-to-voltage converter. Both the voltage produced by the LM2907 and the speed-adjust potentiometer are connected to the input of the LM3524, where they are compared. Any error voltage generated by this comparison is used to adjust the amount of pulse-width modulation applied to the DC motor.

AC Motor Speed Control

In recent years, the AC motor has become increasingly popular in certain areas of industry. Most of the interest in AC motors comes, at least in part, from the advantages it has over DC motors. The AC motor is smaller, and therefore less expensive, than the DC motor of equivalent horsepower rating. Generally, the AC motor is less costly to maintain than the DC motor. In the induction motor, reduced maintenance is due to the absence of a mechanical connection between rotor and stator. Not only does this feature mean lower maintenance costs, but it also means increased safety. Machines with commutators and brushes generate sparks, which could ignite in an explosive atmosphere. It is true that large synchronous motors have slip rings and, therefore, mechanical contacts. However, motors with slip rings produce less sparking and are easier to maintain than mechanically commutated motors.

Several years ago, the AC motor was used predominantly in fixed-speed applications. Recall from Chapter 4 on AC motors that the synchronous stator speed is directly proportional to the line frequency and inversely proportional to the number of magnetic poles. It is relatively easy to change the stator speed by varying the number of poles per phase. This type of speed control, however, results in speed control by steps only. There is no way to adjust the speed proportionally by altering the number of poles.

Speed may also be changed in an AC motor by varying the frequency of the voltage applied to the stator field. Although this type of speed control was known for a number of years, it was not economically sound until the advent of the thyristor in the 1950s. Research and development was soon undertaken to apply the thyristor to adjustable-frequency AC drives.

Universal Motor Speed Control

One circuit that will control the power applied to a universal motor is illustrated in Figure 6.15A. In this circuit, an open-loop control circuit, the ca-

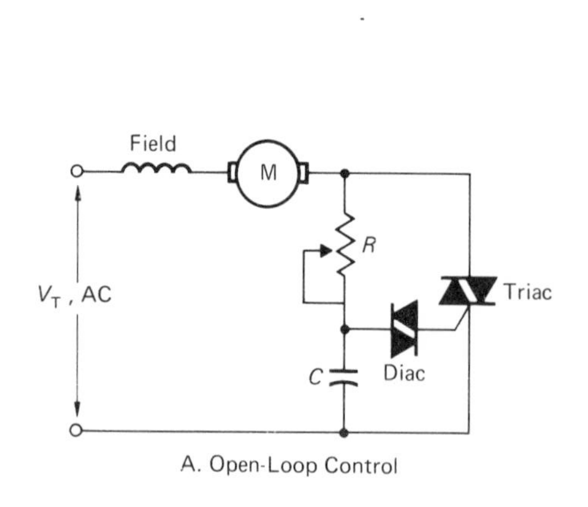

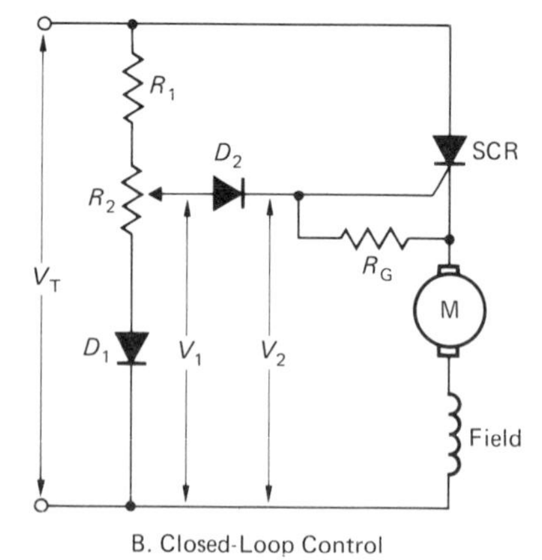

Figure 6.15 Universal Motor Control

pacitor charges up to the firing voltage of the diac in either direction. Once fired, the diac will apply a voltage to the gate of the triac. The triac will conduct and apply power to the motor. Note that the triac will conduct in either direction. Since this device is basically a series DC motor, current flowing in either direction will tend to cause rotation in only one direction. Speed may be changed by varying the resistance of the potentiometer, as discussed previously under phase control.

Closed-loop control for the universal motor is shown in Figure 6.15B. When line voltage is applied, the voltage at R_2 will rise proportionally to the voltage divider network. When the motor is turning at the desired speed, it develops a voltage V_2, which is mostly the cemf of the motor. If V_1 and V_2 are equal, no voltage is applied to the gate of the SCR. In this condition, the reverse-biased diode D_2 blocks the voltage. If the motor's speed decreases, however, the cemf will decrease, causing voltage V_2 to decrease. When V_1 is greater than V_2, a gate voltage is applied to the SCR. This voltage generates torque in the motor. The motor will then pick up speed, returning to near-normal speed.

The firing angle of the SCR is inversely proportional to the load on the motor. As the load on the motor increases, the speed of the motor will decrease. The more the motor speed decreases, the earlier the thyristor fires in the cycle. The earlier the thyristor fires, the more power is applied to the motor. Thus, the universal motor speed is kept relatively constant under varying load conditions.

Synchronous and Squirrel-Cage Induction Motor Speed Control

Changing the frequency of the voltage applied to the stator field will give smooth linear control of motor speed. Frequency control of synchronous and induction motors can be broken down into two methods: rectifier-inverter systems and cycloconverter systems. Figure 6.16A is a block diagram of a *rectifier-inverter system*. Note that the frequency conversion takes place in two steps. First, the fixed 60 Hz line frequency is rectified to DC. Second, the DC is converted to variable-frequency AC by the inverter. Thyristor inverters can supply an output frequency of up to 1 kHz over a range of about 15:1. The *cycloconverter system* is shown in block diagram form in Figure 6.16B. Cycloconverters give output frequencies of up to 25 Hz.

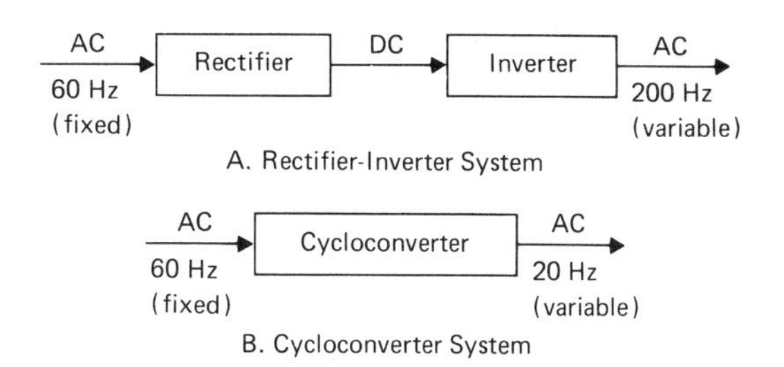

Figure 6.16 Frequency Control Methods for AC Motors

A simplified schematic diagram of a bridge inverter for the rectifier-inverter method is shown in Figure 6.17A. The waveforms associated with this system are illustrated in Figure 6.17B. If we consider the negative DC point of the supply as a reference, during time t_1 with SCR_1 on, V_X equals $+V$ DC. At the same time, with SCR_2 and SCR_4 off, V_Y equals 0 V DC. The total voltage across the load, then, is equal to $+V$ DC. At time t_2, with SCR_2 on and the other thyristors off, the voltage V_Y is equal to $+V$ DC. The total voltage across the load during time t_2 is $-V$ DC. The cycle then repeats itself, with the voltage across the load alternating between $+V$ DC and $-V$ DC. Thus, the inverter changes DC voltage into AC voltage. The frequency may be adjusted by adjusting the triggering and commutation frequencies.

The method of adjusting the stator frequency just described can be improved by pulse-width modulation of the voltage across the load. The waveform for such a system is illustrated in Figure 6.18. Pulse-width modulation may be achieved by varying the on and off times of the thyristors. Instead of turning on only once in a cycle, the SCRs trigger and commutate many times. Pulse-width

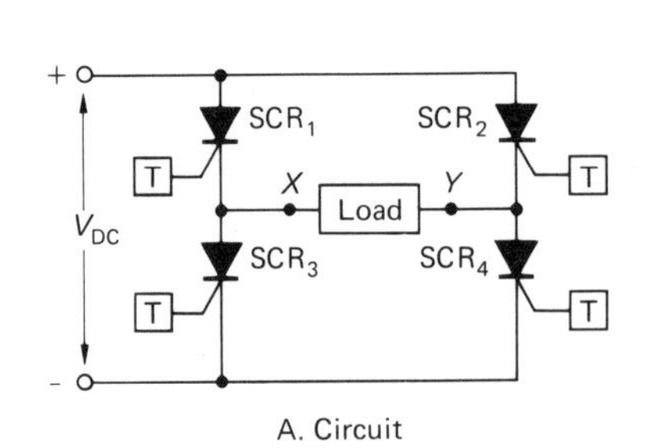

A. Circuit

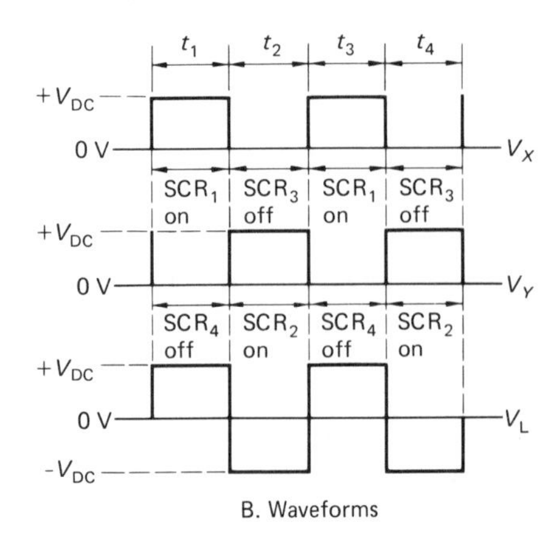

B. Waveforms

Figure 6.17 Rectifier-Inverter Simplified Schematic

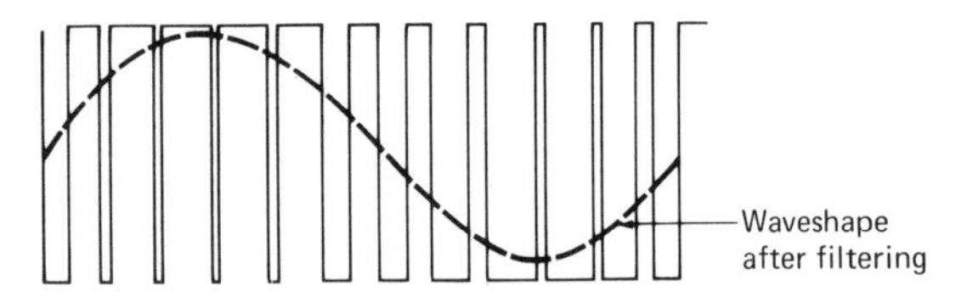

Figure 6.18 Pulse-Width Modulation of Voltage

modulation is used to reduce the amount of undesired harmonic frequencies generated by other methods of inversion.

Although the systems we have shown here use single-phase illustrations, these same techniques apply to polyphase systems as well. We have chosen the single-phase systems as examples because of their simplicity. For example, if an inverter were used in a three-phase system, each phase would require a separate inverter. Although the three-phase system is more complicated in terms of numbers of components, the basic inverter operation is the same in three-phase and single-phase systems.

The cycloconverter method is the second way to provide variable stator frequency control. The cycloconverter changes the fixed-frequency input voltage to a variable-frequency output voltage. The cycloconverter commutates naturally from the input line voltage. Thus, the cycloconverter is less complicated when compared with systems requiring forced commutation, such as the pulse-width modulation inverter discussed above. The cycloconverter has another advantage compared with the inverter. The inverter requires a DC input, which necessitates rectification before it can be used. The cycloconverter, in contrast, operates from the AC input directly.

A simplified circuit diagram of a cycloconverter is shown in Figure 6.19A. During the first input half cycle, SCR_1 and SCR_2 are gated on, producing the first two positive pulses shown in the output of Figure 6.19B. SCR_3 and SCR_4 are gated on during the second input half cycle, producing the next set of positive pulses. The process then repeats. The output voltage, after it is filtered, is sinusoidal in shape. The cycloconverter shown is actually a 7:1 frequency divider.

The cycloconverter finds applications in low-speed synchronous motors where high torques are demanded. Previous to the advent of thyristors, low speeds were achieved by gearing, especially when the application demanded an induction motor. The cycloconverter allows the induction motor to run at a low speed while developing maximum torque.

Before moving on to another topic, we should mention another method of controlling induction motor speed. Varying the voltage applied to the stator winding will give a limited amount of speed

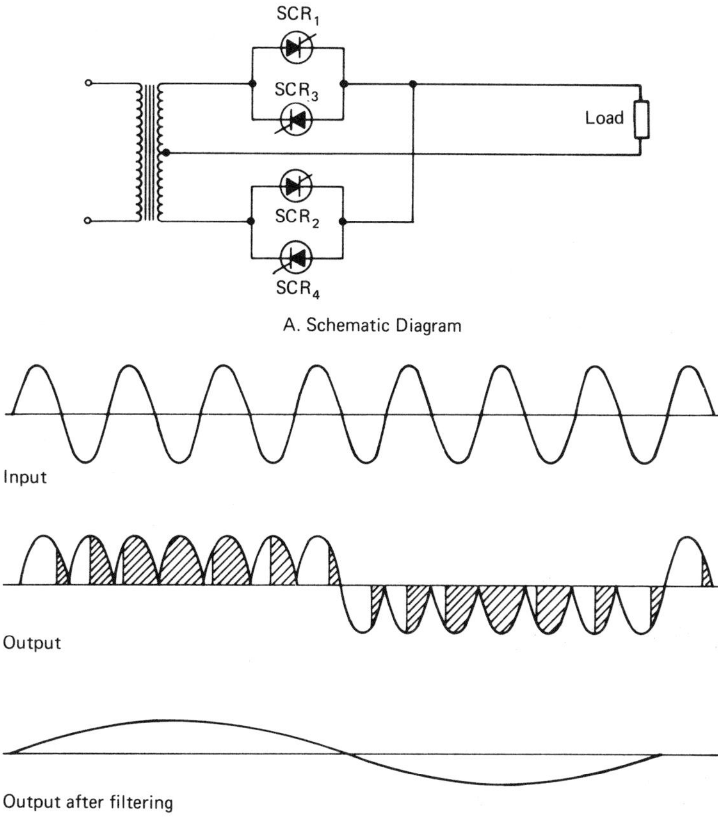

Figure 6.19 Cycloconverter

control while keeping the frequency constant. This type of speed control can be achieved by phase control techniques such as those discussed earlier in the chapter. For example, the circuit illustrated in Figure 6.15A can be used for this type of control. Control is achieved in this case by changing the torque produced by the motor. Recall that the torque produced by an induction motor is directly proportional to the square of the stator voltage. Thus, changing this voltage produces a change in torque. This method of control is not used very much in industry, for several reasons. First, it gives only limited speed control. Second, it changes the breakdown torque of the motor. Third, decreasing the speed by this method can result in the motor overheating.

Transistor and FET Power Control

Before leaving the subject of power control circuits, we should mention that transistors are also used in power applications. Special switching transistors can turn on very quickly and have low collector-to-emitter resistance. They are, therefore, very efficient. Also, transistors do not have the disadvantage that SCRs have in the ability to turn off in DC circuits. In DC circuits, transistors turn off easily by a removal of base current or voltage.

One of the major problems in using transistors in switching circuits involves the base current requirements. Transistors require that base current flow to provide collector current. Normally, power transistors have low current gain (β), some as low as 10. Such a low β means that considerable base current may be needed to keep the device on. This current is not doing useful work. Thus, because of this situation, the transistor is inefficient in power applications.

A recent development that overcomes the base current disadvantage, in large measure, is the *power MOSFET* (metal-oxide semiconductor FET). Recall that, unlike the transistor, the MOSFET is basically a voltage-controlled device. The power MOSFET, which can presently control up to 40 A, draws only nanoamperes of gate current. Consequently, the power MOSFET is a very efficient device. Power MOSFETs are presently being used in motor controls, especially chopper motor controls.

Conclusion

We have seen that the thyristor is a versatile and powerful control device. Its most prominent disadvantage derives from its use in DC circuits. Despite the problems associated with its use, it is very popular in industry. Because of its widespread use, a thorough understanding of the thyristor is a necessity for technicians.

Questions

1. The ramp-and-pedestal control circuit is classified as a ______ control. The ramp-and-pedestal has a greater control over the ______ ______ than other types of phase control.
2. The snap-on effect in some phase controls is due to ______. This effect can be reduced by adding a special diode or by adding another ______ network.
3. The type of switching that only occurs when the supply is at or near 0 V is called ______ switching. This type of switching is appropriate for heating controls but not for ______ controls.
4. The chopper motor control converts pure DC to ______. When the thyristor's on time increases compared with its off time, the ______ voltage at the load increases.
5. Closed-loop, or feedback, control is used to control a motor's speed by using an IC called a ______. This type of control can regulate a motor's speed to within ______% of desired speed.
6. The speed of AC motors is controlled by varying the number of poles and by varying the ______ ______. In the rectifier-inverter speed control, the supply voltage is converted to DC, and the DC is converted to variable-frequency ______.
7. Explain how phase control varies the power delivered to a load.
8. Describe hysteresis, and explain how it affects some phase control circuits.
9. Why is the UJT used as an oscillator to trigger thyristor circuits?
10. Explain how the ramp-and-pedestal circuit works.
11. Explain the advantages and disadvantages of using zero-voltage switching in control circuits. Define zero-voltage switching.
12. Describe basic converter action, and explain how this action changes the power delivered to a load.
13. Define chopper. Describe its operation. How are choppers commutated?
14. Describe how pulse-width modulation can be used to control a DC motor's speed.
15. List the methods by which an AC motor's speed may be controlled for (a) the universal motor and (b) the synchronous and the induction motors.
16. Explain how a closed-loop speed control system works.
17. Explain the operation of the following circuits: (a) rectifier-inverter and (b) cycloconverter.
18. Match the types of conversion with the circuits that perform them.

a. AC to AC	**1.** inverter
b. DC to AC	**2.** rectifier
c. AC to DC	**3.** chopper
d. DC to DC	**4.** converter
	5. cycloconverter

19. List the applications of the cycloconverter. What are its main advantages?
20. What are the advantages of using transistors as power switches?
21. What problems occur if transistors are used as power switches?
22. Describe how the power MOSFET overcomes one of the transistor problems you listed in answer to Question 21.
23. Explain how a PLL can be used as a motor speed controller.

Transducers

CHAPTER 7

Objectives

On completion of this chapter, you should be able to:

- Define a transducer in terms of input, conversion, and output;
- Explain the transduction principle behind individual transducers;
- Identify the schematic symbols of the optoelectronic transducers;
- Classify individual transducers into one of eight basic areas;
- Calculate the unknown resistance in and the sensitivity of a bridge circuit.

Introduction

Control and regulation of industrial systems and processes depend on accurate measurement. It would not be going too far to state that a variable must be measured accurately to be controlled. In industrial systems, the device that does this measurement is the transducer, or sensor. *Transduction* is the process of converting energy from one form to another. We can define a *transducer* as a device that converts a *measurand* (that which is to be measured) into an output that facilitates measurement. In other words, a transducer converts a variable (fluid flow, temperature, humidity) into an analog of the variable.

Let's use an example: the mercury thermometer. A thermometer is a transducer. The measurand is temperature. The transducer converts temperature into an analog of temperature: fluid level in a glass tube. The level of mercury is directly proportional (an analog) to ambient temperature.

In this chapter, we describe several transducers and the principles by which they operate. Our treatment of this topic is in no way exhaustive. Rather, we intend to describe the major sensors in use in industry today. The sensors covered are primarily electrical sensors—that is, sensors that give an electrical output. Sensors giving other outputs are discussed when appropriate. In our treatments, we have classified sensors according to the variable measured. In this chapter, we also briefly discuss bridges, which are common devices for detecting sensor changes.

Temperature

Temperature is undoubtedly the most measured dynamic variable in industry today. Many industrial processes require accurate temperature measurement, because temperature cannot be precisely controlled unless it can be accurately measured.

Before we discuss temperature sensors, we must define temperature. Very simply, *temperature* is the ability of a body to communicate or transfer heat energy. Alternatively, we can define temperature as the potential for heat to flow. Recall that heat will flow from a hotter body to a cooler one.

Temperature is generally measured on three arbitrary scales: Fahrenheit, Celsius, and Kelvin. The *Fahrenheit scale* is referenced to the boiling point of water at 212°F and the freezing point of water at 32°F. The *Celsius scale* references the boiling point at 100°C and the freezing point at 0°C. The *Kelvin scale* is based on the divisions of the Celsius scale, with absolute zero at 0°. To convert from Celsius to Kelvin, you simply add 273°.

Temperature can be measured in many different ways. For purposes of simplicity, we divide temperature sensing into two areas: mechanical and electrical.

Mechanical Temperature Sensing

Mechanical temperature sensing depends on the physical principle that gases, liquids, and solids change their volume when heated. Furthermore, different substances change volume in differing amounts. For example, liquid mercury expands about 0.01% per degree Fahrenheit. Methyl alcohol, on the other hand, expands at about 0.07% per degree Fahrenheit (0.1% per degree Celsius). In this section, we discuss mechanical temperature sensors, devices that convert temperature to position or motion.

Glass-Stem Thermometers

The *glass-stem thermometer* is one of the oldest types of thermometers. Its invention is credited to Galileo in about 1590. As shown in Figure 7.1, it consists of a bulb filled with liquid, a capillary tube, and a supporting glass stem. The capillary tube is quite thin, which helps increase the sensitivity of the device. Some type of magnifying lens is usually included in the glass stem to aid in reading temperature measurement. Linearity of this device is improved by evacuating the remaining air from the tube.

The glass stem has several advantages. It is inexpensive to manufacture and has excellent linearity and accuracy characteristics. It also has several disadvantages. It is fragile and difficult to read, it is not easily used for remote measurement or control, and it allows considerable time lag in temperature measurement. This last disadvantage arises because of the poor thermal conductivity of glass. In spite of these disadvantages, the glass-stem thermometer is still popular in industry today.

Filled-System Thermometers

The *filled-system thermometer* works on the same basic principle as the glass stem thermometer. The bulb, shown in Figure 7.2, is filled with a gas or

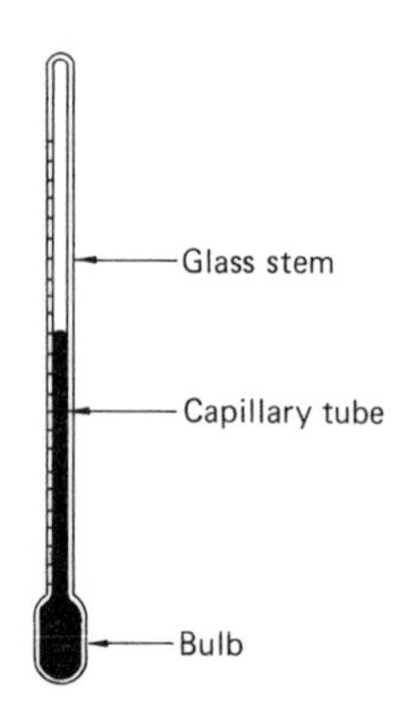

Figure 7.1 Glass-Stem Thermometer

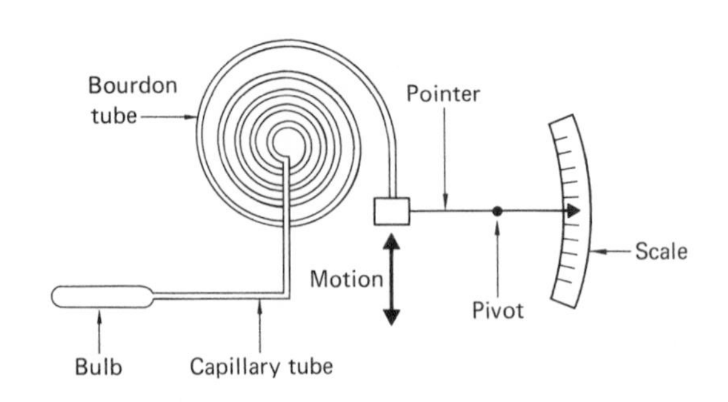

Figure 7.2 Filled-System Thermometer

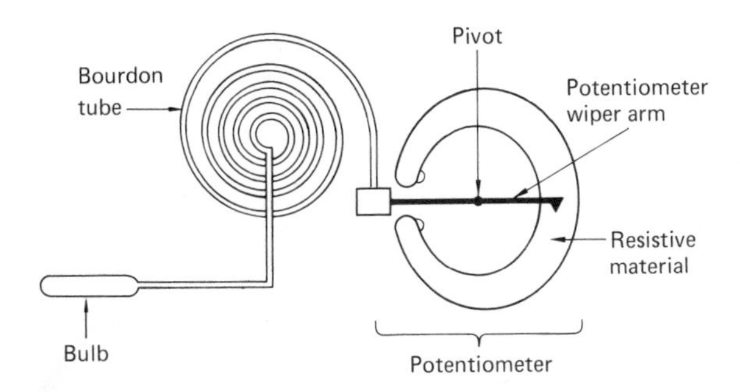

Figure 7.3 Filled System Connected to Potentiometer

a liquid. As the bulb is heated, the gas or liquid expands, exerting pressure on the Bourdon tube (to be described later in the chapter) through the capillary tubing. The Bourdon tube uncoils, moving the pointer. Since this system exerts a pressure, it can be used for chart-recording displays of temperature.

Although this system is mechanical, it can be converted into an electrical transducer by using a potentiometer, as shown in Figure 7.3, or a linear variable differential transformer (LVDT). Potentiometers and LVDTs convert position or displacement into an electrical parameter. These devices are discussed later in the chapter.

Filled systems react very quickly to temperature changes, can be as accurate as 0.5%, and can be used for remote measurement up to 300 ft (100 m). The major disadvantage with filled systems is the need for temperature compensation. Although the sensing element in the filled system is the bulb, the Bourdon tube and capillary tube are also temperature-sensitive. Thus, ambient temperature changes around the Bourdon and capillary tubes can give a false temperature reading. Filled systems can be temperature-compensated by use of bimetal strips or systems using dual elements.

Bimetallic Thermometers

The *bimetallic thermometer* operates on the principle of differential expansion of metals; that is, that metals increase their volume when heated. Different metals expand by different amounts. How much a metal expands when heated is indicated by a parameter called the *linear expansion coefficient*. This parameter shows volume expansion in millionths per degree Celsius. A list of the parameter values for a few metals is given in Table 7.1.

When two metals are joined together and heated, physical displacement occurs. In designing these devices, engineers try to ensure maximum movement with a given temperature change. They do so by joining a metal with a low expansion coefficient to one with a high expansion coefficient. A popular combination is brass and invar (a copper-nickel alloy). From Table 7.1, we see that this combination meets the design criteria.

A bimetallic thermometer is shown in Figure 7.4. The bimetal strips are usually wound in a spiral or helix (coil). The helical type is more sensitive.

The bimetallic thermometer is one of the most popular thermometers in industry. It is relatively inexpensive, has a wide range (800°F, or 400°C), is rugged, and is easily installed and read. Its major disadvantages are that it cannot be used in remote measurement or in analog process control. Bimetallic sensors are used in simple on-off control

Table 7.1 *Linear Expansion Coefficient of Various Metals*

Substance	*Linear Expansion Coefficient (millionths/°C)*
Aluminum	23.5
Brass	20.3
Copper	16.5
Invar (copper-nickel alloy)	1.2
Kovar (copper-nickel-cobalt alloy)	5.9

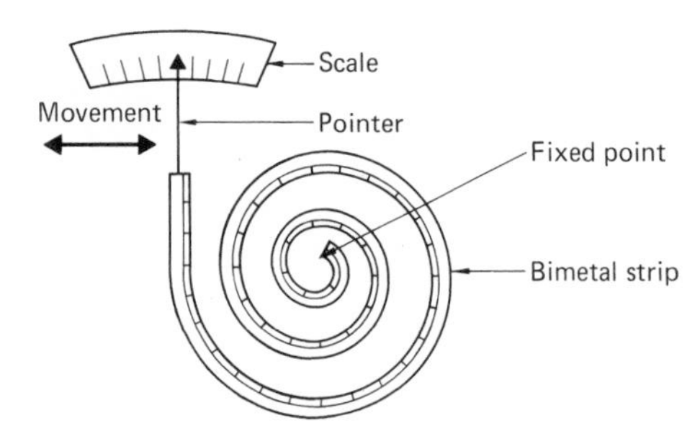

Figure 7.4 Bimetallic Thermometer

systems. For this use, a mercury switch is placed on the bimetal strip, as in home thermostats.

An important calculation when using mercury switches concerns how much the strip moves with a given temperature change. The amount of deflection of a straight bimetallic strip is given by the following equation:

$$y = \frac{3(c_A - c_B)(T_2 - T_1)l^2}{4d} \tag{7.1}$$

where y = amount of deflection

c_A = linear expansion coefficient of metal A

c_B = linear expansion coefficient of metal B

T_1 = lower temperature

T_2 = higher temperature

l = length of strip

d = thickness of strip

For example, suppose we have a 5 cm long, 0.05 cm thick, brass-invar strip that is straight at room temperature (25°C). How much will the strip deflect if the temperature changes 5°C? From Table 7.1, we see that invar and brass have coefficients of 1.2 and 20.3, respectively. Using Equation 7.1, we find that the deflection is 0.036 cm:

$$y = \frac{3(20.3 - 1.2 \times 10^{-6})(5^\circ \text{ C})(5 \text{ cm})^2}{4(0.05 \text{ cm})}$$

$$= 0.036 \text{ cm} \tag{7.2}$$

Note that the amount of deflection is directly proportional to the differences between the temperatures and the coefficients of expansion and proportional to the square of the length. Also, the deflection is inversely proportional to the strip's thickness.

The equation for a spiral strip is similar to the equation for a straight strip:

$$y = \frac{9(c_A - c_B)(T_2 - T_1)rl}{4d} \tag{7.3}$$

where r = spiral's radius

d = length of strip if it were extended

Electrical Temperature Sensing

We have considered several methods of mechanical sensing. Because mechanical rather than electrical principles are used, these sensors are not suitable for use in analog process control systems. So, in this section, we discuss four commonly used electrical sensors: the thermocouple, the thermistor, the resistance temperature detector, and the semiconductor temperature sensor.

Thermocouple

Of the electrical temperature-sensing devices, the *thermocouple* enjoys the widest use in industry. Its discovery dates back to 1821, when Thomas Seebeck, a German physicist, joined two wires made of different metals. He found that when he heated one end, electric current flowed in the loop formed by the wires. This effect is called the *Seebeck effect* and is illustrated in Figure 7.5.

When the circuit was broken, as shown in Figure 7.6, Seebeck observed that there was a

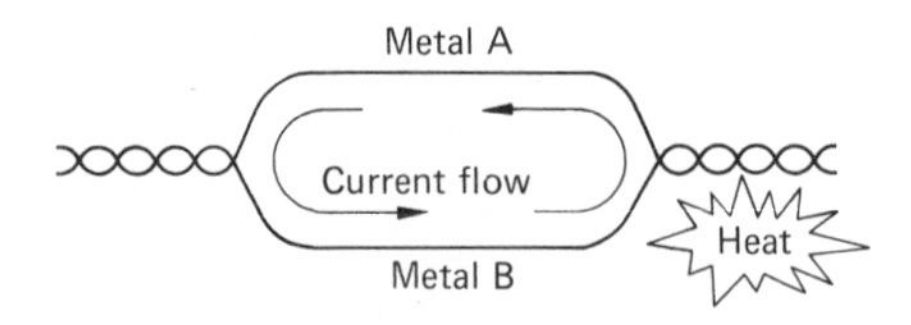

Figure 7.5 Seebeck Effect

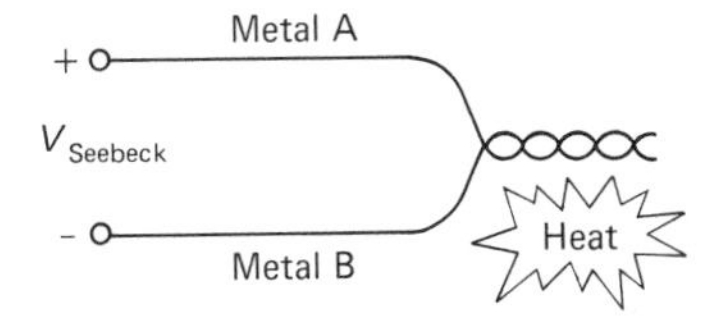

Figure 7.6 Potential Difference across Heated Junction

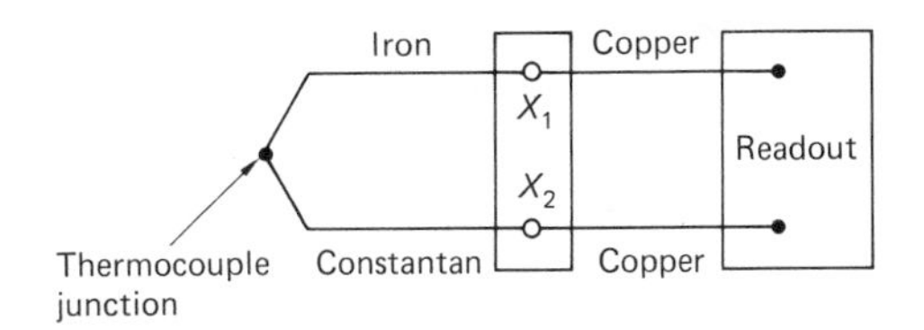

Figure 7.8 Indicating Device Connected to Thermocouple

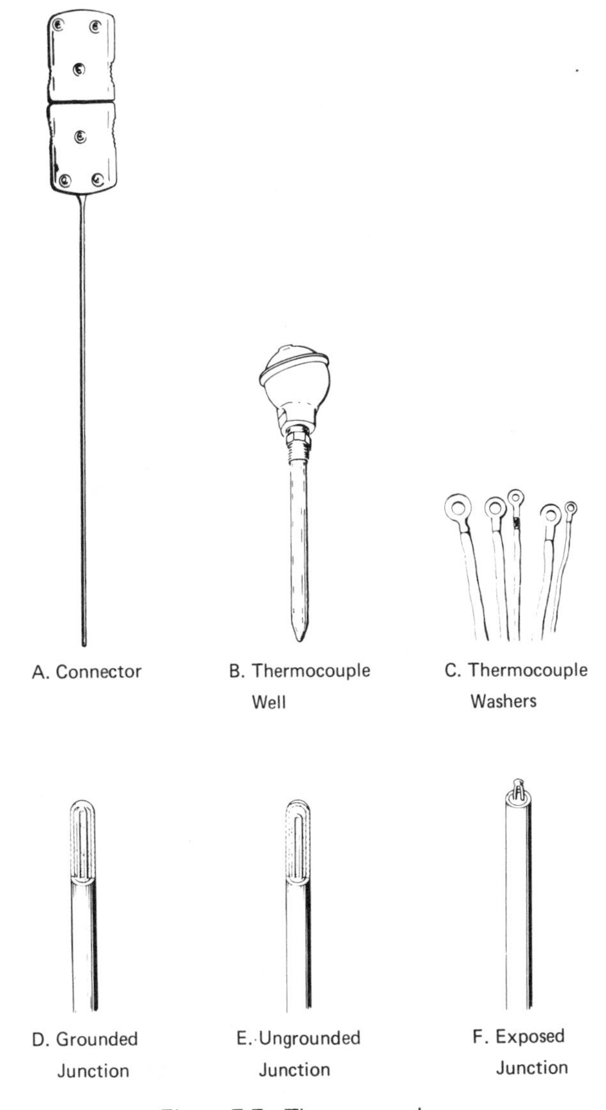

Figure 7.7 Thermocouples

voltage between the two terminals. He discovered that the size of the voltage varied with heat. An increase in heat caused an increase in voltage. He also found that different combinations of metals produced different voltages.

Thermocouples come in many different forms, as shown in Figure 7.7. The thermocouple junction may be exposed, grounded, fashioned into washers, or placed in a well for protection. Each of these methods has its own advantages.

Generally speaking, the thermocouple is simple, rugged, inexpensive, and capable of the widest temperature measurement range (about 4500°F, or 2500°C) of any electrical temperature transducer. On the other hand, it is the least sensitive and stable of the electrical temperature transducers. Perhaps its biggest disadvantage is the need for a reference junction.

Consider the schematic diagram in Figure 7.8. As shown in the figure, we must connect a readout device to a thermocouple to read the voltage produced. But in doing so, we create two more thermoelectric junctions, as indicated in Figure 7.9.

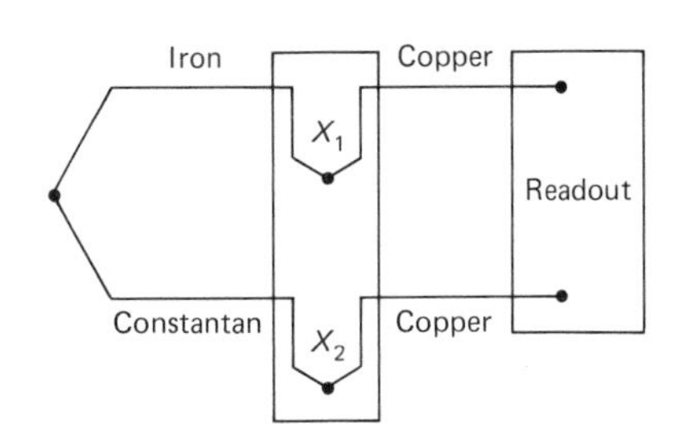

Figure 7.9 Two Additional Thermocouple Junctions Created

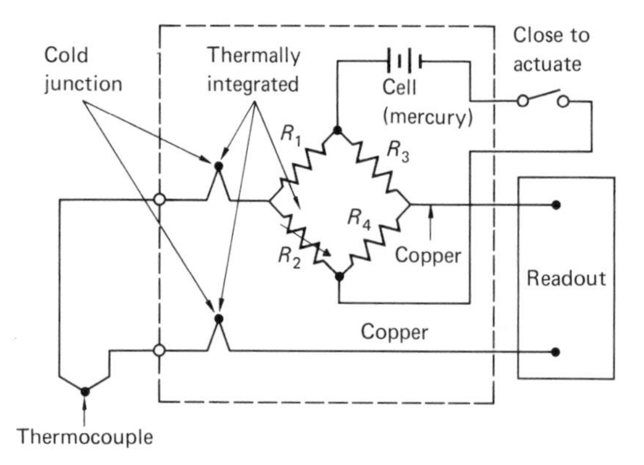

Figure 7.10 Thermocouple System with Thermally Integrated Cold Junctions and Resistors

These extra junctions interfere with the voltage produced by the iron-constantan thermocouple. (Constantan is an alloy of copper and nickel.)

This problem can be overcome by keeping the X_1 and X_2 junctions at a constant reference temperature. If these junctions are held at a constant temperature, it is a simple matter of converting the total voltage produced at the readout device to a temperature. As a matter of fact, most conversion tables are based on a reference junction temperature of 32°F (0°C). This reference temperature came about because for many years the reference junction was immersed in an ice bath to keep its temperature constant.

Because ice baths are often inconvenient to use and therefore impractical, several other methods are used to achieve the same result. One such method is shown in Figure 7.10. Both reference junctions and the bridge resistor R_2 are thermally integrated on a substrate. As the temperature of the cold junction varies, the bridge produces a voltage to cancel out the cold junction potential produced. The only voltage that varies is the one produced by the thermocouple.

Thermocouple temperature versus emf curves are shown in Figure 7.11. The different thermocouples are represented by different letter designations.

Tables are available that give the output voltage of a particular type of thermocouple at various temperatures. These charts usually use a reference junction of 0°C, as mentioned previously. Table 7.2 is an example of a table for a type J thermocouple. Notice that the type J device puts out 5.431 mV at 103°C.

Keep in mind that the voltage shown is a function of the temperature difference between the two junctions. Knowing this fact, we can use this table for references other than 0°C. Using type J as an example, we can find the voltage when our measuring junction is 250°C with the reference at 25°C. At 250°C, the output from the measuring junction is 13.553 mV, while the reference junction's output is 1.277 mV. The thermocouple's output voltage is the difference between the two voltages—that is, 13.553 mV − 1.277 mV = 12.176 mV. Reversing this process allows us to find the temperature given the voltage output of the thermocouple.

At this point, we should note that the temperature of a thermocouple (or any sensor) takes a certain amount of time to react to a temperature change. The thermocouple's response time depends on the mass of the thermocouple, the specific heat of the thermocouple, the coefficient of heat transfer from one boundary to another, and the area of contact between the thermocouple and the material to be measured. These factors are related in the following equation:

$$t_c = \frac{mc}{kA} \tag{7.4}$$

where t_c = thermal time constant
m = mass of sensor, in grams
c = specific heat, in calories per gram-degree Celsius
k = heat transfer coefficient, in calories per centimeter-second-degree Celsius
A = area of contact between sensor and sample, in square centimeters

Just as in a capacitor, the rate of change of a thermocouple's temperature follows the time constant curve. Using Equation 7.4, we can calculate

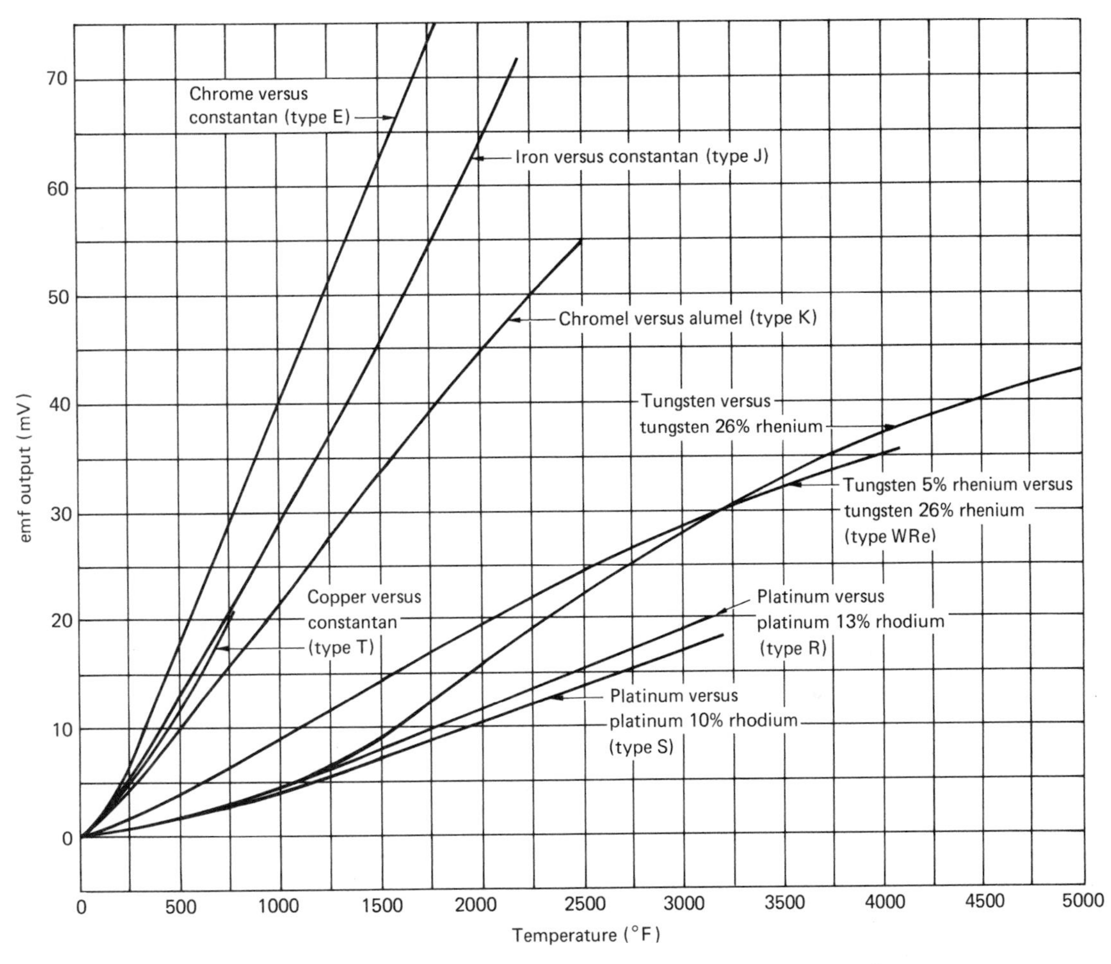

Figure 7.11 Temperature versus emf (Reference Junction at 32° F)

Table 7.2 *Output Voltage for Type J Thermocouple*

DEG C	0	1	2	3	4	5	6	7	8	9	10	DEG C
	THERMOELECTRIC VOLTAGE IN ABSOLUTE MILLIVOLTS											
-210	-8.096											-210
-200	-7.890	-7.912	-7.934	-7.955	-7.976	-7.996	-8.017	-8.037	-8.057	-8.076	-8.096	-200
-190	-7.659	-7.683	-7.707	-7.731	-7.755	-7.778	-7.801	-7.824	-7.846	-7.868	-7.890	-190
-180	-7.402	-7.429	-7.455	-7.482	-7.508	-7.533	-7.559	-7.584	-7.609	-7.634	-7.659	-180
-170	-7.122	-7.151	-7.180	-7.209	-7.237	-7.265	-7.293	-7.321	-7.348	-7.375	-7.402	-170
-160	-6.821	-6.852	-6.883	-6.914	-6.944	-6.974	-7.004	-7.034	-7.064	-7.093	-7.122	-160
-150	-6.499	-6.532	-6.565	-6.598	-6.630	-6.663	-6.695	-6.727	-6.758	-6.790	-6.821	-150
-140	-6.159	-6.194	-6.228	-6.263	-6.297	-6.331	-6.365	-6.399	-6.433	-6.466	-6.499	-140
-130	-5.801	-5.837	-5.874	-5.910	-5.946	-5.982	-6.018	-6.053	-6.089	-6.124	-6.159	-130
-120	-5.426	-5.464	-5.502	-5.540	-5.578	-5.615	-5.653	-5.690	-5.727	-5.764	-5.801	-120
-110	-5.036	-5.076	-5.115	-5.155	-5.194	-5.233	-5.272	-5.311	-5.349	-5.388	-5.426	-110
-100	-4.632	-4.673	-4.714	-4.755	-4.795	-4.836	-4.876	-4.916	-4.956	-4.996	-5.036	-100
-90	-4.215	-4.257	-4.299	-4.341	-4.383	-4.425	-4.467	-4.508	-4.550	-4.591	-4.632	-90
-80	-3.785	-3.829	-3.872	-3.915	-3.958	-4.001	-4.044	-4.087	-4.130	-4.172	-4.215	-80
-70	-3.344	-3.389	-3.433	-3.478	-3.522	-3.566	-3.610	-3.654	-3.698	-3.742	-3.785	-70
-60	-2.892	-2.938	-2.984	-3.029	-3.074	-3.120	-3.165	-3.210	-3.255	-3.299	-3.344	-60
-50	-2.431	-2.478	-2.524	-2.570	-2.617	-2.663	-2.709	-2.755	-2.801	-2.847	-2.892	-50
-40	-1.960	-2.008	-2.055	-2.102	-2.150	-2.197	-2.244	-2.291	-2.338	-2.384	-2.431	-40
-30	-1.481	-1.530	-1.578	-1.626	-1.674	-1.722	-1.770	-1.818	-1.865	-1.913	-1.960	-30
-20	-0.995	-1.044	-1.093	-1.141	-1.190	-1.239	-1.288	-1.336	-1.385	-1.433	-1.481	-20
-10	-0.501	-0.550	-0.600	-0.650	-0.699	-0.748	-0.798	-0.847	-0.896	-0.945	-0.995	-10
0	0.000	-0.050	-0.101	-0.151	-0.201	-0.251	-0.301	-0.351	-0.401	-0.451	-0.501	0

Table 7.2 (*continued*)

DEG C	0	1	2	3	4	5	6	7	8	9	10	DEG C
0	0.000	0.050	0.101	0.151	0.202	0.253	0.303	0.354	0.405	0.456	0.507	0
10	0.507	0.558	0.609	0.660	0.711	0.762	0.813	0.865	0.916	0.967	1.019	10
20	1.019	1.070	1.122	1.174	1.225	1.277	1.329	1.381	1.432	1.484	1.536	20
30	1.536	1.588	1.640	1.693	1.745	1.797	1.849	1.901	1.954	2.006	2.058	30
40	2.058	2.111	2.163	2.216	2.268	2.321	2.374	2.426	2.479	2.532	2.585	40
50	2.585	2.638	2.691	2.743	2.796	2.849	2.902	2.956	3.009	3.062	3.115	50
60	3.115	3.168	3.221	3.275	3.328	3.381	3.435	3.488	3.542	3.595	3.649	60
70	3.649	3.702	3.756	3.809	3.863	3.917	3.971	4.024	4.078	4.132	4.186	70
80	4.186	4.239	4.293	4.347	4.401	4.455	4.509	4.563	4.617	4.671	4.725	80
90	4.725	4.780	4.834	4.888	4.942	4.996	5.050	5.105	5.159	5.213	5.268	90
100	5.268	5.322	5.376	5.431	5.485	5.540	5.594	5.649	5.703	5.758	5.812	100
110	5.812	5.867	5.921	5.976	6.031	6.085	6.140	6.195	6.249	6.304	6.359	110
120	6.359	6.414	6.468	6.523	6.578	6.633	6.688	6.742	6.797	6.852	6.907	120
130	6.907	6.962	7.017	7.072	7.127	7.182	7.237	7.292	7.347	7.402	7.457	130
140	7.457	7.512	7.567	7.622	7.677	7.732	7.787	7.843	7.898	7.953	8.008	140
150	8.008	8.063	8.118	8.174	8.229	8.284	8.339	8.394	8.450	8.505	8.560	150
160	8.560	8.616	8.671	8.726	8.781	8.837	8.892	8.947	9.003	9.058	9.113	160
170	9.113	9.169	9.224	9.279	9.335	9.390	9.446	9.501	9.556	9.612	9.667	170
180	9.667	9.723	9.778	9.834	9.889	9.944	10.000	10.055	10.111	10.166	10.222	180
190	10.222	10.277	10.333	10.388	10.444	10.499	10.555	10.610	10.666	10.721	10.777	190
200	10.777	10.832	10.888	10.943	10.999	11.054	11.110	11.165	11.221	11.276	11.332	200
210	11.332	11.387	11.443	11.498	11.554	11.609	11.665	11.720	11.776	11.831	11.887	210
220	11.887	11.943	11.998	12.054	12.109	12.165	12.220	12.276	12.331	12.387	12.442	220
230	12.442	12.498	12.553	12.609	12.664	12.720	12.776	12.831	12.887	12.942	12.998	230
240	12.998	13.053	13.109	13.164	13.220	13.275	13.331	13.386	13.442	13.497	13.553	240
250	13.553	13.608	13.664	13.719	13.775	13.830	13.886	13.941	13.997	14.052	14.108	250
260	14.108	14.163	14.219	14.274	14.330	14.385	14.441	14.496	14.552	14.607	14.663	260
270	14.663	14.718	14.774	14.829	14.885	14.940	14.995	15.051	15.106	15.162	15.217	270
280	15.217	15.273	15.328	15.383	15.439	15.494	15.550	15.605	15.661	15.716	15.771	280
290	15.771	15.827	15.882	15.938	15.993	16.048	16.104	16.159	16.214	16.270	16.325	290
300	16.325	16.380	16.436	16.491	16.547	16.602	16.657	16.713	16.768	16.823	16.879	300
310	16.879	16.934	16.989	17.044	17.100	17.155	17.210	17.266	17.321	17.376	17.432	310
320	17.432	17.487	17.542	17.597	17.653	17.708	17.763	17.818	17.874	17.929	17.984	320
330	17.984	18.039	18.095	18.150	18.205	18.260	18.316	18.371	18.426	18.481	18.537	330
340	18.537	18.592	18.647	18.702	18.757	18.813	18.868	18.923	18.978	19.033	19.089	340
350	19.089	19.144	19.199	19.254	19.309	19.364	19.420	19.475	19.530	19.585	19.640	350
360	19.640	19.695	19.751	19.806	19.861	19.916	19.971	20.026	20.081	20.137	20.192	360
370	20.192	20.247	20.302	20.357	20.412	20.467	20.523	20.578	20.633	20.688	20.743	370
380	20.743	20.798	20.853	20.909	20.964	21.019	21.074	21.129	21.184	21.239	21.295	380
390	21.295	21.350	21.405	21.460	21.515	21.570	21.625	21.680	21.736	21.791	21.846	390
400	21.846	21.901	21.956	22.011	22.066	22.122	22.177	22.232	22.287	22.342	22.397	400
410	22.397	22.453	22.508	22.563	22.618	22.673	22.728	22.784	22.839	22.894	22.949	410
420	22.949	23.004	23.060	23.115	23.170	23.225	23.280	23.336	23.391	23.446	23.501	420
430	23.501	23.556	23.612	23.667	23.722	23.777	23.833	23.888	23.943	23.999	24.054	430
440	24.054	24.109	24.164	24.220	24.275	24.330	24.386	24.441	24.496	24.552	24.607	440
450	24.607	24.662	24.718	24.773	24.829	24.884	24.939	24.995	25.050	25.106	25.161	450
460	25.161	25.217	25.272	25.327	25.383	25.438	25.494	25.549	25.605	25.661	25.716	460
470	25.716	25.772	25.827	25.883	25.938	25.994	26.050	26.105	26.161	26.216	26.272	470
480	26.272	26.328	26.383	26.439	26.495	26.551	26.606	26.662	26.718	26.774	26.829	480
490	26.829	26.885	26.941	26.997	27.053	27.109	27.165	27.220	27.276	27.332	27.388	490
500	27.388	27.444	27.500	27.556	27.612	27.668	27.724	27.780	27.836	27.893	27.949	500
510	27.949	28.005	28.061	28.117	28.173	28.230	28.286	28.342	28.398	28.455	28.511	510
520	28.511	28.567	28.624	28.680	28.736	28.793	28.849	28.906	28.962	29.019	29.075	520
530	29.075	29.132	29.188	29.245	29.301	29.358	29.415	29.471	29.528	29.585	29.642	530
540	29.642	29.698	29.755	29.812	29.869	29.926	29.983	30.039	30.096	30.153	30.210	540
550	30.210	30.267	30.324	30.381	30.439	30.496	30.553	30.610	30.667	30.724	30.782	550
560	30.782	30.839	30.896	30.954	31.011	31.068	31.126	31.183	31.241	31.298	31.356	560
570	31.356	31.413	31.471	31.528	31.586	31.644	31.702	31.759	31.817	31.875	31.933	570
580	31.933	31.991	32.048	32.106	32.164	32.222	32.280	32.338	32.396	32.455	32.513	580
590	32.513	32.571	32.629	32.687	32.746	32.804	32.862	32.921	32.979	33.038	33.096	590
600	33.096	33.155	33.213	33.272	33.330	33.389	33.448	33.506	33.565	33.624	33.[illegible]3	600
610	33.683	33.742	33.800	33.859	33.918	33.977	34.036	34.095	34.155	34.214	3[illegible].273	610
620	34.273	34.332	34.391	34.451	34.510	34.569	34.629	34.688	34.748	34.807	34.867	620
630	34.867	34.926	34.986	35.046	35.105	35.165	35.225	35.285	35.344	35.404	35.464	630
640	35.464	35.524	35.584	35.644	35.704	35.764	35.825	35.885	35.945	36.005	36.066	640
650	36.066	36.126	36.186	36.247	36.307	36.368	36.428	36.489	36.549	36.610	36.671	650
660	36.671	36.732	36.792	36.853	36.914	36.975	37.036	37.097	37.158	37.219	37.280	660
670	37.280	37.341	37.402	37.463	37.525	37.586	37.647	37.709	37.770	37.831	37.893	670
680	37.893	37.954	38.016	38.078	38.139	38.201	38.262	38.324	38.386	38.448	38.510	680
690	38.510	38.572	38.633	38.695	38.757	38.819	38.882	38.944	39.006	39.068	39.130	690
700	39.130	39.192	39.255	39.317	39.379	39.442	39.504	39.567	39.629	39.692	39.754	700
710	39.754	39.817	39.880	39.942	40.005	40.068	40.131	40.193	40.256	40.319	40.382	710
720	40.382	40.445	40.508	40.571	40.634	40.697	40.760	40.823	40.886	40.950	41.013	720
730	41.013	41.076	41.139	41.203	41.266	41.329	41.393	41.456	41.520	41.583	41.647	730
740	41.647	41.710	41.774	41.837	41.901	41.965	42.028	42.092	42.156	42.219	42.283	740
750	42.283	42.347	42.411	42.475	42.538	42.602	42.666	42.730	42.794	42.858	42.922	750
760	42.922											760
DEG C	0	1	2	3	4	5	6	7	8	9	10	DEG C

Source: *Temperature Measurement Handbook*, 1983, p. T–15. Reprinted with permission from Omega Engineering, Inc. Standford, CT.

the time constant for any thermocouple. For example, let us suppose a thermocouple has a mass of 0.002 g, a specific heat of 0.08 cal/g°C, a heat transfer coefficient of 0.01 cal/cm-s°C, and an area of 0.02 cm^2. The time constant, based on this information, is 0.8 s, as shown in the following calculation:

$$t_c = \frac{(0.002\ \text{g})(0.08\ \text{cal/g°C})}{(0.01\ \text{cal/cm-s°C})(0.02\ \text{cm}^2)} = 0.8\ \text{s} \tag{7.5}$$

When temperature changes quickly, the temperature of a thermocouple at any time is calculated by using the following equation:

$$(T - T_2) = (T_1 - T_2)e^{-t/t_c} \tag{7.6}$$

where T = thermocouple temperature
T_1 = starting temperature
T_2 = final temperature
t_c = thermocouple time constant
t = time from start

For example, suppose a thermocouple at 20°C is placed into a 100°C water bath. With the time constant calculated previously, the temperature after 1 s is about 77°C:

$$(T - 100°\text{C}) = (20°\text{C} - 100°\text{C})e^{-1/0.8} \\ = 77°\text{C} \tag{7.7}$$

We can see from these calculations that the thermocouple does not react instantly to a temperature change. All sensors take some time to respond to a change in the measurand.

Thermistor

A *thermistor* is a thermally sensitive resistor, usually having a negative temperature coefficient. As temperature increases, the thermistor's resistance decreases, and vice versa. Thermistors are made out of oxides of nickel, manganese, cobalt, copper, and other metals. Figure 7.12 shows the almost bewildering variety of packages in which thermistors come.

A typical thermistor's temperature-resistance curve is shown in Figure 7.13. As we can see, there

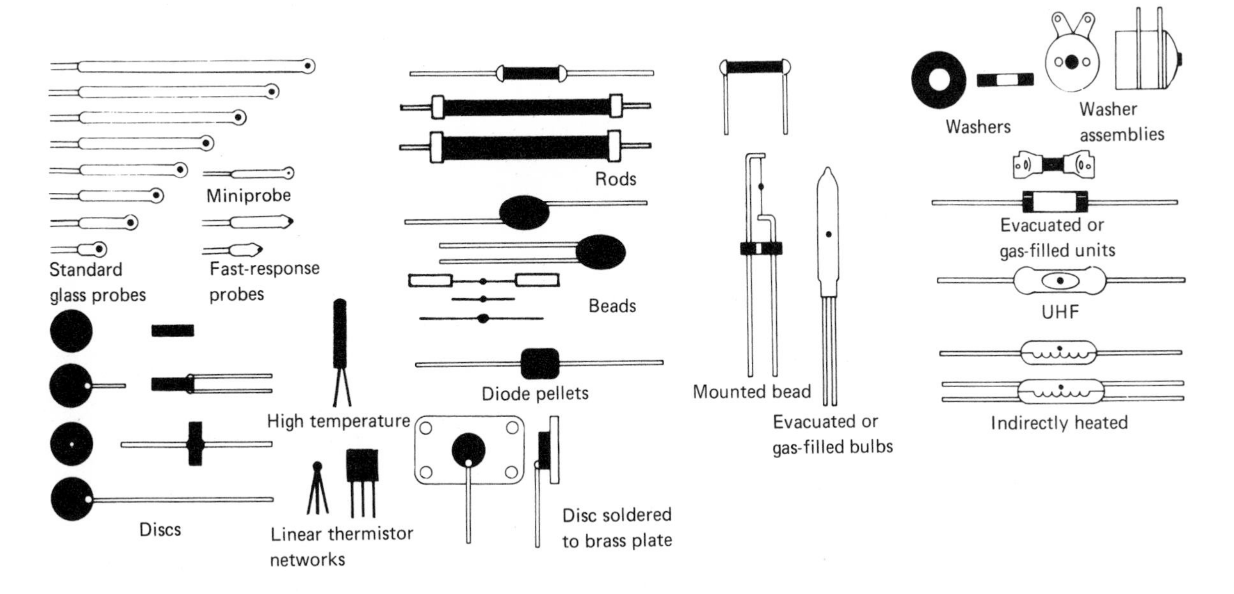

Figure 7.12 Thermistors

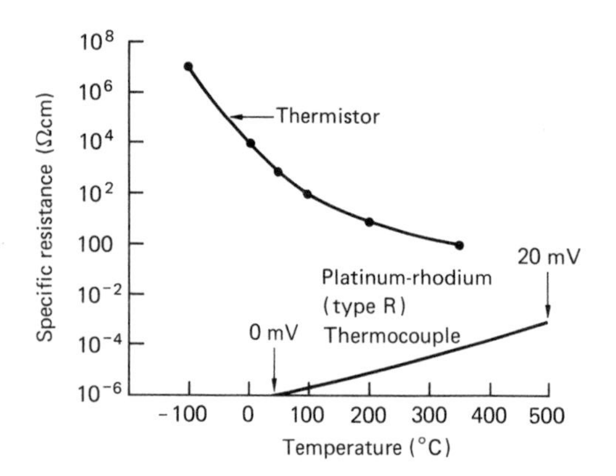

Figure 7.13 Thermistor's Temperature-Resistance Curve and Curve for Thermocouple

are several differences between this curve and that for the thermocouple, which is also shown in Figure 7.13. First, the thermocouple curve is relatively linear. The thermistor's resistance varies exponentially. As a result, the thermistor's thermal sensitivity is very high, as much as 5% resistance change per degree Celsius. Thus, the thermistor is the most sensitive temperature sensor in common use. The thermistor is therefore reasonably linear over a relatively narrow range of temperatures.

Thermistors are often used in conjunction with a parallel resistor to improve linearity, as shown in Figure 7.14A. Resistor R_s (a shunt resistance) is chosen to equal the resistance of the thermistor at the median temperature expected to be measured. Figure 7.14D shows the change in the response curve when a resistor is added.

There are many ways to use thermistors. Figure 7.14 shows several circuits using thermistors. Note the schematic representation of the thermistor as a resistor with a T inside a circle. The circuit shown in Figure 7.14A represents a potentiometric circuit. As temperature increases, the resistance of the thermistor decreases, increasing the current through the ammeter A. Figure 7.14B depicts a more sensitive bridge circuit, where changes in temperature unbalance the bridge, causing a meter indication. Differential temperature is measured with the circuit shown in Figure 7.14C. Bridge unbalance comes only when there is a difference in temperature between the two sensors. Notice that none of these circuits use an amplifier. Because of the large voltage output produced by a typical thermistor bridge, amplification is not normally necessary. A 4000 Ω thermistor in a bridge network at 25°C produces about 18 mV/°C.

The thermistor is gaining popularity as a temperature sensor. It is the most sensitive of the electrical temperature sensors. Some thermistors actually double their resistance with a temperature change of 1°C. Thermistors are relatively inexpensive, react quickly to temperature changes, and require only very simple circuitry. Resistances range from 0.5 Ω to 80 MΩ. Thermistors, on the other hand, are not without disadvantages. They are extremely nonlinear and fragile, and they have a limited temperature range.

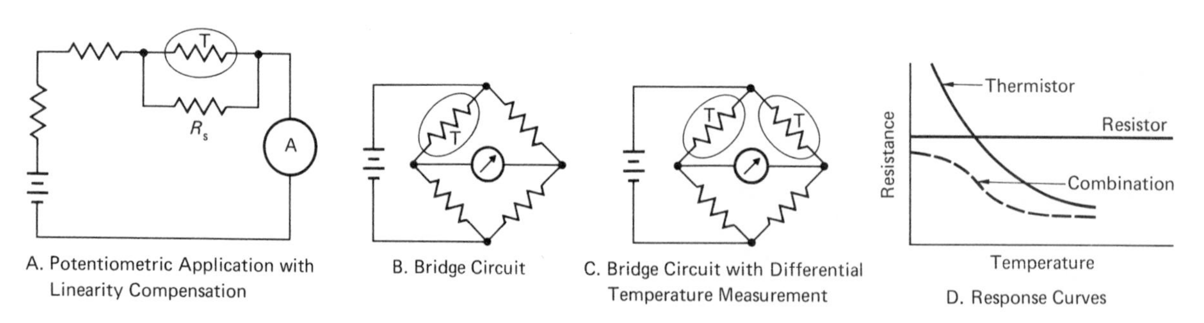

Figure 7.14 Thermistors in Potentiometric and Bridge Circuits

Thermistors find applications in more than temperature measurement and control. They are also used as indicators of liquid level. If a thermistor is placed in a tank, its resistance changes, depending on whether or not the device was immersed in a liquid. If placed in series with a relay, a power supply, and a variable resistor, the thermistor serves as a variable time delay. When current first flows through the system, the thermistor's resistance is high. As current starts to flow, the thermistor heats. Its resistance will then decrease, allowing more current to flow. Eventually, enough current flows to energize the relay.

Resistance Temperature Detector (RTD)

All metals change resistance when subjected to a temperature change. Pure metals, such as platinum, nickel, tungsten, and copper, have positive temperature coefficients. Thus, for pure metals, temperature and resistance are directly proportional. As temperature increases, a pure metal's resistance will increase. This result is the idea behind the *resistance temperature detector* (*RTD*).

RTDs come in many packages; the most popular are shown in Figure 7.15. The helical RTD (Figure 7.15A) consists of a platinum wire wound in a tight spiral (helix) threaded through a ceramic cylinder. A newer construction technique is found in the metal film RTD (Figure 7.15B). Platinum is usually deposited, or screened, on a small flat ceramic substrate. It is then etched with a laser trimming system and sealed. The film RTD offers a substantial reduction in cost over other forms of RTDs and reacts more quickly to temperature changes.

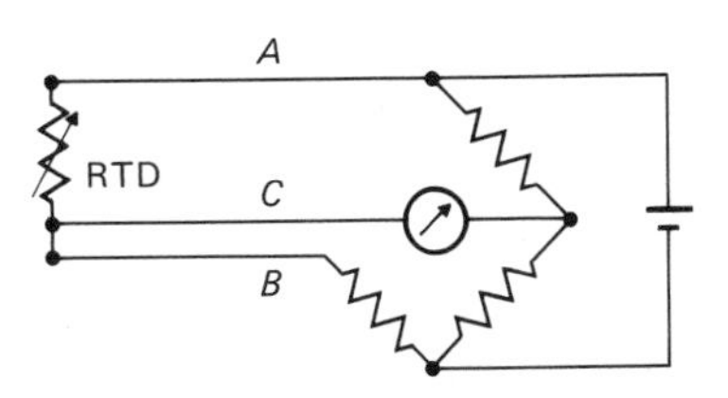

Figure 7.16 Three-Wire Bridge Configuration

Common resistance values for RTDs range (for platinum) from 10 Ω to several thousand ohms. The most common value of resistance is 100 Ω. Because of their low resistance, long lead wires can produce inaccurate temperature measurements. This problem can be solved by using a three-wire bridge configuration, as shown in Figure 7.16. If wires *A* and *B* are identical in resistance, their resistive effects cancel out, because each is in an opposite leg of the bridge. The third wire (*C*) carries no current; it is only a sense lead.

Of the temperature sensors we have presented, the RTD is the most accurate and stable. Furthermore, as can be seen from the curves in Figure 7.17,

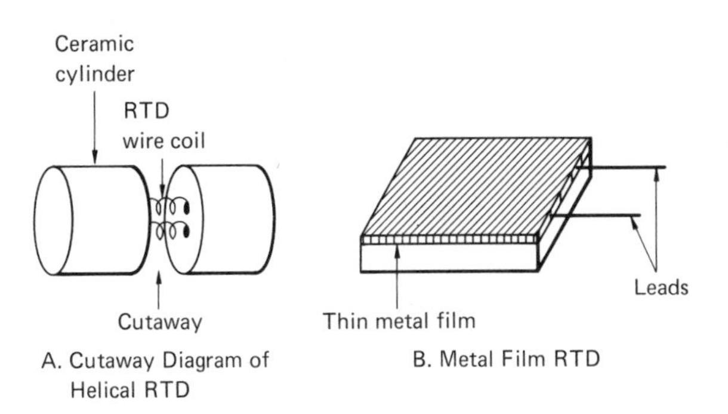

Figure 7.15 Two Popular RTDs

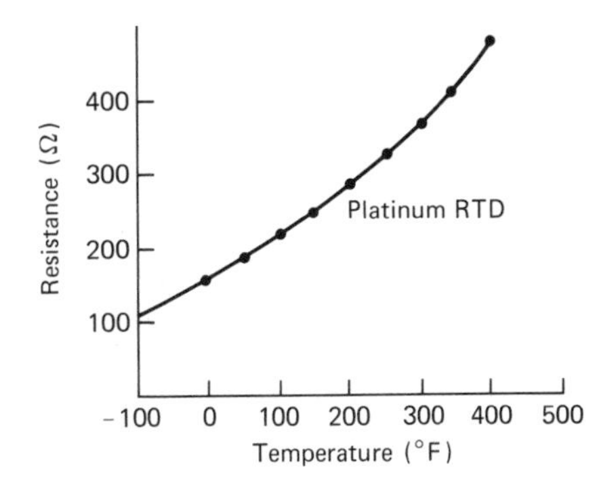

Figure 7.17 Temperature-Resistance Curve for Platinum RTD

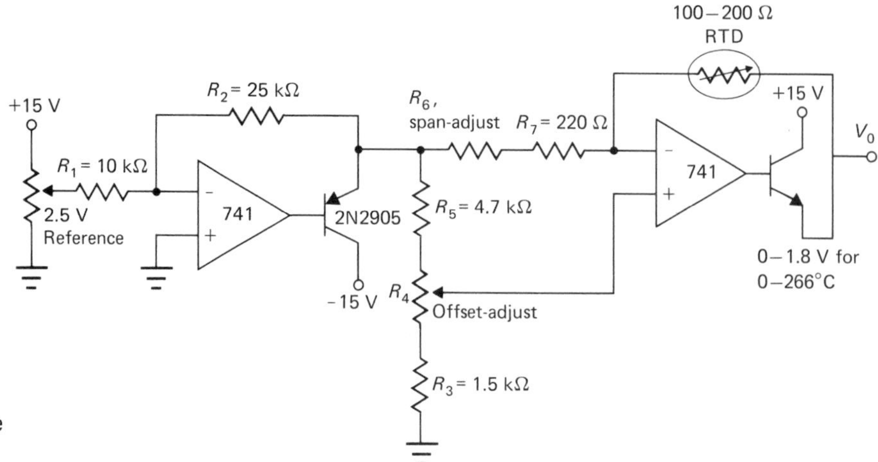

Figure 7.18 Low-Cost Temperature Sensor Using RTD with Fixed Reference and Op Amp

it is very linear and covers a high temperature range. Disadvantages include slowness of response, small resistance changes (usually requiring amplification), and high cost.

A circuit using an RTD is shown in Figure 7.18. It measures temperature between 0° and 266°C, the output ranging between 0 and 1.8 V. The 2.5 V reference is amplified to 6.25 V by the first op amp. The span-adjust sets the output to 1.8 V at 266°C, while the offset is adjusted to 0 V at 0°C.

Semiconductor Temperature Sensors

In the semiconductor area, evenly doped crystals of germanium can be used to sense temperature near absolute zero. The temperature-resistance response curve of the device resembles that of the thermistor. It has a negative temperature coefficient and is very nonlinear.

Silicon crystals are also employed as temperature sensors. They are usually shaped in the form of discs or wafers for measuring surface temperature. Their useful range extends from −67° to 275°F. Unlike germanium crystals and thermistors, silicon crystals have a positive linear temperature coefficient in this range. Below this range, the temperature-resistance response curve becomes negative and very nonlinear.

Recall that a reverse-biased PN junction conducts a small amount of leakage, or minority carrier, current flow. The amount of current flow depends on temperature. As temperature increases, leakage current increases exponentially. This effect is used to measure temperature with transistors and reverse-biased diodes. Normally, germanium semiconductors are used, because their leakage is much greater than that of silicon.

One of the most recent innovations in thermometry is the integrated circuit temperature sensor. The output of the IC (either voltage or current)

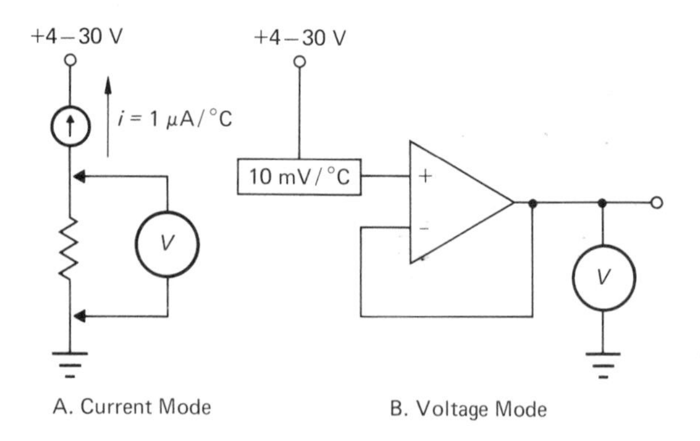

Figure 7.19 IC Temperature Sensor

is directly proportional to temperature. Although limited in temperature range (below +300°F), they produce a very linear output over the operating range. Linearity exceeds that of the RTD. A typical sensitivity value is 1 μA/°C or 10 mV/°C. Figures 7.19A and 7.19B show the voltage and current modes of the AD590 manufactured by Analog Devices.

Generally speaking, semiconductor temperature sensors produce good linearity, are small and low in cost, and produce a high impedance output.

Tables 7.3 and 7.4 list the temperature ranges of the thermometers we have discussed in this section and give a summary of information about them.

Humidity

Many industrial processes depend on accurate assessment and control of humidity. *Humidity* is defined as the amount of water vapor in the air. How much water vapor air can hold depends on several factors. The most important factor is the air temperature. Warm air can hold more water vapor than cold air can. The most common measure

Table 7.3 *Temperature Ranges of Mechanical and Electrical Temperature Transducers*

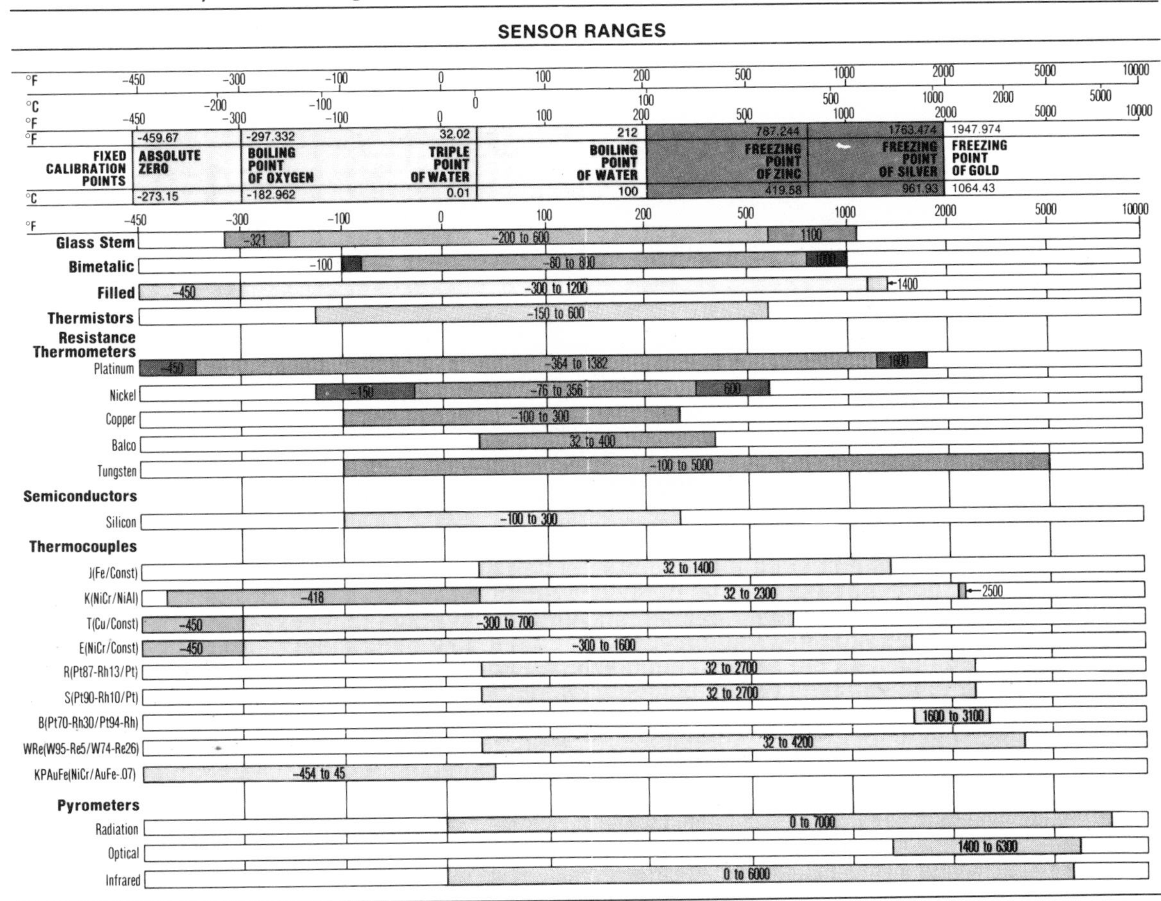

Source: Reprinted with permission from © 1981 *Instrument and Control Systems*, Chilton Company.

Table 7.4 *Summary of Temperature Sensors*

Type/Name	*Temperature Range*	*Linearity*	*Advantages*	*Disadvantages*
Mechanical				
Glass stem	−100°–500°C	Good	Inexpensive Linear Accurate (0.05°C)	Local measurement only Fragile Hard to read Time lag
Bimetallic	−100°–600°C	Good	Inexpensive Wide range Rugged Easy to read and install	Local measurement only
Filled	−200°C–650°C	Fair	Reacts quickly Accurate (0.5%) Remote measurement (300 ft) Can be used with chart recorders	Often needs temperature compensation
Electrical				
Thermocouple	−273°–2000°C	Good	Simple, low cost Rugged Wide range Self-powered	Low sensitivity Reference needed Poor stability
RTD	−200°–800°C	Good	Very stable Very accurate Linear Wide range	Slow response Low sensitivity Expensive Self-heating Limited range Subject to thermal runaway
Semiconductor	−50°–150°C	Poor	Inexpensive Small Output impedance high	
IC	−50°–150°C	Excellent	Excellent linearity Inexpensive Very sensitive	Slow response
Thermistor	−100°–300°C	Very poor	Small size Low cost High sensitivity Fast response	Very nonlinear Poor stability at high temperatures Limited range

of water vapor in air is called relative humidity. *Relative humidity* is the ratio of the amount of moisture air holds at a certain temperature compared with how much it could hold.

In this section, we discuss humidity sensors in two classifications: (1) direct methods through hygrometers; and (2) indirect methods by comparing changes in temperature.

Psychrometers

The *psychrometer*, shown in Figure 7.20, is a common means of measuring relative humidity. Note that its construction is based around two bulbs, one wet and one dry. Air must be passed over the wet bulb, causing the water to evaporate.

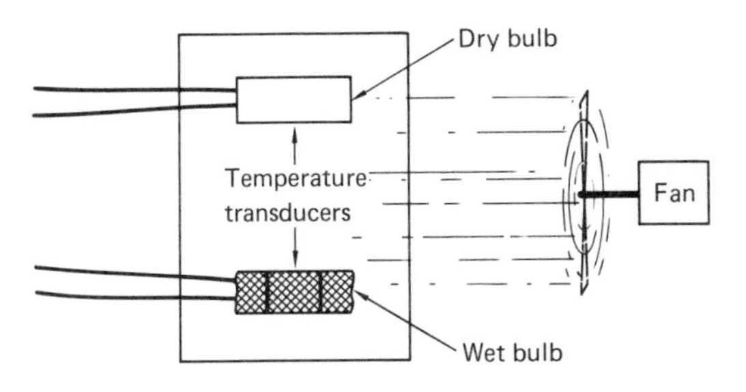

Figure 7.20 Psychrometer Humidity Sensor

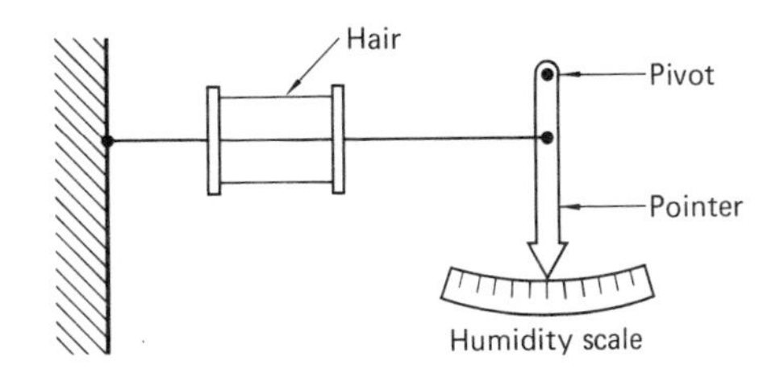

Figure 7.21 Hair Hygrometer

Both bulbs contain temperature sensors. If the air temperature remains constant, the dry bulb sensor remains at a constant temperature. However, the wet bulb's temperature varies with the relative humidity. More evaporation takes place as relative humidity decreases. Thus, as humidity drops, the wet bulb becomes cooler. The difference between the temperature of the two bulbs, then, reflects the relative humidity.

Conversion tables are usually necessary to estimate relative humidity with the psychrometer. However, relative humidity conversion may be accomplished automatically with appropriate hardware or software.

Hygrometers

A *hygrometer* is a device used to measure humidity. These devices contain a material whose properties change when moisture is adsorbed. As we have seen, the psychrometer indirectly measures humidity by comparing the temperatures of two bulbs. The hygrometer, on the other hand, measures humidity directly. In this section, we will describe several common hygrometers in use today.

Hair Hygrometer

The simplest and oldest hygrometer is made from human hair or an animal membrane. A simple *hair hygrometer* is shown in Figure 7.21. Human hair lengthens about 3% from 0% to 100% humidity. This change in length is detected by the pointer, giving a readout in relative humidity.

The hair hygrometer is accurate to within 3%. It is only used for measuring relative humidity between 15% and 90% over a temperature range from 1° to 40°C. Although the hair hygrometer is strictly mechanical, it can be converted to an electrical sensor by attaching the hair to the core of, say, an LVDT or any appropriate position-to-voltage transducer.

Impedance Hygrometer

Several hygrometers use a change in impedance to detect levels of humidity. One, the *resistance hygrometer*, varies its resistance (and, hence, impedance) as humidity changes. The resistance hygrometer, as shown in Figure 7.22, is composed of two electrodes separated by a thin layer of lithium chloride. The lithium chloride film is *hygroscopic*, which means that it adsorbs moisture from the air. As relative humidity increases, the device's resistance decreases. The change is then sensed by potentiometric or bridge methods.

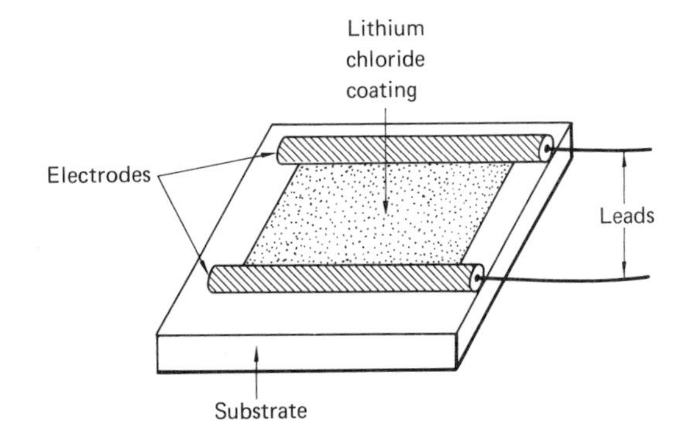

Figure 7.22 Resistance Hygrometer

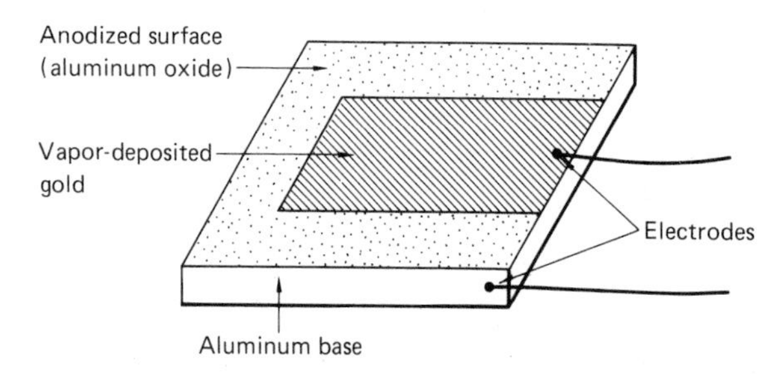

Figure 7.23 Impedance Hygrometer

Another hygrometer using an impedance change is made out of aluminum oxide. This device, sometimes called a *resistance-capacitance hygrometer*, is shown in Figure 7.23. The surface of the aluminum base is anodized to form a layer of aluminum oxide. Gold is then vapor-deposited on top of the oxide coating. Note that one electrode is attached to the gold film and another is attached to the aluminum base. Since the gold is very thin, water vapor diffuses through to the oxide layer. An increase in relative humidity causes the impedance (resistance and capacitance) of the oxide coating to decrease. The change in impedance is detected by an impedance bridge calibrated to read percent relative humidity.

The aluminum oxide element operates over the range from 0% to 100% relative humidity. The element is very sensitive, small in size, and very linear. Condensation of moisture on the surface of the element does not affect it.

Sorption Hygrometer

The *sorption hygrometer* uses the principle of an oscillating crystal to measure humidity. The moisture increases the mass of the crystal and decreases the frequency of oscillation. Frequencies used are normally around 9 MHz. Commercial units use two crystals: One is the sensor, and the other is exposed to a dry gas and acts as a reference.

Capacitive Hygrometer

Capacitive hygrometers work on the principle that the dielectric constant of a capacitor changes in the presence of water. Recall that changes in the dielectric constant vary the capacitance of a capacitor. Measurements of capacitance are made through impedance bridges or oscillator frequency changes.

Microwave Absorption

The absorption of microwave radiation by water vapor is also used in humidity measurements. Microwave radiation is part of the electromagnetic spectrum and has values between 1 and 100 GHz. Water absorbs thousands of times more microwave radiation than a dry gas does. More water vapor in the air increases microwave energy absorption and decreases energy transmitted to a sensor. The

Table 7.5 *Summary of Humidity Sensors*

Name	*Operating Principle*	*Relative Humidity Range*	*Temperature Range*	*Accuracy*
Psychrometer	Evaporation rate of water	2%–98%	32°–212°F	±5%
Hair hygrometer	Hair changes length with humidity	15%–90%	0°–160°F	±5%
Resistance hygrometer	Lithium chloride changes resistance with humidity	1.5%–99%	−40°–160°F	±1.5%
Resistance-capacitance hygrometer	Aluminum oxide changes impedance with humidity	0%–100%	0°–300°F	±2%
Capacitive hygrometer	Dielectric constant changes with humidity	0%–100%	−200°–200°F	±0.5%

level of radiation transmitted is inversely proportional to the relative humidity.

A summary of the humidity sensors discussed in this section is presented in Table 7.5.

Light

In 1887, Heinrich Hertz discovered the photoelectric effect. Albert Einstein received the Nobel Prize in 1921 for his explanation of this effect. Since that time, optoelectronic devices and systems have become an integral part of industrial control. Popularity of optoelectronic devices stems from the fact that they are noncontact sensors. No parts wear out owing to metal fatigue from continuous use. As a technician, you should be familiar with the various devices used to sense light in industry.

Our discussion of optoelectronic sensors is broken down into three areas: photoconductive devices, photovoltaic devices, and photoemissive devices.

Photoconductive Transducers

A *photoconductive transducer* is a transducer that converts a change in light intensity to a change in conductivity. The method used to produce this conductivity change differs with the photoconductive device. We will study two basic types of photoconductive devices: (1) bulk photoconductors, such as the photoresistor, and (2) PN junction photoconductors, such as the photodiode, phototransistor, and photo Darlington.

Photoresistors

Several years before Hertz discovered the photoelectric effect, Willoughby Smith noticed that the resistance of a block of selenium decreased when exposed to light. This same effect is used in the *photoresistor*. The photoresistor, shown in Figure 7.24, is made of either cadmium sulfide (CdS) or cadmium selenide (CdSe). These substances are vapor-deposited on a glass or ceramic substrate and then hermetically sealed in plastic or glass. When light strikes the photoresistive material, it liberates electrons. These electrons are then available as current flow, and the resistance decreases. More light falling on the photoresistor causes resistance to decrease further.

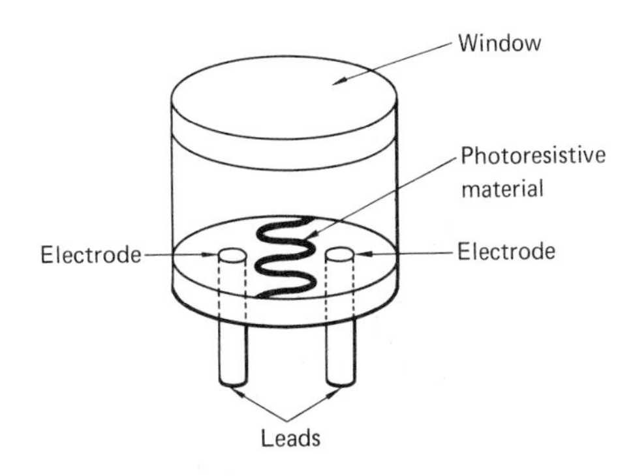

Figure 7.24 Cutaway View of Photoresistor

The spectral response (how the device reacts to different light wavelengths) depends on the type of material used. Notice in Figure 7.25 that the spectral response of the CdS photoresistor closely matches the response of the human eye. The CdSe spectral response is shifted toward the infrared.

The specific material used determines the ratio of dark-to-light resistance, which can range from 100:1 to 10,000:1. The photoresistor, then, is very sensitive to changes in light. It is generally easy to use and inexpensive. Disadvantages include

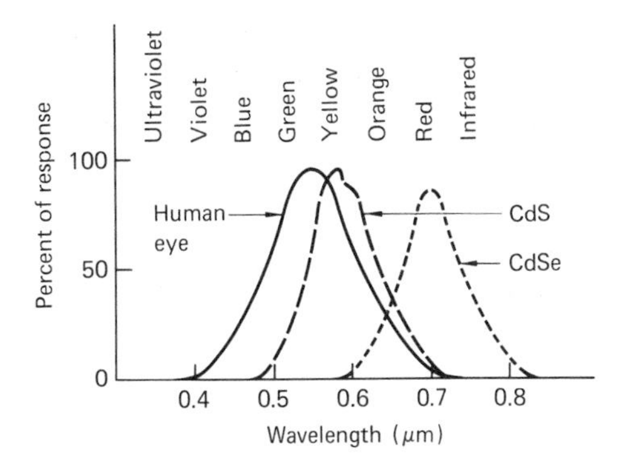

Figure 7.25 Spectral Response of CdS and CdSe Photoresistors Compared with Response of Human Eye

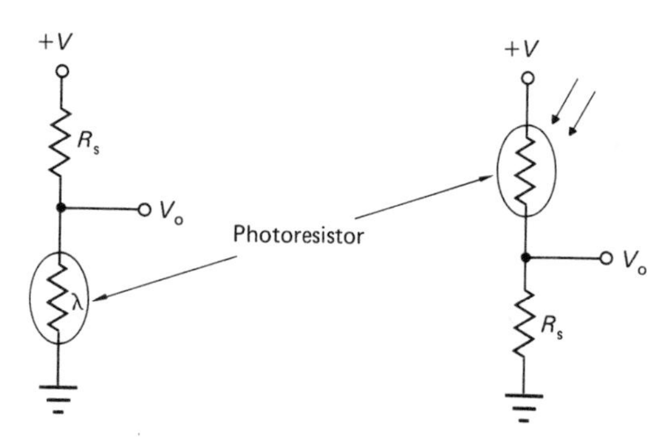

Figure 7.26 Photoresistors Used in Voltage Divider Circuits

narrow spectral response, poor temperature stability, and light history effect. The light history (hysteresis) effect is particularly annoying, since it causes the resistance to depend on past light intensity. In addition, photoresistors are slow to react to light intensity changes. The CdS takes about 100 ms to respond, while the faster CdSe takes about 10 ms.

Because the photoresistor is so sensitive to light changes, it does not normally need amplification. A simple potentiometric circuit, like the ones shown in Figure 7.26, is sufficient. Note the different schematic symbols in common use. The Greek letter λ (lambda), as indicated in Figure 7.26, is often used to symbolize light.

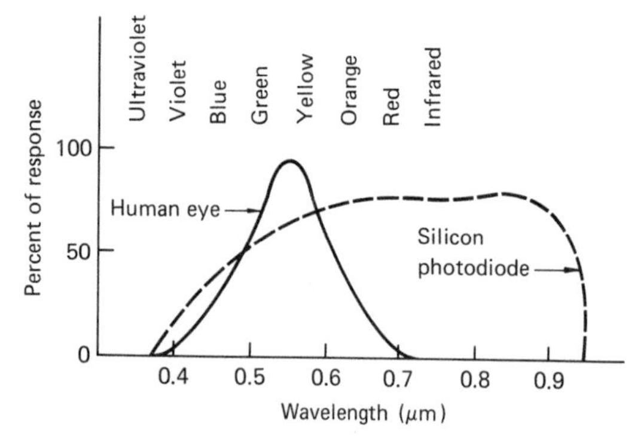

Figure 7.27 Spectral Response of Silicon Photodiode Compared with Response of Human Eye

Photodiodes

Light levels are also detected by PN junctions, like those found in photodiodes and phototransistors. *Photodiodes* are simply PN diodes with their junctions exposed to light.

Early experimenters with PN junctions discovered that minority carrier current flow was influenced by temperature and light levels. Increasing light falling on the junction creates more electron-hole pairs, which increase minority carrier current flow. Since we are dealing solely with minority carrier current, the photodiode is operated with reverse bias.

Note the spectral response curve for the photodiode shown in Figure 7.27. This curve is much wider than that of the photoresistor, covering the visible spectrum as well as the infrared.

Photodiode construction is shown in Figure 7.28. The simple PN device (Figure 7.28A) is nothing more than a PN junction whose structure is optimized for light reception. The PIN device (Figure 7.28B) adds a layer of undoped semiconductor material (I, which stands for "intrinsic"). The layer of undoped material increases the size of the depletion region. Widening the depletion region effectively decreases junction capacitance. This decrease makes the PIN diode respond much faster to light-level changes than the PN photodiode does. PIN photodiode dark currents are much smaller, also.

Photodiodes produce relatively small currents even under fully illuminated conditions. Some manufacturers add an amplifier in the same package to increase sensitivity. Packages for photodiodes without amplifiers are shown in Figure 7.29,

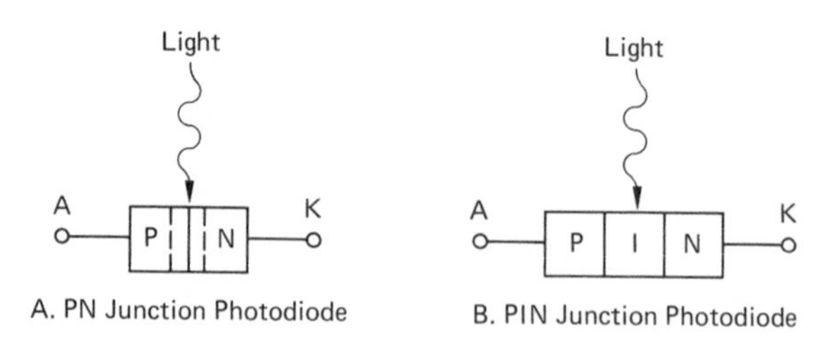

Figure 7.28 Two Types of Photodiodes

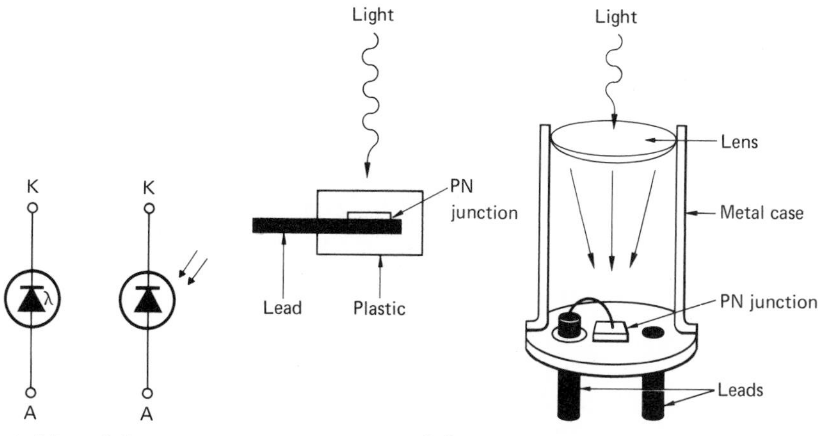

Figure 7.29 Photodiodes

A. Schematic Symbols

B. Packaged Photodiodes

along with common schematic symbols for photodiodes.

Detecting the change in current can be accomplished in several ways. Figure 7.30A shows the photocurrent developing a voltage across an external load R_L. Normally, this method is not sensitive enough to develop small voltages. For better development of small voltages, a differential amplifier may be used, or the Norton amplifier, shown in Figure 7.30B, may be used. In the Norton amplifier, as light level increases, diode D_1 increases its conduction. The increase in current is mirrored in the current flowing through the feedback resistor R_F. Hence, more voltage is produced at the output.

Photodiodes, with their extremely fast response, are ideal for laser detection. They also find applications in ultrahigh-speed demodulation, switching, and decoding.

Phototransistors

As we might expect, if diodes can be made photosensitive, transistors can, too. The *phototransistor* is electrically similar to a small-signal silicon transistor. Structurally, the only difference is that in the phototransistor, the collector-base junction is larger and is exposed to light.

A diagram of the structure of the phototransistor is shown in Figure 7.31. The base material is thin so that impinging light can strike the

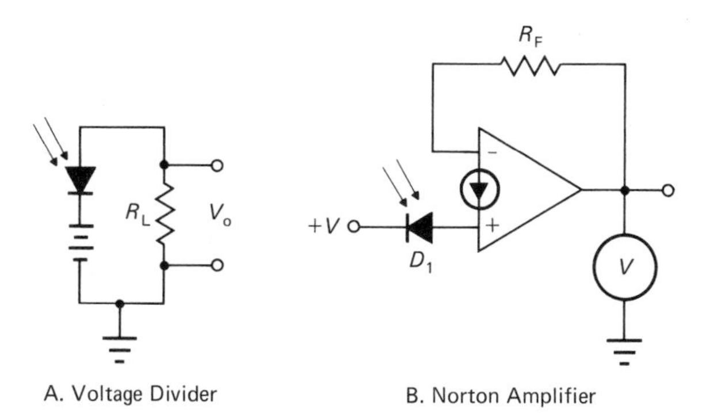

A. Voltage Divider

B. Norton Amplifier

Figure 7.30 Applications of Photodiode

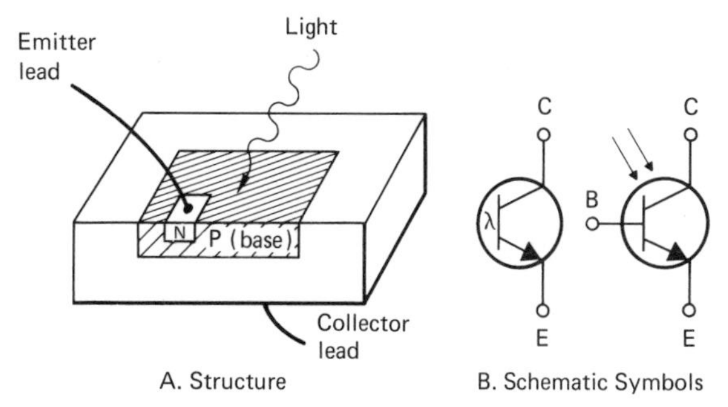

A. Structure

B. Schematic Symbols

Figure 7.31 Phototransistor

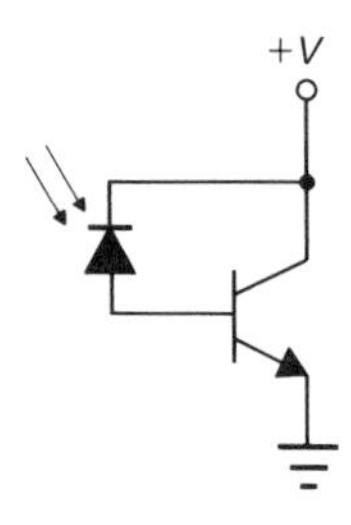

Figure 7.32 Equivalent Circuit for Phototransistor

collector-base (CB) junction. When light strikes the collector-base junction, electron-hole pairs are created, as in the diode. In this way, base current is created and amplified by the current gain of the transistor. The CB junction then acts like a current source, as indicated in Figure 7.32. Although the photodiode needs amplification, the phototransistor does not: The current generated in the phototransistor is automatically amplified in the collector.

The phototransistor is much more sensitive than the photodiode. However, it has higher junction capacitances, which give it poor frequency response compared with the photodiode.

Phototransistors can be either two-lead or three-lead devices (see Figure 7.31B). In the two-lead package, only the emitter and collector are connected. The base is not electrically available for biasing. The only drive for the two-lead device is the light falling on the collector-base junction. Two-lead packages remain the most common form of phototransistor. The three-lead form allows electrical connection to the base for purposes of biasing and decreasing device sensitivity.

Spectral response characteristics of the phototransistor are shown in Figure 7.33. Notice that the response characteristics of the phototransistor extend well into the infrared. For this reason, tungsten lamps are often used to illuminate phototransistors, since they emit radiation in this area.

Because its response is slower than that of the photodiode, and because it is nonlinear, the phototransistor is most often used for computer punch and paper tape readers. When the phototransistor is employed in a digital application, linearity considerations are not important.

Photo Darlington transistors have been developed by several manufacturers. The photo Darlington is characterized by higher sensitivity compared with the phototransistor, but it has slower response time.

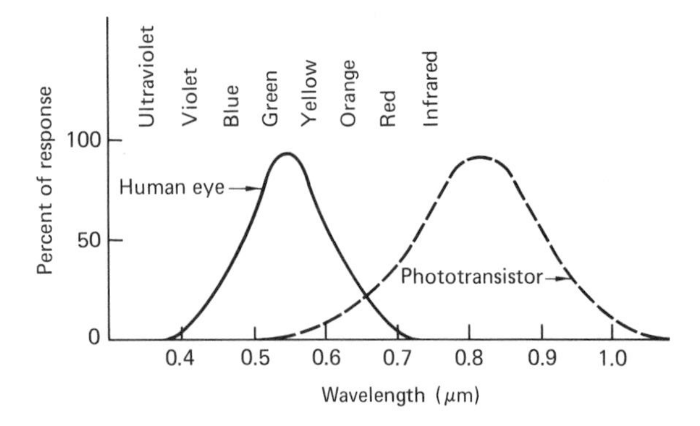

Figure 7.33 Spectral Response of Phototransistor Compared with Response of Human Eye

Photovoltaic Transducers

A *photovoltaic transducer* is a device that generates a voltage when irradiated by light. This phenomenon, called the *photovoltaic effect*, is exhibited by all semiconductors. Recall from semiconductor theory that the area around the PN junction becomes depleted of carriers. The P type region receives an excess of electrons, while the N type region receives an excess of holes. An equilibrium condition is finally reached where no more migration of carriers occurs. The PN junction then blocks any more movement of majority carriers.

The photovoltaic cell is so constructed that when light falls on it, electron-hole pairs are formed, as shown in Figure 7.34. These pairs become minority carriers. Unlike majority carriers, minority carriers can and do move across the junction. Thus, a negative potential is developed at the terminal attached to the N type material. If a load

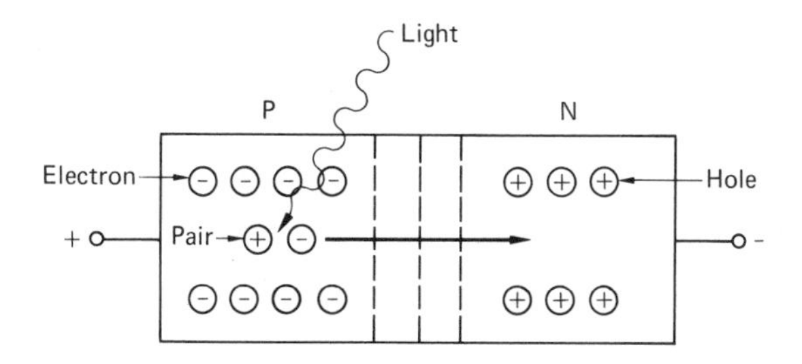

Figure 7.34 Construction of Photovoltaic Cell

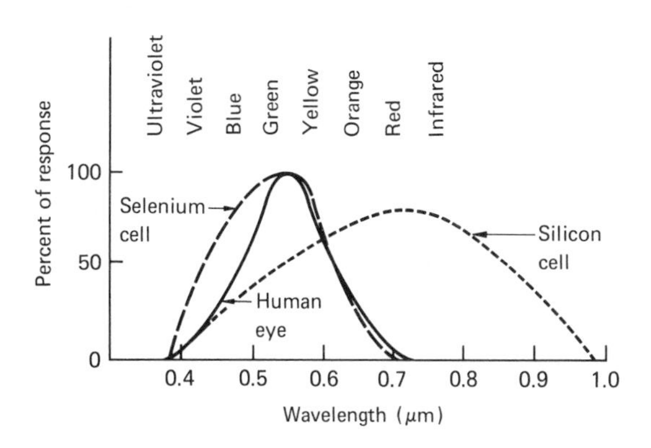

Figure 7.36 Spectral Response of Selenium and Silicon Photovoltaic Cell Compared with Response of Human Eye

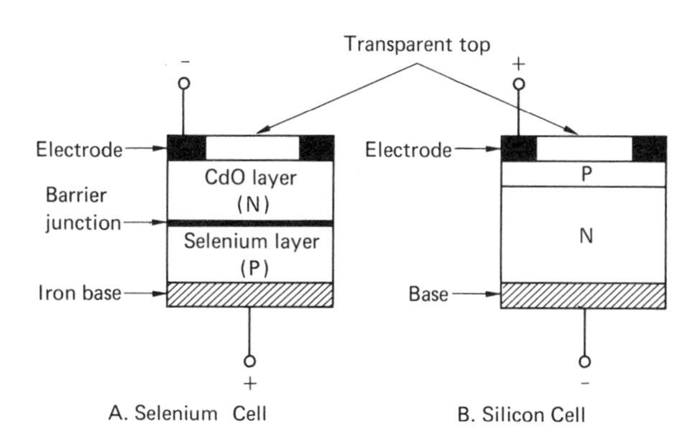

Figure 7.35 Photovoltaic Cells (Photocells)

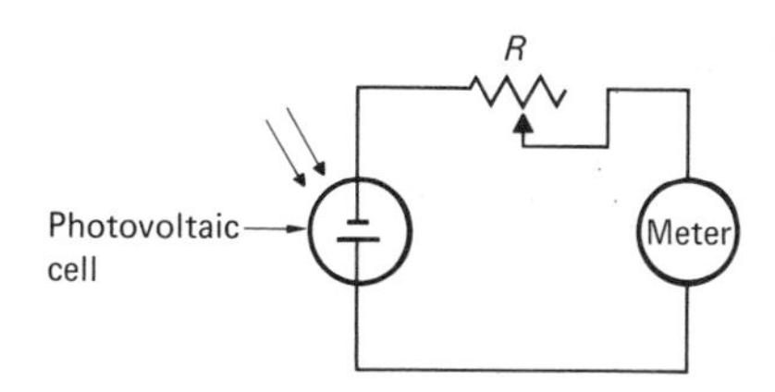

Figure 7.37 Light Meter Using Photovoltaic Cell

is attached to the two terminals, current flows; the current is proportional to the amount of carriers created by light.

Two basic types of photovoltaic devices are in general use today: silicon cells and selenium cells. Pictorial representations of both are shown in Figure 7.35. In the *selenium cell* (Figure 7.35A), a thin layer of selenium is deposited on a metal base. Cadmium is then deposited on top of the selenium. A PN junction forms between the cadmium and selenium. In the *silicon cell* (Figure 7.35B), a thin layer is doped with P type impurities, the remainder with N type.

The photovoltaic cell (photocell) has low internal resistance in order to prevent the voltage generated from being lost within the device under load conditions. This low internal resistance is achieved by heavy doping. A depletion region then forms around the PN junction. The operation of both types of cells is the same: Light falls on the PN junction, creating electron-hole pairs.

Spectral responses of these two types of cells differ, as shown in Figure 7.36. Under high light intensities, silicon cells produce as much as 0.5 V between their terminals and supply as much as 100 mA of current. Selenium cells, although possessing better spectral qualities, are much less efficient as power converters. For this reason, silicon cells are more useful as solar cells. The purpose of a solar cell is to convert radiant energy from the sun into electrical energy. Solar cells are often found in series, parallel, or series-parallel combinations to increase current and voltage output.

Selenium cells are often used in light meters; this application is shown in Figure 7.37. Such circuits are very simple, requiring only a sensitive

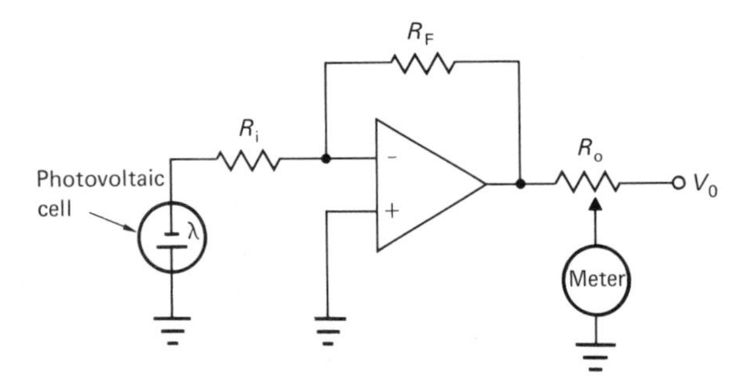

Figure 7.38 Op Amp Combined with Photovoltaic Cell

meter movement, a variable resistor, and the photovoltaic cell.

A more sensitive meter may be obtained by using an op amp, as shown in Figure 7.38. The op amp amplifies the small voltage change from the photovoltaic cell. Note the different schematic symbols used for photovoltaic cells in Figures 7.37 and 7.38. Both symbols are used interchangeably.

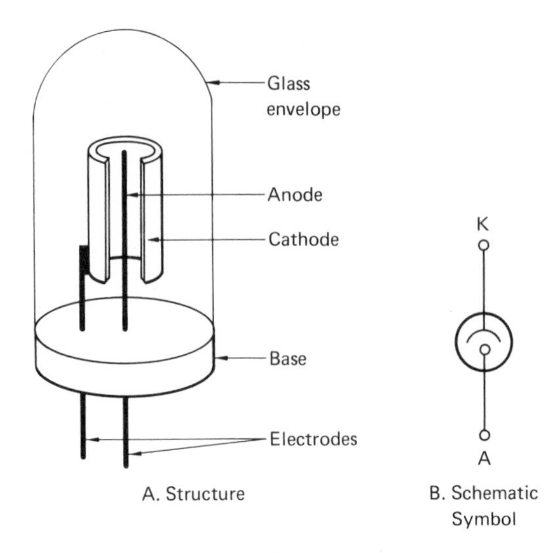

Figure 7.39 Phototube

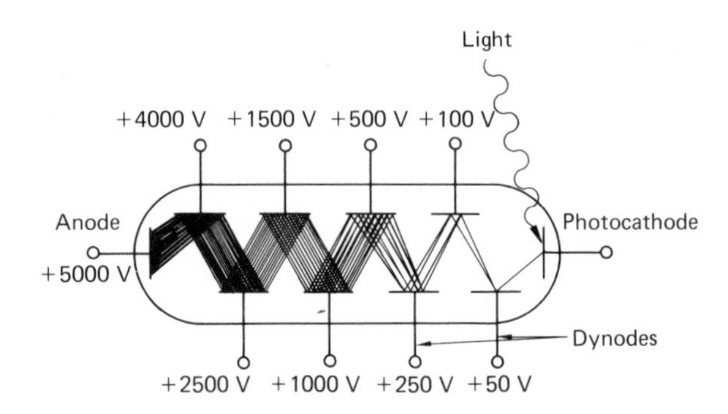

Figure 7.40 Structure and Operation of Photomultiplier Tube

Photoemissive Transducers

Photoemissive transducers emit electrons when struck by light. Most of the transducers using this principle are vacuum tubes. Although most vacuum tube devices have been replaced in industry, two are still being used as photosensors: the phototube and the photomultiplier.

Phototubes

The *phototube*, shown in Figure 7.39, has a peculiar construction. The rod-shaped device is the anode. The cathode, shaped as an open cylinder, is coated with a *photoemissive material.* When light strikes this material, it emits electrons. Increasing light levels increase the anode current.

Although the phototube has been replaced by the photodiode in many applications, it is still used. Its spectral response is very linear, and it has excellent frequency response characteristics. Because current flows are only a few microamperes, the phototube usually needs amplification.

Photomultipliers

The *photomultiplier* works on a principle similar to that of the phototube. It has a photosensitive cathode that emits electrons when light strikes it. In addition to one anode, the photomultiplier has between 10 and 15 electrodes, which are called *dynodes.* The structure of the photomultiplier is shown in Figure 7.40.

Starting with the dynode closest to the cathode, each dynode has an increasingly high positive potential applied to it. When a photon strikes the cathode, electrons are emitted. The emitted electrons are attracted to dynode 1, striking it.

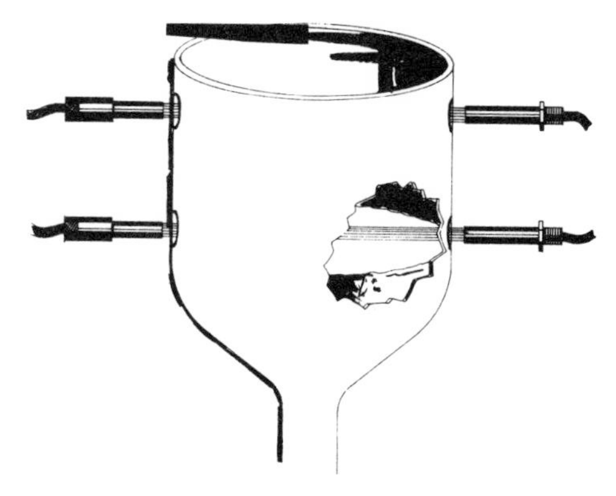

A. Keeping Hopper Fill Level between High and Low Limits

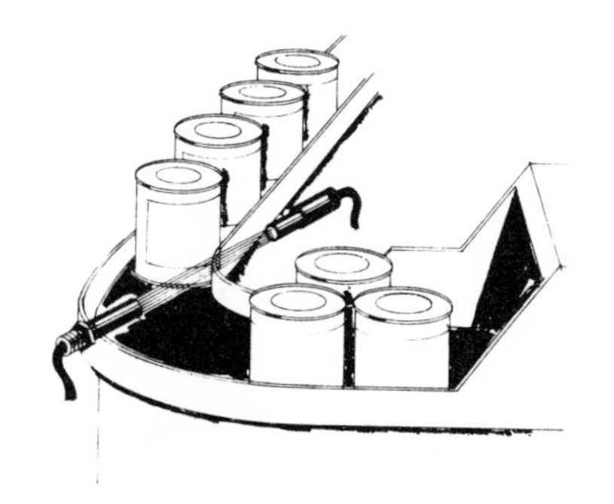

B. Counting Products

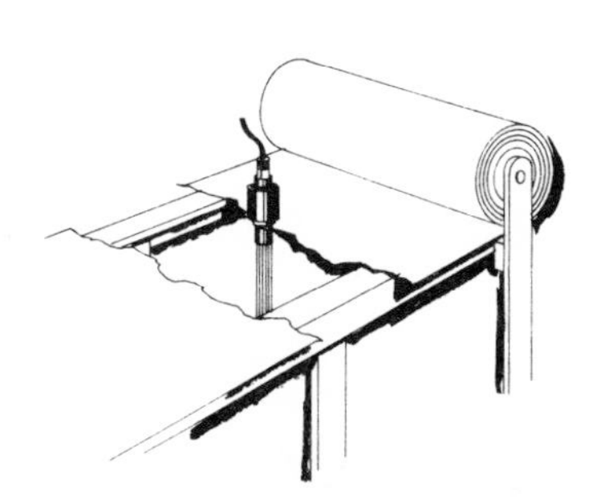

C. Detecting Web Break

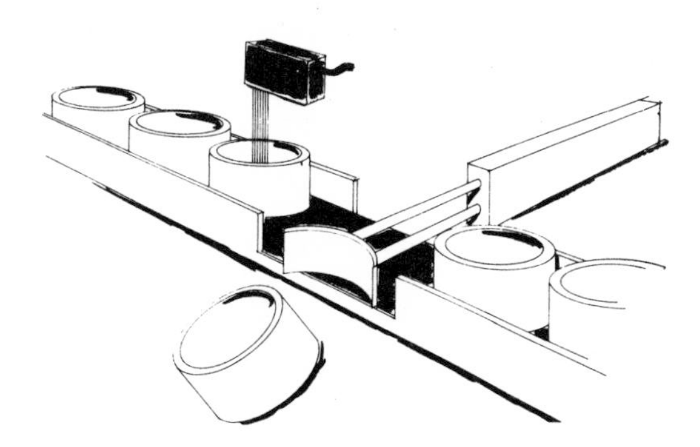

D. Checking Dark Caps for White Liners

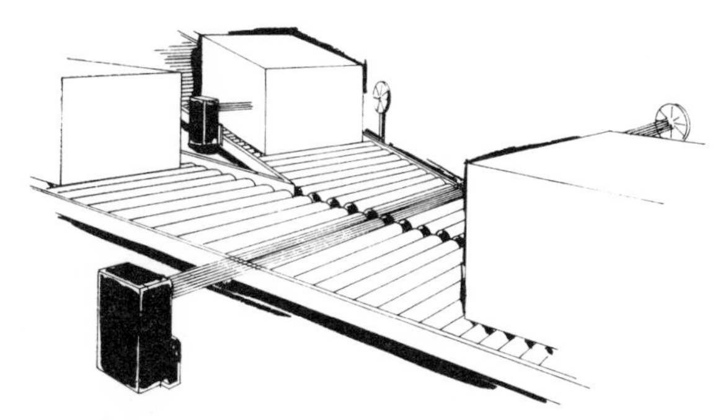

E. Preventing Collisions Where Two Conveyors Merge

F. Detecting Registration Marks

Figure 7.41 Applications of Optoelectronic Sensors

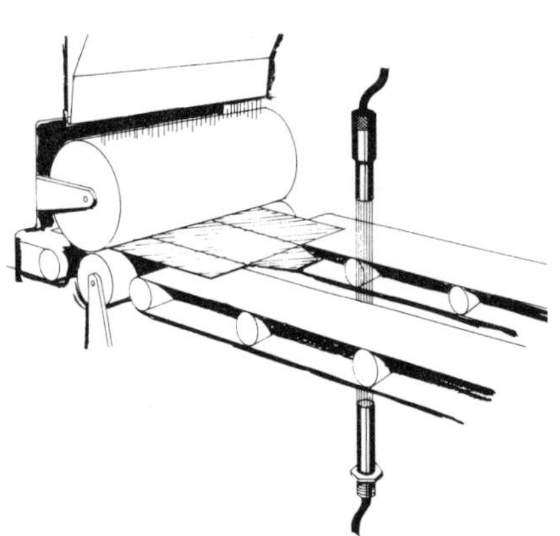
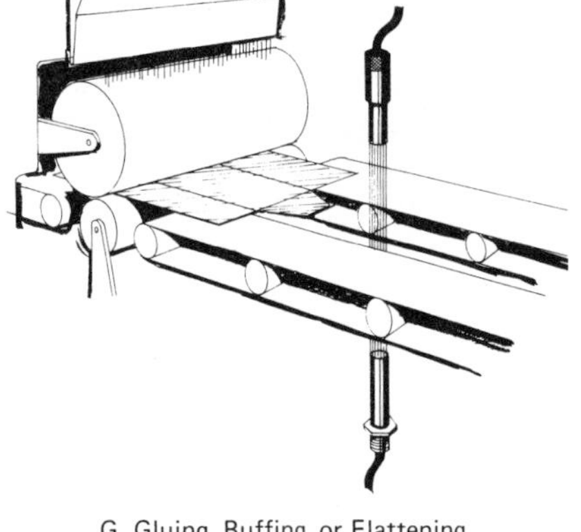

G. Gluing, Buffing, or Flattening

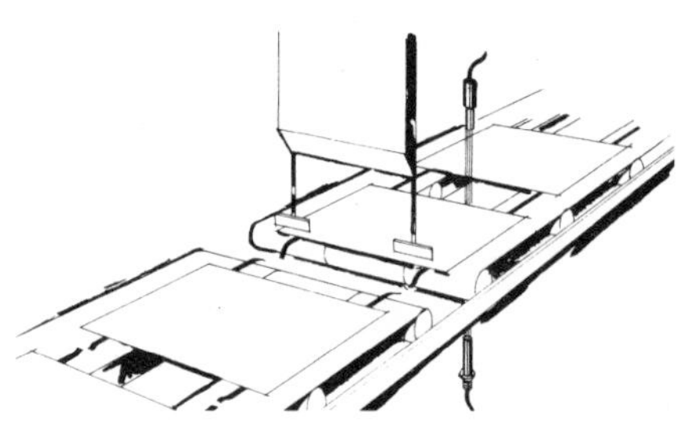

H. Detecting Products and Operating Guillotine Blade

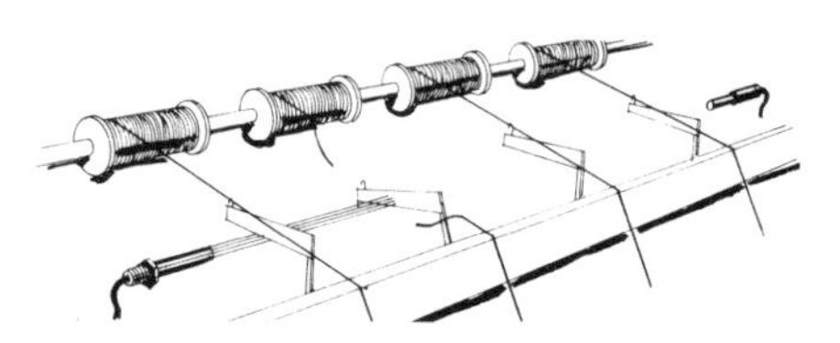

I. Detecting Thread Breaks

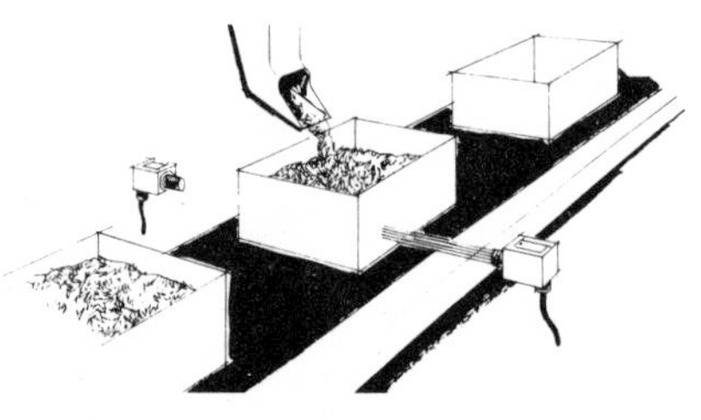

J. Slowing Conveyor and Filling Carton

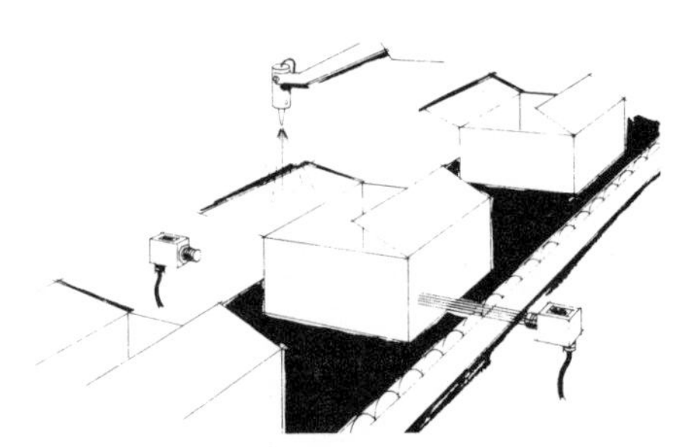

K. Controlling Size of Paper or Fabric Roll

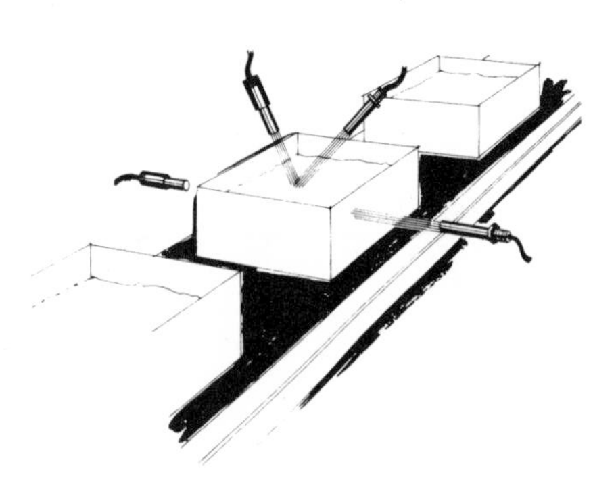

L. Checking Fill Level

M. Turning Glue Nozzle On and Off

Figure 7.41 *(continued)*

Secondary emission occurs from dynode 1, and more electrons are emitted. This cascading process occurs until a great many electrons are emitted from dynode 8. Photomultipliers have gain factors up to 10^9, which makes them one of the most sensitive photodetectors available.

Photosensors have gained a wide following in industry in the past several years. Their popularity is due, in large measure, to their versatility. Figure 7.41 illustrates the photosensor's wide area of application.

Web breaks, that is, the breaks that sometimes occur in a roll of paper or cloth, can be detected with the arrangement shown in Figure 7.41C. Any break will let the light from a source pass through to a detector, thus setting off an alarm or turning off the drive motor. In Figure 7.41E, boxes on a conveyor may be merged without colliding. The box on the right will be stopped until the box on the left passes by. In Figure 7.41F, information from registration marks (reflected back to a sensor) can initiate related operations, such as printing, cutting, or folding. Gluing can be accomplished, as shown in Figure 7.41G, by using the photodetector to detect the presence of a product. Obviously, the manufacturer would not want glue applied when there is no product present to receive it. Guillotine blades can be controlled by photosensors in cutting operations, as shown in Figure 7.41H. A common and important industrial application of photosensors is in detecting fill levels, as shown in Figure 7.41L. The fill detectors (above the box) are turned on by the box detectors (between the box). Without this circuit, the fill inspection pair might mistake the space between the boxes as an improper fill.

A summary of optoelectronic sensors is given in Table 7.6. A summary of the schematic symbols is given in Figure 7.42.

Table 7.6 *Summary of Optoelectronic Sensors*

Type	*Name*	*Spectral Response* (μm)	*Response Time*	*Advantages*	*Disadvantages*
Photoconductive	Photoresistor			Small	Slow
	CdS	0.5–0.7	100 ms	High sensitivity	Hysteresis
	CdSe	0.6–0.9	10 ms	Low cost	Temperature range limited
				Visual range	
	Photodiode	0.4–0.9	1 ns	Very fast Good linearity Low noise Wide spectral range	Low-level output (current)
	Phototransistor	0.25–1.1	1 μs	High current gain Can drive TTL Small size	Low frequency response (500 KHz) Nonlinear
Photovoltaic	Photovoltaic (solar) cell	0.35–0.75	20 μs	Linear Self-powered Visual range	Slow Low-level output (voltage)
Photoemissive	Phototube	0.15–1.2	1 μs	Good stability Fast response High impedance	Fragile
	Photomultiplier	0.2–1.0	1–10 ns	Fast response High sensitivity Good linearity Wide spectral range	Fragile Need high-voltage power supply

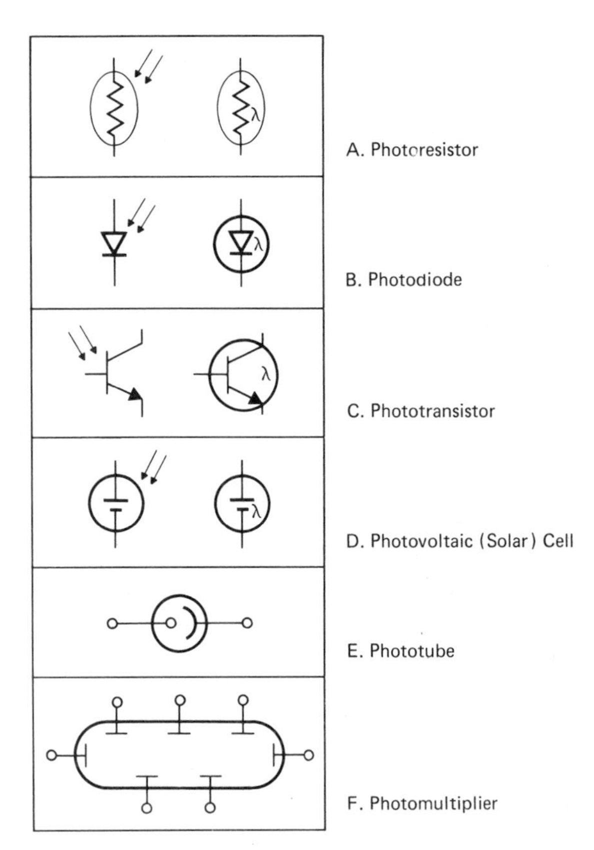

Figure 7.42 Schematic Symbols for Optoelectronic Sensors

Displacement, Stress, and Strain

In industrial process control, it is often necessary to have accurate information about an object's position. *Displacement transducers* provide this kind of information. In measurement systems, displacement and the force that produces the displacement are inseparably linked. We can, then, measure either the displacement or the force (stress) producing the displacement. In some applications, we are interested in measuring the deformation of an object caused by force or stress. This deformation is called strain. In this section, we deal with transducers that can be used to measure each of these concepts.

Displacement Transducers

Displacement is defined as an object's physical position with respect to a reference point. Displacement breaks down into two categories: linear displacement and angular displacement. *Linear displacement* is defined as the position of a body in a straight line with respect to a reference point. *Angular displacement* is the angular position of an object with reference to a fixed point. Industry uses transducers for both types of displacement.

Angular-Displacement Transducers

One of the most common angular-displacement transducers is the potentiometer. You are probably familiar with the basic structure of a potentiometer. It is composed of a resistor shaped in a circle with a wiper sliding on it, as shown in Figure 7.43. If the shaft is rotated clockwise, the resistance between contacts 1 and 2 increases. This increasing resistance can be used to indicate the position of a rotating shaft, say, on a motor. The resistive element may be carbon, conductive plastic, or thin wire wrapped around a nonconductive form.

Linear-Displacement Transducers

Linear displacement can be measured in many ways. For example, if the resistive element in the angular-displacement transducer is straightened

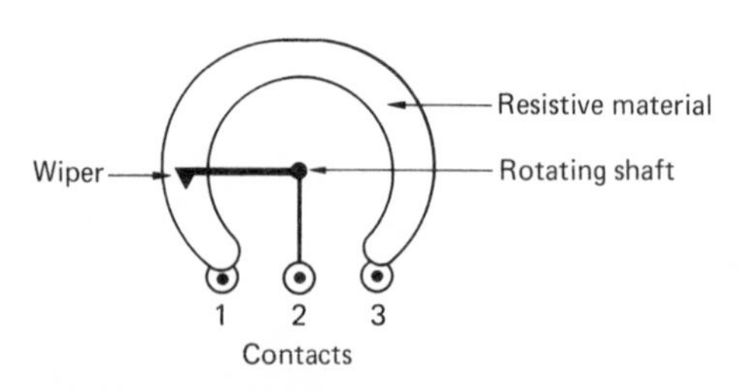

Figure 7.43 Potentiometer as Indicator of Angular Displacement

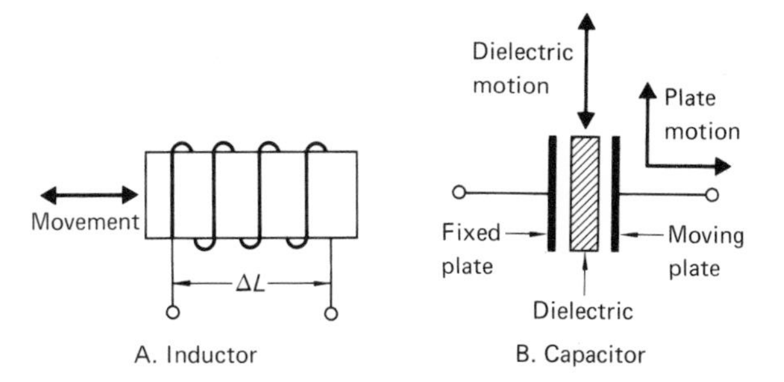

Figure 7.44 Capacitors and Inductors as Indicators of Linear Displacement

out, we have a linear-displacement transducer. The position of the tap or wiper is directly controlled by linear displacement.

Variations of capacitance and inductance are also used to indicate linear displacement. Inductance is normally changed by withdrawing the core from the windings, as shown in Figure 7.44A. Capacitance is varied by moving the plates or by withdrawing or inserting the dielectric, as shown in Figure 7.44B.

By far, the most common displacement transducer is the *linear variable differential transformer (LVDT)*. The LVDT is basically a transformer with two secondaries; see Figure 7.45. A movable core is connected to the shaft. An object attached to the shaft moves the core. The primary winding is excited by an AC frequency between 50 Hz and 15 kHz with an amplitude of up to 10 V. The secondary windings are usually connected series opposing so that when the core is centered, there is no output from the secondary. If the core material moves in either direction, an output voltage is produced, because the mutual inductance changes. Although the LVDT output is usually AC, the AC output can be rectified to produce DC output.

LVDTs are used extensively in industry as displacement transducers. Displacements of as little as ten-thousandths of an inch are detected by LVDTs. In addition to displacement, LVDTs are used to measure weight and pressure in combination with mechanical transducers.

Stress and Strain Transducers

Stress. strain, and force are included in this section on displacement because displacement is the primary means of measuring these quantities. *Force* is defined as the quantity that changes the motion in a body. *Stress*, a similar concept, is the force acting on a solid's unit area. *Strain*, on the other hand, is the change in shape or form resulting from stress. The change in shape may be a change in length or width. Thus, the force applied to a solid material is the stress, and the deformation of that material is the strain.

In this section, we discuss two popular types of strain gauges: the wire gauge and the semiconductor gauge.

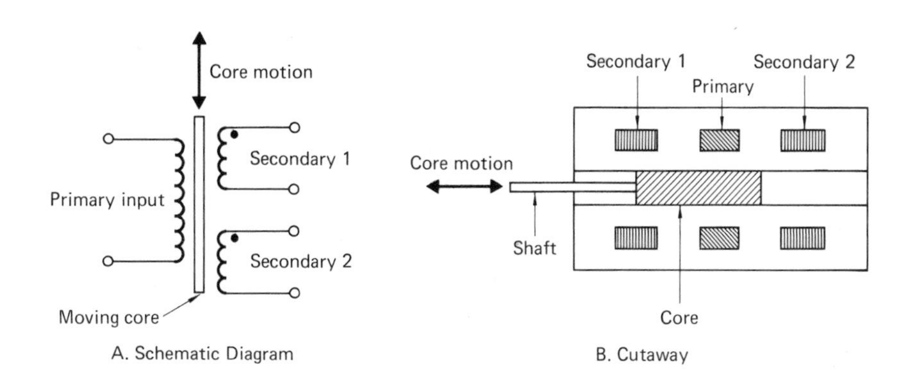

Figure 7.45 LVDT as Indicator of Linear Displacement

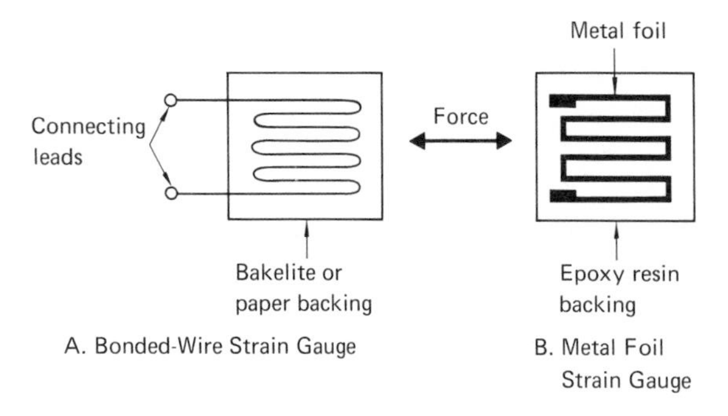

Figure 7.46 Pictorial Representations of Wire and Foil Strain Gauges

Bonded-Wire Strain Gauge

Probably the most popular strain-sensing device is the *bonded-wire strain gauge*. Strain gauges are made of metal, either in the form of wire (about 0.001 in. in diameter) or foil (about 3 μm thick). Strain gauges are usually shaped in a serpentine fashion, as shown in Figure 7.46.

The wire strain gauge is firmly cemented to a paper or Bakelite backing. The metal foil type is normally photoetched on an epoxy resin backing, as a printed circuit board is. Both react to force with a change in resistance. Recall that the resistance of a conductor depends on length and cross-sectional area, among other factors. As length increases and diameter decreases, resistance increases. This principle is the one on which the wire and foil strain gauges work.

As mentioned in the introduction to this section, the change in an object's length compared to its original length is called strain ($\Delta l/l_0$). Strain is often represented by ε (Greek letter epsilon). How much the resistance of a strain gauge changes depends on several things. Any resistance change is directly proportional to the original resistance, the elasticity of the material, and the strain, or deformation. These factors are related in the following equation:

$$\Delta R = R_0 G \varepsilon \qquad (7.8)$$

where ΔR = change in resistance
R_0 = original resistance
G = gauge factor
ε = strain

The gauge factor contains a constant called Poisson's ratio, which is an index of elasticity.

Dividing both sides of this equation by the original resistance and the strain, we see that the *gauge factor* G is a measure of the fractional change in resistance per unit of strain:

$$G = \frac{\Delta R/R}{\varepsilon} \qquad (7.9)$$

Or we can say that the gauge factor tells us the percent change in resistance of the gauge compared to its percent change in length. Most metals have a gauge factor between 1.8 and 5.0.

Stress and strain are related through a law called *Hooke's law*. Hooke pointed out that for many materials, there is a constant ratio between

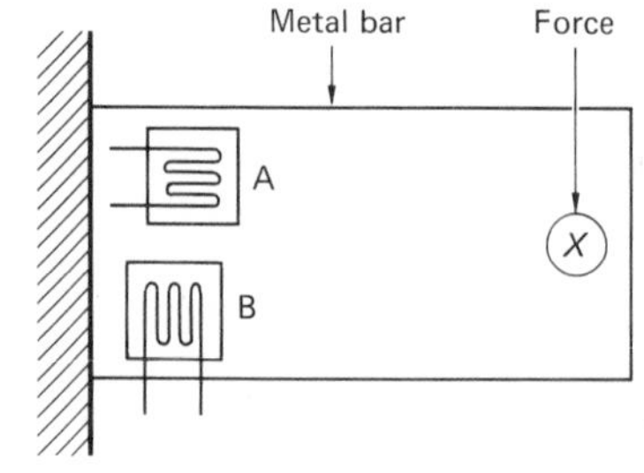

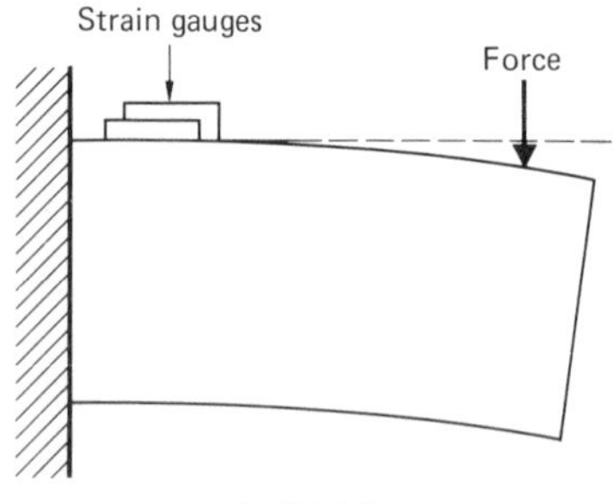

Figure 7.47 Strain Gauges Mounted on Bar and Subjected to Stress

stress and strain. The constant of proportionality between stress and strain is called *Young's modulus*. This relationship is expressed mathematically as follows:

$$E = \frac{\sigma}{\varepsilon} \tag{7.10}$$

where E = Young's modulus

σ = stress, in force per unit area

ε = strain

Using these equations, we can find how much stress a material is receiving. We will use these equations for the system shown in Figure 7.47.

Strain gauge A in Figure 7.47 is the sensor. As a force is applied down at point X, the metal bar bends. So does strain gauge A. In bending, the strain gauge wire increases its length and decreases its cross-sectional area. Thus, its resistance increases. (Strain gauge, unstrained, resistances range from 10 to 10,000 Ω.) This change in resistance is related to strain and the stress (or force) that produced it.

One problem with this type of strain gauge is its inherent lack of sensitivity. The resistance changes only a few percent over its total range. For example, a 100 Ω resistor may only change by 5 Ω. Measuring so small a resistance change usually requires a bridge, as shown in Figure 7.48A. The unbalance in the bridge caused by the strain gauge resistance change is detected by a differential amplifier. Resistances R_A and R_B are designed to be greater than ten times the strain gauge resistance in order to make the circuit a constant-current source for the strain gauge. The design current is generally 1 mA in order to keep I^2R heat to a minimum and still have the strain gauge voltage high enough to be above the background noise level.

Let us say that strain gauge A has a resistance of 1000 Ω. We have measured this resistance and found it to have increased by 2 Ω under stress. Let us further specify that Young's modulus E is 10^{12} dyn/cm^2 and the gauge factor G is 2. Solving Equation 7.9 for strain, we find it to be 0.001:

$$\varepsilon = \frac{\Delta R/R}{G} = \frac{2/1000}{2} = 0.001 \tag{7.11}$$

To find the stress, we solve Equation 7.10 for stress. Substituting the values for Young's modulus and the strain, we find stress to be 10^{-2} dyn/cm^2:

$$\sigma = E\varepsilon = (10^{12}\ \text{dyn/cm}^2)(0.001) = 10^{9}\ \text{dyn/cm}^2 \tag{7.12}$$

Up to this point, we have not talked about the other strain gauge in Figure 7.47, strain gauge B. Another problem in strain gauge measurements results from resistance changes due to ambient temperature fluctuations. Strain gauge B (called the dummy) is used for temperature compensation. Figure 7.48B shows both strain gauges in a balanced bridge. If temperature rises, both strain gauges change resistance by the same amount, creating no imbalance. Only the change in strain gauge A is proportional to strain.

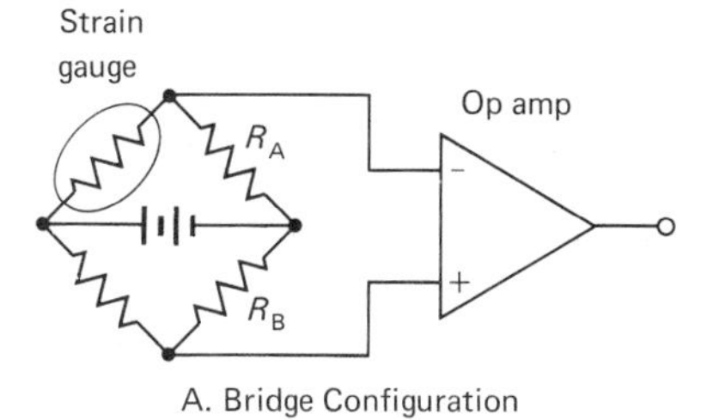

A. Bridge Configuration

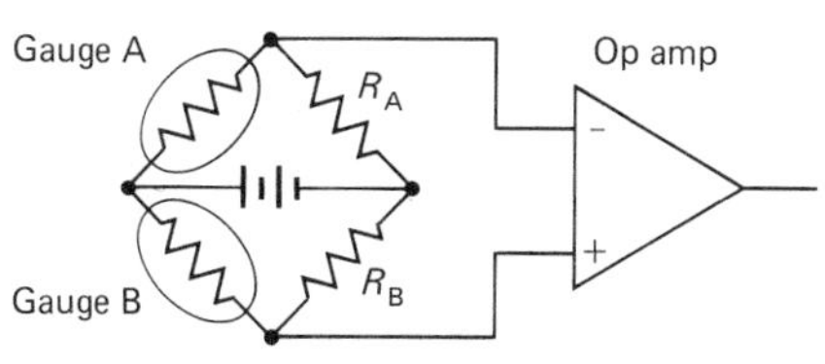

B. Strain Gauge B Used to Compensate for Ambient Temperature Changes

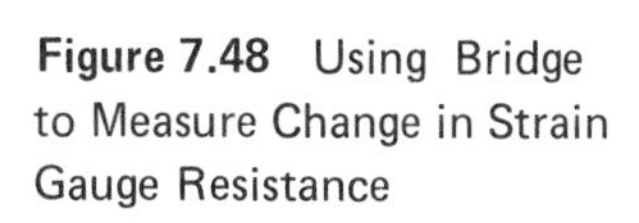

Figure 7.48 Using Bridge to Measure Change in Strain Gauge Resistance

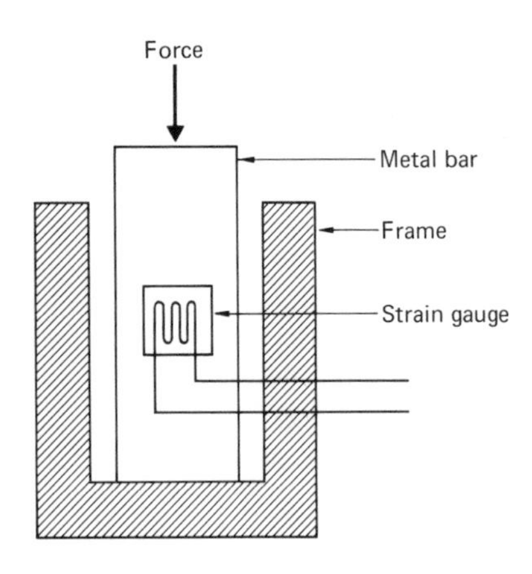

Figure 7.49 Load Cell

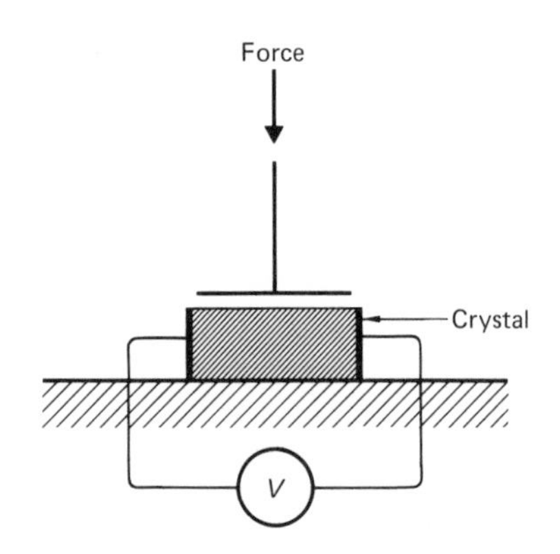

Figure 7.50 Piezoelectric Crystal Generating Voltage

Strain gauges are also found in load cells, which are commonly used to measure force. The *load cell* consists of one or more strain gauges mounted on some form of metal beam or bar, as shown in Figure 7.49. The gauges are usually wired in a bridge configuration. The imbalance is then taken from the load cell directly.

Semiconductor Strain Gauge

A more sensitive device (but one that is less linear) is the *silicon semiconductor strain gauge*. Silicon is a *piezoresistive substance*; that is, it changes its resistance under pressure. The semiconductor strain gauge is typically used in a bridge (with temperature compensation), as wire or foil strain gauges are. It is 30 to 50 times more sensitive than the comparable bonded-wire or foil strain gauge. Over wide ranges, linearity sometimes is a problem. So, these devices are commonly used over narrow ranges. Where the application calls for wide ranges and high sensitivity, these devices are combined with linearity-compensating devices or circuits.

Another device used for strain or stress measurement is the *piezoelectric crystal*. Such crystals (commonly quartz, Rochelle salts, or tourmaline) exhibit a potential difference when compressed; see Figure 7.50. The voltage that appears at the edges of the crystal can be high in voltage value but is normally low in the amount of current it provides. Also, the charge leaks off very quickly under static conditions. Therefore, this device is usually used to measure forces or stresses that change rapidly, such as vibrations.

Semiconductor strain gauges have gauge factors as high as 200. They are, however, much less linear than wire strain gauges.

Acceleration Transducers

We have seen how displacement transducers can be used to measure stress and strain. The displacement transducer can also be used to measure acceleration. Such a device is called an *accelerometer*.

Most accelerometers use the same principle, which is illustrated in Figure 7.51. As acceleration starts, the mass, which is free to move, goes backward. The mass moves backward because all mass has inertia, or resistance to motion. The higher the acceleration, the more backward movement that results. Any displacement, force, strain, or pressure transducer can be used to detect the motion or backward force.

Many accelerometers use the piezoelectric crystal to convert the force to electricity. Accelera-

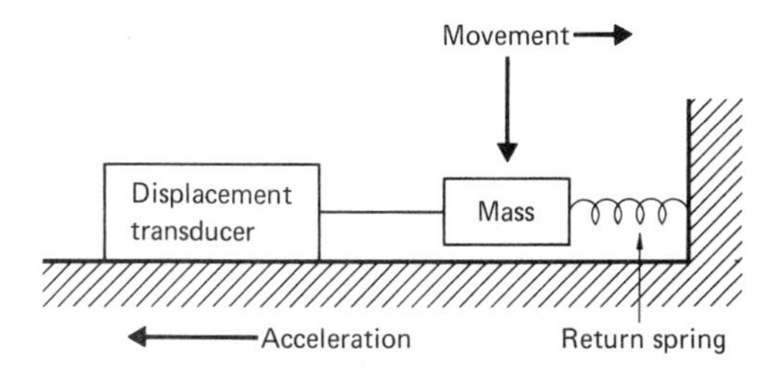

Figure 7.51 Basic Accelerometer

tion in these cases is calculated by using the following equation:

$$a = \frac{V}{s} \tag{7.13}$$

where a = acceleration

V = peak voltage, in millivolts

s = sensitivity of the sensor, in millivolts per g

Recall that 1 g is the acceleration due to the earth's gravity, which is about 981 cm/s^2. Using this equation, we can calculate the acceleration of a system with a sensitivity of 20 mV/g and an output of 45 mV peak. We find the acceleration to be 2.25 g:

$$a = \frac{V}{s} = \frac{45\text{ mV}}{20\text{ mV}/g} = 2.25\ g \tag{7.14}$$

Magnetism

Recall that magnetic fields are distributed spatially around an object as lines of force. Sensors that detect the strength of magnetic fields are used in measuring devices called gauss meters. Other common applications of magnetic field sensors are in ammeters. Two devices, in particular, are used in industry to detect the presence of magnetic fields: Hall effect devices and magnetoresistors. We will describe each type in turn.

Hall Effect Devices

The most popular device for detecting magnetic fields is the *Hall effect sensor*. The transduction principle used in the Hall effect device was discovered in 1879 by Edward H. Hall at Johns Hopkins University. He found that when a magnetic field was brought close to a gold strip in which a current was flowing, a voltage was produced. This effect is named the *Hall effect*.

Today, semiconductors are used in Hall devices instead of the gold with which Hall worked. Semiconductors produce the highest Hall voltages of any solid material.

The Hall effect is illustrated in Figure 7.52. Note that there is no distortion of current carrier paths in Figure 7.52A. When a magnet is brought close to the semiconductor material, as in Figure 7.52B, electrons are forced to the right side of the material. A differential voltage is then felt across the sides of the device.

Hall effect devices are typically four-terminal devices. Two terminals are used for excitation and two for output voltage. Hall effect devices come in two basic functional classifications: linear and digital.

Linear Hall effect devices are used for gauss meters, with a sensitivity of 1.5 mV/G, and for DC current probes. Normally, to measure DC current, we must break the circuit and insert an ammeter. But breaking the circuit is not necessary with the

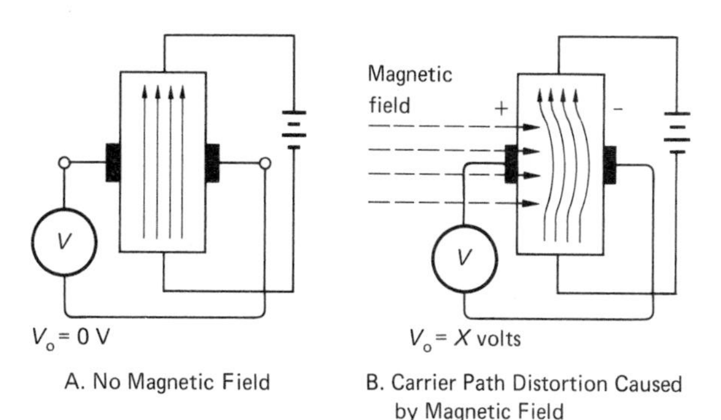

Figure 7.52 Hall Effect

Hall probe. The Hall effect DC current probe senses the magnetic field strength around a wire, which is proportional to current flow. Other applications include cover interlocks and ribbon speed monitoring in computer printers. Hall effect devices also find uses in brushless DC motors in computer disc and tape drives. When a permanent-magnet rotor and a stator of coil windings are used, Hall sensors can replace brushes.

Two further applications are illustrated in Figures 7.53A and 7.53B. The diagram in Figure 7.53A shows a means of keeping tabs on a conveyor operation. The Hall effect sensor is mounted on the frame. A magnet is mounted on the drum. The Hall effect sensor actuates every time the magnet passes by. This system is used for speed control or for informing the operator that the conveyor is still in operation. Figure 7.53B shows a beverage dispenser. Magnets are attached to the buttons. When the buttons are pushed, the magnets actuate the sensors, dispensing the correct drink. The unit can be completely sealed for ease of cleaning.

Digital Hall effect devices are used in counting, switching, and proximity sensors. Proximity is sensed by placing a magnet on the object whose proximity is to be detected.

Hall effect devices typically produce about 10 mV potentials at the output terminals. So, many manufacturers include an amplifier in the package with the Hall effect device.

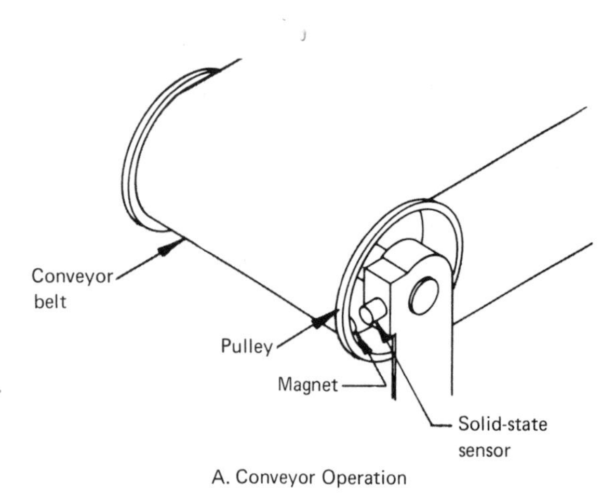

A. Conveyor Operation

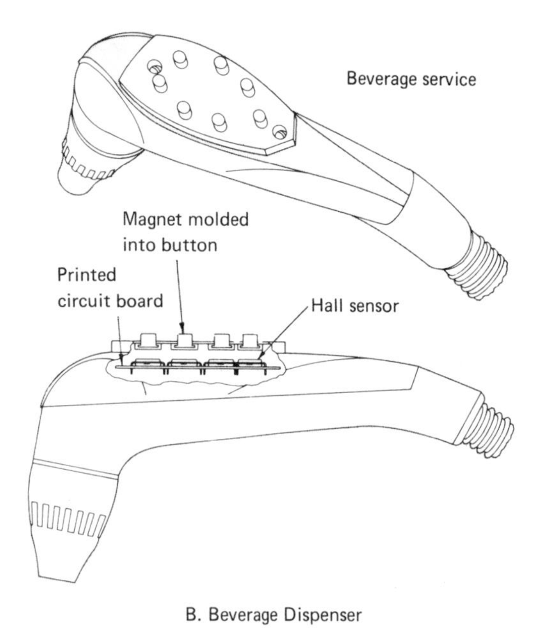

B. Beverage Dispenser

Figure 7.53 Applications of Hall Effect Sensors

Magnetoresistors

Magnetoresistors, although not as popular as Hall effect devices, are also used to sense the presence of a magnetic field. Magnetoresistors look like photoresistors in structure. In these devices, indium antimonide is deposited in a serpentine pattern on a substrate, as shown in Figure 7.54. When a magnetic field impinges on the semiconductor, the carrier paths are distorted, as in the

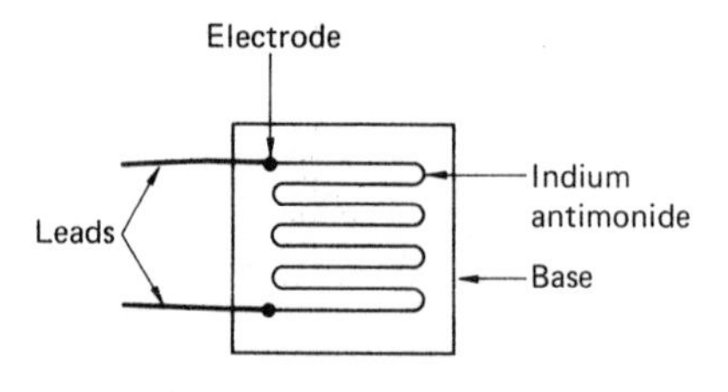

Figure 7.54 Magnetoresistor Construction

Hall effect device. However, in the magnetoresistor, carrier path distortion effectively narrows the cross-sectional area of the conductor, thereby increasing resistance.

Note that the magnetoresistor is a two-terminal device whose resistance varies with magnetic field proximity. As field strength increases, magnetoresistor resistance increases.

Magnetoresistors, being two-terminal devices, are used to replace resistors in low-voltage circuits, which Hall effect devices cannot be used for. Magnetoresistors are also used to sense mechanical position (proximity), current, and magnetic fields, as Hall effect devices are. An advantage of the magnetoresistor is its sensitivity. In a bridge circuit, magnetoresistors produce up to 1 V output. The Hall effect devices typically produce less than 10 mV.

Pressure

Pneumatic and hydraulic systems abound in industry, even in the electronic age. There are many advantages to such systems, advantages that are discussed in Chapter 8. Suffice it to say here that these systems are still used. And it is often necessary to measure the pressures exerted at different points in these systems.

Pressure can be defined as the action of one force acting against an opposing force. Pressure is normally measured as a force per unit area, such as pounds per square inch (commonly abbreviated as psi rather than lb/in.^{2} in industry). In the English system, if the pressure is referenced to ambient pressure, the measurement is called gauge pressure (symbolized by psig). If the pressure measurement is referenced to a vacuum, the measurement is called absolute pressure (symbolized by psia). The metric system unit for absolute pressure is the pascal (Pa) and is equal to 1 N/m^2. Also, 1 Pa = 1.45×10^{-4} psi. Because this unit (pascal) is so small, the kilopascal (kPa) is often used.

Pressure-measuring devices will be broken down into two basic classifications: (1) pressure-to-position transducers, such as the manometer, Bourdon tube, bellows, and diaphragm, and (2) pressure-to-electrical transducers, such as the piezoresistive transducer.

Although the transducers discussed in this section are, for the most part, pressure-to-position transducers, they can be easily converted to produce an electric output. As we proceed within this section, we will point out methods to achieve this conversion.

Manometers

The U tube manometer, shown in Figure 7.55, is representative of the liquid-column pressure transducers. The *U tube manometer* is a differential pressure–measuring device constructed from a glass or plastic tube bent in a U shape. The tube contains an amount of liquid, whose makeup depends on the application. Note that pressure p_1 is greater than pressure p_2 in the figure. The difference in pressure causes the liquid to move into the left leg until all pressures are in equilibrium. The difference in the height (Δh) of the liquid is proportional to the difference in pressures.

The major disadvantage of manometers is that they are strictly mechanical in nature. They can, however, be converted to electrical systems with a float device attached to a displacement transducer, such as the LVDT.

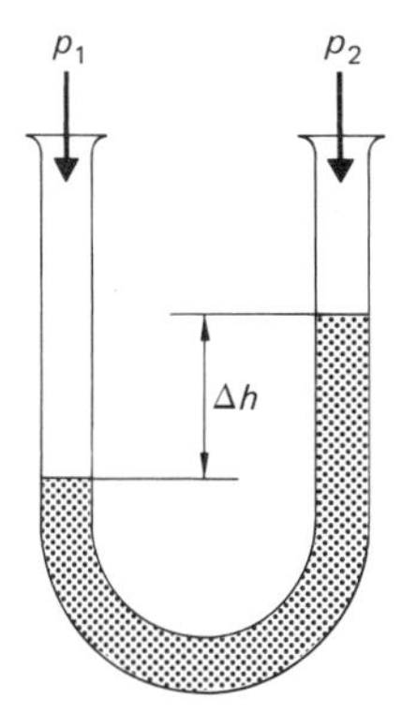

Figure 7.55 U Tube Manometer

Elastic Deformation Transducers

All *elastic deformation pressure transducers* use the same transduction principle. Pressure causes a bending, or deformation, of the transducer material, usually a metal. This deformation results in a deflection or displacement. We might say, then, that these transducers are pressure-to-position converters. As we will see, the indicators can be purely mechanical, or they can use electrical transducers to convert the displacement to an electrical parameter. We will discuss three types of elastic deformation transducers, all based on the same principle but each different in physical shape and size. The transducers we will consider are the Bourdon tube, the bellows, and the diaphragm.

Bourdon Tube

The *Bourdon tube* is one of the oldest and most popular (even today) pressure transducers. It was invented in 1851 by Eugene Bourdon and has been used in industry ever since. Bourdon tubes come in three basic shapes: C tube, spiral tube, and helical tube; see Figures 7.56A, 7.56B, and 7.56C.

All three devices work in essentially the same fashion. They are all composed of a flattened metal tube, usually made of brass, phosphor bronze, or steel, as shown in cross section in Figure 7.56D. The pressure at the input is communicated to the inside of the device. Since there is more area on the outside of the tube than on the inside, the tube unwinds, much like a paper-coil party toy. The unwinding of the Bourdon tube is not linear. Therefore mechanical pointers fastened to the moving tips have displacement applied through a gearing system. This gearing system compensates for the nonlinearity of movement. Nonlinearity may also be compensated for by using a nonlinear pointer scale.

The Bourdon tube is simple to manufacture

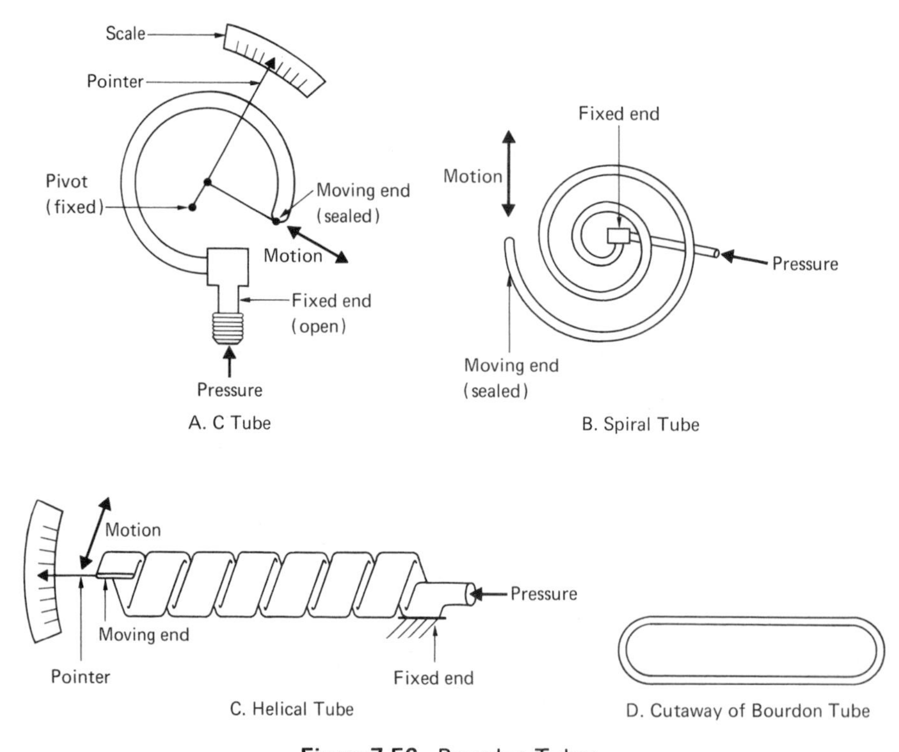

Figure 7.56 Bourdon Tubes

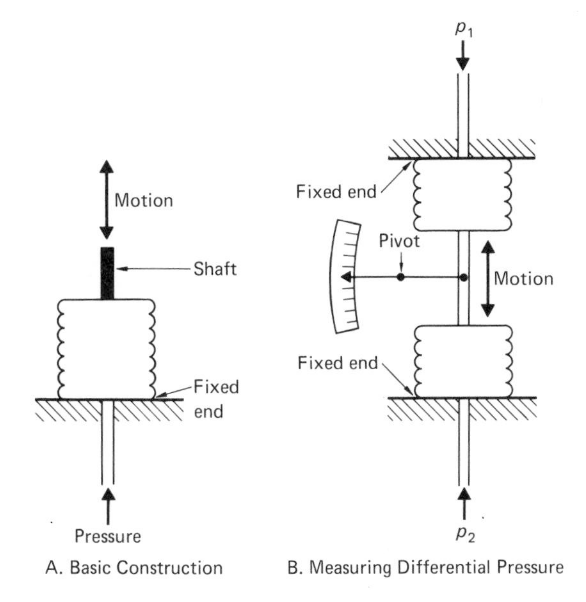

Figure 7.57 Pressure Bellows

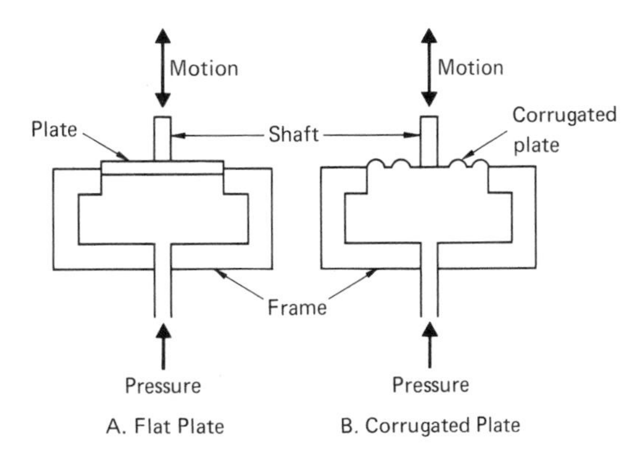

Figure 7.58 Pressure Diaphragms

(and low in cost), is accurate, and measures pressures up to 100,000 psi (700,000 kPa). The C-shaped Bourdon tube is the least sensitive to pressure changes, and the helical tube is highest in sensitivity. These devices do not work well with pressures under 50 psi (750 kPa). Bourdon tubes are converted very simply to electrical transducers by using electrical displacement transducers such as potentiometers and LVDTs.

Bellows

The *pressure bellows* is a cylindrical-shaped device with corrugations along the edges, as shown in Figure 7.57A. As pressure increases, the bellows expands, moving the shaft upwards. The shaft may be attached to a pointer or an electrical displacement transducer like an LVDT.

In Figure 7.57B, the bellows is connected to indicate differential pressure. If p_1 is greater than p_2, the pointer moves up. The amount of displacement is proportional to the difference in pressures.

Bellows are more sensitive to pressures in the 0–30 psi (0–210 kPa) range than are Bourdon tubes. Thus, they are more useful at lower pressures.

Diaphragms

The *diaphragm* is nothing more than a flexible plate, as indicated in Figure 7.58. The plates are usually made of metal or rubber. The flat (or curved) metal plates are generally less flexible (and less sensitive) than the corrugated type. Increases in pressure will cause the shafts to move up. Diaphragms are used in the 0–15 psi (0–105 kPa) pressure range.

Piezoresistive Transducers

Both National Semiconductor and the Micro Switch Division of Honeywell manufacture very sensitive and linear piezoresistive pressure transducers. These sensors use the piezoresistive effect and have their own built-in amplifiers and temperature compensation networks. National's line of products can sense pressures ranging from 10 to 5000 psi (70 to 35,000 kPa) with linearity less than 1%. Figure 7.59 shows one of National's transducers.

Fluid Flow

Along with temperature, fluid flow is a very important measure. Many industrial processes, especially biomedical applications, depend on accurate

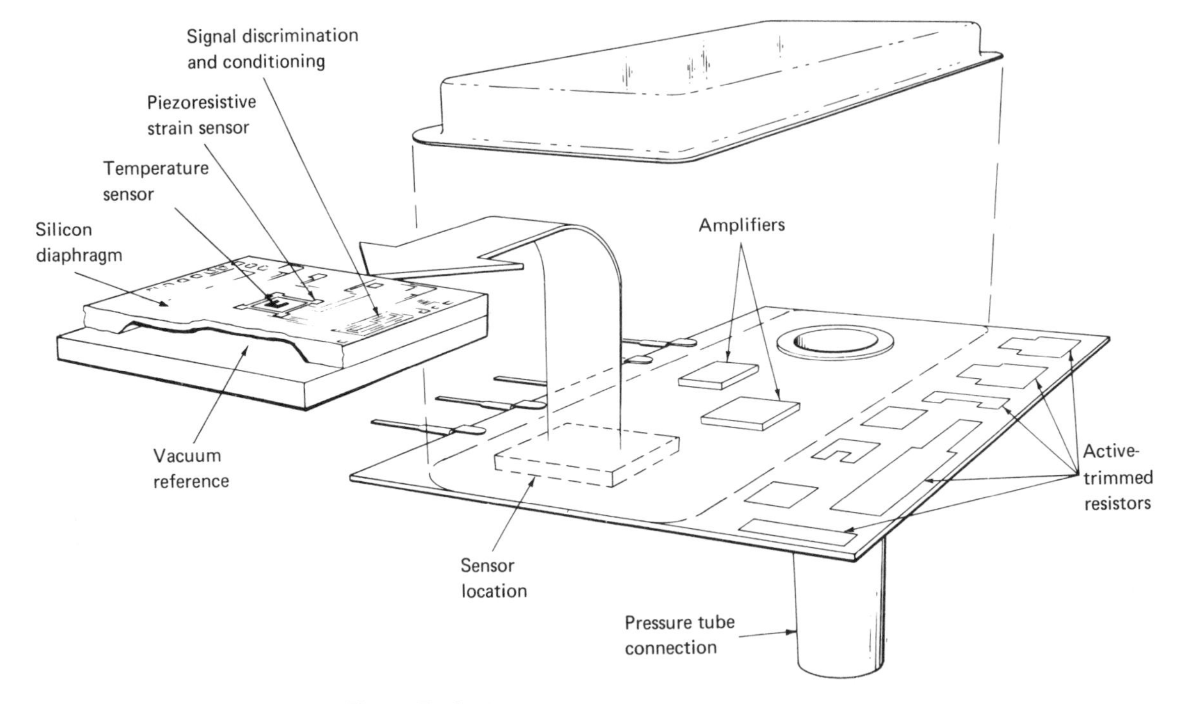

Figure 7.59 Piezoresistive Pressure Transducer

assessment of fluid flow. Fluid flow measurements are an integral part of chemical, steam, gas, and water treatment plants.

There are many different ways to measure fluid flow. As with the other transducers we have discussed, we limit our treatment to only the more common fluid flow measuring devices: differential pressure flowmeters, variable-area flowmeters, positive-displacement flowmeters, velocity flowmeters, and thermal heat mass flowmeters.

Differential Pressure Flowmeters

The most common industrial flowmeters come from this group. Most *differential pressure flowmeters* use a pressure difference caused by a restriction as a transduction principle. Each flowmeter has a different method of producing this pressure differential. The principle involved here is called the Bernoulli effect. The *Bernoulli effect* states that as the velocity of a fluid (liquid or gas) increases, the pressure decreases, and vice versa. We say that an inverse relationship exists between the pressure and velocity of a fluid. In this section, we consider five of the commonly used differential pressure flowmeters.

Orifice Plate

We will use the most common of the differential pressure flowmeters, the *orifice plate flowmeter*, as an example. It is shown in Figure 7.60. This flowmeter consists of a plate with a hole in it. If we place liquid columns all along the tube to indicate pressure, we see a *pressure gradient*, or pressure distribution, as pictured in Figure 7.60. Pressure increases slightly just before the restriction and falls off quickly after the restriction. The pressure decreases sharply because of the increase in velocity at and just after the restriction. A manometer or

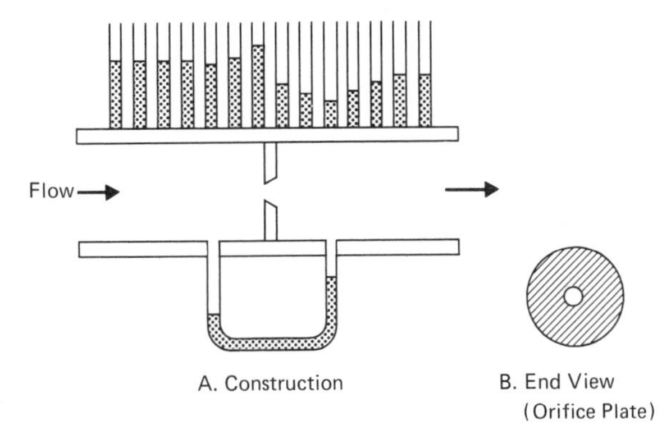

Figure 7.60 Static Pressure Gradient in Orifice Plate

other differential pressure indicator can be calibrated to read the flow rate. As fluid flow increases, the pressure difference increases also.

The orifice plate flowmeter is the most widely used of all the flowmeters. It is simple in structure, low in cost, and easy to install. It suffers from the disadvantages of poor accuracy, limited range, and inability to be used with slurries (liquids with suspended particles of solid matter).

Venturi Tube

Another popular flowmeter is the *Venturi tube*, shown in Figure 7.61. The Venturi tube employs the same principle as the orifice, differential pressure. The pressure difference is caused by the restriction in the tube.

The Venturi has two major advantages over the orifice. First, it creates less turbulence than the orifice because of the gently sloping sides. Thus, the final pressure loss is less than that in the orifice. This loss is called *insertion loss*. Up to 90% of the initial pressure can be recovered with the Venturi. Second, the Venturi can be used for slurries because there are no sharp edges or projections for solids to catch on. Its major disadvantage is cost. It is expensive to buy and install.

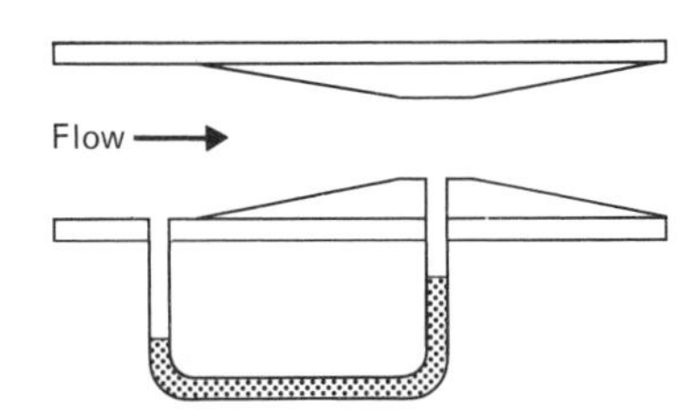

Figure 7.61 Venturi Tube Flowmeter

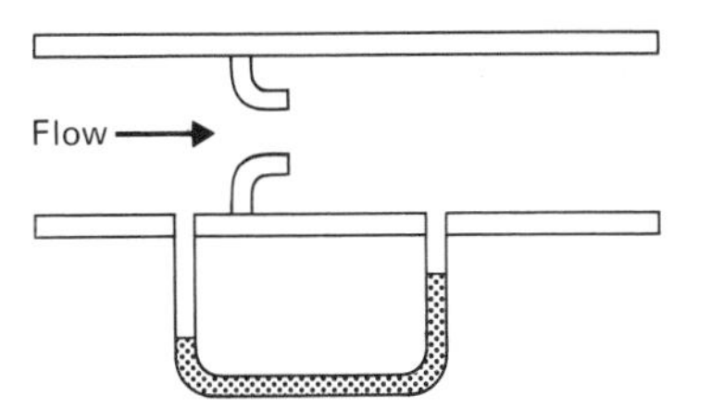

Figure 7.62 Flow Nozzle Flowmeter

Flow Nozzle

The *flow nozzle*, shown in Figure 7.62, falls somewhere between the Venturi and the orifice plate in cost and pressure loss. In the flow nozzle, the edges of the restriction are somewhat rounded. Thus, the flow nozzle permits up to 50% more flow rate than the orifice. An additional advantage is its ability to be welded into place, which cannot be done with the other differential pressure flowmeters.

Dall Tube

Of all the differential pressure flow meters, the *Dall tube* has the least insertion loss. The Dall tube, shown in Figure 7.63, consists of two cones, the shorter one upstream of the restriction, the larger cone downstream. It is generally cheaper to purchase and install compared with the Venturi tube.

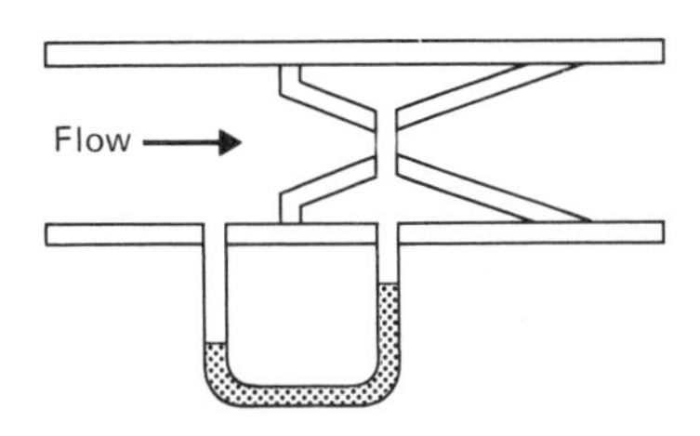

Figure 7.63 Dall Tube Flowmeter

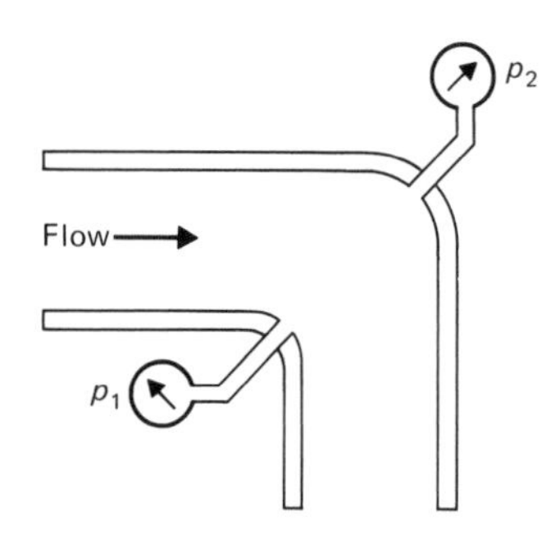

Figure 7.64 Elbow Tube Flowmeter

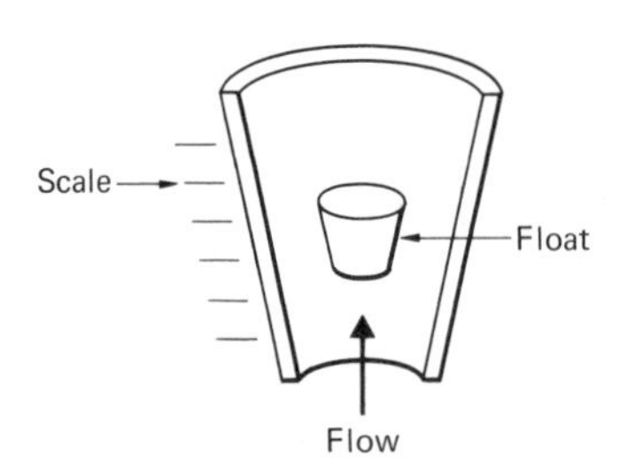

Figure 7.65 Rotameter

However, it cannot be used with solids and slurries, as the Venturi can.

Elbow

The *elbow flowmeter* is a differential pressure flowmeter that does not use the Bernoulli effect. It is shown in Figure 7.64. The differential pressure is developed by the centrifugal force exerted by the fluid flowing around the corner. In this case, p_2 is greater than p_1. Since elbows (corners) usually already exist in pipes, installation of elbows poses no problems. There is no additional pressure loss because there is no restriction. However, the elbow flowmeter, at best, has only $\pm 5\%$ accuracy.

The devices we have discussed thus far are called *primary elements.* They are so named because, in a primary sense, they convert fluid flow to pressure differential. *Secondary elements* are needed to convert this pressure difference into a usable parameter. This conversion can be accomplished with any of the pressure transducers discussed previously. In addition, National Semiconductor manufactures a differential pressure transducer that is used as a secondary element to measure flow.

Variable-Area Flowmeters

The *variable-area flowmeter* is a flowmeter that controls the effective flow area and, therefore, the flow rate within the meter. An example of the variable-area flowmeter is the rotameter, also called the float or rising ball. Shown in Figure 7.65, the *rotameter* consists of a float that is free to move up and down in a tapered tube. The tube is wider at the top and narrows toward the bottom. Fluid flows from the bottom to the top, carrying the float with it. The float rises until the area between it and the tube wall is just large enough to pass the amount of fluid flowing. The higher the float rises, the higher is the fluid flow rate.

Normally, the rotameter is used for visual flow measurement. It cannot be used for automatic process control.

Positive-Displacement Flowmeters

Positive-displacement flowmeters divide the flowing stream into individually known volumes. Each volume is then counted. Adding each volume increment gives an accurate measurement of how much fluid volume is passing through the meter. Positive-displacement meters are accurate to less than 1%, simple, inexpensive, and reliable. Furthermore, pressure losses are generally lower when compared with other flowmeters.

The *nutating-disc flowmeter*, shown in Figure 7.66, is a common example of a positive-displacement flowmeter. The nutating-disc meter is used in residential water metering as well as in industrial metering. As the liquid enters the left chamber, it fills quickly. Because the disc is off center, it has unequal pressures on it, causing it to wobble, or nutate. This action empties the volume of liquid from the left chamber into the right. The process

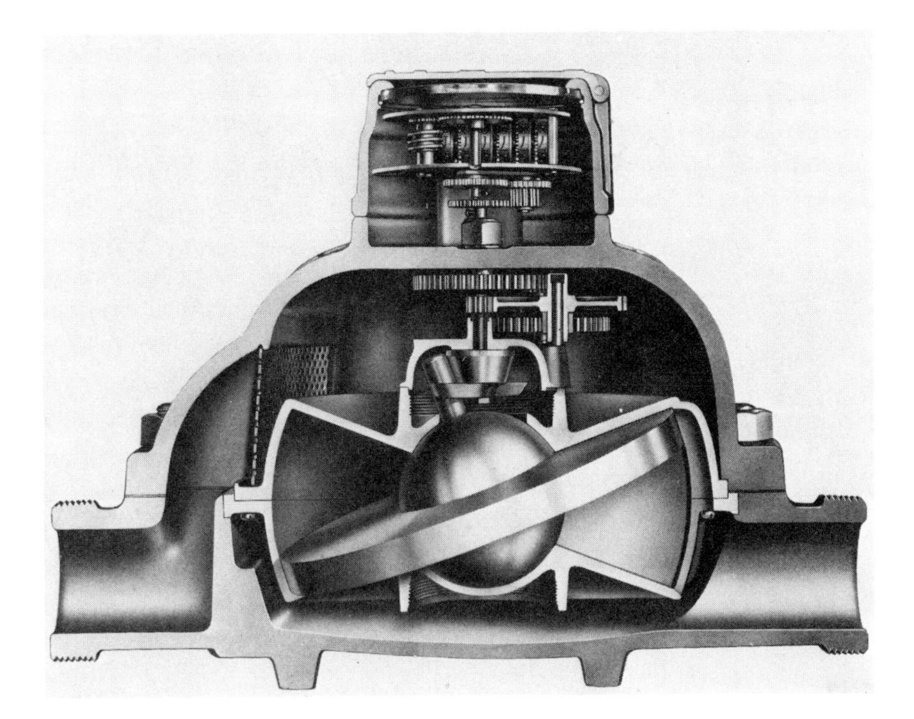

Figure 7.66 Nutating-Disc Meter

repeats itself, driven by the force of the flow stream. Each nutation of the disc is counted by a mechanical or electromagnetic counting mechanism.

Because of the mechanical structure of the nutating-disc flowmeter, it is not suitable for measuring the flow of slurries.

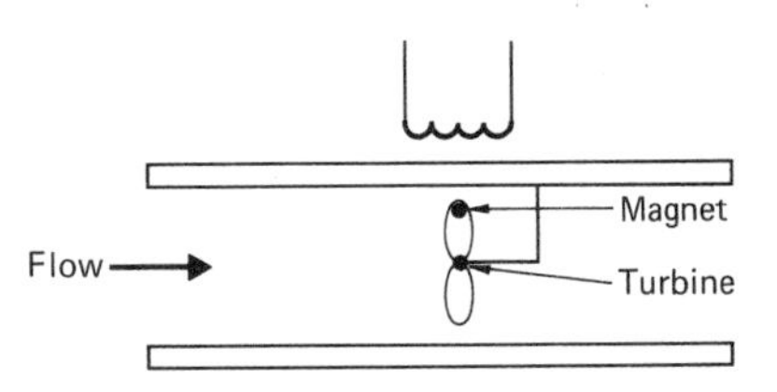

Figure 7.67 Turbine Flowmeter

Velocity Flowmeters

Velocity flowmeters (sometimes called volumetric rate flowmeters) give an output that is proportional to the velocity of fluid flow. We will consider three of the most commonly used velocity flowmeters: turbine, electromagnetic, and target.

Turbine Flowmeter

The *turbine flowmeter* is shown in Figure 7.67. The fluid flowing past the turbine exerts a force on the blades. The blades then turn at a rate directly proportional to the fluid velocity. The turbine blades can be magnetized to produce pulse as voltage is induced in the coil. Turbine meters are useful because of their good linearity, range, and accuracy. They are one of the most accurate flowmeters available.

Electromagnetic Flowmeter

The *electromagnetic flowmeter* operates on the principle of electromagnetic induction. The *principle of electromagnetic induction*, known as Faraday's law of electromagnetic induction, states that

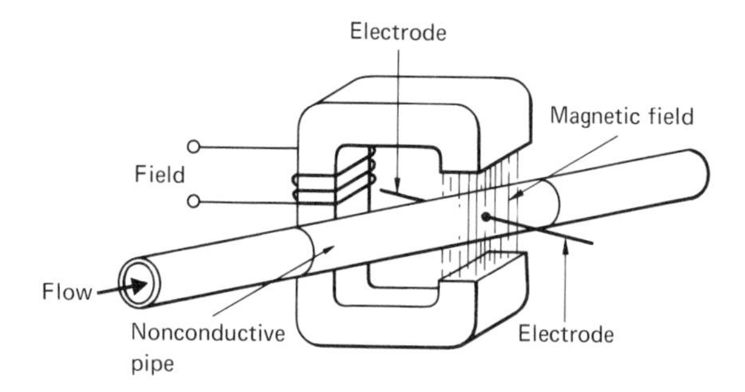

Figure 7.68 Electromagnetic Flowmeter

a conductor passing through a magnetic field (at right angles) produces a voltage. The voltage produced is proportional to the relative velocity of fluid flow.

The electromagnetic flowmeter, shown in Figure 7.68, generates a magnetic field through an electromagnet. The conductor is the flowing fluid. As we might expect, the fluid must be conductive. A potential is then produced between the electrodes, which are imbedded in a piece of nonconductive pipe. Since no obstructions exist, the pressure loss is small. Additionally, the lack of obstructions makes measuring slurries easy.

Target Flowmeter

The *target flowmeter*, shown in Figure 7.69, presents an obstruction to fluid flow. The fluid exerts a pressure on the target that is directly proportional to the fluid velocity. The force on the target is sensed by a strain gauge.

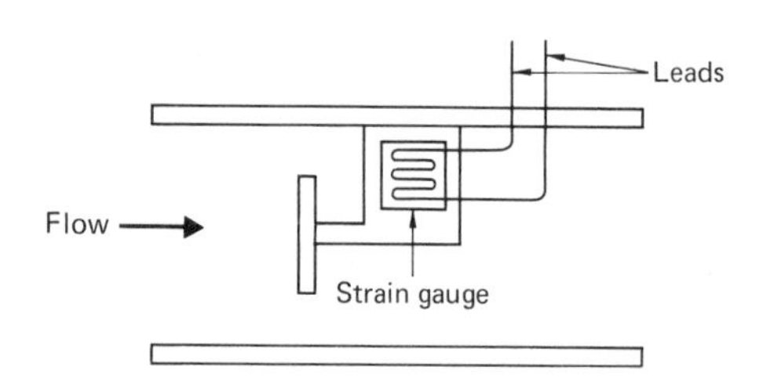

Figure 7.69 Target Flowmeter

Thermal Heat Mass Flowmeters

The *thermal heat mass flowmeter* is an accurate device for measuring fluid flow. In this device, heat is applied to the flowing fluid. A temperature sensor is then used to measure the rate at which the fluid conducts heat away. This rate is directly proportional to the flow rate. One of the foremost advantages of the thermal heat mass system is its lack of obstructions to fluid flow.

Table 7.7 summarizes the fluid flow measuring devices discussed in this section.

Liquid Level

The methods of measuring liquid level are as numerous and varied as those of fluid flow. Also, there are as many reasons for wanting to know the level of a liquid (or solid) as there are reasons for wanting to know the rate of fluid flow. Liquid level is an important part of process control and accounting.

Liquid-level measurement is divided into two basic types: point and continuous. The *point level* is just what its name implies: measurement of a level at a point. *Continuous-level measurement* is analog in nature. The output of a continuous-level sensing instrument is directly proportional to the level of the liquid. Some sensors are more suited to one or the other of these types of measurement. In our description of the sensors, we point out which application is more appropriate. For purposes of discussion, we classify liquid-level sensors into sight, force, pressure, electric, and radiation sensors.

Sight

Sight level sensors are among the oldest and simplest. The *dipstick* is possibly the oldest, and it is still used in some parts of industry. To make a

Table 7.7 *Summary of Flowmeters*

Type	*Name*	*Range*	*Operating Principle*	*Accuracy*	*Advantages*	*Disadvantages*
Differential pressure	Orfice plate	3:1	Variable head	1%–2%	Lowest in cost Easy installation	Highest permanent-head loss Cannot be used with slurries
	Flow nozzle	3:1	Variable head	1%–2%	Can be used at high velocity Can be used with slurries	Expensive Difficult to remove
	Venturi tube	3:1	Variable head	1%–2%	Most accurate of these meter types Can be used with slurries Low permanent-head loss	Difficult to remove and install Costly
	Dall tube	3:1	Variable head	1%–2%	Lowest permanent-head loss	Costly
	Elbow	3:1	Centrifugal force	5%–10%	Saves space No head loss No obstructions	Low accuracy Small differential pressure created
Positive displacement	Nutating disc	5:1	Measured volume	0.1%	High accuracy Simple to install	Costly to install and maintain Cannot be used with slurries
Velocity	Turbine	10:1	Spinning turbine or propeller	0.5%	High accuracy Low head loss Wide temperature range	Cannot be used with slurries Hard to install
	Electromagnetic	30:1	Electromagnetic voltage generation	1%	Linear No obstructions No head loss	Must have conductive fluid High cost
	Target	3:1	Fluid pressure on target	0.5%–3%	Compact Easy to install Low cost	Subject to temperature variations
Variable area	Rotameter	10:1	Balance between gravity and fluid pressure	1%–2%	Little head loss Linear response	Expensive Fragile
Mass flow	Thermal heat mass flowmeter	100:1	Temperature rise to mass flow rate	2%	Measures low flow rates Low head loss Fast response	Only useful for gases Costly and complex

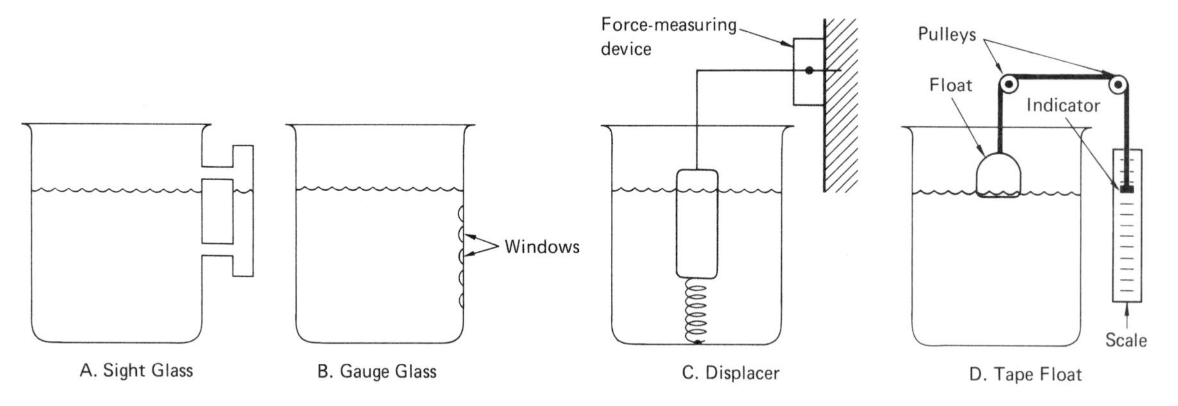

Figure 7.70 Liquid-Level Measurement by Sight

measurement, you insert a stick (sometimes calibrated with lines, as in a car oil dipstick) into the container. You then withdraw it and observe the level of fluid on the dipstick.

Related to the dipstick method are the sight glass and the gauge glass. The *sight glass* is similar to the coffee urns found in cafeterias. The sight glass allows you to see the liquid level in the tube, as shown in Figure 7.70A. A *gauge glass* is similar (Figure 7.70B). It has windows in its sides to allow you to see the liquid level.

The *displacer* works on a different principle entirely. It uses Archimedes' principle. Archimedes found that a body immersed in water loses weight equal to the amount of water displaced. In Figure 7.70C, the weight of the displacer object is proportional to the liquid level. As the liquid rises, the displacer object has less weight. The weight measurement is made mechanically and displayed on a pointer, or electric force sensor. A strain gauge works nicely here.

The *tape float* (Figure 7.70D) works in a similar fashion. The float rides on top of the liquid. It is attached to a tape, at the end of which is an indicator. As the level rises, the tape moves down the scale.

All the sight methods have the same disadvantage: They require human operators to see and

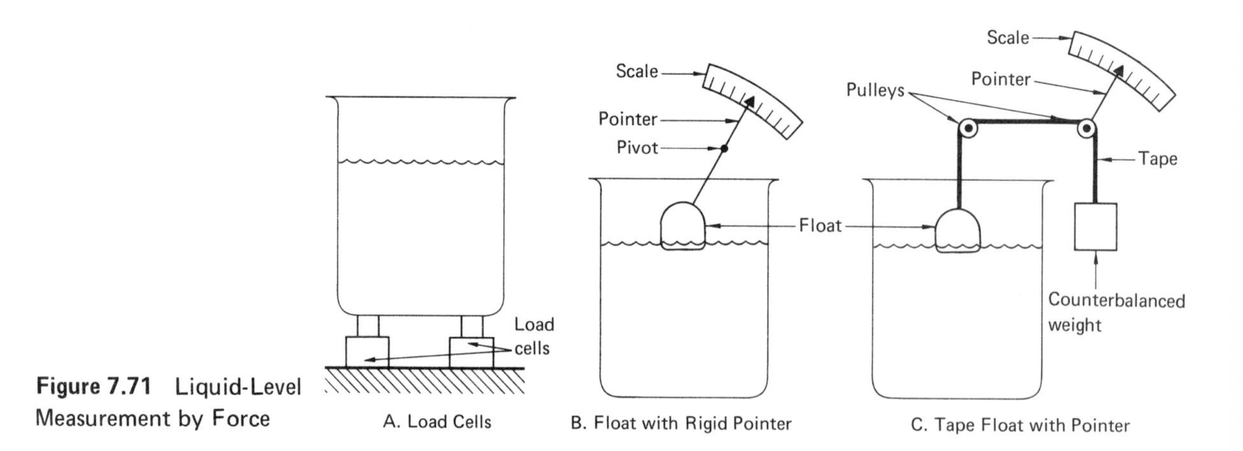

Figure 7.71 Liquid-Level Measurement by Force

record the levels. However, these methods are very inexpensive. And all are suited to continuous measurement.

Force

The weight of a container (force) is often used to indicate liquid levels. When more liquid (or solid) enters the container, it weighs more. The weight is detected by *force* or *strain sensors*, as in the load cell of Figure 7.71A. Alternatively, the buoyant force of a float moves a rigid rod or flexible tape, cable, or chain, as in Figures 7.71B and 7.71C.

All these methods are suited to continuous measurement of liquid level. The float methods shown may be converted to an electrical output by placing a potentiometer at the pivot point. Changes in the linear displacement of the float are then changed to angular displacement by the pulley or pivot. The potentiometer changes angular displacement into an electrical resistance change.

Pressure

Pressure is another variable that is affected by liquid level. As liquid level increases, more pressure is exerted at the bottom of the tank. Devices that use this method of sensing level are called *hydrostatic-head devices*. In the *diaphragm type*, shown in Figure 7.72A, more pressure is exerted on

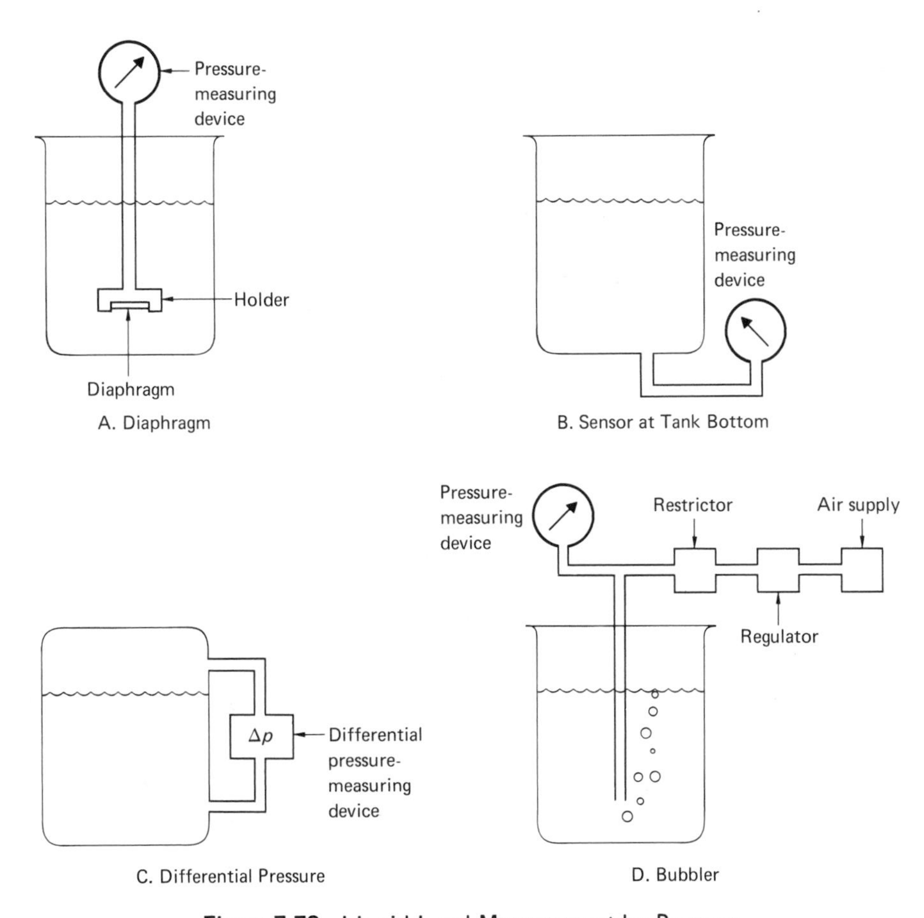

Figure 7.72 Liquid-Level Measurement by Pressure

the diaphragm as level increases. The air pressure inside the tube then increases and registers a pressure on the meter. Figure 7.72B shows a similar device mounted on the bottom of the tank.

Other hydrostatic-head devices use differential pressure, as shown in Figure 7.72C. The pressure difference between the bottom and top of the tank depends on liquid level. The difference is sensed by a differential pressure–measuring device, such as bellows or National Semiconductor's IC differential pressure transducers.

The *bubbler system*, shown in Figure 7.72D, is one of the oldest and simplest devices for measuring pressure. Air flows in through the tube placed in the tank. If enough pressure is exerted in the tube, bubbles are forced out of the tube. The pressure required to force the bubbles out varies proportionally with the liquid level. This pressure is indicated by the pressure-measuring device.

Electric

Beyond the mechanical-to-electrical conversions mentioned previously, certain sensors give direct electric outputs with changing liquid levels. One interesting device, manufactured by Metritape, changes the resistance of a wire helix as pressure is applied; see Figure 7.73. This device is suited to continuous-level measurement.

Electrodes are also used to sense electrical changes. In Figures 7.74A and 7.74B, the capacitance changes as the liquid level fluctuates. The change in capacitance is caused by the changing dielectric constant between the liquid and the air. The changes in capacitance are usually sensed with an oscillator or an AC bridge.

Note the resistive electrode sensor in Figure 7.74C. With conductive liquids, the electrodes in Figure 7.74C provide a path for current flow

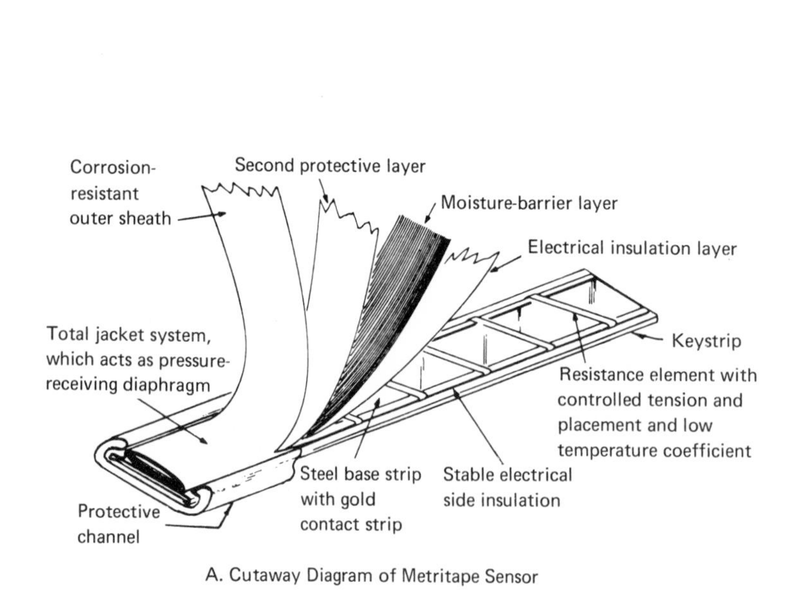

A. Cutaway Diagram of Metritape Sensor

B. Electrical Equivalent

Figure 7.73 Resistance Changes with Changes in Liquid Level

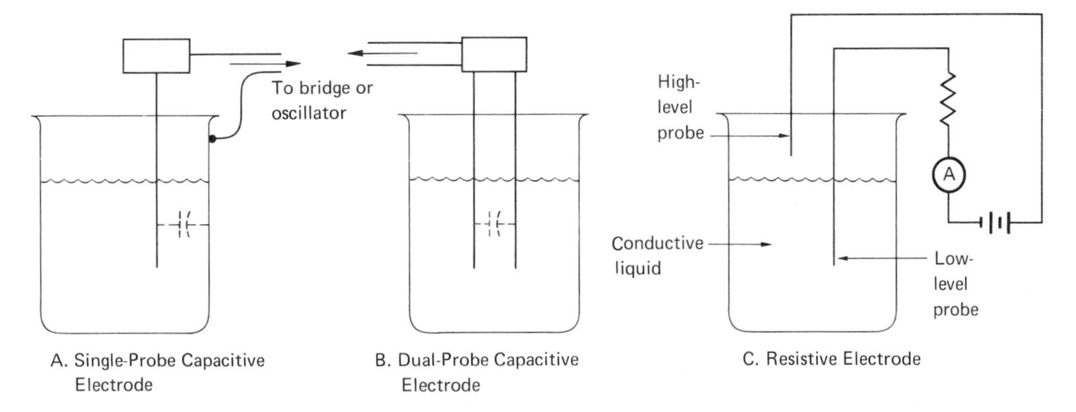

Figure 7.74 Liquid-Level Measurement by Electrical Means

through the series circuit. This type of system is more suited to a point-sensing application.

Radiation

Most of the level measurement devices discussed previously have elements that are in contact with the liquid. But in some industrial measurements, the level-sensing device must be used with corrosive liquids or liquids under high pressures. Hence, contact with the liquid might destroy the device. In this section, we will discuss one class of sensors that do not physically contact the liquid at all: radiation sensors. We divide these sensors into two classifications: sonic (including ultrasonic) and nuclear radiation.

Sonic

Sonic and ultrasonic beams (7.5–600 kHz) use the *echo principle* to measure liquid level. In Figure 7.75A, the sonic beam is transmitted from the transducer to the liquid surface. It is then reflected back to the transducer from the surface. The time it takes for the beam to travel from the transducer to the surface and back to the transducer depends on the liquid level. The lower the level, the more time it takes for the beam to travel the distance. The time interval is then inversely proportional to the liquid level.

Nuclear

Nuclear radiation is sometimes used to make liquid-level measurement for point-sensing applications. In Figure 7.75B, radiation is emitted from

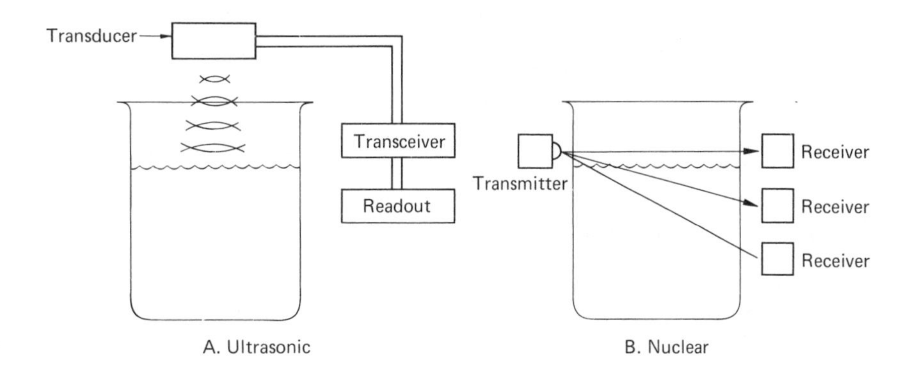

Figure 7.75 Liquid-Level Measurement by Radiation

Table 7.8 *Summary of Liquid-Level Sensors*

Type	*Name*	*Use**	*Temperature Range*	*Pressure* (lb/in.^{2})	*Accuracy*	*Range*
Sight	Tape float	C	to 149°C	300	$\pm 1/8$ in.	to 60 ft
	Displacer	P, C	to 260°C	300	$\pm 1/4$ in.	10 ft
	Sight glass	P, C	to 260°C	10,000	$\pm\frac{1}{2}\%$–1%	6–8 ft
Pressure	Diaphragm	P, C	to 459°C	Atmospheric	± 1 in.	Unlimited
	Differential pressure	P, C	to 649°C	to 6000	$\pm\frac{1}{2}\%$–1%	—
	Bubbler	C	Dew point	Atmospheric	$\pm 1\%$–2%	Unlimited
Electrical	Resistive electrode	P	−26°–82°C	to 5000	$\pm 1/8$ in.	to 66 ft
	Capacitance	P, C	−26°–982°C	to 5000	$\pm 1\%$	to 20 ft
Radiation	Sonic	P, C	−40°–149°C	to 150	$\pm 1\%$–2%	to 125 ft
	Ultrasonic	P, C	−26°–60°C	to 320	$\pm 1\%$–2%	to 12 ft
	Nuclear	P, C	to 1648°C	Unlimited	$\pm 1\%$–2%	Unlimited

* P = point; C = continuous.

the transmitter in the form of gamma rays. The receiver is usually a Geiger-Müller tube. The radiation is absorbed by the liquid or solid. The output of the receiver is then dependent on whether or not the liquid is between the sensor. Nuclear sensors are expensive compared with other electrical methods of sensing level. They also require extensive in-plant safety precautions.

Table 7.8 summarizes the different types of liquid-level sensors and some pertinent information about them.

Measurement with Bridges

Many of the transducers discussed previously use bridges to transform their resistance or impedance change to a voltage change. The *bridge* can transform a small change in resistance or impedance to a large change in voltage. The bridge circuit is capable of detecting changes on the order of 0.1%.

The basic four-arm DC bridge is shown in Figure 7.76. No current flows through the meter when the following resistance ratio is met:

$$\frac{R_1}{R_3} = \frac{R_2}{R_4} \tag{7.15}$$

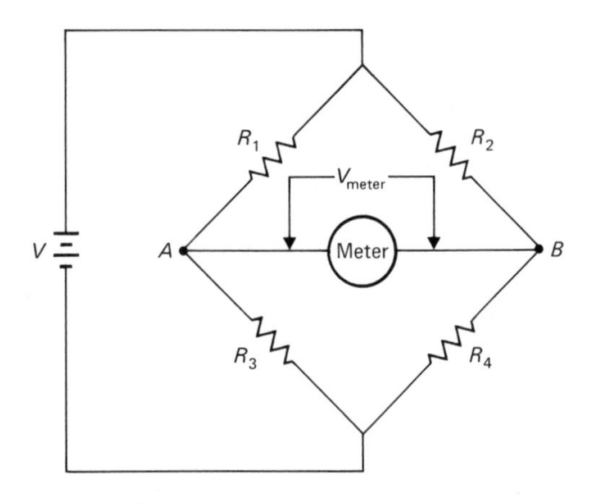

Figure 7.76 Four-Arm DC (Wheatstone) Bridge

When no current flows through the meter, the bridge is said to be in a *null state.*

In transducer applications, one of the resistors of Figure 7.76 is replaced by the transducer. Since the three remaining resistors are constant, any change in the bridge null state is caused by the transducer. Furthermore, the resistance of the transducer can be calculated at null by using Equation 7.15, solving the equation for the unknown resistance.

If R_4 is the unknown resistance, it is related to the other resistances in the bridge by the following equation:

$$R_4 = \frac{R_2 R_3}{R_1} \tag{7.16}$$

For example, if $R_1 = 75\ \Omega$, $R_2 = 100\ \Omega$, and $R_3 = 150\ \Omega$, then R_4 is $200\ \Omega$.

A resistance change causes a bridge to become unbalanced. The imbalance is in the form of a voltage between points A and B. This difference in potential causes current to flow through the meter. The potential difference (V_{meter}) between A and B can be approximated by the following equation:

$$V_{\text{meter}} = V\left[\frac{R_2\,\Delta R_4}{(R_2 + R_4)^2}\right] \tag{7.17}$$

where V = supply voltage
ΔR_4 = change in resistance of R_4
R_4 = original resistance of R_4

Let us suppose that the bridge is balanced with $R_1 = 100\ \Omega$, $R_2 = 200\ \Omega$, $R_3 = 400\ \Omega$, and $R_4 = 800\ \Omega$. The supply voltage is 100 V. If R_4 changes by $10\ \Omega$, then V_{meter}, previously at 0 V, will be at about 0.2 V, as shown in the following calculation:

$$\begin{aligned} V_{\text{meter}} &= V\left[\frac{R_2\,\Delta R_4}{(R_2 + R_4)^2}\right] \\ &= 100\ \text{V}\left[\frac{(200\ \Omega)(10\ \Omega)}{(200\ \Omega + 800\ \Omega)^2}\right] \\ &= 0.2\ \text{V} \end{aligned} \tag{7.18}$$

Another helpful relationship to know is the bridge's sensitivity. The *sensitivity* of a bridge is the change in output voltage compared with the transducer's change in resistance. That is, the sensitivity (s) of a bridge (in volts per ohm) is equal to the change in voltage between points A and B (V_{meter}) divided by the change in the transducer's resistance (ΔR_4):

$$s = \frac{\Delta V_{\text{meter}}}{\Delta R_4} = V\left[\frac{R_2}{(R_2 + R_4)^2}\right] \tag{7.19}$$

The sensitivity of the bridge in our example is $0.2\ \text{V}/10\ \Omega$, or $20\ \text{mV}/\Omega$. In other words, for every ohm that R_4 changes, the voltage V_{meter} changes about 20 mV.

Conclusion

We have covered a great many sensors in this chapter. All sensors convert the variable to be measured to an analog of that variable. As we have seen, the most common conversion is to an electrical quantity, because electric control systems predominate in industry.

You should have a firm grasp of the transduction principles behind each sensor. Although you will probably not be asked to choose a sensor for a system, you will have to work with existing sensors in control systems. Thus, you will need a knowledge of the concept behind the sensor.

In the next chapter, the sensor is assumed to be an integral part of a process control system.

Questions

1. A transducer is a device that converts a ______ into an output that facilitates measurement. The mercury thermometer converts temperature to ______.
2. The mechanical temperature transducers operate on the principle that gases and liquids change ______ when heated. The potential produced by the junction of two dissimilar metals is used in the ______ temperature sensor.
3. The psychrometer assesses relative humidity by comparing the ______ of a wet and a dry bulb. The hair hygrometer element changes ______ with varying relative humidity.
4. The ______ photoresistor's spectral response closely approaches that of the human eye. Of the optoelectronic sensors discussed, the ______ is the most sensitive.
5. Because the phototransistor's response is so nonlinear, it is used more in ______ systems than analog systems. The ______ cell increases its voltage output with increasing light levels.

6. The displacement sensor with a primary wound between two secondary windings is called a(an) ______. The bonded-wire strain gauge changes its ______ with strain.
7. Two devices used to measure magnetic field strength are the Hall effect device and the ______.
8. The ______ pressure transducer is more suited to measuring high pressures. A bellows converts pressure to ______.
9. The most commonly used differential pressure flowmeter is the ______ ______. Differential pressure flowmeters use the ______ effect to generate a pressure difference.
10. The two types of liquid-level measurement are called continous and ______. The electronic level probe is more suited to ______ measurement.
11. Define temperature.
12. Describe the function of a transducer.
13. Compare sensors that are active (give an output voltage with input energy) and those that are passive (give no output voltage). Give an example of each of the two types of sensors from among the temperature transducers and optical transducers.
14. Name the parts of a filled-system, temperature-measuring device.
15. Describe the behavior of a thermistor when it is heated. What type of temperature coefficient does it have? Is its response linear?
16. Describe how a thermistor can be used in (a) a fluid flow system, (b) a liquid-level system, and (c) time delay. *Hint*: Does the thermistor change its resistance immediately when heated?
17. How can a thermistor be made more linear? Name two methods.
18. Of the temperature sensors discussed in this chapter, which is the most sensitive?
19. Describe how a thermocouple is used to measure temperature.
20. Describe the operational principle behind the bimetallic strip.
21. Describe how the following humidity sensors work: (a) hair hygrometer, (b) resistance hygrometer, and (c) microwave transceiver.
22. What is the relationship between gain and sensitivity in optoelectronic sensors?
23. Define the gauge factor for a strain gauge.
24. Describe two types of differential pressure transducers.
25. Describe measurement of liquid level and fluid flow using differential pressure transducers.
26. Compare the operation of the Bourdon tube with that of the bellows.
27. List the maximum ranges of the bellows, diaphragm, and Bourdon tube.

Problems

1. A type J (iron-constantan) thermocouple measures 6.907 mV across the device. Find the temperature of the test junction if the reference junction is at (a) 0°C and (b) 25°C.
2. The temperature of a type J thermocouple is 400°C. Find the voltage across the thermocouple if the reference junction is at 0°C.
3. A straight bimetal strip is 20 cm long, 0.1 cm thick, and clamped at one end. How much would it move if it was made of brass-invar and heated to 15°C?
4. A spiral bimetal strip has a radius of 10 cm and a thickness of 0.2 cm, and it is made of metals with coefficients of 17 and 5 millionths/°C. The spiral is 50 cm long, and it deflects by 1 cm. What is the heat change that produced this deflection?
5. A thermocouple has a mass of 0.008 g, a specific heat of 0.08 cal/g°C, a heat transfer coefficient of 0.05 cal/cm-s°C, and an area of 0.05 cm^2. What is its thermal time constant?
6. A strain gauge has an initial resistance of 200 Ω. Under a strain of 10^{-4}, resistance increases to 402 Ω. What is the gauge factor?

Process Control

Objectives

On completion of this chapter, you should be able to:

- Define process control;
- List the elements of a process control system;
- Define process load, process lag and stability, and describe how they relate to a process control system;
- List and give examples of two types of process control systems;
- Distinguish between feedback and feedforward control;
- Describe the following controller modes: on-off, proportional, integral, and derivative;
- Contrast electrical controllers and pneumatic controllers in terms of the advantages and disadvantages of each.

CHAPTER 8

Introduction

A *process control system* can be defined as the functions and operations necessary to change a material either physically or chemically. Process control normally refers to the manufacturing or processing of products in industry.

Every process has one or more controlled, or dynamic, variables. The *controlled (dynamic) variable* is a variable we wish to keep constant. Processes also have one or more manipulated variables, or control agents. A *manipulated variable* is a variable that we change to regulate the process. Specifically, the manipulated variable enables us to keep the controlled variable constant. Examples of controlled or manipulated variables are pressure, temperature, fluid flow, and liquid level. Finally, each process control system has one or more disturbances; *disturbances* tend to change the controlled variable. The *function* of the process control system is to regulate the value of the controlled variable when the disturbance changes it.

An example may help to identify the components of a process control system. Figure 8.1 illustrates a system in which milk is heated to pasteurizing temperature. The temperature of the milk is the controlled variable. Steam flows through the pipes, transferring its heat to the milk. The flow of steam controls the temperature of the milk. Therefore, the steam flow is the manipulated variable. The ambient temperature surrounding the tank could be considered a disturbance. For

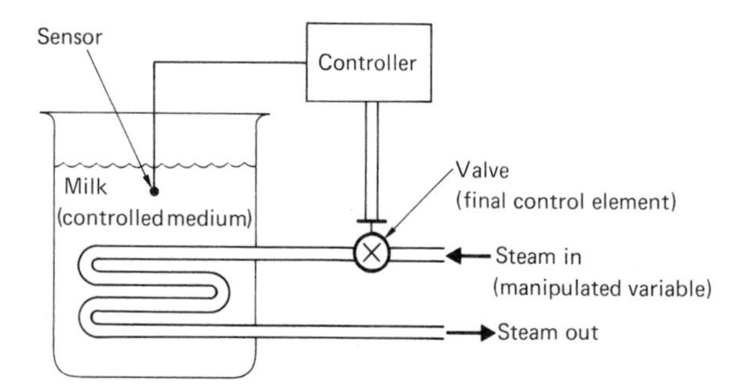

Figure 8.1 Process Control System for Pasteurizing Milk

instance, if the ambient temperature decreased, the temperature of the milk would eventually decrease.

In an automatic control situation, the temperature of the milk would be monitored. If it decreased beyond allowable limits, the controller would make adjustments to bring the temperature under control. The *controller* is that part of a process control system that decides how much adjustment the system needs and implements the results of that system. For example, the controller might open the steam control valve, causing the milk to be heated and returning it to the required temperature. This valve is the final control element in a process control system.

In this chapter, we will present a basic outline of process control systems. Since this subject is rather complex, we will present only a general overview of these systems. We will first discuss the characteristics that are found in most process control systems. Next, we will consider general types of process control and the ways in which controllers operate. Finally, we will give a short summary of those final control elements we have not discussed in previous chapters. Recall that we have already covered motors in Chapters 3 and 4.

Well-designed process control systems can save an enormous amount of time and money, reduce error (which improves the quality of a product), and provide greater operator safety. Because of these advantages, sophisticated process control systems are frequently found in the manufacturing industry. The task of preventive and corrective maintenance varies directly with the complexity of the system. Therefore, it is wise for technicians to have a firm grasp of the concepts behind process control systems.

Characteristics of Process Control Systems

All process control systems have three characteristics in common. First, each automatic process control system makes a *measurement* of the controlled variable. This measurement is usually made by a sensor or transducer. As we saw in Chapter 7, transducers change the controlled variable into another form, usually electrical. In our milk example, a thermistor could have been used to change the temperature of the milk into a corresponding electrical resistance. As the temperature of the milk changed, so would the sensor's resistance. Since the sensor's output must be evaluated by the controller, the output must be in a suitable form. For example, if the controller is a computer, the sensor's output must be digital. Thus, the sensor's output may need to be converted, or conditioned, to an appropriate form for evaluation.

Second, as mentioned above, the controller needs to *evaluate* the information from the sensor. It does so by comparing the sensor measurement with a reference called the *set point*. If the sensor measurement differs from the set point, an *error condition* occurs. The controller then decides whether or not this amount of difference is acceptable. If it is not acceptable, the controller initiates action to reduce the error.

Finally, each process control system must have a final control element. A *final control element* makes those adjustments that are necessary to bring the controlled variable back to the set point value.

Process Characteristics

Every process exhibits three basic characteristics that are important in the understanding of process control systems. These characteristics are process load, process lag, and stability.

Process Load

Process load can be defined as the total amount of control agent needed to keep the process in a balanced condition. In the milk pasteurization example, we need a certain amount of steam (the control agent) to keep the milk at the correct temperature. Suppose the ambient temperature around the tank decreased. Then, we would need a greater rate of steam flow to keep the milk at a constant temperature. Thus, a decrease in the ambient temperature constitutes a change in the process load.

The process load is directly related to the setting of the final control element. Any process load change will cause a change in the state of the final control element. The final control element's adjustment is what keeps the process balanced.

Process Lag

Process lag is the time it takes the controlled variable to reach a new value after a process load change. This time lag is a function of the process and not the control system. The control system has time lags of its own. Process lags are caused by three properties of the process: capacitance, resistance, and transportation time.

Capacitance can be defined as the ability of a system to store a quantity of energy or material per unit quantity of a reference. For example, a large volume of water has the ability to store heat energy. As a result, we could say that the water has a large thermal capacitance.

A large capacitance in a process means that it takes more time for process load changes to occur. From one viewpoint, large capacitance is a desirable characteristic, because it is then easier to keep the controlled variable constant. Small disturbances do not have very much effect on the process load. On the other hand, large capacitance also means that it is more difficult to change the controlled variable back to the desired point once it has been changed.

Resistance in a process can be defined as opposition to flow. In the pasteurization example, we would notice thermal resistance in the walls of the steam pipes. In other words, the material of the pipes slows down the transfer of heat to the liquid. Sometimes, gases and liquids surround the pipes in layers or films, increasing the thermal resistance compared with pipes without the layers or films. These layers have a slowing effect on the transfer of heat energy and thus have resistance. Large resistances will, therefore, increase the process lag by opposing the change in the controlled variable.

Often, the capacitance and resistance of a system are combined into a factor called *RC delay*, or the *RC time constant*. Recall that, in an electric system, the time it takes to charge a capacitor to 63% of its final voltage is called one time constant. The rate at which the capacitor charges will depend on the capacitance of the capacitor and the resistance through which it must charge. The product of the resistance and the capacitance (*RC*) gives one time constant, in seconds.

Process control systems have the same *RC* lag. In the pasteurization example above, the product of the thermal capacitance, in joules per degree Celsius, and the thermal resistance, in degrees Celsius per joule per second, will give one time constant, in seconds. In practice, one system may have many *RC* lags caused by many individual resistances and capacitances.

The third component of process lag, *transportation time*, or *dead time*, can be defined as the time it takes for a change to move from one place to another in a process. Or we can define transportation time as the time between the application of the disturbance and the changing of the process load. Dead time is most easily illustrated in applications where fluids or solids are moving, as in a flowing fluid or a conveyor belt application. In a fluid flow application, let us say that a temperature change is made some distance upstream from the sensor. It will take some time for the temperature change to be transported downstream to the sensor. In the conveyor illustration, the same principle applies. Any change in the condition of the product (weight, for example) will not be detected until that change is communicated to the sensor. This time delay is called dead time or transportation time.

Note that any controller action is delayed by the amount of time delay present. To find the time delay, we use the equation $d = rt$, which can be rewritten as follows:

$$t = \frac{d}{r} \tag{8.1}$$

where t = transportation (delay) time, in minutes
d = distance, in centimeters
r = flow rate, in centimeters per minute

In the fluid flow example, suppose that fluid is flowing at a rate of 100 cm/min. And say the temperature sensor is located 50 cm from the temperature change. Then, transportation (delay) time is as follows:

$$t = \frac{d}{r} = \frac{50 \text{ cm}}{100 \text{ cm/min}}$$
$$= 0.5 \text{ min or } 30 \text{ s} \tag{8.2}$$

In other words, it will take 30 s for the temperature change to reach the sensor. Thus, there will be a delay of 30 s *before* the controller can even start to react.

Process engineers are especially interested in keeping dead time to a minimum. Large dead times make accurate process control difficult.

Stability

An important consideration when examining control systems using feedback is the stability factor. We say that a process control system is *stable* if it can return the controlled variable to a steady-state value. Typically, an unstable system will cause the controlled variable to oscillate above and below the desired value.

If the controlled variable oscillates above and below the desired value, three things will happen. First, the strength of the oscillations will increase in amplitude as time increases. This result occurs when the feedback is in phase and the loop gain is greater than 1. (Recall the Barkhausen criteria for oscillators discussed in Chapter 2.) Second, if the feedback is in phase with the oscillations and the loop gain is 1, the oscillations will have a constant amplitude. Third, if the loop gain is less than 1 and the feedback is out of phase with the oscillations, then the oscillations will gradually die out. The amount of time that it takes for the oscillations to die out, or damp, is called the *settling time*. A good process control system will reduce settling time to a minimum.

Types of Process Control

Generally speaking, process control can be classified into two types: open-loop and closed-loop control. Closed-loop control can be further divided into feedback and feedforward control.

Open-Loop Control

Open-loop control involves a prediction of how much action is necessary to accomplish a process. That is, in an open-loop system, no check is made during the process to see whether corrective action is necessary to accomplish the end result.

Figure 8.2 shows a block diagram of a typical open-loop system. A washing machine is a good example of such a system. In this case, the person

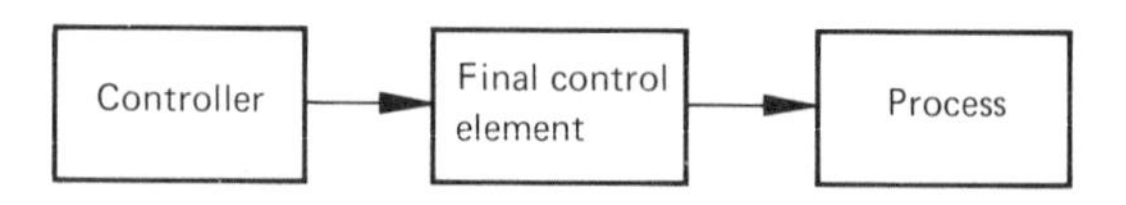

Figure 8.2 Block Diagram of Open-Loop Control System

operating the machine takes a measurement (the clothes are dirty), compares this information with a reference level (clean clothes), and makes a prediction (a cycle is chosen and soap added). The operator then starts the machine and attends to other business.

The operator assumes that the prediction made will accomplish the desired objective (clean clothes). If the prediction is absolutely correct in all aspects (amount of soap, amount of hot and cold water, and so forth), the clothes will be absolutely clean. Therefore, open-loop control is capable of perfect control. However, if the prediction was wrong for any reason, the objective will not be reached. Open-loop control does not always give perfect control. Since the washing machine does not make any measurements or comparisons, any mistakes in the prediction will produce an undesirable outcome.

Closed-Loop Control

In a *closed-loop control system*, a measurement is made of the variable to be controlled. This measurement is then compared to a reference point called a set point. If a difference or error exists between the actual and desired levels, an automatic controller will take the necessary corrective action.

A typical block diagram of a closed-loop control system is shown in Figure 8.3. A comparison of Figures 8.2 and 8.3 shows the difference between open-loop and closed-loop systems. In the open-loop system, no actual measurement is made on the process. In contrast, in closed-loop control, the variable to be controlled is measured and compared to a reference (the set point), and corrective action is taken.

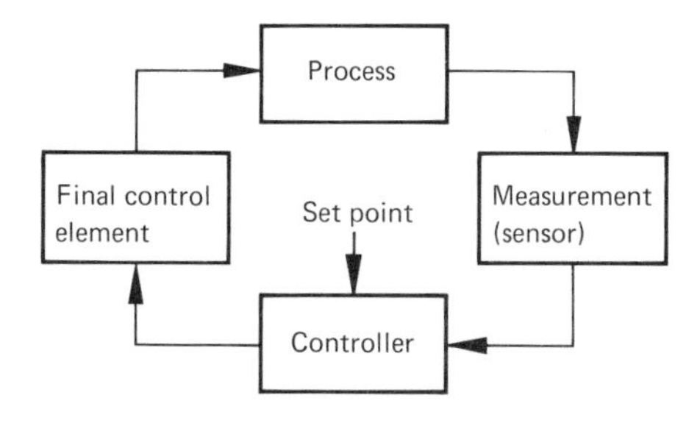

Figure 8.3 Block Diagram of Closed-Loop Feedback Control System

The milk pasteurization example is a good illustration of a closed-loop system. This system is often called a *feedback control system*, since sensor information is fed from the output back to the input.

Figure 8.3 shows a block diagram for a closed-loop feedback control system. The measurement of the dynamic variable is taken by the sensor. The controller compares the information from the set point and the measurement. It then decides whether or not to take corrective action on the basis of its evaluation. The controller can be any circuit or system capable of evaluation and decision making, from a simple op amp comparator to a complex digital computer. In the block diagram, the controller also performs any signal conditioning needed to make the sensor measurement compatible with the input to the evaluation circuit. The controller, if it decides action is to be taken, adjusts the final control element to bring the dynamic variable back under control.

We can relate this block diagram to the pasteurization example discussed at the beginning of the chapter. The measurement is taken by a temperature sensor. The controller, on the basis of this measurement, decides whether to adjust the final control element (a valve) or to leave it unchanged. Recall that the valve controlled the flow of steam in the pipes (the manipulated variable).

Feedforward Control

Closed-loop feedback control has disadvantages in two areas of process control systems. Feedback control is not satisfactory when there are large disturbances in the process load and when there are large process lags. These disadvantages can be dealt with by using a type of control called feedforward (or predictive) control. *Feedforward control* is defined as a closed-loop system that feeds a

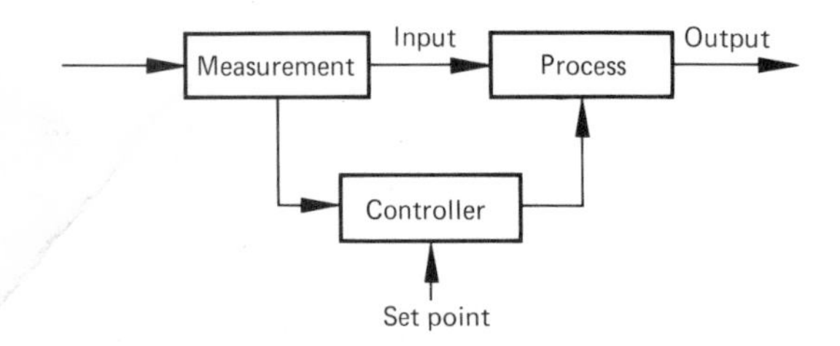

Figure 8.4 Block Diagram of Feedforward Control System

correction signal forward to the controller based on a measurement of the disturbance. A simple feedforward control system is shown in Figure 8.4.

There is one basic difference between feedforward control and open-loop control: Open-loop control makes an assumption about the variables in the system. In contrast, feedforward control makes a measurement of a disturbance (or the manipulated variable). From this measurement, the controller estimates what action needs to be taken to keep the process from changing. In other words, the controller decides what change to make in the manipulated variable so that the change, when combined with the disturbance, will produce no change in the controlled variable. The controller anticipates the effect the disturbance will have on the process.

Feedforward control, like open-loop control, relies on a prediction. It differs from open-loop control in that it does not rely on a fixed program. Feedforward control is a type of closed-loop control.

Note the difference between feedback control and feedforward control. In feedback control, a measurement of the dynamic variable is taken. In feedforward control, a measurement of the disturbance is taken.

Like feedback control, feedforward control is not without its problems. The prediction made in feedforward control assumes that all significant disturbances are known. It also assumes that there are sensors present to measure these disturbances and that we know exactly how the disturbances will affect the process. However, the engineering and technical know-how needed to accomplish this type of control is immense. Thus, feedforward control is used only where the process load and the disturbances to it are well understood and can be accurately predicted.

Basic Control Modes

As we discussed previously, the controller performs evaluation and decision-making operations in our system. It takes the difference or error signal from the comparator and determines the amount of change the manipulated variable needs to bring the controlled variable back to normal. The controller may contain the *error detector* (or comparator), a signal-conditioning device, a recording device, and a telemetering device. The entire system with all these possible combinations is shown in Figure 8.5.

Controllers may be classified in several different ways. For instance, they may be classified

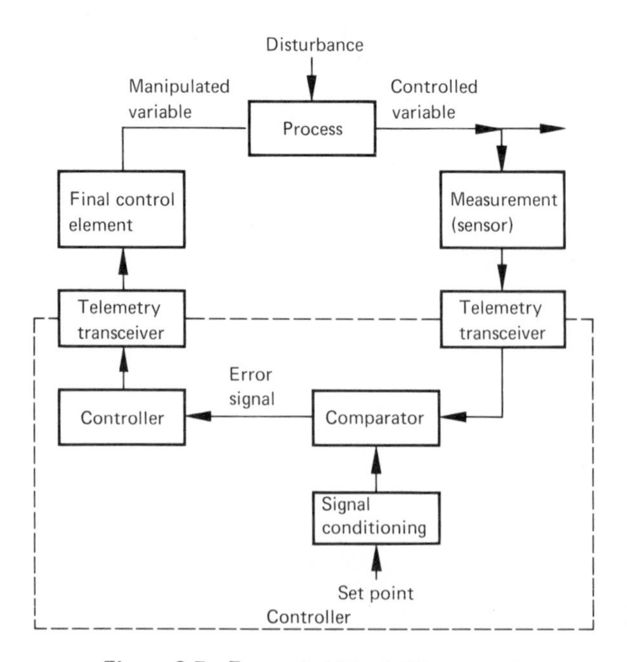

Figure 8.5 Expanded Block Diagram of Closed-Loop Feedback Control System

according to the type of power they use. Two common types in this category are electric and pneumatic controllers. *Pneumatic controllers* are decision-making devices that operate on air pressure. *Electric* (or *electronic*) *controllers* operate on electric signals. You will see both types used widely in industry. Pneumatic controllers are used in the chemical and petrochemical industries for reasons of safety. They are less expensive and simpler than comparable electric controllers. However, pneumatic controllers are difficult to interface with digital computers, a great disadvantage in light of the increasing popularity of computers in process control. Electric controllers suffer no such disadvantage.

Controllers are also classified according to the type of control they provide. In this section, we will discuss five types of control in this category: on-off, proportional, proportional plus integral, proportional plus derivative, and proportional plus integral plus derivative.

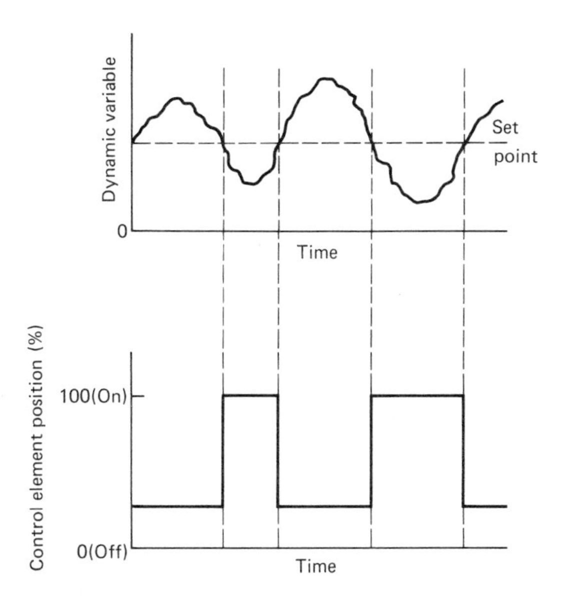

Figure 8.6 Plot of Measured Variable and Final Control Element Position with Respect to Time in an On-Off Controller

On-Off Control

In *on-off control,* the final control element is either on or off. This controller is also called bang-bang from the speed of the response of the on-off state. In the pasteurizing example, if the controller were an on-off controller, the valve would be either open or closed. This type of control is sometimes also called two-position control, since the final control element is either in the open or the closed position. That is, the controller will never keep the final control element in an intermediate position.

On-off control is undoubtedly the most popular mode in industry today. On-off control also has domestic applications. For example, most home heating systems use the on-off control mode. If the room temperature goes below a predetermined point, the heater is turned on. When the room temperature goes above that point, the heater shuts off. Thus, the control is on-off.

On-off control is illustrated in Figure 8.6. As soon as the measured variable goes above the set point, the final control element is turned off. It will stay off until the measured variable goes below the set point. Then, the final control element will turn on. The measured variable will oscillate around the set point at an amplitude and frequency that depends on the process's capacity and time response.

The on-off control mode is chosen by the process engineer under the following conditions:

1. Precise control must not be needed.
2. The process must have sufficient capacity to allow the final control element to keep up with the measurement cycle.
3. The energy coming in must be small compared with the energy already existing in the process.

On-off control is most often found in air-conditioning and refrigeration systems. It is also widely used in safety shutdown systems to protect equipment or people. In this application, on-off control is called a shutdown or cutback alarm.

Differential-gap control, shown in Figure 8.7, is similar to on-off control. Notice that a band, or

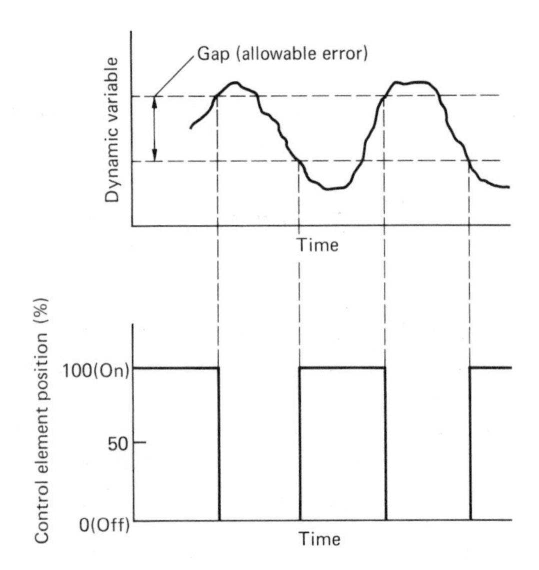

Figure 8.7 Plot of Measured Variable and Final Control Element Position with Respect to Time in Differential-Gap Controller

gap, exists around the control point. When the measured variable goes above the upper boundary of the gap, the final control element closes. It will stay closed until the measured variable drops below the lower boundary. Some home heating systems use this type of control mode rather than on-off.

Differential-gap control does not work as well as on-off, but it does save wear and tear on the final control element. In industry, differential-gap control is often found in noncritical level-control applications, such as keeping a tank from running dry or from flooding.

The gap in differential-gap control is often called a *dead zone*. Engineers normally choose a dead zone of about 0.5% to 2.0% of the range of the final control element.

On-off control is almost always the simplest and least expensive controller mode. This type of control works equally well with electric systems and with pneumatic systems. However, many industrial processes need better, more sophisticated control. Proportional control was developed to meet this need.

Proportional Control

In *proportional control*, the final control element is purposely kept in some intermediate position between on and off. Proportional control is a term usually applied to any type of control system where the position of the final control element is determined by the relationship between the measured variable and the set point.

An example of a proportional control mode using a filled-system temperature sensor is shown in Figure 8.8. The bulb of the filled system senses the temperature of the fluid in the line. Any change in the temperature causes the valve position to change.

The *gain* of the controller is determined by the relationship of change in temperature to change in valve position. If the valve moves a great deal for a given temperature change, the gain is high. If the valve does not move very much for a given temperature change, the controller gain is small. The valve is normally set so that when the temperature is in the center of its range, the valve is one-half open. In this system, the set point is

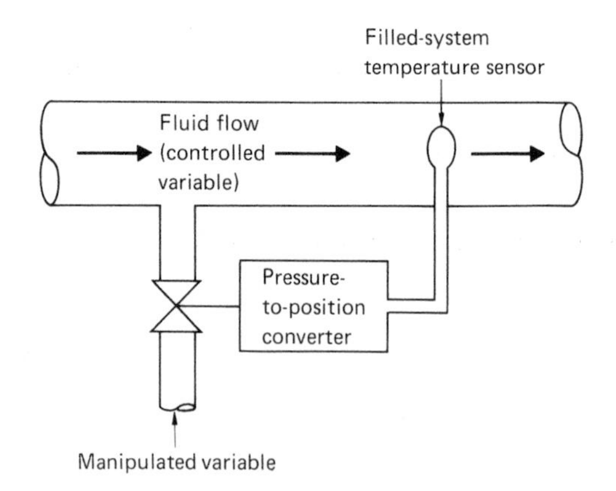

Figure 8.8 Proportional Control System Using Filled-System Temperature Sensor

determined by the initial setting of the valve (50%). Once this system is installed, the controller gain is also fixed. More sophisticated proportional control systems have variable gain or set points.

The final control element setting when the measured variable equals the set point is called the *bias*. In our example, the bias is 50%. As the measurement deviates from the set point, the final control element setting will change from 50%. The amount of change in output (final control element) for a given change in input is a function of the gain of the controller's amplifier, as shown in the following equation:

$$\text{controller gain} = \frac{\Delta\text{output}}{(\text{set point} - \text{measurement})} \quad \textbf{(8.3)}$$

Most industrial controllers have a gain adjustment that is expressed in percent of the proportional band. *Proportional band* is the amount of change in the dynamic variable that causes a full range of controller output. Or we can say that the proportional band is equal to the range of values of the dynamic variable that corresponds to a full or complete change in controller output. Proportional band is also called throttling range. Normally, the proportional band is expressed as a percentage:

$$\%\ \text{proportional band} = \frac{1}{\text{gain}} \times 100 \quad \textbf{(8.4)}$$

Then:

$$\text{gain} = \frac{100}{\%\ \text{proportional band}} \quad \textbf{(8.5)}$$

Note the relationship between gain (or sensitivity) and proportional band. Systems with large proportional bands are less sensitive (have less gain) than narrow-band systems.

The equation for determining output for the proportional controller in Figure 8.8 is as follows:

$$\text{output} = \frac{100}{\%\ \text{proportional band}} \times (\text{set point} - \text{measurement}) + \text{bias} \quad \textbf{(8.6)}$$

Thus, for a fixed gain (proportional band setting) and a fixed bias, we can calculate the output. To do so, we must know the measurement and the set point. For example, let's say the proportional band is 100% (gain = 1) and the bias is 50%. When the measurement equals the set point, the output (final control element setting) is 50%. When the measurement exceeds the set point by 10%, the output is 40%. When the measurement is below the set point by 10%, output is 60%.

The higher the gain, the more the output will move for a given change in either measurement or set point. Figure 8.9A shows the effect on the input-output relationship for a gain change.

Another way of showing the effect of changing the proportional band is illustrated in Figure 8.9B. Each position in the proportional band produces a controller output. The wider the band, the more the input signal (set point − measurement) must change to cause the output to swing from 0% to 100%. Changing the bias shifts the proportional band so that a given input signal will cause a different output level.

Proportional control tries to return a measurement to the set point after a disturbance has occurred. However, it is impossible for a proportional controller to return the measurement so that it equals the set point. By definition, the output must equal the bias setting (normally 50%) when the measurement equals the set point. If the loading conditions require a different output, a difference between measurement and set point must exist for this output level. Proportional control may reduce the effect of a load change, but it can never eliminate it.

The resulting difference between measurement and set point, after a new balance level has been reached, is called *offset*. The amount of offset may be calculated from the following equation:

$$\Delta\text{offset} = \frac{\%\ \text{proportional band}}{100} (\Delta\text{measurement}) \quad \textbf{(8.7)}$$

The change in measurement is the change required by load upset.

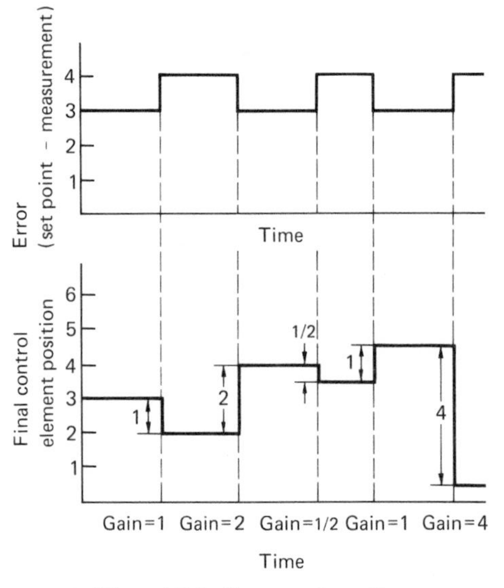

A. Effect of Gain Change on Input-Output Relationships

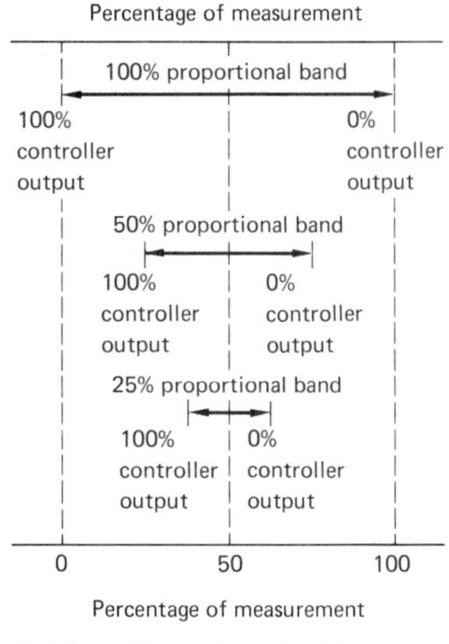

B. Effect of Proportional Band Change on Controller Output and Measurement

Figure 8.9 Plots of Controller Gain and Proportional Band Changes

From this equation, we see that as the proportional band goes toward zero (gain approaches infinity), offset will approach zero. This result seems logical, because a controller with infinite gain is, by definition, an on-off controller. We know that we cannot have an offset in an on-off controller. On the other hand, as proportional band increases (gain decreases), proportionately more offset will exist.

Proportional plus Integral Control

Often, in industrial applications, the offset caused by proportional control cannot be tolerated. Process engineers solve this problem by adding another control mode. Recall from Chapter 2 that the op amp integrator integrates any voltage present at the input. *Integral control* (sometimes called reset action) in a process control system will integrate any difference between the measurement and the set point. The controller's output will change until the difference between the measurement and the set point is zero.

The response of a pure integral controller is shown in Figure 8.10A. Note that the controller output changes until it reaches 0% or 100% of scale (or until the measurement is returned to the set point). This figure also assumes that an open-loop condition exists where the controller's output is not connected to the process.

Figure 8.10B shows an open-loop, proportional plus integral controller's response to a step change. *Proportional plus integral control* (*PI*) combines the characteristics of both types of control. Reset time is the amount of time required to treat the amount of change caused by proportional action. In Figure 8.10B, reset time t is the amount of time required to repeat the amount of output change y.

Another way to visualize PI action is shown in Figure 8.11. In this example, a 50% proportional band is centered around the set point. If a disturbance comes into the system, the measurement will deviate from the set point. Proportional

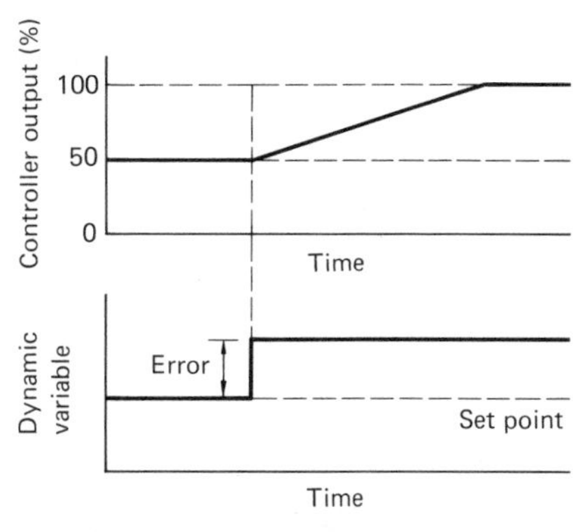

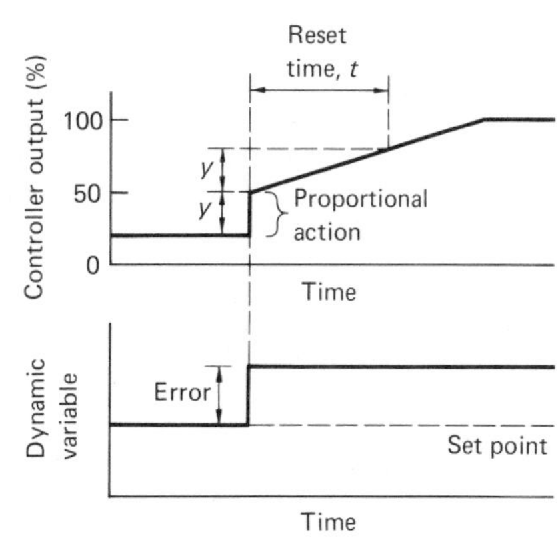

Figure 8.10 Plots of Controller Output and Dynamic Variable Measurement versus Time

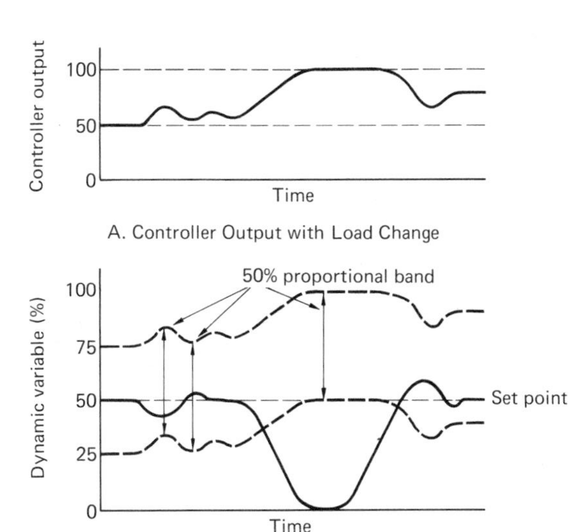

Figure 8.11 Shifting of Proportional Band in PI Controller When Error Changes

response will be seen immediately in the output, followed by integral action.

We may think of integral action as forcing the proportional band to shift. This shift causes a new controller output for a given difference between measurement and set point. Integral control will continue to shift the proportional band as long as there is a difference between the measurement and the set point.

Integral control action does exactly what an operator would do by manually adjusting the bias in the proportional controller. The width of the proportional band stays constant. It is shifted in a direction opposite to that of the measurement change. Thus, an increasing measurement signal results in a decreasing output, and vice versa. Because integral control acts only on long-term or steady-state errors, it is seldom used alone.

The PI control mode is used in situations where changes in the process load do not happen very often, but when they do happen, changes are small. These small changes may occur for a long time before they finally go beyond the allowable error limits. In many industrial processes, even small amounts of error that persist for long periods of time are undesirable. Consequently, PI control is a very popular mode of control in industry today.

Proportional plus Derivative Control

Some process control systems have errors that change very rapidly. This situation is especially true in processes that have a small capacitance.

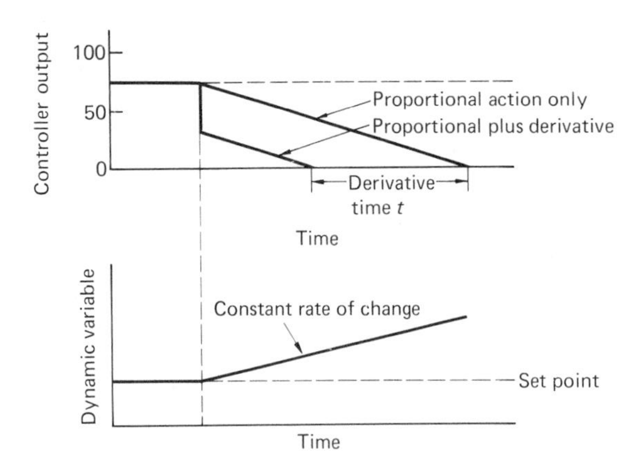

Figure 8.12 Plot of Measurement and Controller Output versus Time in PD Controller

Neither proportional control nor PI control responds well to errors that change rapidly. Recall from Chapter 2 that the differentiator circuit responded to the input rate of change, or slope. The same concept applies to the type of controller mode called *derivative*, or *rate, control*. By adding derivative control to proportional control, we get a controller output that responds to the measurement's rate of change as well as to its size.

Proportional plus derivative control (*PD*) is illustrated in Figure 8.12. When a measurement changes, derivative action differentiates the change and maintains a level as long as the measurement continues to change at a given rate.

Derivative control is never used alone because it can only react to measurements when they are changing. It cannot react to steady-state errors.

Proportional plus Integral plus Derivative Control

All three controller modes we have discussed can be combined into one mode, *proportional plus integral plus derivative* (*PID*) *control*. PID control has all the advantages of the three types of control. The proportional mode portion produces an output proportional to the difference between the measurement and the set point. The integral control action produces an output proportional to the amount and the length of time the error is present. The derivative section produces an output proportional to the error rate of change. The PID mode of control applies to systems that have transient as well as steady-state errors.

Controllers

As we have seen, the controller is the part of a control system that compares the measurement of the controlled variable with the set point. The controller also directs the action of the final control element, which corrects or limits the deviation.

Controllers in industry may be classified in several ways. For example, they may be classified as either electric (electronic) or pneumatic. Also, electric and pneumatic controllers can both be broken down into two popular types: analog and digital. So, in this section, we will discuss both analog and digital types of electric and pneumatic controllers.

Electric Controllers

Many electric (or electronic) analog controllers use the op amp as a basic building block. Recall from Chapters 1 and 2 that the op amp can perform the processes of comparison, summation, integration, and differentiation. These are precisely the building blocks we need to put together the types of contoller modes we have just discussed. Process engineers choose the types of control to suit the particular process application. All the controller modes we discussed above can be accomplished with the op amp.

While each manufacturer of electronic analog process controllers produces a different system, many of these systems have common factors. For example, many electronic controllers have similar

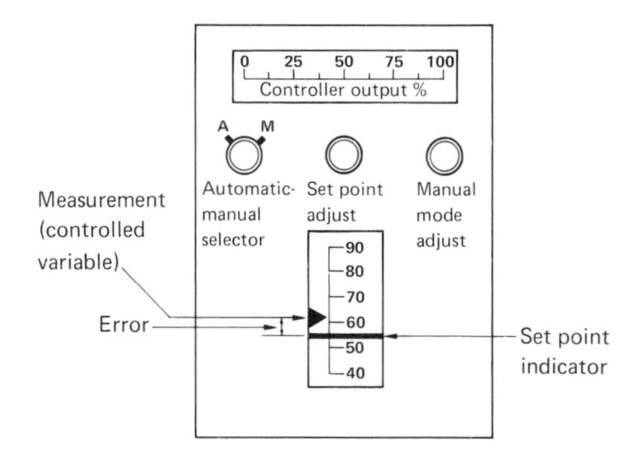

Figure 8.13 Front Panel of Typical Electronic Controller

front panels, with metering systems similar to the one shown in Figure 8.13. This meter displays two types of information, the set point and the measured variable values. The set point is normally adjustable from the front panel by means of a potentiometer. Thus, it is simple to change the set point for different applications. The scale is usually calibrated to read from 0% to 100% of signal variation. In addition, many controllers give operators the ability to switch back and forth from manual to automatic mode. The manual mode is often used when the process is started up and when it is shut down.

Digital electronic control is also found in industry. The computer, of course, can be classified as a digital controller. Because of the ability of the computer to make decisions and initiate action, it is ideally suited to process control applications. The most popular digital controller today is the programmable controller, which is a microprocessor-based system capable of a wide range of tasks. The programmable controller will be discussed in Chapter 12.

Pneumatic Controllers

Because of their low cost, low maintenance, and safety advantages, pneumatic controllers are still popular in industry. Pneumatic controllers, like electric controllers, come in both analog and digital versions.

The analog pneumatic controller uses the pressure bellows and the flapper-nozzle combination to effect its control. A simple *flapper-nozzle amplifier* is shown in Figure 8.14A. Constant air pressure is supplied to the nozzle through a restrictor whose diameter is about 0.010 in. (0.254 mm). The nozzle itself has a diameter of about 0.020 in. (0.5 mm). The flapper is positioned against the nozzle opening. The position of the flapper is determined by the transducer output and the set point. The nozzle's back pressure is inversely proportional to the distance between the nozzle opening and the

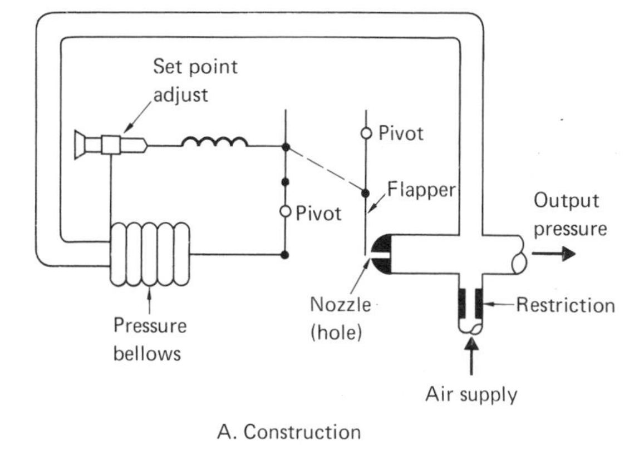

A. Construction

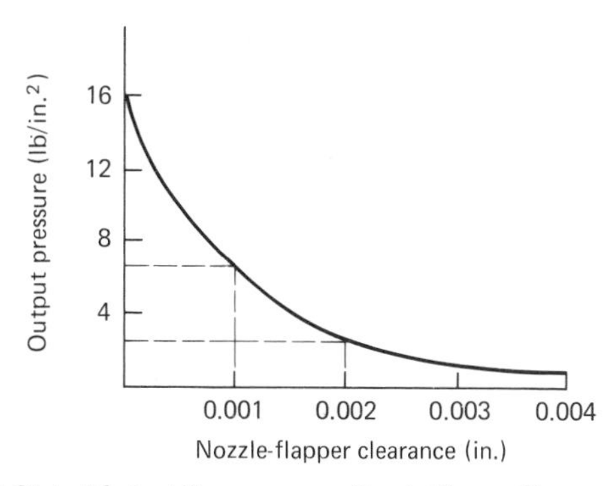

B. Plot of Output Pressure versus Nozzle-Flapper Clearance

Figure 8.14 Pneumatic Flapper-Nozzle Amplifier

flapper. As shown in Figure 8.14B, a flapper motion of about 0.002 in. (0.05 mm) is sufficient to provide a full range of output. The sensitivity of this system can be adjusted by moving the flapper pivot point.

The flapper-nozzle amplifier shown in Figure 8.14A can be used as a proportional controller. When the controlled variable (output pressure) increases, the bellows expand, causing the flapper to move away from the nozzle. Air then escapes through the nozzle into the atmosphere. This action reduces the output pressure to the set point value. Conversely, a decrease in output pressure causes the output pressure to increase to the set point value.

A proportional plus derivative pneumatic controller is shown in Figure 8.15. The addition of a variable restrictor results in a delayed negative feedback. If the controlled variable increases or decreases suddenly, the transducer causes the flapper to open or close. This change in flapper position causes the output pressure to change suddenly, but the pressure in the bellows can only change slowly. The slower change in the bellows is due to the variable restriction and the capacity of the bellows. This slower change delays and reduces the amount of negative feedback. Since the feedback is negative, the output pressure is higher and leads

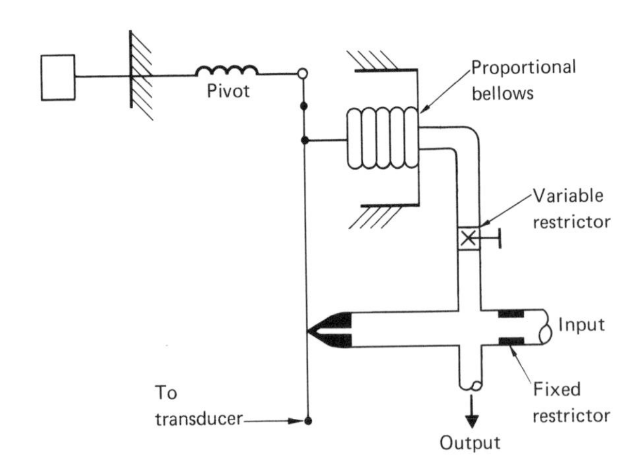

Figure 8.15 Proportional Plus Derivative Pneumatic Controller

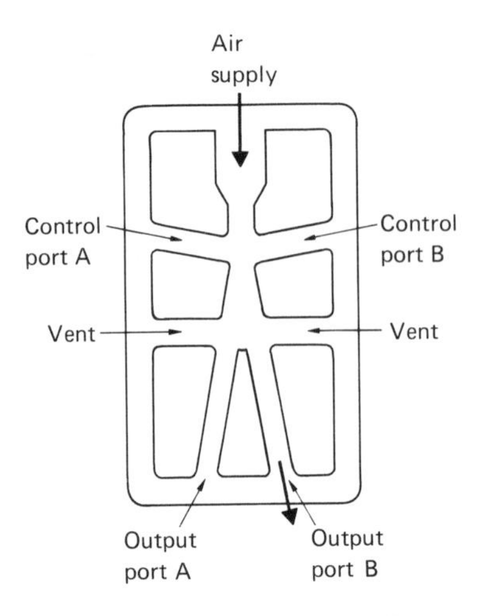

Figure 8.16 Wall Attachment Effect in Fluidic Device

the transducer signal. Therefore, the delayed negative feedback produces a derivative response.

These examples show that control can be accomplished by analog pneumatic means. However, digital pneumatic control is also available through fluidic devices. A *fluidic device* is a digital logic element that operates by air flow and pressure.

One of the most popular fluidic devices uses the *Coanda*, or *wall attachment, effect*. This effect is illustrated in Figure 8.16. A fluid flowing past a wall has a tendency to attach itself to that wall. When air pressure flows from the supply, the stream of air will attach itself to one of the two walls. The stream will stay attached until a jet of air from the control port detaches it from the wall.

You may recognize this fluidic device, functionally, as a bistable flip-flop. By a rearrangement of the input ports and the addition of a restriction to one of the output legs, both AND/NAND and OR/NOR gates may be formed. The logic symbols are shown in Figure 8.17. With these three logic elements, a simple digital pneumatic computer may be made. It can be used for simple controlling and decision-making functions in the same way an electronic digital computer is used.

Fluidic systems have several advantages over electric systems. Since no electric potentials are present, there is less likelihood of explosive gases

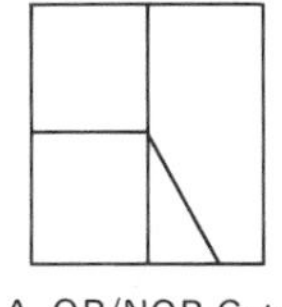

A. OR/NOR Gate

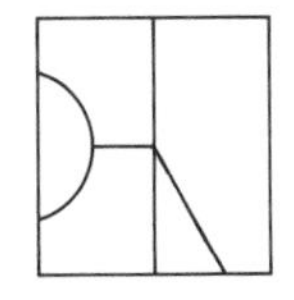

B. AND/NAND Gate

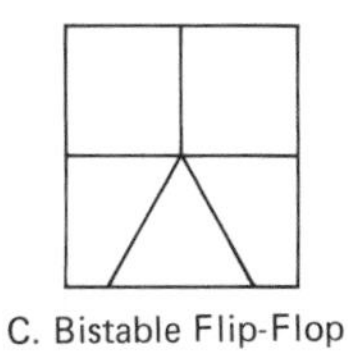

C. Bistable Flip-Flop

Figure 8.17 Schematic Diagrams of Fluidic Logic

being ignited. This advantage is especially important in the chemical and petrochemical industries. Also, the fluidic system can be completely flushed out with cleaning solvents at regular intervals. It is difficult to do such cleaning in electric equipment. This advantage is particularly important in the food-processing industry, where bacterial contamination must be kept low.

There are some disadvantages to fluidics, however. Generally, fluidic systems are much slower than comparable electric systems. Also, fluidic systems take up much more space, since large-scale integration is not available with pneumatic systems.

Final Control Elements

Previously, we defined the final control element as a device that corrects the value of the controlled variable. For instance, in an assembly line, the controlled variable might be conveyor belt speed. The conveyor belt is driven by a motor, which is the final control element. In a proportional control system, an increase in the load on the belt would tend to slow the belt down. This speed change would be sensed and compared to a set point. The controller would increase the belt speed back to normal by varying the speed of the motor, the final control element. In the pasteurization example, recall that the steam heated the milk to the proper temperature. Therefore, the valve controlling the flow of steam is the final control element. If the liquid were heated by a resistance element, that element would be the final control element.

In many process control applications, the final control element is a control valve. The *control valve* is a mechanism that regulates the flow rate of a fluid, either a gas, a liquid, or a vapor. The control valve can be classified on the basis of its flow characteristics or on the basis of its body style.

In electrical terms, a valve can be considered analogous to a resistance. Flow through a valve is then proportional to two things: (1) the area of the valve opening and (2) the pressure drop across the valve. The following formula expresses this relationship:

$$Q = KA\sqrt{\Delta p} \tag{8.8}$$

where Q = quantity of fluid flow
K = constant of proportionality for conditions of flow
A = area of valve opening
Δp = pressure drop across valve

Figure 8.18 is a diagram of the percent of valve travel, or position, plotted against the percent of flow. Note that the *quick-opening valve* shows a large increase in flow rate with a small change in valve opening. This type of valve is used in on-off control situations. The *linear valve* has a linear response curve. That is, the valve position is linearly related to the flow rate.

The most commonly used valve type is the *equal-percentage valve.* Note that a change in the valve position produces an equal-percentage change in flow. If you were to plot this relationship

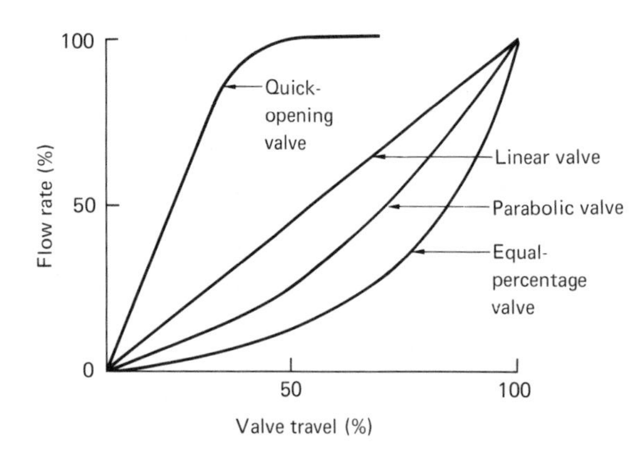

Figure 8.18 Plot of Valve Travel versus Flow Rate for Four Control Valve Designs

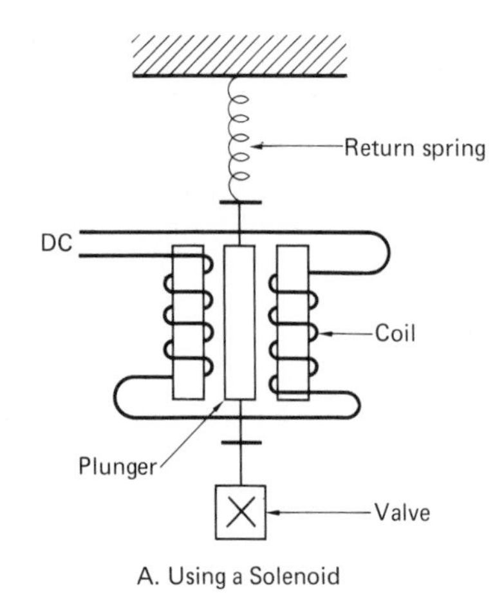

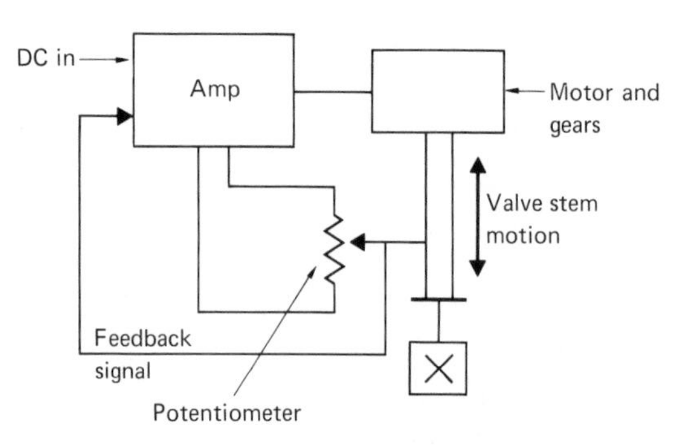

Figure 8.19 Electrical Valve Control

on a logarithmic scale, the relationship would be a linear one. The equal-percentage valve is designed to be used between a certain minimum and maximum flow rate. The maximum flow rate (Q_{max}) divided by the minimum flow rate (Q_{min}) is called the *rangibility* (R) of the valve. The following formula expresses the rangibility of the valve:

$$R = \frac{Q_{max}}{Q_{min}} \quad \textbf{(8.9)}$$

Most commercial valves have a rangibility between 30 and 50.

Generally speaking, valves are controlled in industry in two ways: electrically and pneumatically. Electrical actuation of valves may be done by a solenoid, such as the one illustrated in Figure 8.19A. When current flows through the coil, a magnetic field is generated, which moves the plunger. Such a device is used in conjunction with on-off controllers.

Another method of controlling valve actuation electrically is by motors or servomechanisms, as illustrated in Figure 8.19B. The DC signal from the controller is amplified and operates a geared motor, driving the valve stem. The valve position is converted to an electric signal that feeds back to the amplifier circuit. Here, it is compared to the DC signal. Any difference is amplified to drive the valve to a position exactly proportional to the original DC signal.

Valves may also be actuated pneumatically. Pneumatic actuation is shown in Figure 8.20. In this case, the diaphragm moves proportionally to the amount of air pressure at the inlet. Air pressure may be used to either open or close the valve.

Although control valves are the most common final control elements used in industry, there are others in use. In Chapters 3 and 4, we discussed DC and AC motors. These devices are used in industrial velocity and position control systems, such as conveyor belts and mixing operations. The speed of DC motors is easily controlled and is capable of reversible operation. Speed control is possible in AC motors as well, although not as conveniently.

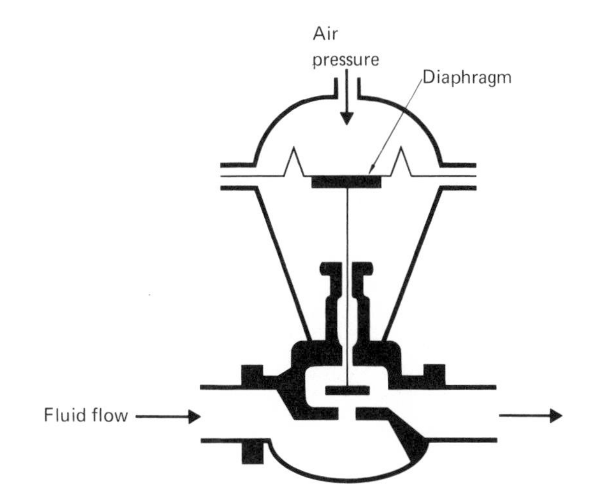

Figure 8.20 Pneumatic Valve Control

Conclusion

At this point, we have discussed most of the components of a process control system. These parts should be starting to fit together into a coherent whole for you. All the previous chapters have been building up to this point—a clear understanding of process control systems and their components.

Questions

1. The function of a process control is to keep the controlled variable constant when it tends to be changed by a ______
2. The process of comparing the information from a sensor to a reference is called ______. The difference between the sensor and the set point reference is called the ______.
3. ______ control has no feedback from output to input. Feedback control is sometimes unsuited to control applications, since large ______ or ______ cannot be dealt with effectively by feedback control.
4. ______ control is sometimes called anticipatory control because it anticipates the effect of a disturbance and tries to correct it before it happens. ______ control is suited to systems that have large capacitances.
5. In proportional control, the final control element action is directly related to the difference between the measurement and the ______. The amount of final control element change for a given input change is called ______.
6. Integral control provides controller action that changes as a function of ______ ______ the error is present. Integral control is also called ______ control.
7. Derivative control provides controller action that varies as a function of the ______ at which the error changes. Derivative control is also called ______ control.
8. Define the following terms: (a) dynamic variable and (b) manipulated variable.
9. Draw a block diagram of a closed-loop control system with blocks for measurement, evaluation, and control.
10. Define process load.
11. Define process lag, and list the three properties of a process control system that cause it.
12. Give an example of an open-loop control system.
13. Draw a block diagram of a feedforward control system, and describe the system's operation.
14. Is it necessary to have a sensor in an open-loop control system? Why or why not?
15. Is a sensor necessary in a closed-loop control system? Why or why not?
16. What is an error signal? How is it generated?
17. Describe and draw response curves for the following control modes (open loop): (a) on-off, (b) proportional, (c) integral, and (d) derivative.
18. As the controller gain increases, what happens to the controller output for a given change in the measurement?
19. Define offset. ~~How may offset be calculated?~~
20. Explain the advantages and disadvantages of (a) proportional control, (b) integral control, and (c) derivative control.
21. Can integral and derivative control be used alone? Why or why not?
22. Explain how control can be achieved by pneumatic means. Describe digital and analog pneumatic controllers and how they work.
23. Define final control element.
24. What is the relationship between controller gain and proportional band? Is it ever possible to have a 0% proportional band? Why?

Problems

1. A process control system with a DC motor as a final control element has a proportional band of 100%. What is the gain of the controller? If the proportional band were 50%, what would the gain be?
2. In Problem 1, suppose a system had a gain of 3, a set point of 70°C, and a measurement of 50°C. How much would the controller output change over this difference?
3. Calculate the proportional band of a pneumatic controller that has a total range of 0 –200 lb/in.^{2} with its set point at 50 lb/in.^{2}. A deviation of $\pm$10 lb/in.^{2} will cause the output of the controller to vary from 0 to 100 lb/in.^{2}.
4. A pneumatic-to-electric transducer provides the following information:

Pounds/in.^{2}	V_o
0.5	1.00
7.5	1.50
10.0	1.75

What is the gain of the transducer using the values in the first and second rows?

Pulse Modulation

CHAPTER

9

Objectives

On completion of this chapter, you should be able to:

- Define pulse modulation;
- Determine the minimum sample rate for pulse modulation;
- Modulate and demodulate the four types of analog pulse modulation;
- List the advantages and disadvantages of the pulse modulation types;
- Generate and demodulate pulse-code modulation;
- Determine quantizing noise values for pulse-code modulation.

Introduction

Pulse modulation has been known to humanity for a long time. Modulated drum beats and smoke signals are two examples of this process. However, in this chapter, we will be concerned with a more recent development: human-generated electric pulse modulation.

Pulse modulation is a process in which an analog signal is sampled at regular periods of time. Information contained in the analog signal is transmitted only at the sampling time and may also have synchronizing and calibrating pulses. At the receiver, the original waveform may be reconstructed from the information contained in the samples. If the original samples are taken frequently enough, the analog signal can be reproduced with minimal error or distortion.

This chapter will deal with the various types of pulse modulation and demodulation and some examples of circuits using these pulse modulation methods. Pulse modulation is fundamental to the understanding of telemetry, the subject of Chapter 10.

Electric Pulse Communication

Electric pulse communication had its origin in 1837 when Samuel Morse invented the telegraph. *Telegraphy* is the process of sending written messages from one point to another in the form of code. A few years later, in 1845, a Russian general, K. I. Konstantinov, and Dr. Poulié constructed a telemetry system. A *telemetry system* performs measurements on distant objects. The original system recorded and analyzed the flight of a cannonball. This system automatically reported the course of the cannonball as it passed through screens and recorded the electric impulses and a timing impulse on a manually turned recording drum loaded with graph paper. This system was one of the first written records concerning telemetry.

More recently, with the development of television and radar, an additional type of pulse communication called data communication has come into widespread use. *Data communication* is the process of transmitting pulses, which are the output of some data source, from one point to another. Examples of data communication are computer-to-computer transmission, data collection, or telemetry and alarm systems. Other commercial uses are financial/credit information, travel and accommodation booking services, and inventory control for stores.

Facsimile is also considered to be a form of data communication. *Facsimile* is the process whereby fixed graphic material, such as pictures, drawings, or written material, is scanned and converted into electric pulses, transmitted, and, after reception, used to produce a likeness (facsimile) of the original. This process can be considered similar to the transmission of a single television picture, but here the picture is recorded on paper. Facsimile has been used by news services to transmit newspaper photos and by ships for reception of up-to-date weather charts or maps.

In recent years, the volume of pulse communication has increased greatly. There are at least three factors contributing to this increase in pulse communication: (1) Much of the information to be transmitted is in pulse form to begin with, such as computer data or, to some extent, picture elements; (2) a more error-free transmission process is possible, as, for example, in distant space probe picture transmission; and (3) the advent of large-scale integration has greatly simplified the necessary electronic circuitry. In addition, large-scale integration has permitted the use of complex coding systems that take the best advantage of channel capacities.

Telegraphy and telemetry can properly be considered subsets of data communications, but historically they have been kept separate, as indicated in Figure 9.1. Telegraphy has been considered a separate branch since it is the transmission of messages, which are not considered to be data. Telemetry has been considered a separate branch because its first major use employed pulse modulation and radiotelemetry and did not use public transmission networks, such as the telephone line, for its transmission. Since the inception of data communications, however, these distinctions have become less clear.

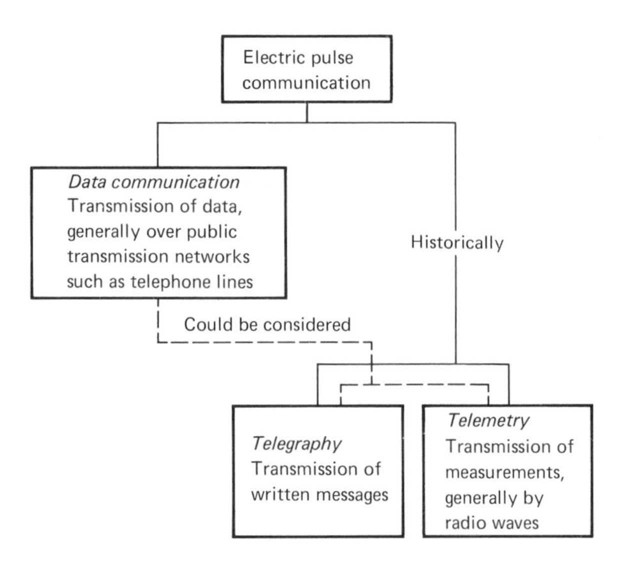

Figure 9.1 Subsets of Electric Pulse Communication

It is instructive to understand the relationships among telegraphy, telemetry, and data communications. This background will be beneficial when you are studying the next chapter, which is concerned with telemetry.

Pulse Modulation Types

Pulse modulation can be divided into two major categories: analog and digital. In *analog pulse modulation*, a characteristic of the pulse, such as height or width, may be infinitely variable and varies proportionally to the amplitude of the original waveform. In *digital pulse modulation*, a code is transmitted; the *code* indicates the sample amplitude to the nearest discrete level.

All modulation systems sample the information waveform to be telemetered, but they all have different ways of indicating the sampled amplitude. Each modulation system also has its advantages and disadvantages, as will be discussed in the following sections.

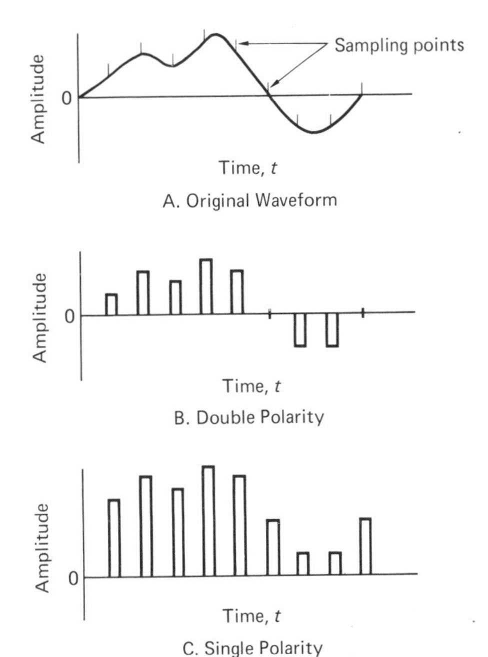

Figure 9.2 Pulse-Amplitude Modulation

Analog Pulse Modulation

There are four major types of analog pulse modulation: pulse-amplitude modulation, pulse-width (or duration) modulation, pulse-position modulation, and pulse-frequency modulation. We will discuss each type in detail in this section.

Pulse-Amplitude Modulation (PAM)

Pulse-amplitude modulation (*PAM*) is illustrated in Figure 9.2. PAM is a process in which the signal is sampled at regular intervals, and each sample is made proportional to the amplitude of the signal at the instant of sampling. As shown in Figure 9.2, there are two types of PAM. *Double-polarity PAM* can have pulse excursions both above and below the reference level (Figure 9.2B). *Single-polarity PAM* has a fixed DC level added to the pulses so that the pulse excursions are always positive. (Figure 9.2C).

Noise

Since the amplitude of the pulse contains the information, it is very important that the amplitude be a true representation of the original level. However, in the transmission of the signal, noise can be added to the pulse, causing reconstruction errors. *Reconstruction errors* are errors that cause the reconstructed waveform to differ from the original waveform. These errors may be caused by noise or they may result from other phenomena.

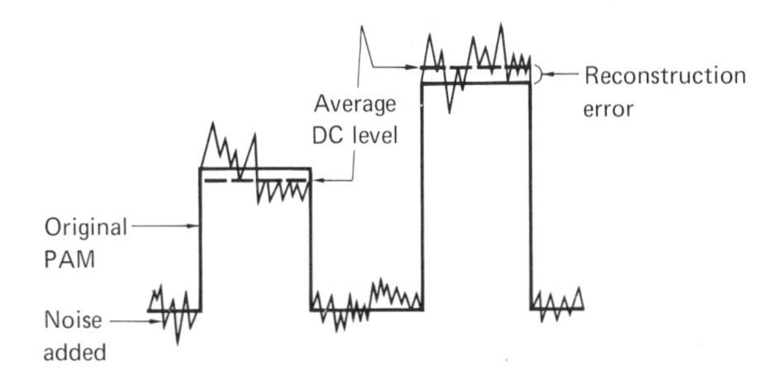

Figure 9.3 Effect of Noise on PAM

Figure 9.3 illustrates reconstruction errors due to noise. Noise and the other reasons for error are discussed in more detail in Chapter 10. The added noise depends on many factors, such as the strength of the signal, the distance the signal is to travel, and the environment the signal has to pass through. Because PAM is greatly affected by noise, it is not used as often as other forms.

Aliasing

Another problem that can exist with any form of sampled telemetering is called aliasing. *Aliasing* is the reconstruction of an entirely different signal from the original signal. The different signal is a result of the sampling rate being low in comparison with the rate of change of the signal sampled. If, for example, the signal to be sampled varies at a rate of 10 Hz, the signal may be sampled ten times during its period (sampling rate of 100 Hz). This sampling rate will not result in severe aliasing. However, if the sampling rate were reduced to 9 Hz, or less than once each signal period, a certain amount of aliasing would occur. Figures 9.4 and 9.5 illustrate these ideas.

In Figure 9.4, the sampling rate is ten times the frequency of the waveform to be sampled. As shown, the resulting, or recovered, waveform is approximately equal to the original signal. In Figure 9.5, the original waveform frequency remains the same as in Figure 9.4, but the sampling rate is reduced by more than ten times the sample rate of Figure 9.4. The resulting, or recovered, waveform in Figure 9.5 no longer resembles the original waveform. Therefore, considerable aliasing is produced.

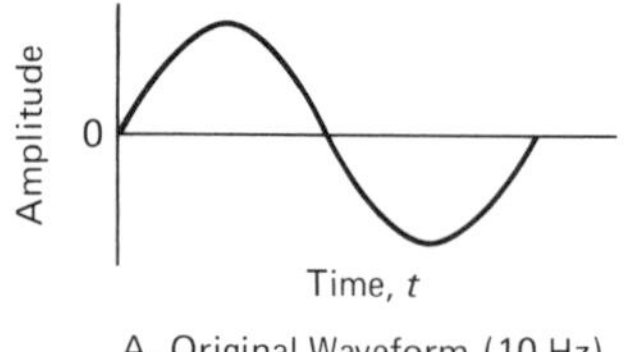

A. Original Waveform (10 Hz)

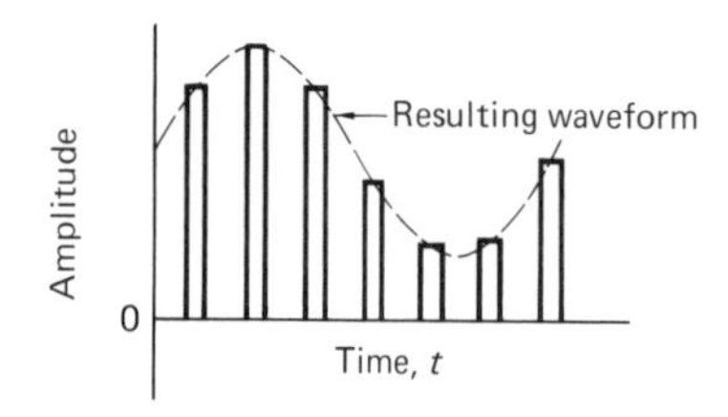

B. Pulses Used to Reconstruct Original Waveform

Figure 9.4 Sample Rate Resulting in Minimum Aliasing

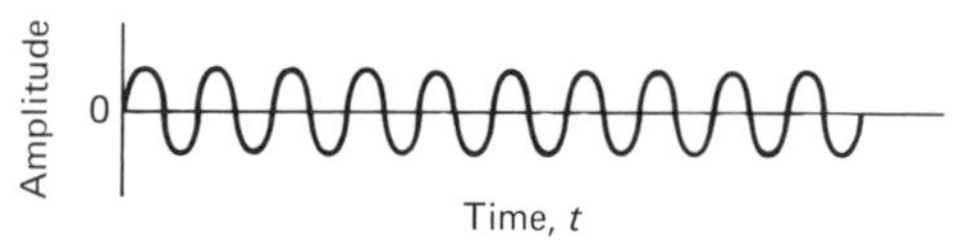

A. Original Waveform (10 Hz)

B. Pulses Used to Reconstruct Original Waveform

Figure 9.5 Sample Rate Resulting in Considerable Aliasing

Sampling Theorem

The aliasing problem associated with sample rate has been investigated by many researchers and has resulted in a theorem known as the sampling theorem.

Sampling Theorem:
If the sampling rate in any pulse modulation system exceeds twice the maximum signal frequency (or Nyquist rate), the original signal can be reconstructed in the receiver with vanishingly small distortion.

As a result of this theorem, most pulse modulation sampling rates for speech over standard telephone channels, whose bandwidth is 300–3400 Hz, is standardized at 8000 samples per second. The 8000 samples per second is slightly more than twice 3400 Hz, and, therefore, the sampling theorem is satisfied.

To illustrate the theorem, we will assume Figure 9.6A represents a small portion of a speech transmission. The portion of the waveform from point *A* to point *B* represents the steepest slope, or greatest rate of change, of the wave. Since it is telephonic speech, this rate of change corresponds to the frequency of 3400 Hz. A 3400 Hz sine wave (the dotted line) whose peaks are at points *A* and

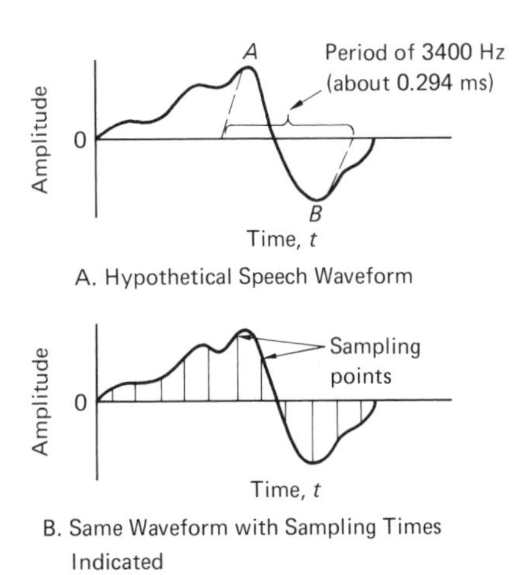

Figure 9.6 Waveform Illustrating Sampling Theorem

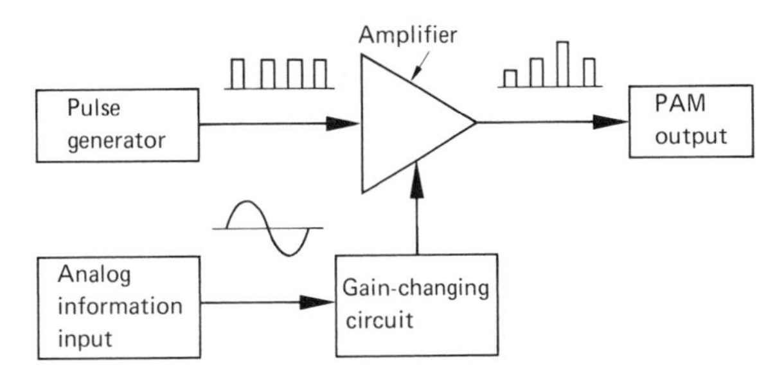

Figure 9.7 Block Diagram for Producing PAM

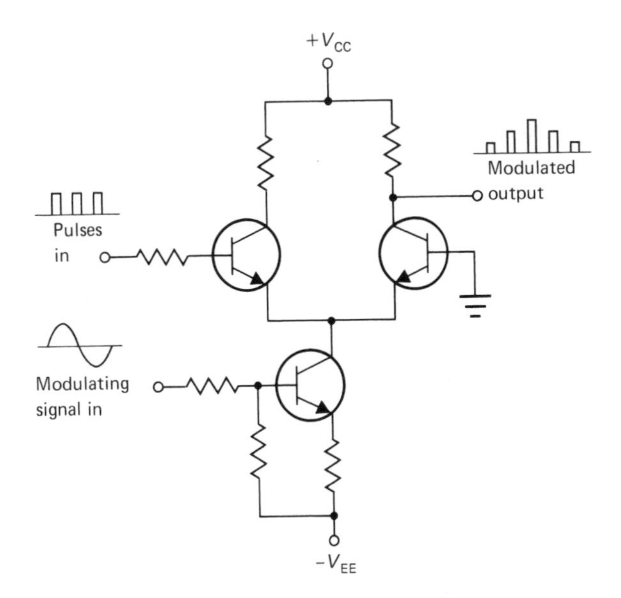

Figure 9.8 Variable-Gain Differential Amp

B is shown for comparison. The period of the 3400 Hz sine wave is also shown. According to the sampling theorem, the samples are taken 8000 times per second and therefore have a pulse period of 0.125 ms.

The waveform in Figure 9.6B is the same as the waveform in Figure 9.6A, but it shows how far apart the samples are taken to indicate 8000 pulses per second. This rate is the slowest pulse rate required to reproduce the original signal with "vanishingly small distortion."

Generation and Demodulation of PAM

In most electronic systems, there are many different circuits that can be used to produce the same outcome. So it is here. The circuits in this chapter represent only a few of the circuits that can be used to produce pulse modulation.

Pulse-amplitude modulation can be produced by first generating a pulse, then varying the gain of the amplifier to which the pulse is going. A block diagram for producing PAM is shown in Figure 9.7. A National Semiconductor LM555 timer connected in the astable configuration or a CD4047 low-power, monostable/astable multivibrator in the astable mode can be used as the pulse generator. The CD4047 is CMOS (complementary metal-oxide semiconductor).

The variable-gain amplifier employed here is often used to generate analog amplitude modulation (AM). The basic circuitry is that of a differential amplifier, shown in Figure 9.8. The pulses are input into one of the differential transistors, while the modulating signal is applied to the normally constant current transistor. The modulating signal thus causes the amplifier gain to vary. (The

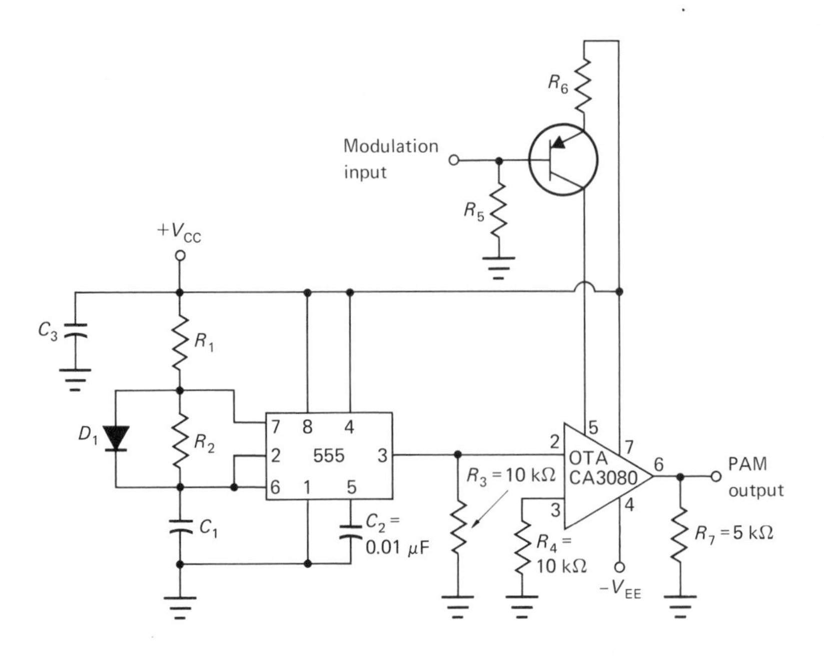

Figure 9.9 Circuit for Generating PAM

constant-current amplifier was discussed in Chapter 1.) The output is now pulse-amplitude–modulated. This result is essentially what the transconductance amplifier, discussed in Chapter 2, produces.

Other classes of circuits that can be used for the variable-gain amplifier are Motorola Semiconductor's MC1595L linear four-quadrant multiplier, National Semiconductor's LM1596 balanced modulator-demodulator, or RCA's CA3080 operational transconductance amplifier (OTA).

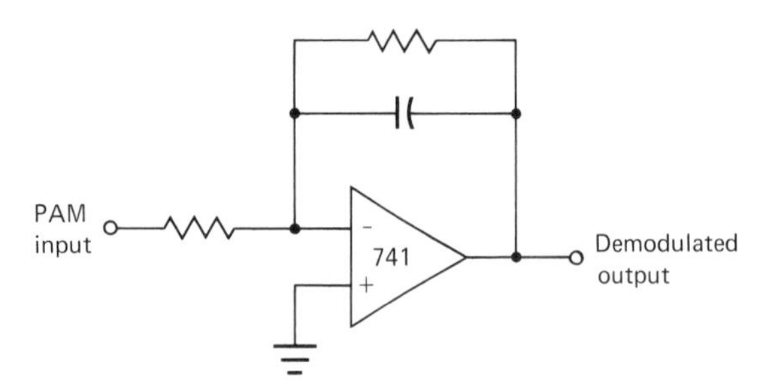

Figure 9.10 Low-Pass Filter for Demodulating PAM

The complete circuit for PAM is shown in Figure 9.9.

Demodulation, or signal recovery, of pulse-amplitude modulation can be accomplished with a low-pass filter, as shown in Figure 9.10. Additional filtering or signal smoothing may be required.

Pulse-Width Modulation (PWM)

Pulse-width modulation (*PWM*), sometimes referred to as *pulse-duration modulation* (*PDM*), is shown in Figure 9.11. In PWM, the signal is sampled at regular intervals, but the pulse width is made proportional to the amplitude of the signal at the instant of sampling. As shown in Figure 9.11B, the pulses are of equal amplitude and the leading edges of the pulses are the same time apart. However, the trailing edges of the pulses are not.

One concern in PWM is that the pulse-width variations do have practical limits. Obviously, the pulse width cannot exceed the spacing between pulses. If it did, there would be overlapping, and the proportional relationship between pulse width and signal amplitude would no longer be true. And

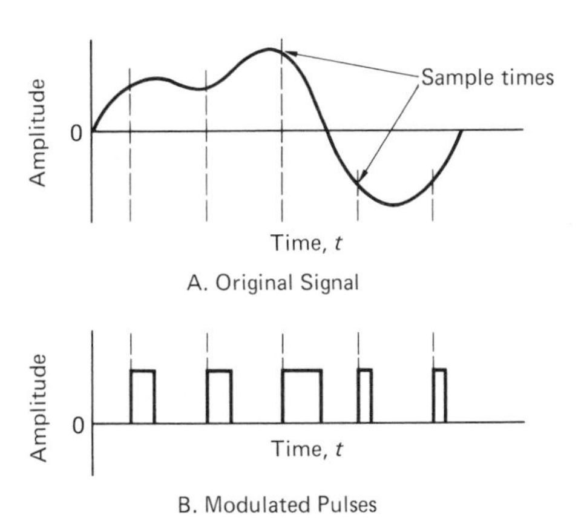

Figure 9.11 Pulse-Width Modulation

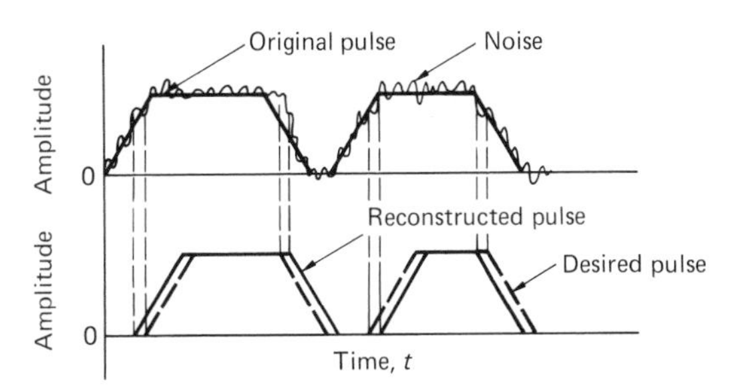

Figure 9.13 Exaggerated Effect of Noise on PWM

since a negative pulse width makes no sense, the most negative swing of the signal to be sampled can, at most, produce zero pulse width. In practice, these limits are more restricted. As a rule of thumb, the widest pulse width should not exceed 80% of maximum; the narrowest pulse width should not be less than 20% of maximum. Figure 9.12 illustrates these limitations.

Pulse-width modulation is less affected by noise than pulse-amplitude modulation. However, it is not completely immune. Figure 9.13 shows the possible error due to noise for PWM.

Generation of PWM can be produced by again employing the LM555. National Semiconductor's LM555 specification sheet contains a circuit similar to the one shown in Figure 9.14. In this circuit, two LM555s are used. However, the circuit can be made more compact by using an LM556, which is a dual LM555.

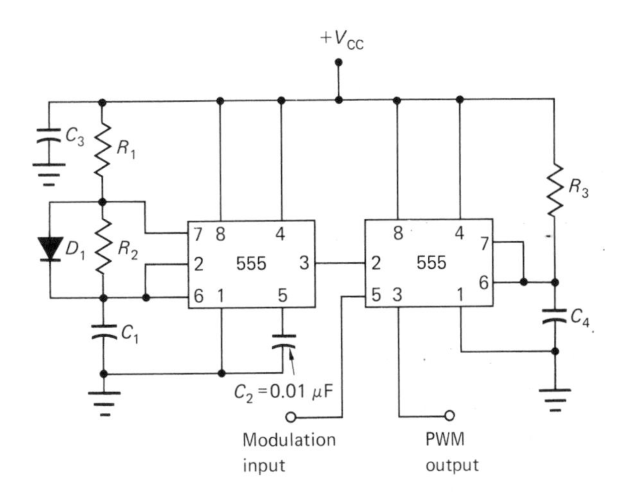

Figure 9.14 Circuit for Generating PWM

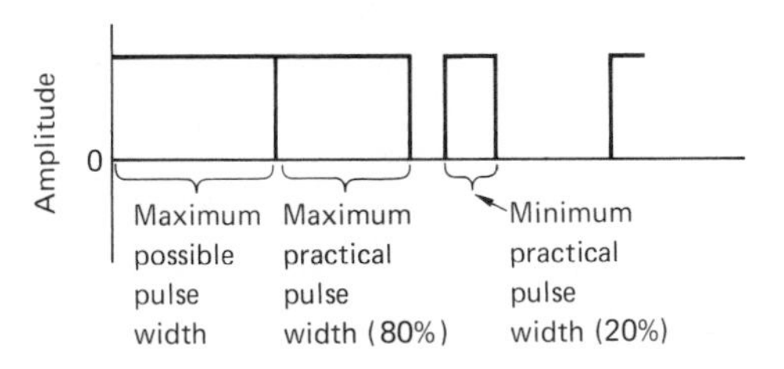

Figure 9.12 Practical Pulse-Width Limitations

Demodulation of PWM can be accomplished in the same manner as for PAM (see Figure 9.10). A low-pass filter will demodulate pulses that vary in either amplitude or width.

Pulse-Position Modulation (PPM)

Pulse-position modulation (PPM), along with a method of generating it, is shown in Figure 9.15. The original signal is first pulse-width–modulated, then differentiated and clipped. The resulting

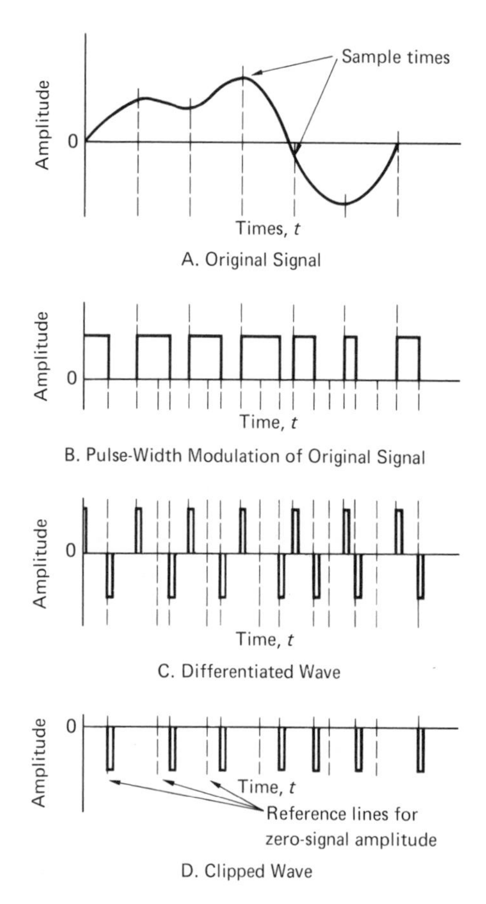

Figure 9.15 Generation of Pulse-Position Modulation

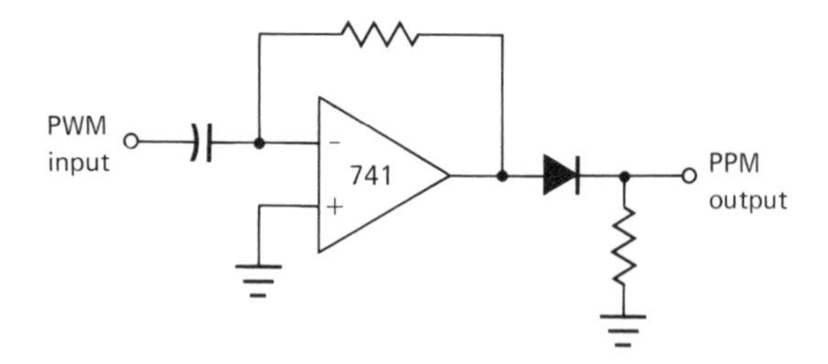

Figure 9.16 Circuit for Producing PPM from PWM

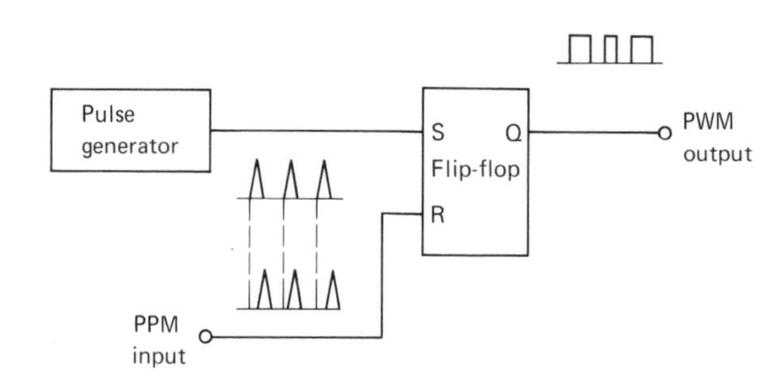

Figure 9.17 Converting PPM to PWM

pulses vary their position about a reference line, which is established by the zero-signal pulse position. As shown in Figure 9.15D, the PPM pulses are either on the reference line, lag the reference, or lead it. The amount of lead or lag from the reference line is proportional to the amplitude of the original signal above or below the zero line. Note that the leftmost pulse of Figure 9.15D is on the reference line; the second, third, and fourth pulses from the left lag the reference lines; the fifth and sixth pulses lead the reference lines; and the last pulse is on the reference line.

PPM is as susceptible to noise on the pulse as PWM is, but it is not as bad as PAM. The same cause of error shown in Figure 9.13 can affect PPM. But, of course, for *PWM*, the position of the pulse is affected; its width is not important.

As shown in Figure 9.15, in the generation of PPM, the signal is first converted into PWM. The circuit of Figure 9.14 shows generation of PWM. The output from the circuit of Figure 9.14 is then input to the circuit of Figure 9.16. The PWM is thus differentiated, and the negative pulses are clipped. Because the differentiator is an inverter, the PPM output is now a positive-going pulse.

PPM can be demodulated in at least two different ways. In the first method, PPM is converted back to PWM, and then the output is filtered. Figure 9.17 illustrates the conversion to PWM. Here, a flip-flop is used for the conversion to PWM, which then can be filtered to the original signal. This circuit has a drawback, however: The pulse generator input must somehow be synchronized with the PPM input. This synchronization calls for more circuitry and is more difficult than necessary. Therefore, it will not be discussed further.

The second method for demodulating PPM does not require synchronization and is easier to do than the first. This method requires the use of a phase-locked loop (PLL) circuit, which can be used to demodulate any frequency-modulated signal. The circuit is shown in Figure 9.18. The two diodes, D_1 and D_2, are used to limit the input signal to a safe range in case it is large. The input of the LM565 responds to zero or reference crossings only, and therefore the input signal does not have to be a sine wave but can be any waveform.

The voltage-controlled oscillator (VCO), free-running frequency f_o of the 565 is given by the following equation:

$$f_o = \frac{1}{4R_1C_1} \tag{9.1}$$

This free-running frequency should be adjusted to be near the center frequency of the input PPM frequency range. In Equation 9.1, R_1 should be between 2 kΩ and 20 kΩ, with 4 kΩ being the optimum value. The value of C_1 is not critical. Resistor R_2 in Figure 9.18 is not necessary but can be used to decrease the lock range of the 565. The output of the 565 (pin 7) will need filtering and amplification.

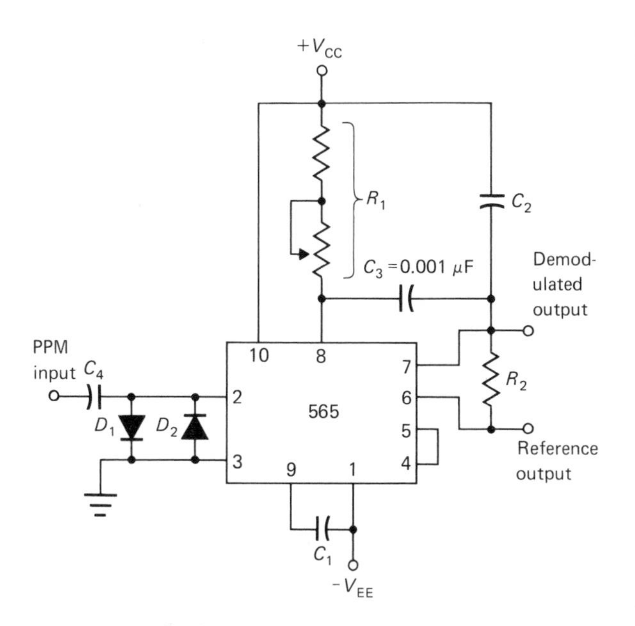

Figure 9.18 Demodulation of PPM Using a Phase-locked Loop

Pulse-Frequency Modulation (PFM)

Pulse-frequency modulation (*PFM*) may look very much like PPM, but it is generated in a different way. While PPM has only one pulse in each sample space, PFM can have more than one pulse in each sample space. (See Figure 9.21E.)

Of the four pulse modulation systems discussed up to this point, PFM is the least affected by noise. PFM does not depend on pulse width or pulse location but on the number of pulses per second. Thus, noise spikes would have to be very large in order to affect frequency.

One method for generating PFM uses a single LM555 connected as shown in Figure 9.19. This circuit is shown in National Semiconductor's specifications for the LM555 as PPM. However, according to the definitions in this chapter, Figure 9.19 is the circuit for PFM, not PPM.

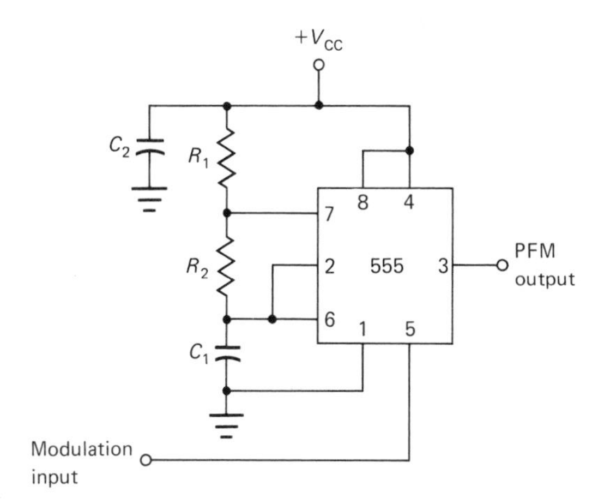

Figure 9.19 Generation of PFM Using the LM555

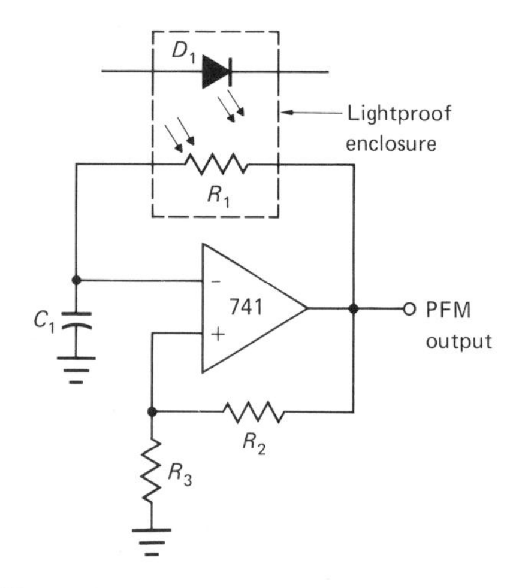

Figure 9.20 Generation of PFM Using the LM741

Another circuit for generating PFM very simply is that of Figure 9.20. Here, the LM741 operational amplifier is used as a relaxation oscillator. The R_1C_1 combination determines the rate of change of voltage at the inverting input to the op amp. The R_2R_3 combination provides positive feedback to the op amp and is the comparator-with-hysteresis portion of the circuit.

Some means of varying the value of R_1 would cause the output square wave to change its frequency. Therefore, if the modulation signal is applied such that R_1 varies its resistance at the modulation rate, then the output square wave will vary its frequency at the modulation rate. Figure 9.20 shows an LED (D_1) and a photoresistor (R_1) enclosed in a lightproof enclosure. The circuitry associated with the LED causes it to vary its intensity, which, in turn, varies the resistance of R_1. The output of the 741 is now PFM.

Demodulation of PFM can be done in the same way that demodulation of PPM is done, that is, with the phase-locked loop of Figure 9.18. The PLL *chip* (the semiconductor material that makes up the integrated circuit) is a very handy circuit. You should get to know it well.

Summary of Analog Pulse Modulation

Figure 9.21 gives a visual summary of the analog pulse modulation types. As we have seen, these pulse modulation types can be easily implemented by using the simple circuits illustrated in this chapter. These circuits have been given to facilitate understanding of the modulation types and also to illustrate the problems associated with each modulation type, such as modulation, demodulation, and noise. An industrial grade modulation/demodulation system would necessarily be more

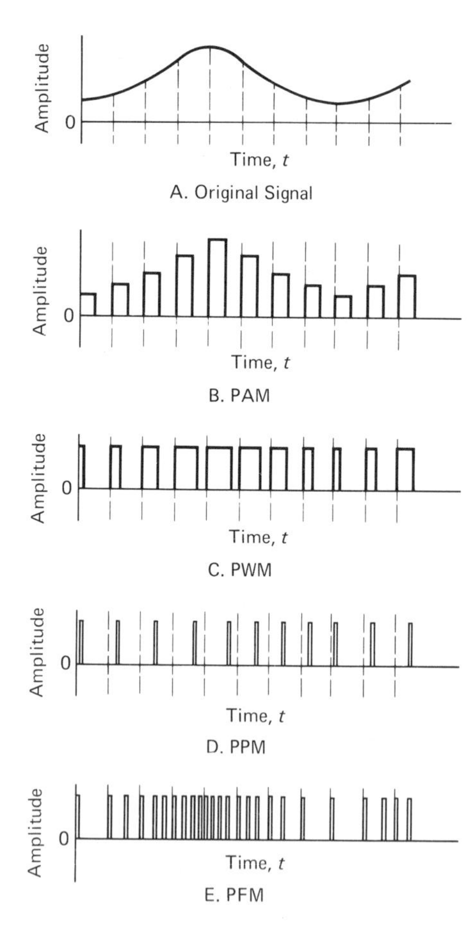

Figure 9.21 Analog Pulse Modulation Types

complex, but the principles involved would be the same.

Digital Pulse Modulation

Pulse-code modulation is the major type of digital pulse modulation in use today. In this section, we will describe pulse-code modulation in detail and briefly discuss some other digital systems found in industry.

Pulse-Code Modulation (PCM)

Pulse-code modulation (PCM), a form of digital pulse modulation, is very different from analog pulse modulation. Analog pulse modulation requires that some characteristic of the pulse, such as amplitude, width, position, or frequency, vary in accordance with the amplitude of the original signal. Pulse-code modulation, in contrast, transforms the amplitude of the original signal into its binary equivalent. This binary equivalent represents the approximate amplitude of the signal sampled at that instant. The approximation can be made very close, but it still is an approximation.

To illustrate PCM, let's suppose a signal could vary between 0 and 7 V. This voltage range is divided into a number of equally spaced levels, or *quanta*, which can be a representation of the integer values of voltages, as shown in Figure 9.22A.

When the signal is sampled (the vertical lines in Figure 9.22A), the digit produced depends on the level, or quantum, the signal is in at that instant. For instance, the leftmost sample time in Figure 9.22A finds the signal in the range between 1/2 and 1 1/2 V. This range corresponds to the decimal number 1 and the binary number 001. The process of determining which level the signal is in is called *quantization*. This quantized digit is represented in binary code by 1s and 0s (Figure 9.22A) and sent as pulses (Figure 9.22B), generally in reverse order to make decoding easier.

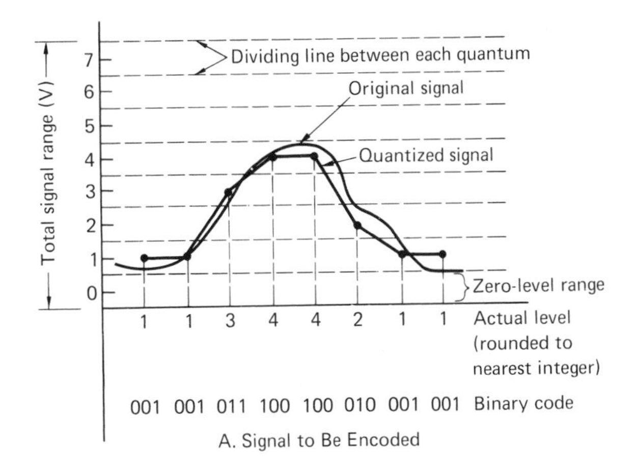

A. Signal to Be Encoded

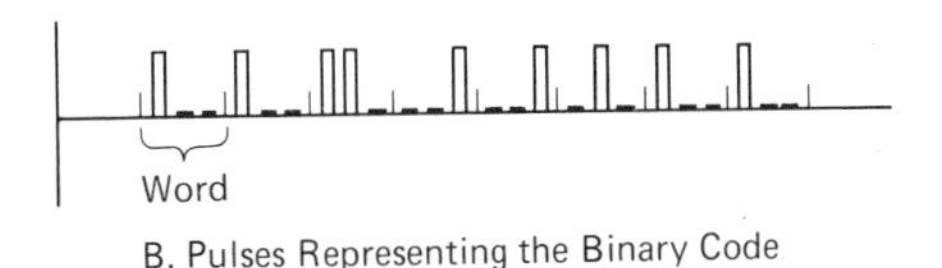

B. Pulses Representing the Binary Code

Figure 9.22 Encoding a Signal for Pulse-Code Modulation

As shown in Figure 9.22B, it may be difficult to determine when one group of pulses, called a *word*, ends and the next group or word starts. For this reason, a *supervisory*, or *synchronizing*, *bit* is generally added to each coded word to separate them. This bit is distinguishable from the other pulses in some way, such as having a longer pulse duration or a greater pulse amplitude. This bit allows easier decoding.

You probably immediately noticed in Figure 9.22 that the quantized signal does not look like the original signal. This difference is called *quantizing noise*. It is called noise because the errors are randomly produced. The largest quantizing error that can occur is equal to one-half the amplitude of the levels into which the signal is divided. In Figure 9.22, the largest quantizing error is 1/2 V.

For reduction of this quantizing noise, the obvious solution is to increase the number of quantizing levels used to encode the original signal.

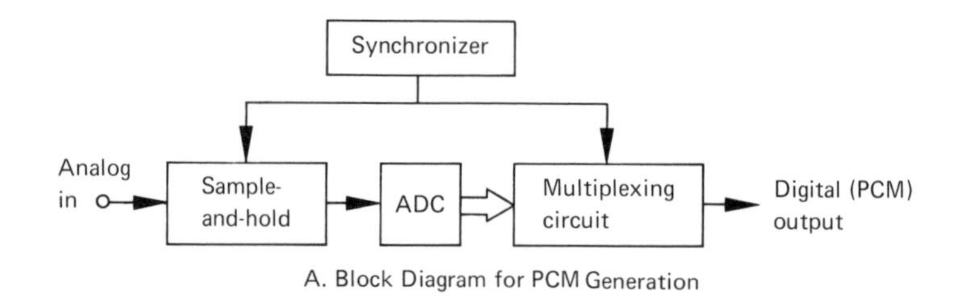

A. Block Diagram for PCM Generation

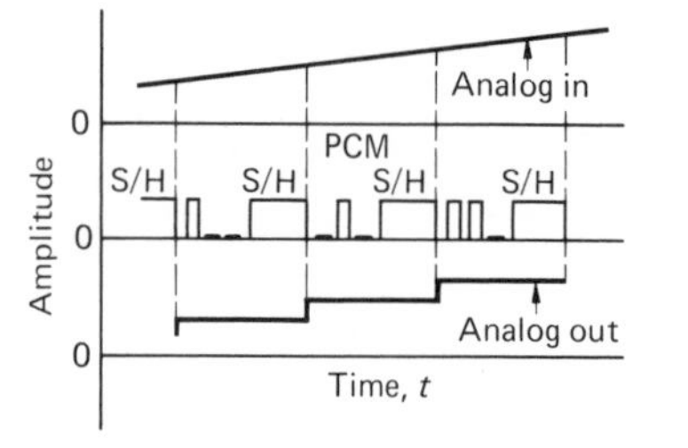

B. Signal Being Encoded and Then Decoded

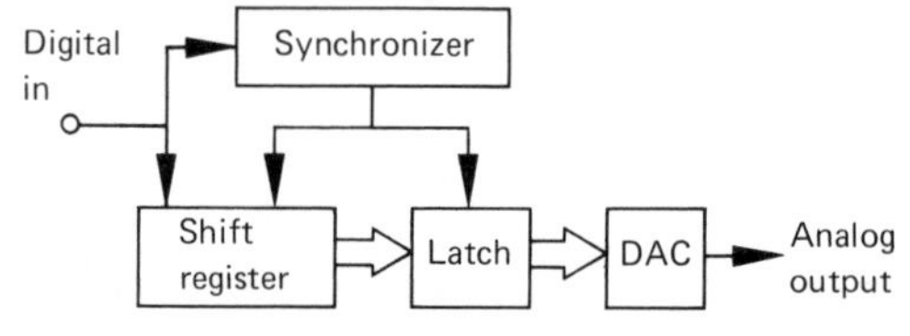

C. Block Diagram for PCM Demodulation

Figure 9.23 PCM Generation and Demodulation

For example, 1/2 V intervals instead of 1 V intervals could have been used. In the case of Figure 9.22, then, 4 bits (16 voltage levels) instead of 3 bits (8 voltage levels) would have to be used to represent each level. The increased number of levels results in more pulses per second at the output of the PCM encoder and, therefore, an increase in the frequency of the signal to be transmitted. The advantage of increasing quantizing levels is that quantizing noise is reduced; the disadvantage is that a greater bandwidth is required to transmit the signal. In practical systems, 128 levels for speech are considered adequate.

Generation and Demodulation of PCM

Generation and demodulation of PCM are much more complex than they are for analog pulse modulation. The block diagrams of Figures 9.23A and 9.23C give some idea of the complexity involved. Figure 9.23B shows the original input analog signal at the top, the analog signal transformed into PCM in the middle, and the demodulated PCM on the bottom line. The middle diagram of Figure 9.23B also shows when the sample-and-hold (S/H) circuit is sampling and holding (to the end of the S/H block). The dotted lines indicate the precise time

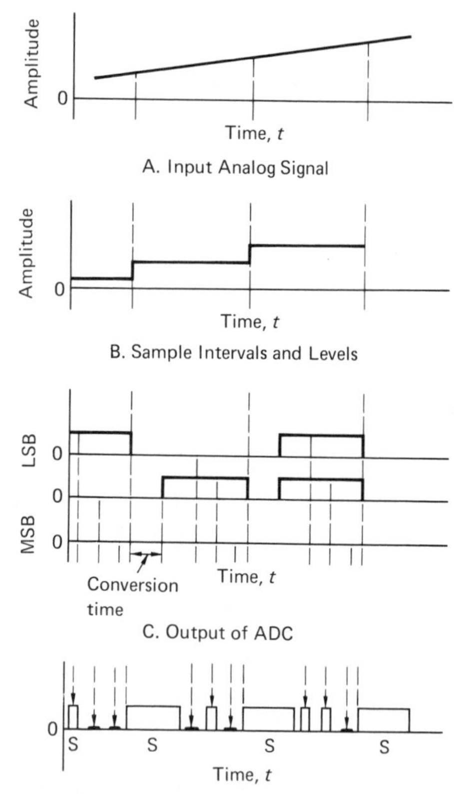

Figure 9.24 Signal Process for Generating PCM

the input signal is held to be converted to digital code, which follows the S/H block.

Figure 9.24 shows the encoding process for generating PCM. For encoding of the incoming analog signal into a serial pulse train, the signal is first sampled at regular intervals, as shown in Figure 9.24A. This sampled voltage is held at a constant level for a short period of time by a sample-and-hold circuit (Figure 9.23A). An example of a sample-and-hold circuit is shown in the bottom left corner of Figure 9.25. The components involved are the CD4016 bidirectional switch (the *B* part), the 1 μF capacitor, and the 741 voltage follower.

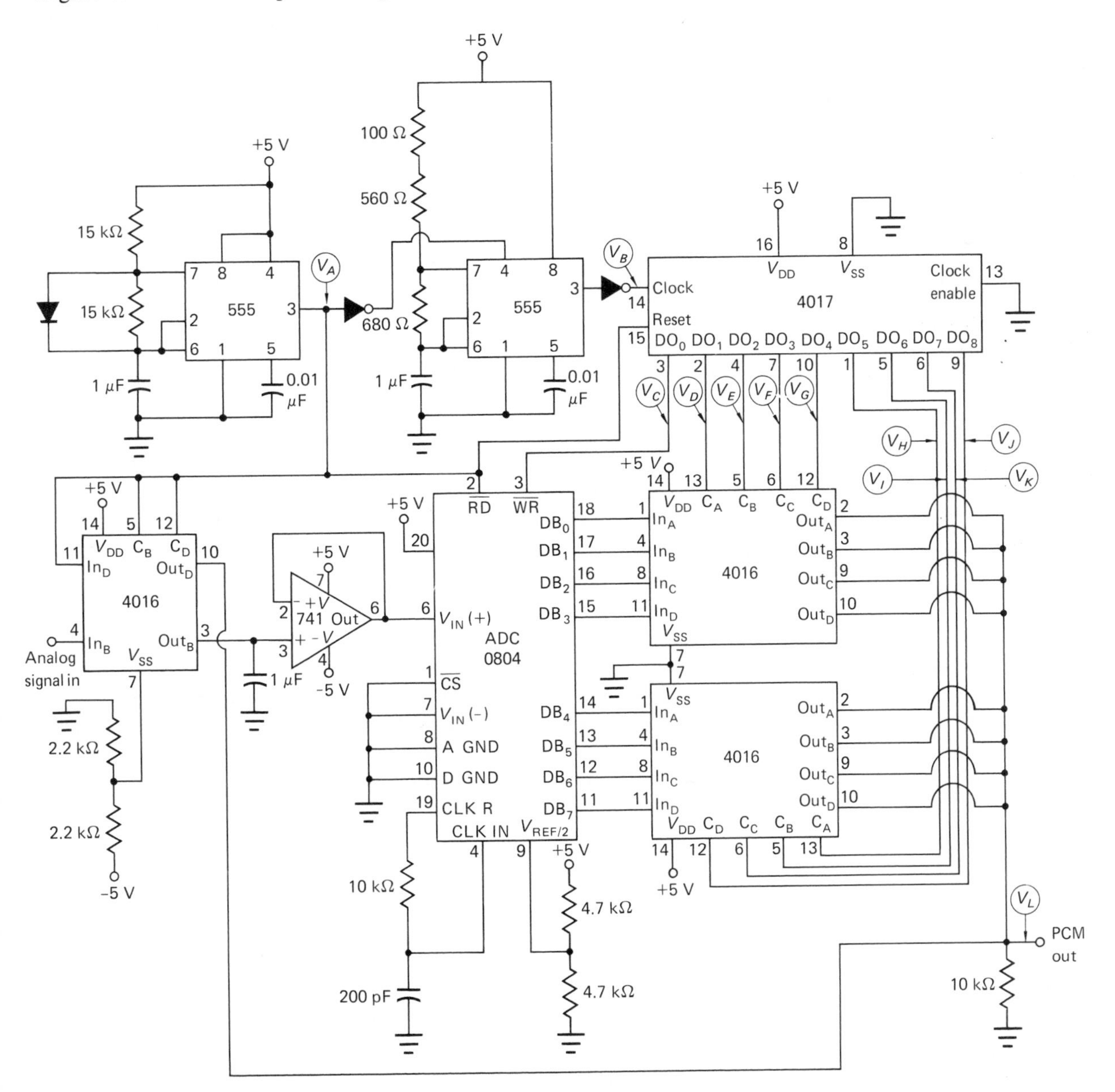

Figure 9.25 Schematic of PCM Modulator

The next step in the process is to convert the sampled signal into a digital representation. This conversion is done by the *analog-to-digital converter* (ADC, or A/D, shown in Figure 9.23A). The ADC in Figure 9.25 is the ADC0804. The specifications and description of this device are available from the manufacturer. Because the ADC does not operate instantaneously, the input signal must be sampled and held at a steady-state voltage long enough for the digital conversion to occur. In addition, the ADC converts the single input into a number of parallel outputs, each output representing a bit location. The thick arrow out of the ADC in Figure 9.23A represents this parallel output. Figure 9.24B illustrates the sampled intervals, and Figure 9.24C shows the output of the ADC.

Lastly, the outputs of the ADC are multiplexed onto the output line (Figure 9.23A), one at a time and each in turn. Multiplexing starts with the *least significant bit* (*LSB*) and ends with the *most significant bit* (*MSB*); see Figures 9.24C and 9.24D. At the same time, a synchronizing pulse (S) is added in order to separate each word and allow for easier demodulation. In Figure 9.25, the multiplexing circuit is on the right side and is composed of the CD4017 and two CD4016s. There may or may not be a space between bits, depending on the system used. The resulting signal is shown in Figure 9.24D.

All these operations must occur repeatedly and within a given time frame. This job belongs to the synchronizer shown in Figure 9.23A. In Figure 9.25, the synchronizer is the pair of 555s at the top.

The encoded signal is now ready to be sent to the transmitter. The transmitter will convert the pulses into amplitude modulation (AM), frequency modulation (FM), single sideband (SSB), or whatever type of modulation is desired. The signal does not necessarily have to be sent by radio, however. Light, sound, or telephone line transmission may be employed, depending on transmission distance and other factors.

At the receiver, the transmitted signals are converted back into pulses, then demodulated into the original analog signal (or a close approximation). Figures 9.23C and 9.26 show the PCM demodulation process.

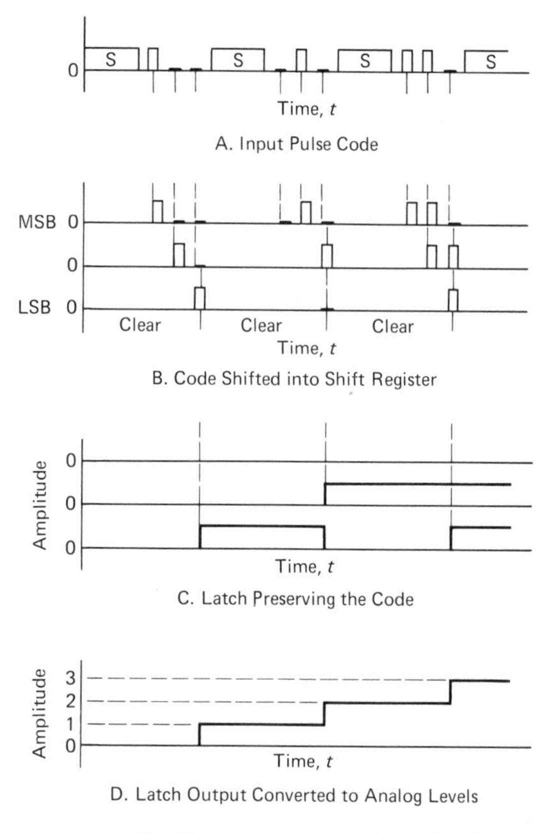

Figure 9.26 Signal Process for Demodulating PCM

The PCM signal is first input into a *shift register* (Figures 9.23C and 9.26B), which converts the serial pulses of each word into a parallel output. The first bit of the word is shifted into the portion of the register we will call the MSB location. As each bit, in turn, is shifted into the MSB location, the preceding bit in that location is shifted into the next bit location. Finally, when the last bit of the word is shifted in, the register contains the serial pulse code displayed in parallel fashion. This process is illustrated in Figures 9.26A and 9.26B. The shift register is shown in Figure 9.27; it is the 74164.

Also at this time, the output of the shift register is latched into the latch (Figures 9.23C and 9.26C) to preserve the pulse code and present it to the *digital-to-analog converter* (DAC, D to A, or D/A, shown in Figure 9.23C). *Latching* is the process of

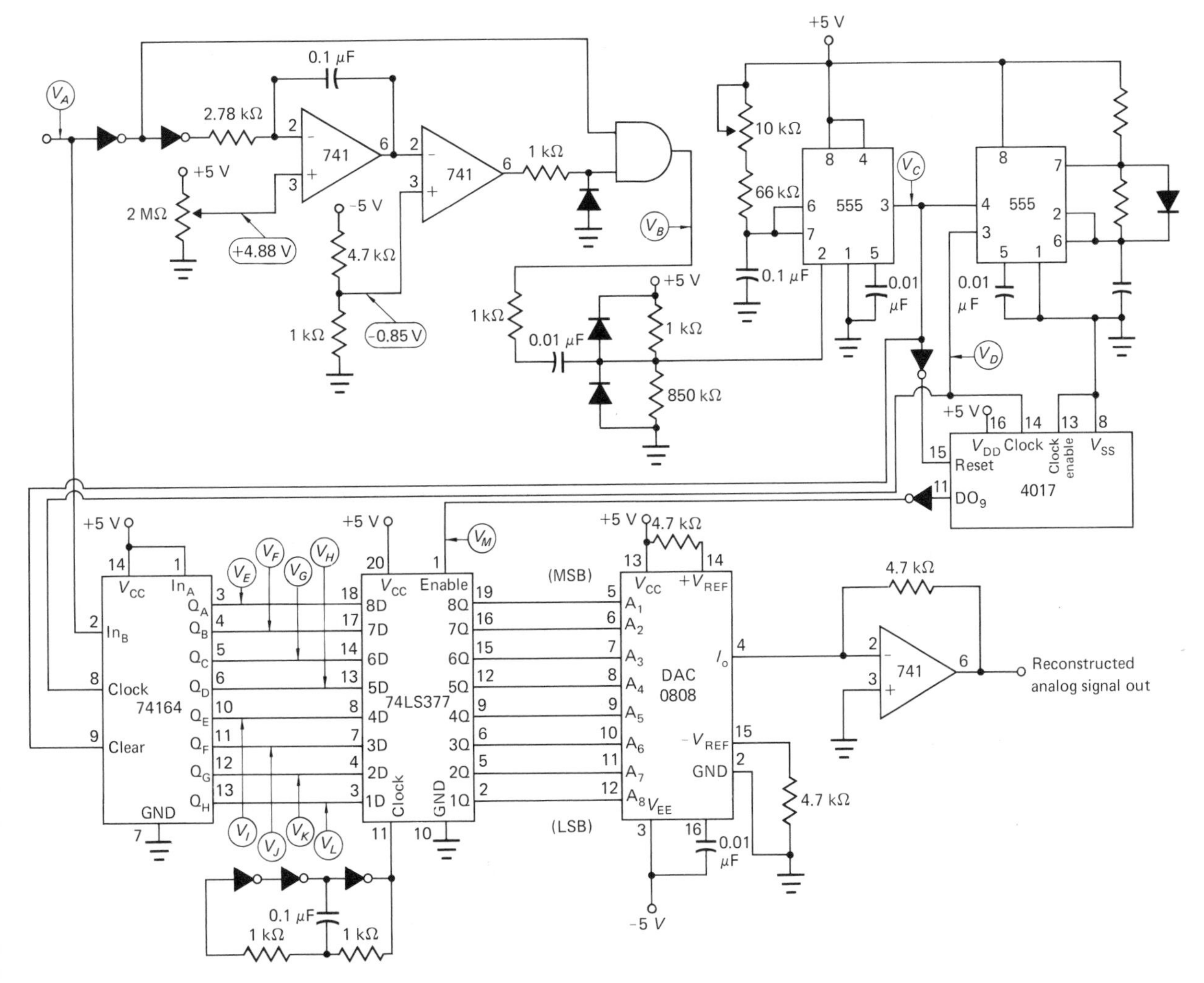

Figure 9.27 Schematic of PCM Demodulator

locking the input signal onto the output and holding it there regardless of what the input signal does following the latching process. The shift register is cleared during each synchronizing pulse time. The latch is shown in Figure 9.27; it is the 74LS377.

The DAC (Figure 9.23C) converts the code into the appropriate analog level, and the process is complete. Figure 9.26C shows the latched digital code, and Figure 9.26D shows the resulting analog output. The DAC converter in Figure 9.27 is the DAC0808. If this signal is now applied to a low-pass filter, the high-frequency components of the steps between voltage levels can be reduced, and the waveform will look more like the original analog signal.

If we look again at the block diagram in Figure 9.23C, we see that the PCM signal was input into the synchronizer as well as the shift register. The synchronizer detects the synchronizing pulse and synchronizes the shifting and latching process described above. In Figure 9.27, the entire top half of the circuit is the synchronizing pulse detector and synchronizer. This part of the circuit is a very critical part. If the synchronizer is off slightly, the latching process can occur too early or too late to give correct information. Most of the problems associated with this circuit will involve synchronizer timing.

We have gone through the process of generation and demodulation of PCM in detail. Certainly, there are many other ways to do the same thing. However, the ideas presented here are valid. Detailed schematics of a working PCM modulator and demodulator were presented in Figures 9.25 and 9.27. These circuits are not designed for efficiency but as teaching aids to illustrate PCM modulation and demodulation. With the ideas presented here, you can design your own system or improve on the circuits given.

Example of PCM

Figure 9.28 shows several *synchrograms*, which are diagrams showing waveforms synchronized in time. (Another name for synchrogram is timing diagram.) The diagrams in Figure 9.28 are synchrograms of the voltage waveforms for the voltages specified by the letters in the PCM modulator schematic of Figure 9.25. The data word represented in Figure 9.28 is 00000010, which corresponds to an analog input of 0.05 V.

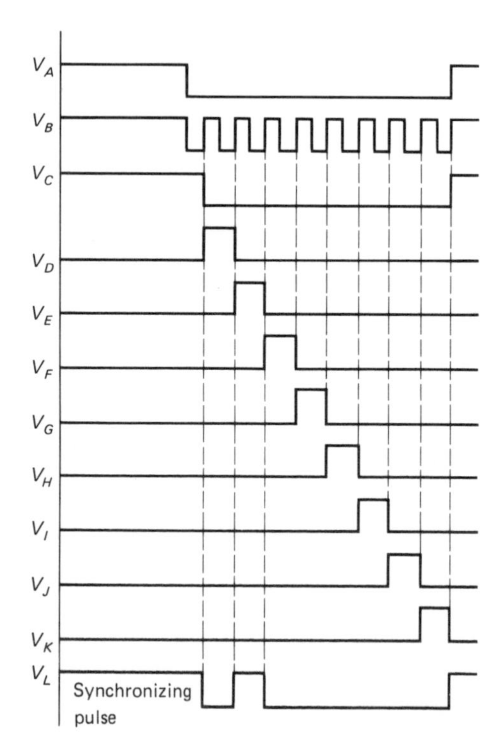

Figure 9.28 Synchrogram for PCM Modulation

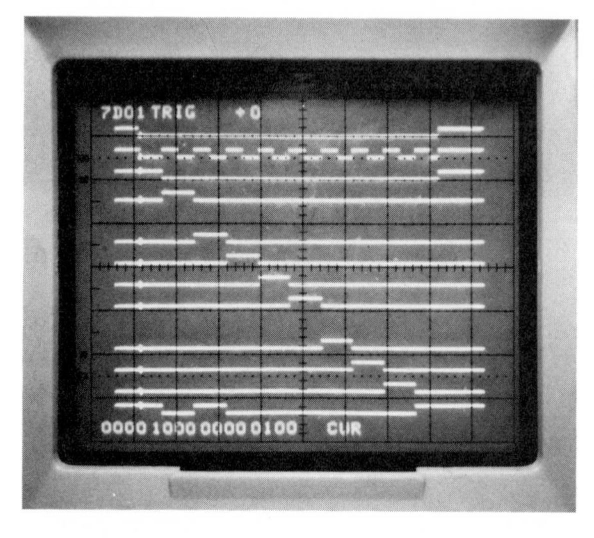

A. Data Word 00000010, Representing Analog Input of 0.05 V

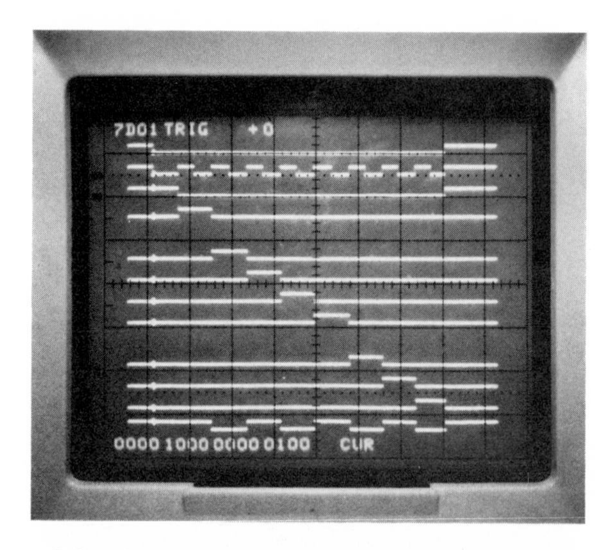

B. Data Word 01010101, Representing Analog Input of 1.57 V

Figure 9.29 Synchrograms for PCM Modulator

The photograph in Figure 9.29A presents the same information as the synchrogram in Figure 9.28. The voltage waveforms for V_A through V_L (the letters in the schematic in Figure 9.25) are consecutive in each photograph in Figure 9.29, from top to bottom.

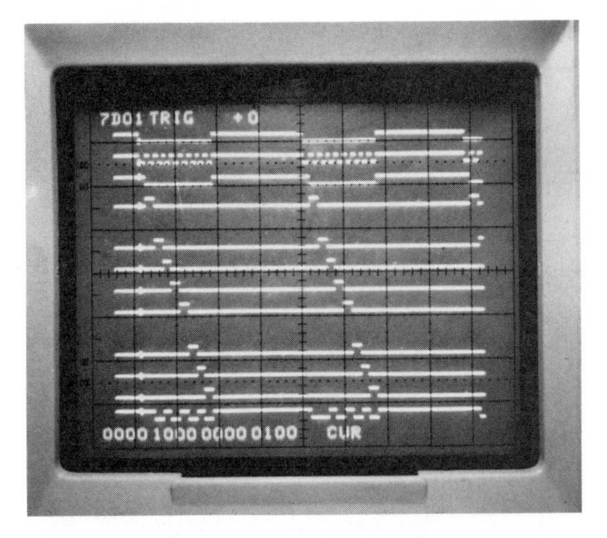

C. Two Data Words Representing Same Information as in Part B

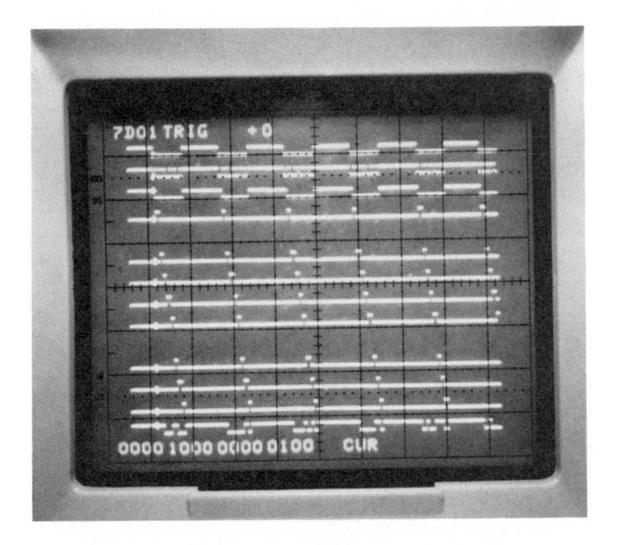

D. Five Data Words Representing a Time-varying Signal

Figure 9.29 (*continued*)

Figure 9.30 shows synchrograms for the PCM demodulator whose schematic was given in Figure 9.27. The voltage waveforms for V_A through V_M (the letters in the schematic) are consecutive in each photograph of Figure 9.30, from top to bottom. The last line of the photographs is unused.

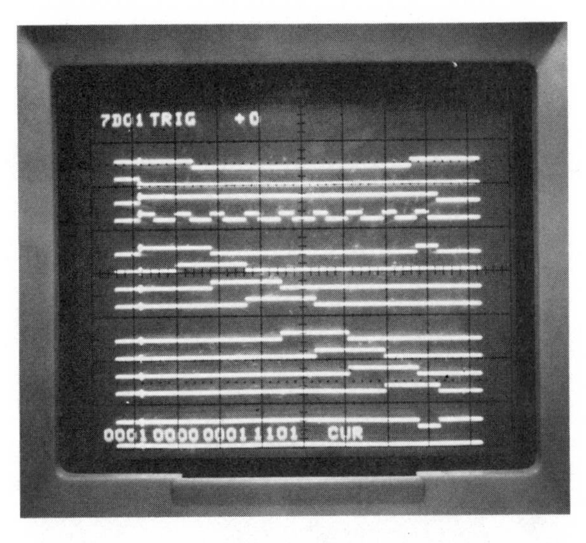

A. Input Data Word 00000001, Representing Analog Input of 0.03 V

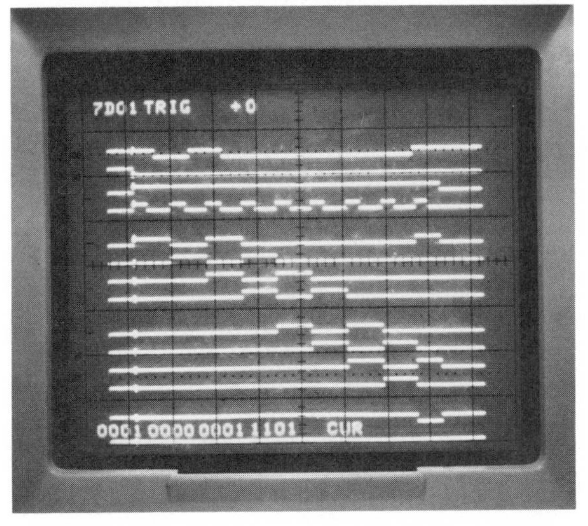

B. Input Data Word 00000010, Representing Analog Input of 0.05 V

Figure 9.30 Synchrograms for PCM Demodulator

Each photograph in Figure 9.31 shows the PCM signal (V_A, at the top of each photograph) and the output of the integrator (bottom). (The integrator is also called the synchronizing pulse detector.) The bottom signal in each photograph is the input to the next 741 (the comparator) and triggers it when its input signal goes to -0.85 V. Notice the effect of the PCM code on the bottom trace in each photograph. The worst-case condition shown in Figure 9.31C has an input code of 11111111. However, the result of the integration at the end of the coded part is still not low enough to reach the -0.85 V level and trigger the comparator.

Go through the schematics and synchrograms to verify for yourself how the system operates.

The PCM system we have been discussing was designed to operate at very low frequencies. The photograph in Figure 9.32A shows a 1 Hz analog input and the resulting PCM demodulated output.

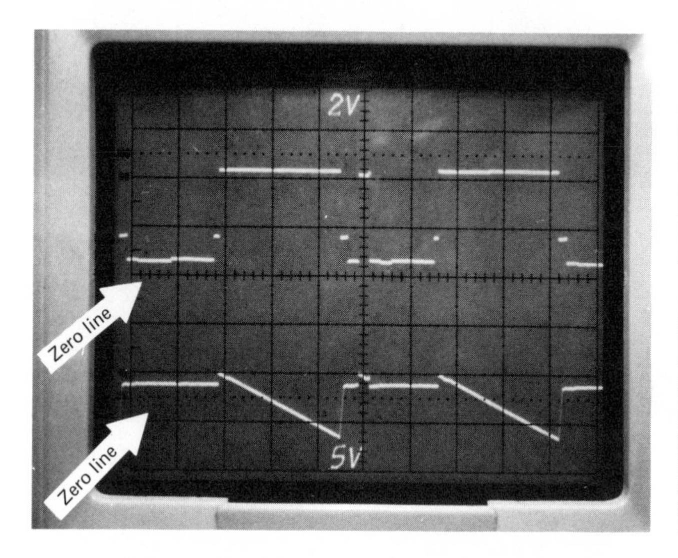

A. Zero-Level Input Signal and 00000010-Level Input

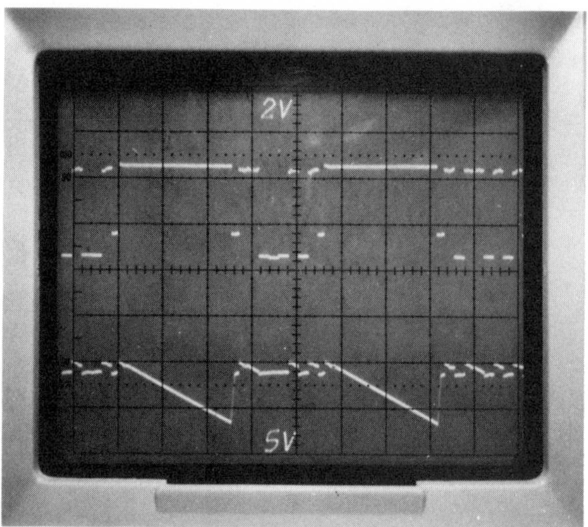

B. Time-varying Input

Figure 9.31 Results of Processing the Input Signal in Order to Synchronize the Demodulator (Figure 9.27)

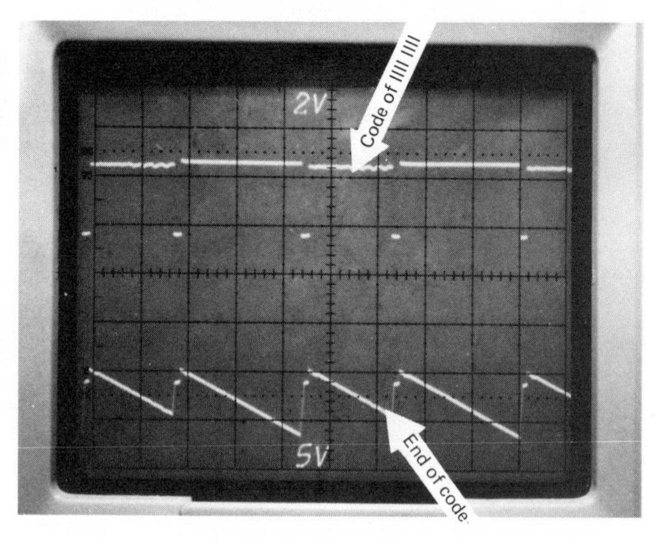

C. Worst-Case Condition

The input is the top waveform; the output is the bottom waveform. As the input frequency increases (Figure 9.32B), the steps in the output become much more visible, and the output becomes increasingly phase-shifted. The output is being sampled approximately 44 times during one cycle of the input signal. This sampling rate is 22 times the Nyquist rate for this input signal. (Recall that the Nyquist sampling rate is twice the highest frequency expected. See Figure 9.32D.) The phase shift of Figure 9.32B is a direct result of the *frame time*—the time from synchronizing pulse to synchronizing pulse—compared with the input signal frequency.

In the circuit for Figure 9.32, the frame time is about 23 ms. The input analog signal is sampled

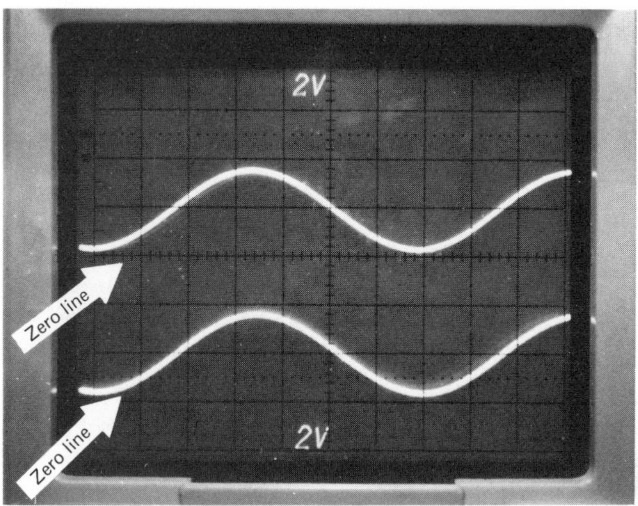

A. Well-reconstructed Waveform

B. Approximately 22 Times the Nyquist Sampling Rate

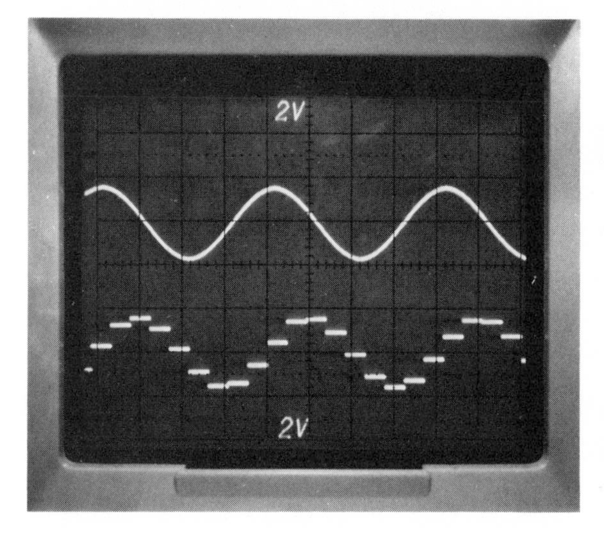

C. Approximately 5 Times the Nyquist Sampling Rate

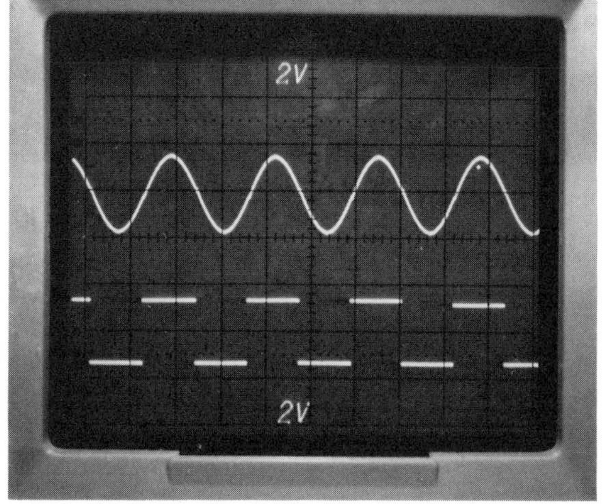

D. Highest Input Frequency (Nyquist Sampling Rate)

Figure 9.32 Waveforms of Input and Output Signals of Circuits in Figures 9.25 and 9.27

at the beginning of the synchronizing pulse, but it is not converted back into an analog signal until the end of the frame in the demodulator. Therefore, a phase shift results as the input analog period approaches the frame time. Figure 9.32A through 9.32D illustrate this progression.

Figure 9.32D shows the highest input frequency possible, resulting in two samples per period (Nyquist rate) of the input in order to be within the specifications set forth by the sampling theorem. In this case, filtering would be necessary to recover the original signal.

Companding

Often, we must reduce quantizing error for different-amplitude signals. To do so, we employ the process called *companding*, which is the compression of large signals at the transmitter followed by expansion of the signals at the receiver. This process is called *companding* since it *com*presses the large signals (at the transmitter) and then ex*pands* those that were compressed (at the receiver). The following discussion demonstrates why this process is helpful.

As was previously shown, the maximum quantizing error for a PCM system is one-half the amplitude of a quantizing level. The PCM system of Figure 9.25 has a signal amplitude range of 5 V and uses eight bits for each word. Thus, the maximum quantizing error in this system is as follows:

$$\frac{1}{2^8} \times 5\text{ V} = \frac{1}{256} \times 5\text{ V} = 19.5\text{ mV} \tag{9.2}$$

For an input analog signal of 5 V peak to peak, 19.5 mV represents about a 0.4% error. However, for an input signal of 0.1 V peak to peak, 19.5 mV is almost a 20% error. Thus, small-signal inputs can produce large errors at the output because of quantizing.

One way to reduce this problem is to have more quantizing levels for small signals, or, in other words, to *taper* the quantizing levels. Figure 9.33 illustrates a nonlinear quantizing-level arrangement. At the receiver, the levels would be corrected by a corresponding amount so that the amplitude of the original signal is returned.

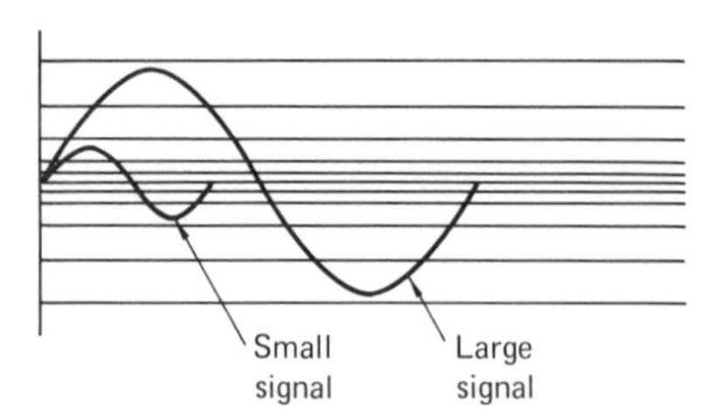

Figure 9.33 Tapered Quantizing Level

The circuitry necessary to produce nonlinear quantizing levels is much more complex than that required for linear levels. However, a suitable alternative can be employed. The large signals can be attenuated proportionally more than small signals before they are encoded. Then, at the receiver, the large signals can be amplified more than the small signals. The results here would be the same as those for the nonlinear circuitry, without greatly complicating the circuitry.

Integrated Circuit CODEC

At this point, we can appreciate the complexity of designing a PCM modulator and demodulator system. National Semiconductor now has two integrated circuits that perform these functions. They are called *PCM CODECs* (pulse-code modulation coders/decoders). They are the TP3001 μ-law CODEC and the TP3002 A-law CODEC. The *μ-law* and *A-law* designations refer to the two different transfer functions used in companding. The μ-law function is used in North America and parts of the Far East. The A-law function is used in most of the remaining countries. For more applications and information, see National Semiconductor's application note 215. A reprint of the application note is given in National Semiconductor's *Special Functions Databook*.

Advantages of PCM

Unless the noise signal completely obliterates the PCM pulse or pulses, PCM is not affected by amplitude noise, as PAM is. Nor is PCM affected by rising or falling edge noise, as PWM, PPM, and PFM are. Thus, PCM is less affected by noise than any of the other pulse modulation methods so far presented. Insensitivity to noise is a great advantage, especially in systems where the signal must be relayed from point to point before it reaches its destination. Other signals are degraded slightly after each relay process; PCM is not.

If noise rejection is such an advantage, why is PCM not the only system in use today? Other systems are used because they were developed first, because PCM circuitry is much more complex, and because PCM requires a much larger bandwidth than the other systems.

PCM was patented in 1937 by Alex H. Reeves of Great Britain. Even though PCM was well documented, it did not come into its own until the early 1960s, when integrated circuits could be used to implement the circuitry. As noted above, integrated circuits make implementation of complex PCM much easier.

The bandwidth requirement for PCM will also become less of a problem as technology advances. With the steady increase in the use of lasers and fiber optics, noise-free transmission of wide-bandwidth signals can be realized.

PCM is becoming more popular as circuit complexity is solved. In 1965, the *Mariner IV* space probe transmitted pictures by using PCM. The system required 30 min to transmit each picture, but the transmitter was over 200,000,000 km away, and the transmitter power was only 10 W! Today, PCM is being used to produce noise-free recordings on phonograph records (digital recordings) and on video recorder systems (laser discs). PCM will become more widespread as technology continues to develop.

Other Digital Systems

Other digital pulse modulation systems have been proposed, but none is as popular as PCM. Two of the other digital modulation systems are briefly discussed below.

One digital system in use is called *differential PCM*. Differential PCM is very similar to regular PCM. The difference is that regular PCM quantizes the absolute amplitude of the signal, whereas differential PCM quantizes the difference in amplitude, positive or negative, between one sample and the previous sample. The idea here is that most signals do not change much from their previous level. Thus, it would take fewer pulses to represent a change in amplitude than it would take to represent an absolute amplitude. If this technique could be implemented, then a bandwidth reduction could be realized. Again, complex encoding and decoding circuits have kept differential PCM from becoming widely accepted.

The second and more popular digital system is called *delta modulation*. Delta modulation in its simplest form is similar to differential PCM. However, delta modulation changes only one bit, positive or negative, per sample. Figure 9.34 illustrates delta modulation.

One of the problems associated with delta modulation is *slope clipping*. Here, the input signal is changing so rapidly that the delta modulator cannot keep up with it. Slope clipping is illustrated in Figure 9.35.

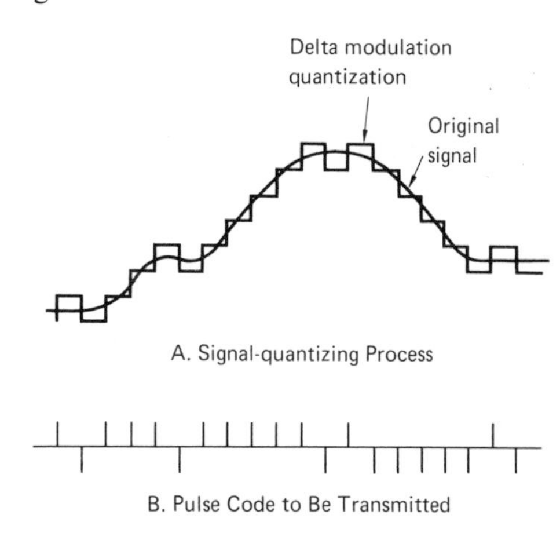

Figure 9.34 Delta Modulation

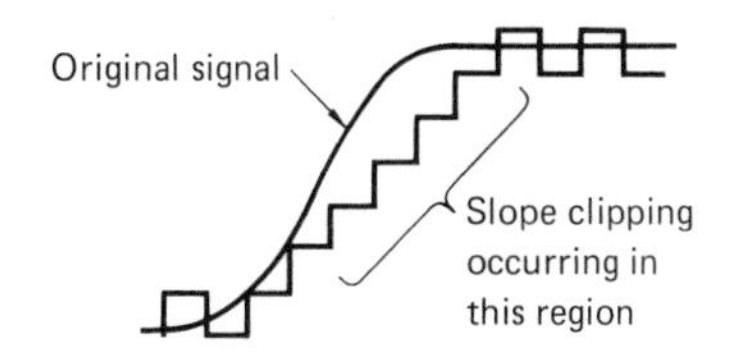

Figure 9.35 Slope Clipping in Delta Modulation

Conclusion

This chapter has been concerned with pulse modulation, the basis for telemetry. Of all the pulse modulation types, pulse-code modulation is probably going to become the most widely used. PCM's advantage of excellent noise rejection is beginning to overshadow its disadvantage of complex circuitry, especially now that single integrated circuits are acting as encoders and decoders. Already, the industrial, consumer, and government markets have made extensive use of the process.

Questions

1. ______ is the process for performing measurements on distant objects.
2. News services transmit photographs electronically by a process known as ______.
3. The ______ measurement technique is easier to use over short distances. The ______ technique is more noise-free over long distances.
4. The four types of analog pulse modulation are ______, ______, ______, and ______.
5. The subclassification of PAM in which pulse excursions occur both above and below the reference level is called ______.
6. In PWM, the widest pulse width should not exceed ______% of maximum, and the narrowest pulse width should not be less than ______%.
7. Pulse-______ modulation is differentiated and clipped PWM.
8. Pulse-______ modulation is a form of digital pulse modulation.
9. In PCM, the analog signal voltage is divided into levels called ______.
10. Name one of the five modulation types that is most affected by noise.
11. Identify the reconstruction phenomenon that may occur when the sampling rate of the analog signal is too slow.
12. Name the type of modulation produced by an amplifier whose gain can be controlled by an analog signal.
13. Name the type of filter with which both PAM and PWM can be demodulated.
14. The PLL can be used to demodulate what type of modulation?
15. Does the input to the PLL have to be a sine wave for the PLL to work properly?
16. Name the term used for a group of pulses in PCM representing an analog voltage level.
17. Discuss why quantizing noise is random noise.
18. Discuss why the largest quantizing error is one-half the amplitude of the levels into which the analog signal is divided.
19. Discuss the use of a synchronizing pulse in PCM.
20. Define the term synchrogram.
21. Discuss the need for companding.
22. Explain the term CODEC.
23. Discuss two reasons that are keeping PCM from becoming more common.
24. Define slope clipping.

Problems

1. Observing the rules of the sampling theorem, determine the minimum sampling rate for high-fidelity voice (20 kHz).
2. Draw to scale the PPM waveform for a straight line that has a slope of 0.2 and a length of 20 cm, and goes through zero at the 10 cm length point. The pulse width (of PWM) at the lower end of the straight line is 20% of its maximum, and the pulse width (of PWM) at the upper end of the straight line is 80% of its maximum pulse width. Use eleven equally spaced sample points from one end of the 20 cm point to the other.
3. Determine the VCO free-running frequency of the 565 given in Figure 9.18. The value of R_1 is 4 kΩ, and the value of C_1 is 0.1 μF.
4. In a PCM system with six bits in each word, the maximum input signal swing is 10 V. For quanta that are equally spaced, determine the maximum quantizing error.
5. In Figure 9.20, suppose $C_1 = 0.01$ μF, $R_1 = 1$ kΩ, $R_2 = 4.6$ kΩ, and $R_3 = 5.4$ kΩ. What is the approximate output frequency of the circuit if the power connections on the 741 are ± 10 V?
6. If R_1 in Problem 5 changes to 1.5 kΩ, what is the new output frequency?

Telemetry

CHAPTER 10

Objectives

On completion of this chapter, you should be able to:

- Describe frequency-division and time-division multiplexing;
- Determine the relationship among data bandwidth, frequency deviation, and modulation index;
- Relate pulse rise time to data bandwidth;
- List the sources of error in multiplexed transmission;
- Determine the classification of multiplexing systems;
- Calculate frame time, frame rate, sample time, and commutation rate for pulse modulation systems.

Introduction

Telemetry was discussed briefly at the beginning of the preceding chapter because it relies heavily on pulse modulation. Now that pulse modulation has been presented, telemetry can be discussed in more detail.

Telemetry can be defined as the science involved in performing a measurement at a remote location and transmitting the data to a central location for processing or storage. The telemetry system can be divided functionally into three parts: transducers, transmission system, and display and interpretation system. See Figure 10.1.

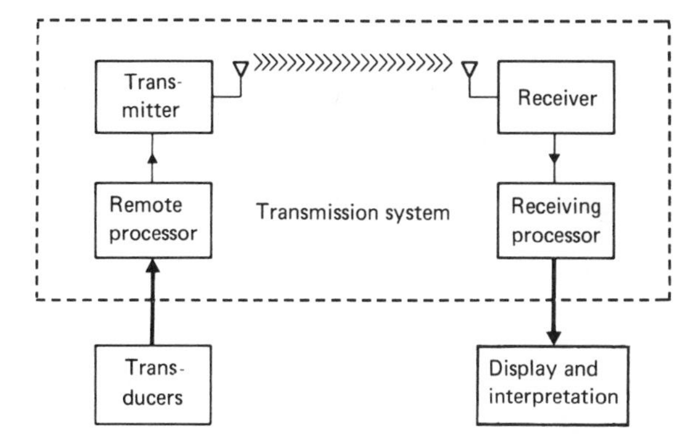

Figure 10.1 Functional Diagram of Telemetry System

The transducers are located at the remote station and transform the physical quantities to be monitored into electrical signals. The transmission system consists of a device for transforming the electric signals from the transducers into signals suitable for transmission to the receiving station; the transmission and receiving devices; and a device for transforming the output of the receiver to forms suitable for display and interpretation. The display and interpretation system consists of the devices that calculate the desired parameters and display them for final interpretation by automatic or human means.

Transducers were discussed in detail in Chapter 7 and that discussion need not be repeated here. Our attention in this chapter will focus on the portion of the telemetry system called the transmission system and, in particular, on the remote and receiving processor portions in Figure 10.1. These devices transform the signals for transmission and later transform them back to signals suitable for display.

Applications of Telemetry

Telemetry is considered a necessary topic in a text on industrial electronics because of the need in industry to transmit measurements over a distance. Typical applications, to name only a few, have been the monitoring and automatic adjustments of electrical power transmission grids or networks and other utility systems; the monitoring of meteorological (weather) conditions in atmospheric, subsurface, or severe climate locations; the monitoring of seismic (earthquake) conditions; the monitoring or control of satellite, space probe, missile, or ordnance systems; easier monitoring of vital signs of animals or humans who are ambulatory (able to move about); and the monitoring or control of rotating machines, such as the blades of a turbine.

Of the telemetry applications listed above, most of the measurements to be telemetered are analog (or continuously variable) and may be transmitted by analog methods. A simple example of an analog method is a voltmeter (or ammeter) with long leads. Many factors must be considered when telemetering analog measurements, such as ground loops, line loss, and noise. Nevertheless, most of these applications are relatively easy to realize with telemetry when compared with pulse modulation, especially over short distances.

If analog transmission is easy and pulse modulation is more complicated, why is pulse modulation used so much? Some of the advantages have already been listed in Chapter 9. To summarize: (1) Some of the information is in pulse form to start with; (2) pulse transmission is a more error-free transmission process; and (3) large-scale integration has reduced circuit complexity. In addition to these advantages are two more: Transmitters can operate on very low duty cycles (percent of time the transmitter is on), and the time intervals between pulses can be filled with samples of other data. This last advantage illustrates how a number of different sets of data can be *multiplexed*, that is, transmitted over the same channel. We will discuss multiplexing in detail in this chapter.

Multiplexing

In most telemetry applications, it is desirable to perform a number of different measurements. Although a separate transmission line or link could be used for each measurement, the problems of efficiency in terms of power, bandwidth, weight, size, and cost normally prevent us from doing so. Therefore, it is common to send many measurements over a single transmission channel. This process, as mentioned above, is called multiplexing. Multiplexing can be used in systems for purposes other than telemetering, but it is so useful in telemetering that it will be the subject of the remainder of this chapter.

There are two types of multiplexing in common use today: frequency-division multiplexing (FDM) and time-division multiplexing (TDM). These two types of multiplexing are illustrated in Figure 10.2.

Figure 10.2A is a representation of the space available for communication. This communication space can be divided into different frequency channels at the same time, for *frequency-division multiplexing*, or into different time slots for the same frequency, for *time-division multiplexing*. Of

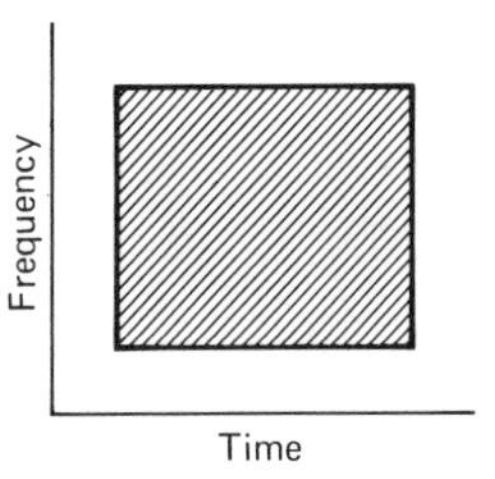

A. Space Available for Communication

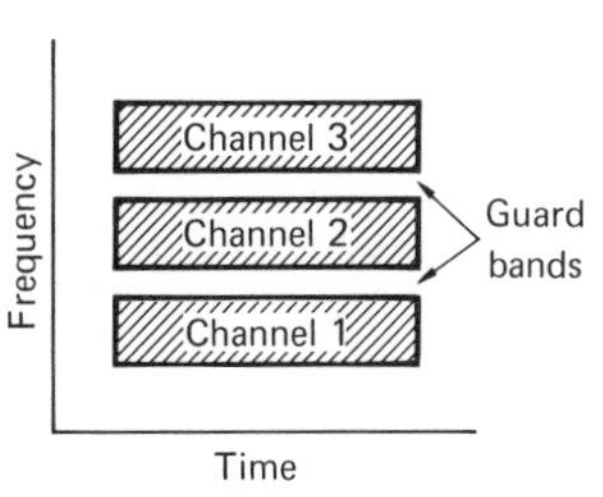

B. Communication Space Divided into Frequency Channels for FDM

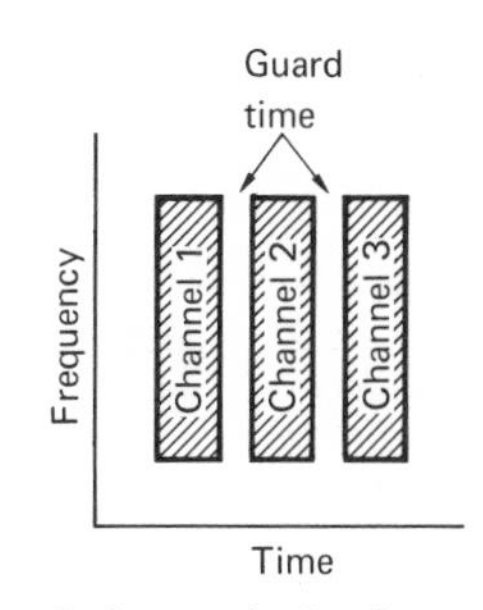

C. Communication Space Divided into Time Slots for TDM

Figure 10.2 Frequency-Division and Time-Division Multiplexing

course, combinations of these two methods could also be used. Notice that each type has a *guard space* to separate channels. Generally, the guard space is used to reduce cross talk (interference between channels) and therefore reduce circuit cost and complexity.

Frequency-Division Multiplexing (FDM)

A familiar example of FDM is radio broadcasting. The signals received by a radio antenna contain many different programs traveling together but occupying different frequencies on the radio spectrum. The tuning circuits in the radio receiver allow the selection of one station from all the others.

The use of the radio frequency spectrum for telemetry and other applications is regulated by various government agencies. The Inter-Range Instrumentation Group (IRIG) publishes a set of telemetry standards that are revised periodically and that specify all pertinent details of telemetry applications.* Table 10.1 is an excerpt from these telemetry standards.

Twenty-one data channels, with subcarrier center frequencies ranging from 400 Hz to 165 kHz (column 2 of Table 10.1), have been designated for telemetry use. *Subcarriers* are carrier frequencies in telemetry systems that are modulated by the measurement or intelligence signal. These center frequencies are designed with a $\pm 7.5\%$ frequency deviation. Thus, Table 10.1 is used for frequency modulation (FM) purposes.

The *nominal frequency response* (or data bandwidth) (column 5 of Table 10.1) designates the highest frequencies at which the data can vary for each channel. For example, channel 1's data cannot vary any faster than 6 Hz in order to stay in channel 1, assuming a modulation index of 5. The *modulation index* for FM is defined as the maximum frequency deviation of the signal divided by the modulating frequency. While modulation indices of 5 are recommended, indices as low as one or less may be used. For these low indices, low signal-to-noise ratios, increased harmonic distortion, and cross talk must be expected.

The relationship among the data bandwidth (data BW), the frequency deviation of the center frequency (400 ± 30 for channel 1), and the modulation index is expressed by the following equation:

$$\text{data } BW = \frac{\text{frequency deviation}}{\text{modulation index}} \quad \textbf{(10.1)}$$

For channel 1:

$$\text{data } BW = \frac{30}{5} = 6 \text{ Hz} \quad \textbf{(10.2)}$$

* A copy of the standards may be requested from Secretariat, Inter-Range Instrumentation Group, White Sands Missile Range, New Mexico 88002.

Table 10.1 *Proportional Bandwidth FM Subcarrier Channels*

Channel	Center Frequencies (Hz)	Lower Deviation Limit* (Hz)	Upper Deviation Limit* (Hz)	Nominal Frequency Response (Hz)	Nominal Rise Time (ms)	Maximum Frequency Response† (Hz)*	Minimum Rise Time† (ms)
			±7.5% CHANNELS				
1	400	370	430	6	58.00	30	11.700
2	560	518	602	8	42.00	42	8.330
3	730	675	785	11	32.00	55	6.400
4	960	886	1,032	14	42.00	72	4.860
5	1,300	1,202	1,398	20	18.00	98	3.600
6	1,700	1,572	1,828	25	14.00	128	2.740
7	2,300	2,127	2,473	35	10.00	173	2.030
8	3,000	2,775	3,225	45	7.80	225	1.560
9	3,900	3,607	4,193	59	6.00	293	1.200
10	5,400	4,995	5,805	81	4.30	405	0.864
11	7,350	6,799	7,901	110	3.20	551	0.635
12	10,500	9,712	11,288	160	2.20	788	0.444
13	14,500	13,412	15,588	220	1.60	1,088	0.322
14	22,000	20,350	23,650	330	1.10	1,650	0.212
15	30,000	27,750	32,250	450	0.78	2,250	0.156
16	40,000	37,000	43,000	600	0.58	3,000	0.117
17	52,500	48,562	56,438	790	0.44	3,938	0.089
18	70,000	64,750	75,250	1050	0.33	5,250	0.067
19	93,000	86,025	99,975	1395	0.25	6,975	0.050
20	124,000	114,700	133,300	1860	0.19	9,300	0.038
21	165,000	152,624	177,375	2475	0.14	12,375	0.029
			±15% CHANNELS‡				
A	22,000	18,700	25,300	660	0.53	3,330	0.106
B	30,000	25,500	34,500	900	0.39	4,500	0.078
C	40,000	34,000	46,000	1200	0.29	6,000	0.058
D	52,500	44,625	60,375	1575	0.22	7,875	0.044
E	70,000	59,500	80,500	2100	0.17	10,500	0.033
F	93,000	79,050	106,950	2790	0.13	13,950	0.025
G	124,000	105,400	142,600	3720	0.09	18,600	0.018
H	165,000	140,250	189,750	4950	0.07	24,750	0.014

Source: Reprinted from *IRIG Telemetry Standards*, May, 1973, with permission from Inter-Range Instrumentation Group, White Sands Missile Range, NM.

*Rounded off to nearest Hz.

†The indicated data for maximum frequency response and minimum rise time are based on the maximum theoretical response that can be obtained in a bandwidth between the upper and lower frequency limits specified for the channels.

‡Channels A through H may be used by omitting adjacent lettered and numbered channels. Channels 13 and A may be used together with some increase in adjacent channel interference.

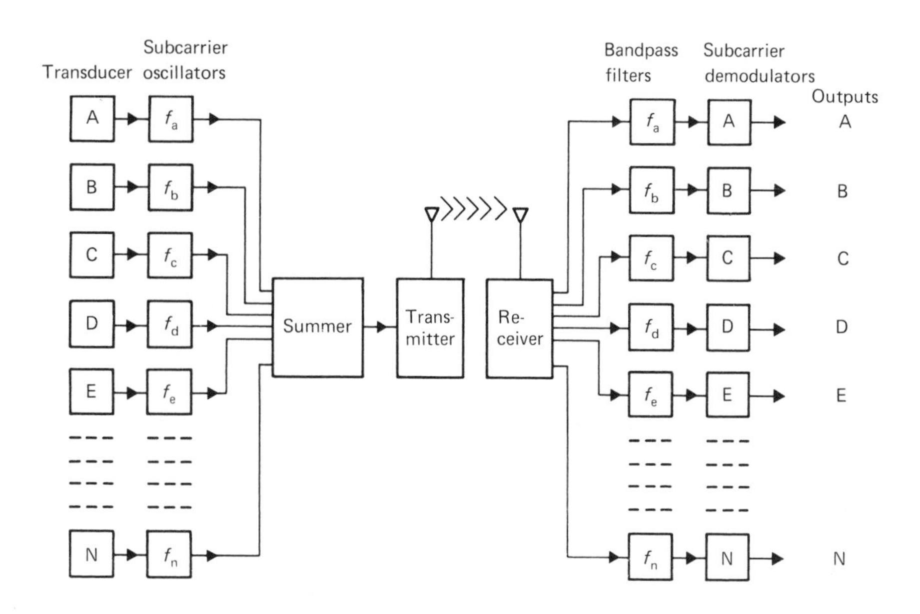

Figure 10.3 FDM System

In Table 10.1, the center frequency deviation and the maximum frequency response are identical, since the maximum frequency response column is based on a modulation index of one. The *rise times* t_r (nominal and minimum) given in Table 10.1 are related to the data *BW* by the following equation:

$$t_r = \frac{0.35}{BW} \quad \textbf{(10.3)}$$

Again, nominal rise time is based on a modulation index of 5, while minimum rise time is for an index of one. The rise time equation can now relate pulse rise time to data bandwidth and therefore expands the use of Table 10.1 beyond FM applications to pulse applications.

The 21 bands were chosen to make the best use of present equipment and the frequency spectrum. There is a ratio of approximately 1.3:1 between center frequencies of adjacent bands, except between 14.5 and 22 kHz, where a larger gap was left to provide for a compensation tone for magnetic tape recording. The deviation has been kept at $\pm 7.5\%$ for all bands, with the option of $\pm 15\%$ deviation on the five higher bands to provide for transmission of higher-frequency data. When this option is used on any of these five bands, certain adjacent bands cannot be used, as noted in a footnote to Table 10.1.

The basic operation of an FDM system is illustrated in Figure 10.3. The measurement signals from the transducers are used to modulate subcarrier oscillators tuned to different frequencies. The outputs of the subcarrier oscillators are then linearly summed (linearly mixed), and the resulting composite signal is used to modulate the main transmitter. At the receiving site, the composite signal is obtained from the receiver demodulator and input to a number of bandpass filters, which are tuned to the center frequencies of the subcarrier oscillators.

An example might be helpful at this point. Let's assume that the system given in Figure 10.3 has four input transducers and therefore four outputs at the receiver. We will also assume that transducer A varies no faster than 6 Hz, transducer B no faster than 8 Hz, transducer C no faster than 11 Hz, and transducer D no faster than 14 Hz. This arrangement allows us to use channels 1, 2, 3, and 4 from Table 10.1. The subcarrier oscillator frequencies will be 400, 560, 730, and 960 Hz, respectively. Figure 10.4 is a diagram of this arrangement.

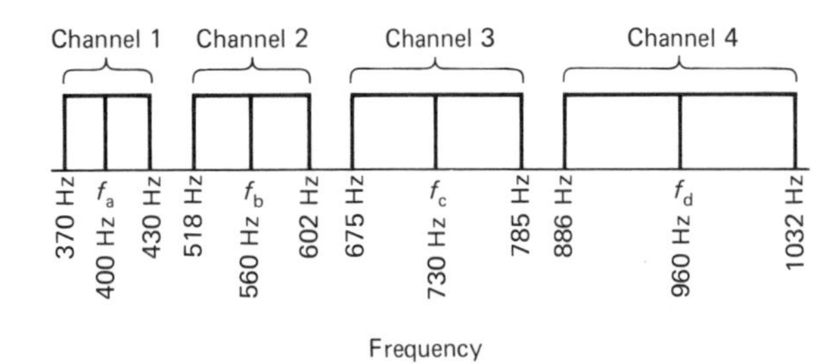

Figure 10.4 FDM System Using Channels 1, 2, 3, and 4 of Table 10.1

The system bandwidth now varies from 370 to 1032 Hz, with guard bands in between each channel. This data bandwidth of 662 Hz is now sent to the transmitter to be transmitted. If we maintain a modulation index of 5, the main transmitter frequency deviation about its center frequency should be 3.31 kHz (622 Hz × 5, from Equation 10.1). The receiver will demodulate these frequencies back into their original data signals. Both the subcarriers and the main carrier frequencies were frequency-modulated in this example.

All types of modulation can be used for both the subcarrier and the main carrier. The types of modulation are frequency modulation (FM), phase modulation (PM), amplitude modulation (AM), suppressed-carrier amplitude modulation (SC), and single-sideband amplitude modulation (SS or SSB). FDM systems are normally designated by listing the type of modulation used by the subcarriers followed by the type of modulation used by the main carrier. Thus, FM/AM indicates an FDM system in which the subcarriers are frequency-modulated by the measurement and the main carrier is amplitude-modulated by the composite subcarrier signals. Almost all combinations of subcarrier and main-carrier modulation techniques have been used in the past. However, FM/FM is by far the most common technique in use today.

The principal sources of errors in an FDM transmission system are drift, bandlimiting, cross talk, distortion, and radio frequency (RF) link noise.

The errors due to drifts (slowly changing circuit parameters due to heat, age, and so on) are those associated with modulation and demodulation of the subcarriers. If the drifts are slow, their effect can be greatly diminished by use of calibration signals. Their reduction depends on the circuit design.

Errors due to bandlimiting (limited frequency spectrum) occur throughout the system whenever the data are dynamic (changing). This error source cannot be eliminated regardless of the circuit design.

Cross talk errors are of two kinds. The first is caused by nonlinearities in the summing amplifier before the main transmitter or in the modulation or demodulation process of the main carrier. This type of cross talk error can be eliminated. The second type of cross talk error is due to overlap of frequencies from the bandpass filters of adjacent channels. Since no filter is perfect, some overlap will always be present in other channels. This error cannot be eliminated.

Distortion error is due to nonlinearities in the subcarrier modulator and demodulator (not to be confused with cross talk errors due to nonlinearities in the main-carrier modulator or demodulator). Calibration signals can help reduce or eliminate this error.

RF link noise (the RF link is the transmitter, receiver, and medium in between) causes errors that are generally random and come from three sources: receiving station front-end noise, noise from space, and technological interference. These types of noise will always be present but can be reduced by proper circuit design.

Time-Division Multiplexing (TDM)

The major alternative to FDM is time-division multiplexing (TDM). Here, the time available is divided up into small slots, and each of these slots is occupied by a piece of one of the signals to be

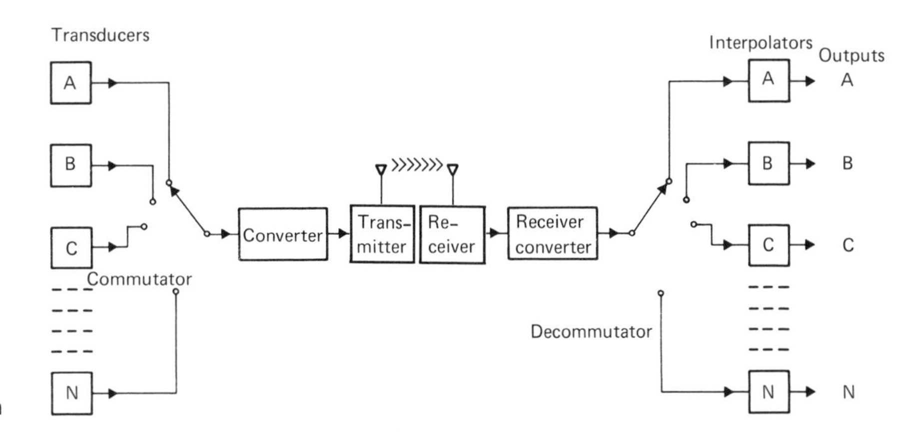

Figure 10.5 TDM System

sent. The multiplexing apparatus scans the input signals sequentially. Only one signal occupies the channel at one instant. TDM is thus quite different from FDM, in which all the signals are sent at the same time but each occupies a different frequency band.

The operation of a TDM system from a functional standpoint is illustrated in Figure 10.5. The signals from the transducers are input to a *commutator* (a switching device), which samples the channels sequentially. Thus, the output of the commutator (also referred to as the multiplexer) is a series of pulses, the amplitudes of which correspond to the sampled values of the input channels from the transducers. This train of pulses is then passed through a device that converts it to a form suitable for modulating at the transmitter.

At the receiving station, the process is reversed. The demodulated output from the receiver is passed through a converter, which reproduces the pulse train that existed at the commutator at the transmission site. This pulse train can then be decommutated to produce pulses with values corresponding to samples of the original measurement signals.

Let's look at the commutator section a little more closely. Figure 10.6A is a diagram of a mechanical commutator connected to three transducers. As the commutator arm or rotor rotates, it makes contact with conductors A, B, C, and sync. Each time a conductor is contacted, a voltage is output. Between each conductor is an insulator. When the commutator rotor is on an insulator, there is no output. The insulator contact corresponds to the space between signals in the output pulse train diagram of Figure 10.6B.

Figure 10.6B shows a synchrogram of the operation occurring in Figure 10.6A. Providing the values of the voltages from the instruments are not varying too rapidly compared with the rotation time of the rotor, the individual inputs can be reconstructed from the composite signal.

For separation of the signals when they are received, a commutator similar to that illustrated in Figure 10.6A might be used, but with the input and output reversed. The receiving commutator must be exactly synchronized with the transmitting commutator. That is exactly what the sync pulse is for, to synchronize the receiver commutator with the transmitter commutator. The commutators used today operate at very high speeds and are electronic instead of mechanical.

As in FDM, in TDM, the modulation of the transmitter may take any form, that is, AM, FM, PM, and so on. However, the principal distinction among TDM systems lies in the form of the processor used (see Figure 10.1). For the processor, any of the pulse modulation methods discussed in

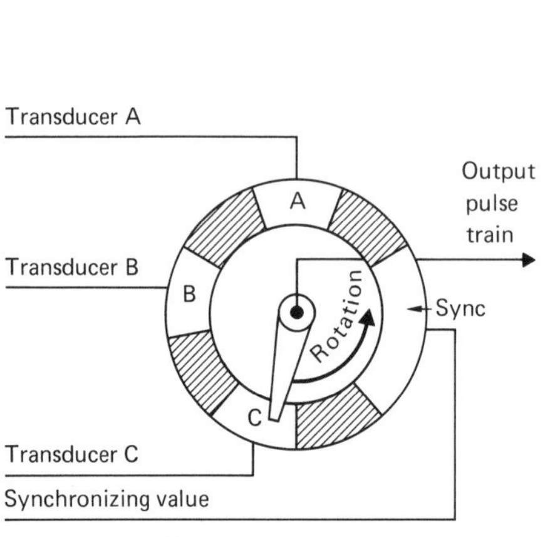

A. Mechanical Commutator

B. Construction of Composite Signal Using PAM Transmission

Figure 10.6 TDM Commutation

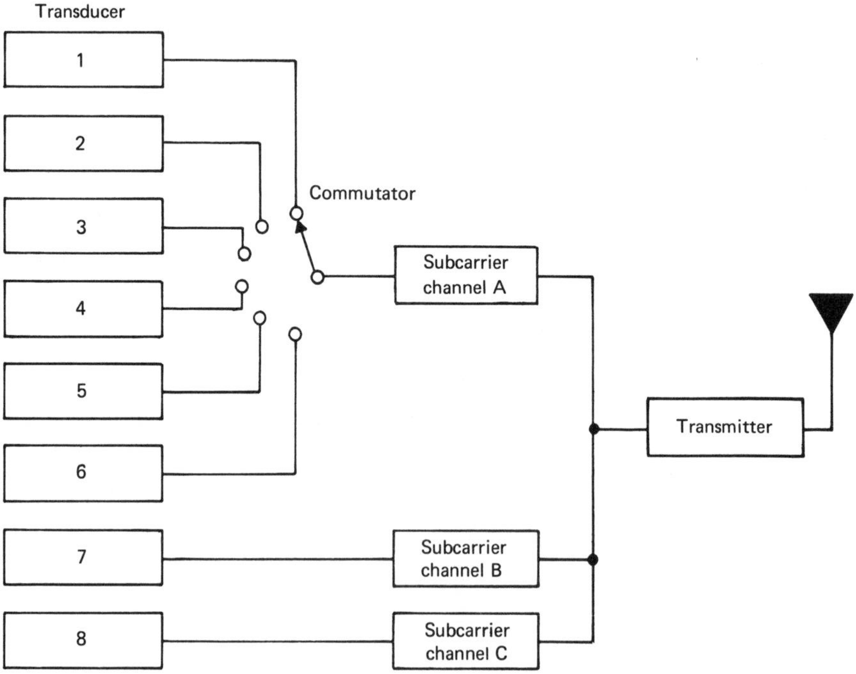

Figure 10.7 Combination of TDM and FDM in One System

Chapter 9, such as PAM, PWM, PPM, PCM, and delta modulation, will work. The system designation normally lists the type of processor (or converter) first, followed by the type of main-carrier modulation. That is, a PWM/PM system has a pulse-width modulation processor, the output of which is used to phase-modulate the main carrier.

Submultiplexing

In FDM transmission systems, it has been common to submultiplex some of the wider-band subcarrier channels. *Submultiplexing* usually involves the use of a TDM waveform as the modulation of one of the subcarrier channels, as shown in Figure 10.7. The use of submultiplexed channels of this nature allows the total number of measurements handled by the transmission system to be increased considerably. The frequency content of the submultiplexed measurements must be low relative to the frequency content that could normally be transmitted over the subcarrier channel.

The most common types of submultiplexing are PAM/FM/FM and PWM/FM/FM, although almost all varieties have been used for some applications. Combined telemetry systems have not been widely used to date. However, there is some indication that they may receive increased attention in the future.

Pulse-Amplitude TDM

Let's return our attention now to pure TDM. The first modulation type we will consider is pulse-amplitude modulation (PAM). A PAM system is one in which the output of the commutator is used to directly modulate the main transmitter. Thus, the processor is simply a pair of wires. So, PAM is the simplest of the TDM systems.

The PAM wave trains may take several forms, two of which are illustrated in Figure 10.8. The principal difference here lies in the duty cycle system (50% and 100%). The length of time necessary to sample all channels (including frame sync and calibration pulses) is normally referred to as *frame time*. For identification of the sample at the receiving station, it is necessary to insert frame synchronization. Several different methods can be used to designate a frame. The one illustrated in Figure 10.8 consists of forcing several consecutive

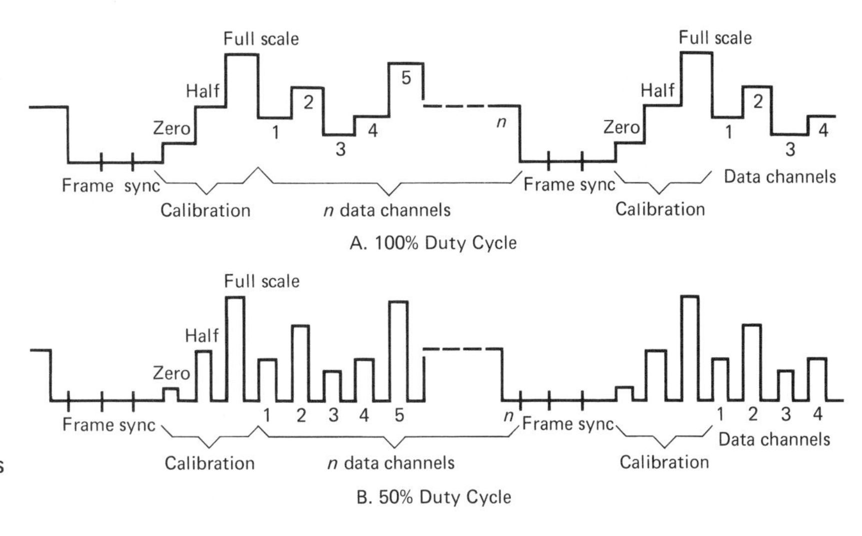

Figure 10.8 PAM Waveforms in a TDM System

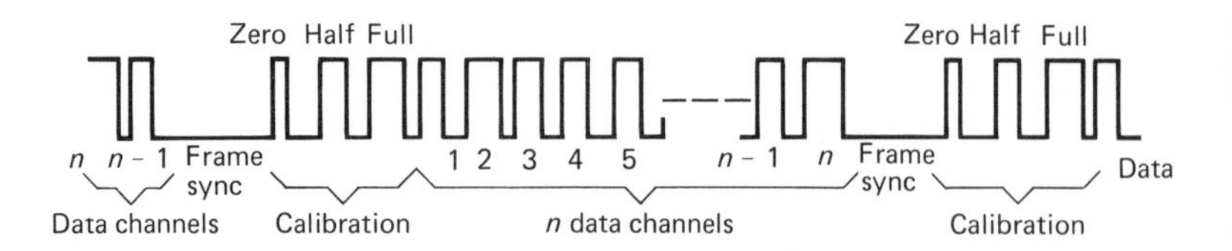

Figure 10.9 PWM Waveform in a TDM System

channels to a level below the minimum allowable data value. Since drifts and nonlinearities in the system cause error directly, it is also common to transmit calibration pulses (zero, half, and full scale), as shown. The data is offset in such a fashion that channels with signal outputs that are capable of both positive and negative polarities are centered about the half-scale value. The frame sync pulses and calibration pulses are sometimes referred to as *housekeeping pulses*, since they are not part of the data but are used to synchronize and calibrate the receiver.

In general, the 100% duty cycle system (Figure 10.8A) requires a smaller frequency spectrum than that of the 50% duty cycle system (Figure 10.8B). However, the 50% system was used to a greater extent in the past because of the relative ease of synchronizing at the receiving station. In addition, the dead time available in the 50% system allows the use of circuitry to get rid of transients (undesirable short-duration signals) and therefore reduce cross talk between successive channels.

Pulse-Width TDM

In pulse-width modulation (PWM), the width of pulses is varied in proportion to the modulation signal. The PWM pulse train, then, consists of a string of pulses with different widths, as illustrated in Figure 10.9.

Guard time is allowed at both the beginning and the end of each pulse to reduce the difficulties associated with interchannel cross talk. Since the information is carried in terms of pulse width, drifts and nonlinearities in the system do not have as great an effect on data accuracy as they do in an equivalent PAM system.

Pulse-Position TDM

Pulse-position modulation (PPM) is similar to PWM. Here, though, only the trailing edge of the pulses, rather than the entire duration of the pulse, is transmitted to identify the pulse width. See Figure 10.10. PPM is used principally in connec-

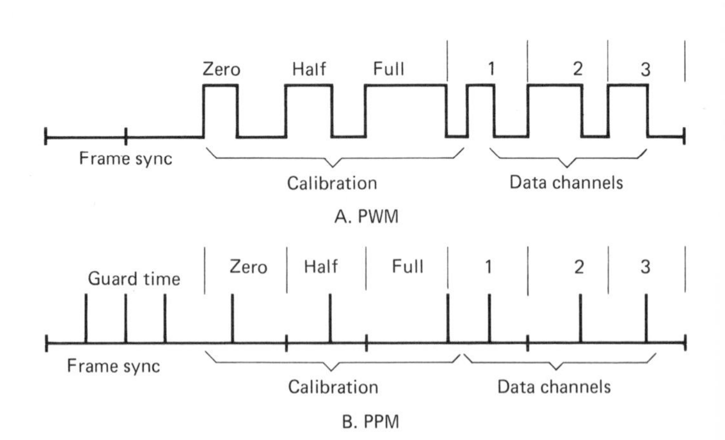

Figure 10.10 PPM Waveform Compared with PWM Waveform in TDM System

tion with an amplitude-modulated main carrier since its greatest advantage is in the small percentage of time that pulses are present. Thus, the transmitter operates at relatively high peak powers and low average powers. The wider system bandwidths required (as compared with PWM and PAM) makes its use in connection with a frequency-modulated main carrier undesirable, since the reduction in average power for the FM system could not be realized.

A typical PPM waveform is shown in Figure 10.10 along with the equivalent PWM waveform. The synchronization pulses shown may vary from system to system. Although pulses corresponding to the leading edge are not transmitted in the system illustrated, they may be in order to reduce synchronization problems at the receiving station.

Pulse-frequency modulation (PFM) systems will not be discussed here since they are so similar to PPM systems.

Analog TDM Errors

The principal sources of error in analog TDM systems are closely related to those in FDM systems. These error sources are drift and nonlinearity, bandlimiting, cross talk, interpolation errors, and RF link noise.

With the exception of PAM, only drifts and nonlinearities associated with equipment prior to (and including) the remote converters (processors) and after (and including) the receiver converter are of primary concern to analog TDM systems. As in FDM systems, this drift and nonlinearity is entirely dependent on circuit design. Calibration signals can be of considerable help here.

Bandlimiting errors in TDM systems are produced only in the equipment preceding the multiplexer. Errors due to bandwidth restrictions in other parts of the system are usually classified under different names. Bandlimiting errors depend on the phase and the amplitude response of the filters involved, as well as on the character of the data.

Cross talk in a TDM system normally occurs from bandlimiting of the pulse signal, which causes transients from channel pulses to affect the pulses of the channels following. The relationship between bandlimiting and cross talk varies from system to system.

Interpolation error comes about when an attempt is made to reconstruct a continuous waveform from the sampled values of the original waveform available to the receiving station. The amount of interpolation error depends on the characteristic of the data that is sampled, the method of interpolation, and the sampling rate.

The RF link errors cannot be eliminated. They depend on the received signal power at the receiving station and many other system parameters.

Pulse-Code TDM

Pulse-code modulation (PCM) comes in many varieties. The most common is binary modulation, the type shown in Figures 9.26 through 9.32 in Chapter 9. In *binary modulation*, the transmission of a 1 or 0 value during a bit interval is done by the presence or absence of a pulse. In actual practice, this waveform can have many different forms. Figure 10.11 shows a number of the different waveforms that can be used.

The nonreturn-to-zero (NRZ) waveform is probably the most commonly used of those shown. However, all waveforms shown in the figure either have been used or are to be used in present-day systems. When the nonreturn-to-zero or return-to-zero (RZ) waveforms are used, it is common practice to employ a premodulation filter to round off the corners of the modulating waveform to limit the bandwidth requirement.

The arrangement of the code sequence in a binary transmission system is called the *format*. In general, the format is defined by the arrangement of *bits*, or binary digits, in each code group representing a sample value (which is sometimes called a *word*) and the arrangement of these code groups with respect to one another.

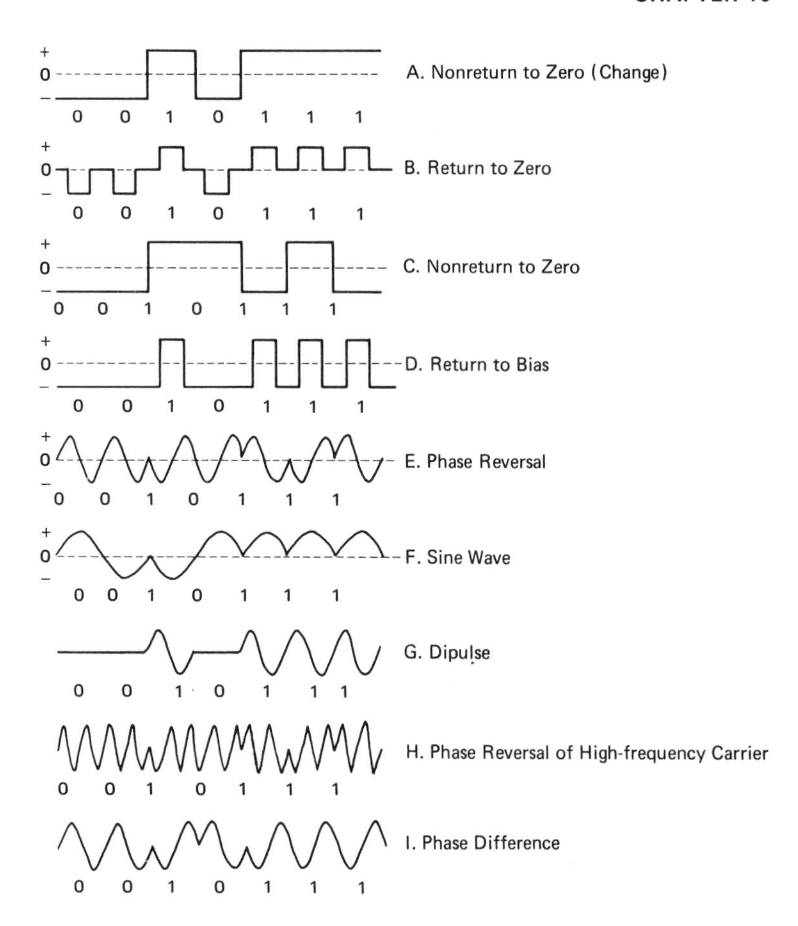

Figure 10.11 Different Ways to Represent 1 or 0 in Binary Modulation

The bits within a particular coded word may consist of information bits, parity bits, or synchronization (sync) bits. *Information bits* appear in a binary code that represents the sample value. A *parity bit* is either present or not present in order to make the total number of one bits per word either even or odd (depending on whether even or odd parity is to be used). Parity bits allow the detection of words with a single bit in error. Parity bits are generally not used in radio transmission systems since it is much more likely that two or more bits will be missing. However, in ground data-handling systems, parity bits are nearly always used since single-bit errors are most likely. The *sync bits* are timing bits. They may appear after every word or after a certain small sequence of words, and they are usually called *word synchronization bits.* Either the parity bits or the sync bits, or both, may be missing, depending on the requirements of the system. Today, virtually all operational binary PCM systems include some form of word synchronization, although some systems presently under development do not.

In the simplest cases of time division multiplexing of PCM, the information channels are sampled sequentially (as in Figure 10.5) with fixed word lengths. In this case, each word represents the code for a separate information channel, appearing sequentially in time. The sequence is repeated each time the commutator completes a cycle. This entire cycle is normally called a *frame.*

In most recent PCM systems, the programming of channels within a frame has been quite complex, with some channels being sampled more often than others. This system can be best visualized by considering a commutator with a very large number of terminals and with some of the terminals tied together so that some channels are sampled more often than others.

In most PCM systems, it is also necessary to supply timing information to designate the start of the frame. Timing information is usually supplied by transmitting a unique or identifiable code pattern in one or more word positions.

Digital TDM Errors

Errors in a digital system can be categorized as analog errors (drifts, bandlimiting, nonlinearity, and so on, in analog portions of the system), digital dropouts, quantization errors, interpolation errors, and RF link errors.

Analog errors due to drifts, bandlimiting, and so on, are the same as those types discussed previously. Since most digital systems contain analog circuitry, they also have analog errors.

The errors associated with data after it has been digitized are called *dropouts*. In a binary-coded system, the principal sources of dropouts are in the tape and disc recording processes and are due to tape or disc imperfections.

Recall from Chapter 9 that the largest quantization error occurring is equal to one-half the amplitude of the levels into which the signal is divided. Interpolation and RF link errors are the same as those previously discussed.

Multiplexing Sample Problems

We begin this section by presenting a few definitions that were given previously or were implied.

A *frame* is one set of data. The *frame time* is the time from the start of one sync pulse to the start of the next. Its unit of measure is second. The *frame rate* is the reciprocal of frame time; its unit of measure is hertz. A *sample* is the individual data input into the frame. The *sample time* is the time allotted for each sample. *Commutation* is the process of changing from one sample to the next. The *commutation rate* is the rate of changing from sample to sample; it is also the reciprocal of sample time. Its unit of measure is hertz.

Note that the circuits in this section are only proposed circuits. Some have been constructed, and some have not.

Now, let's work some problems. In a certain PAM TDM system, there are 14 data inputs, 3 calibration pulses, and a sync pulse that takes three sample times. The sample time divisions are all equal and are at a 100% duty cycle (see Figure 10.8). All data inputs are sampled only once per frame. The fastest data input changes no faster than 10 Hz, and it is sampled 5 times during its period. Determine the following: frame time, frame rate, sample time, and commutation rate.

The first thing to do is draw a diagram to clarify the problem. The diagram is shown in Figure 10.12.

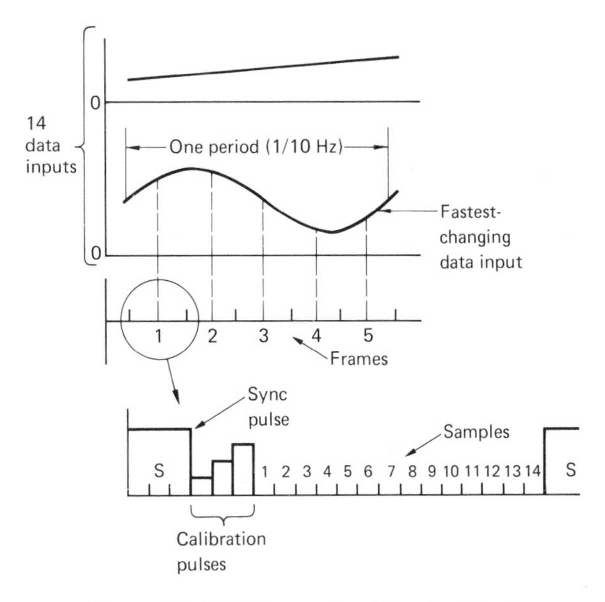

Figure 10.12 Diagram for First Problem

If the fastest data input changes no faster than 10 Hz, then this sample will determine the frame time. The period of the fastest-changing data input is then 0.1 s. Five frames occur during this time, so frame time is 0.1 s divided by 5, or 0.02 s. Therefore, the frame rate is 50 Hz. Since there are 14 samples, 3 calibration pulses, and a sync pulse that takes three sample times, then the sum of these yields 20 sample times occurring within the frame. Sample time is the number of samples divided into the frame time, or 0.001 s. Commutation rate is then 1 kHz (the reciprocal of sample time).

For the second problem, we want to design a single-channel, voice PCM system. Voice sample rate is generally set at 8000 samples per second. We will use eight bits for each word, a 50% duty cycle, and a sync pulse whose amplitude is twice that of a data bit but has the same sample time. Determine the following: pulse width, commutator frequency, frame time, and frame rate.

The diagram for this problem is shown in Figure 10.13. Since there are 8000 samples per second, frame time, which is the reciprocal of 8000 samples per second, is 125 μs, and frame rate is 8 kHz. A 50% duty cycle will require the equivalent of 18 sample spaces per frame because there is one sample space between bits and one-half sample space at each end of the frame. Thus, pulse width is frame time divided by 18, or 6.94 μs. Commutator frequency is 144.1 kHz.

If the circuit of Figure 9.25 (in Chapter 9) were to generate the output desired in this problem, the first 555 would be set to 8 kHz and the second 555 to 144.1 kHz.

For our third problem, let's design a physiological monitor of some type. It should be lightweight, portable, and inexpensive. As an example, we will make a heartbeat monitor. TDM is generally used for slowly changing data, so we will use PFM for this problem.

The PFM generator of Figure 9.20 (Chapter 9) would seem to be a prime candidate for this problem. And if we want an inexpensive system, we could use an FM radio as the receiver. The bandwidth for FM radio is from 88 to 108 MHz, but the frequency limit of the LM741 is 1 MHz, so this device won't do. However, other op amps may be used if they can operate at high frequencies.

An example of an op amp that might meet the frequency requirements is the LM733 differential video amp. It has a bandwidth of 120 MHz and selectable gains of 10, 100, and 400, and no frequency compensation is required. Figure 10.14 shows how the transmitter might be connected.

An alternative might be to use a lower-frequency transmitter of the same design and a phase-locked loop IC as the receiver/demodulator. Since the IC is low power, the transmission distances would be necessarily short.

For the fourth problem, we will make a single-channel oscilloscope into a multichannel

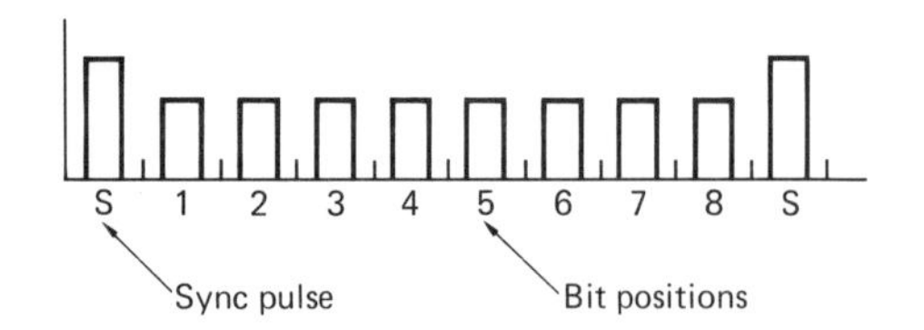

Figure 10.13 Diagram for Second Problem

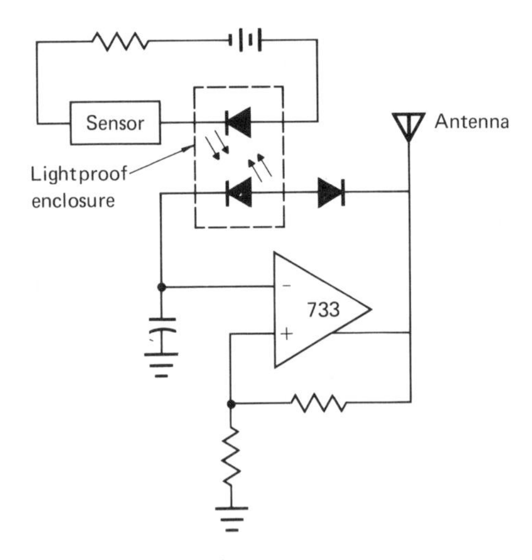

Figure 10.14 Proposed Transmitter for Heartbeat Monitor

oscilloscope. To do so, we will use TDM principles but no transmitter or receiver. (The oscilloscope will be the receiver.)

The idea in this problem is to divide the inputs (called channels here) into PAM and give each input channel a different DC offset. These PAM TDM signals will be input to the single-channel oscilloscope, which will display all the inputs on the screen, multiplexed in time. If the multiplexing is done fast enough, the display will appear to be that of a multichannel oscilloscope.

The schematic of Figure 10.15 can be used to produce the desired results. The input resistors R_1 and R_2 are used to attenuate the input signals to keep them within the voltage limits of the circuitry. Notice that the input resistors for the three channels have the same designation (R_1 and R_2). Thus, all R_1's are the same resistance and all R_2's are the same resistance. The input zener diodes, Z_1 and Z_2, protect the ICs if the maximum voltages are still exceeded. The first line of op amps gives each channel a different DC level offset so that each input channel will be displayed on a different level on the screen. Resistors R_7, R_8, and R_9 adjust the individual channel gains, while R_{10} adjusts the overall gain. The final op amp sums all the signals and prevents each channel from interfering with the others.

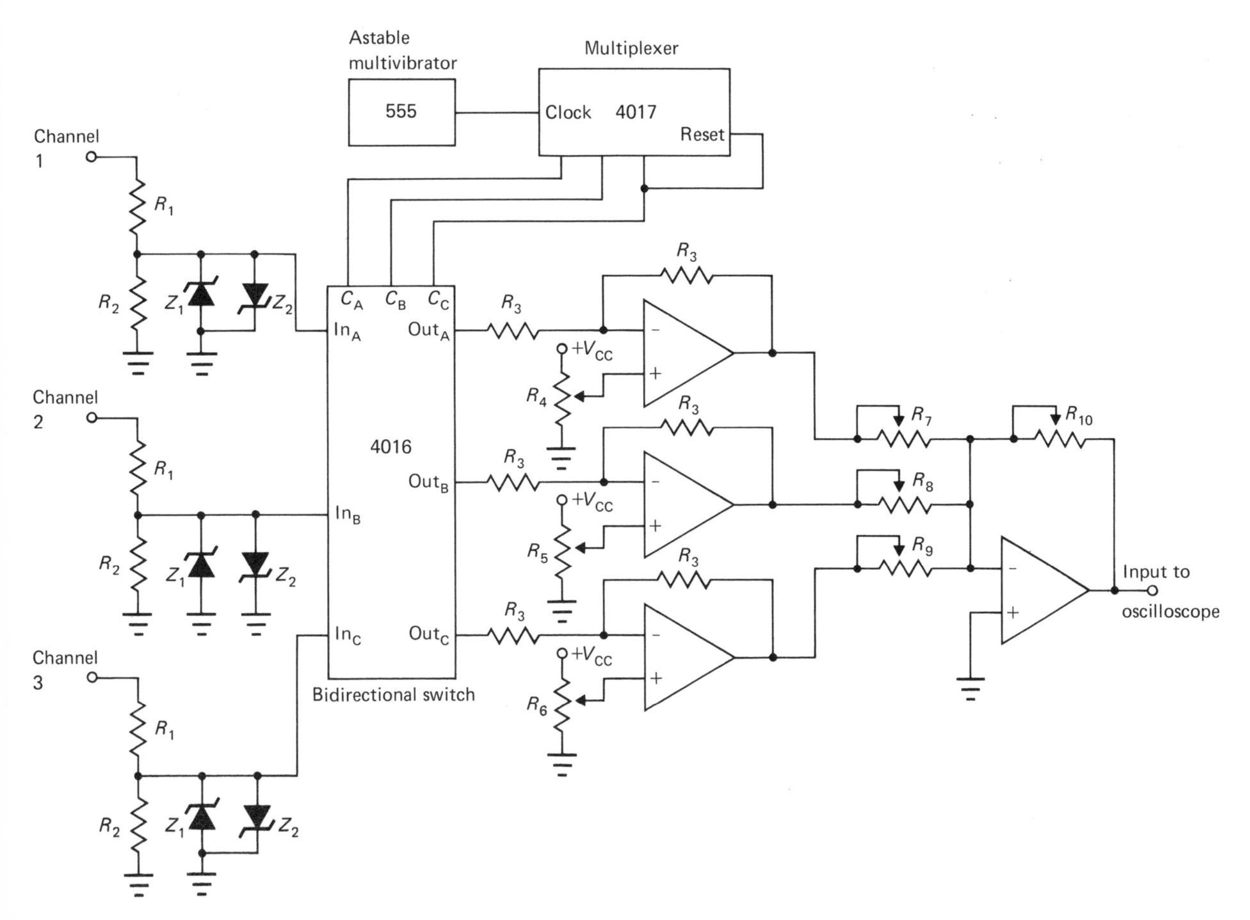

Figure 10.15 Basic Schematic for Changing Single-Channel Oscilloscope into Three-Channel Oscilloscope

The final problem is to make a three-channel multiplexer suitable for a radio-controlled airplane. The solution is represented in Figures 10.16, 10.17, 10.18, and 10.19.

Figures 10.16 and 10.18 are the schematics for the modulator and demodulator. Figures 10.17 and 10.19 are the synchrograms for the schematics. The receiver in the schematic of Figure 10.18 is producing pulses whose pulse widths correspond to the pulse widths being transmitted. The output from each channel can be attached to appropriate mechanical actuators in the airplane.

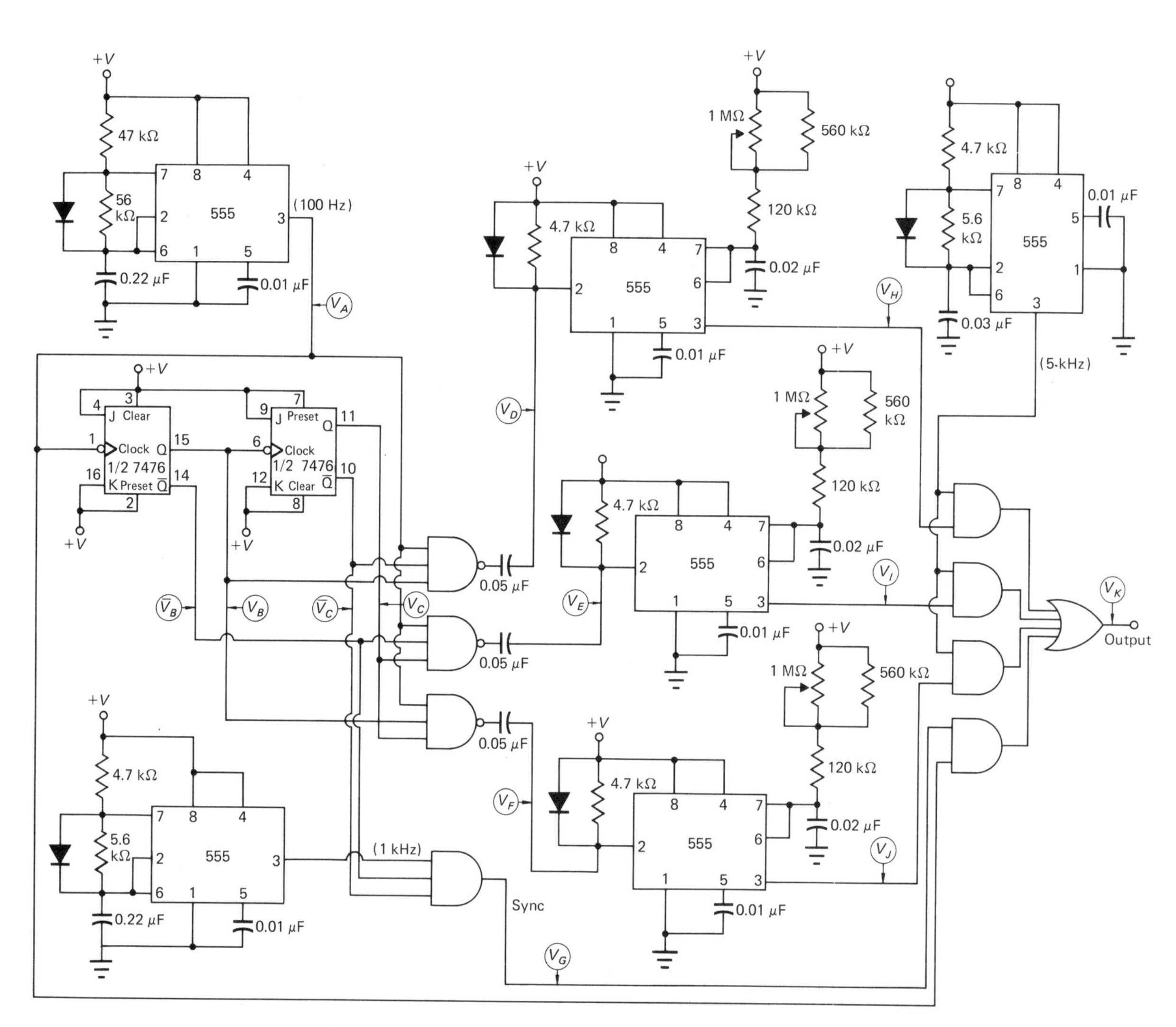

Figure 10.16 Three-Channel Multiplexer Using PWM

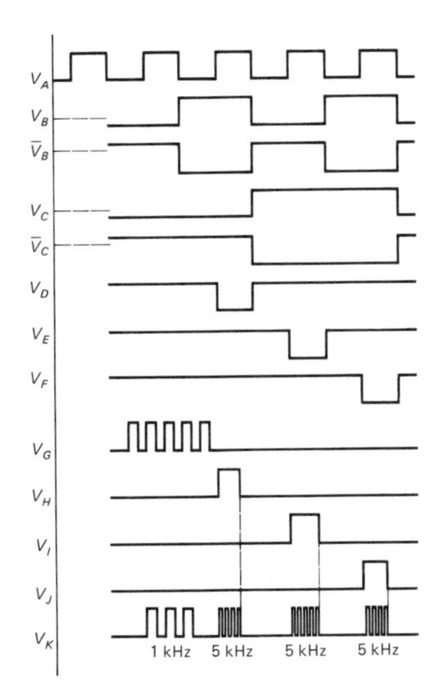

Figure 10.17 Waveforms for Schematic of Figure 10.16

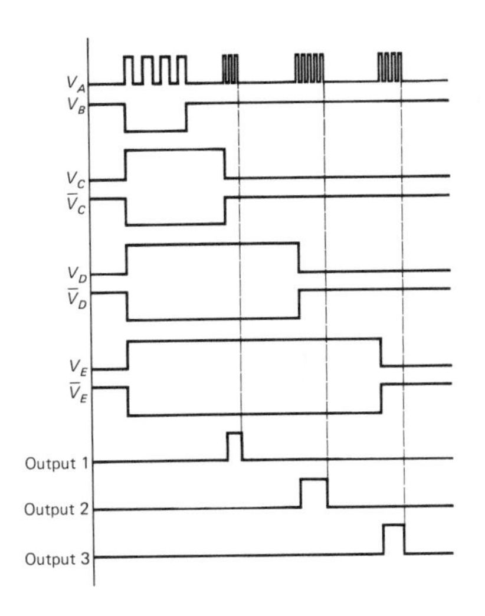

Figure 10.19 Waveforms for Schematic of Figure 10.18

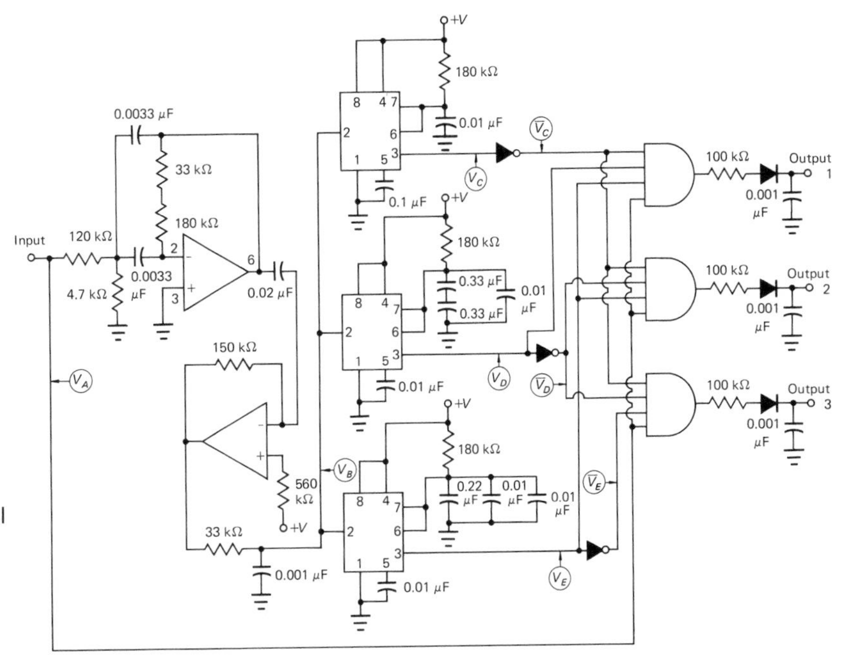

Figure 10.18 Three-Channel Demultiplexer Using PWM to Demultiplex Signal from Circuit of Figure 10.16

Conclusion

As shown in this chapter, the principles of telemetry can be applied to many industrial applications. And, as in the case of most present-day electronics applications, only the competence and imagination of the investigator will limit the results.

Questions

1. A telemetry system can be divided into three parts: The ______, which are at remote locations and transform the physical quantities to be monitored into electrical signals; the ______ system, which transforms the transducer signals, transmits and receives signals, and transforms receiver signals for display; and the ______ and ______ system, which displays received signals for interpretation.
2. According to Table 10.1, in order to keep a low signal-to-noise ratio while transmitting telemetry signals, we need a modulation index of ______.
3. An FM/AM multiplexed telemetry system designation indicates that the subcarriers are ______-modulated and the main carrier is ______-modulated.
4. ______/______ is by far the most common modulated telemetry technique in use today.
5. A 100% duty cycle telemetry system uses a ______ frequency spectrum than the 50% duty cycle system uses.
6. In Question 11, the 50% duty cycle system ______ cross talk between successive channels.
7. Pulse-position TDM uses the ______ edge of the pulse to carry the information.
8. Pulse-code TDM can only transmit 1 and 0 information by the ______ or ______ of a pulse.
9. Define multiplexing.
10. Name two types of multiplexing in common use.
11. Define subcarrier.
12. Define frame time.
13. What are housekeeping pulses?
14. Define binary transmission system format.
15. Define pulse-code TDM word.
16. Discuss when parity bits are and are not used.
17. If a data signal varies at a rate of 75 Hz, what would the lowest standard frequency band be that could be used to carry this information?
18. Band 14 would use a subcarrier of what frequency?
19. To demodulate an FM data signal on band 10, what discriminator center frequency would you use?
20. What is the lowest frequency the subcarrier can deviate to and still be in band 8?
21. Measurement of a signal indicates that its minimum rise time is 7 ms. What channels could be used to carry this information?

Problems

1. It is desired to transmit voice (20 kHz) on an FM system that has a modulation index of 5. What will be the maximum frequency deviation of this transmitter?
2. A telemetry system has seven different sensors to sample. Two additional sample spaces are set aside for the high-level and low-level references, and one sample is set aside for a sync pulse. In this telemetry system, the highest frequency input variable has a maximum frequency of 10 Hz. It is desired to sample this variable ten times in each waveform period. Each of the seven inputs is sampled only once during each frame. Determine the following: sample time, frame time, frame rate, and commutation rate.

Microprocessors

CHAPTER 11

Objectives

On completion of this chapter, you should be able to:

- Describe the five structural blocks of a computer system;
- Describe the basic architecture of a microprocessor;
- Describe the different microprocessor and one-chip microcomputer categories;
- List some microprocessor application areas and typical examples;
- Determine single-loop and multiloop control in microprocessor controllers.

Introduction

Computers and calculating machines have been around a long time. Probably the earliest calculating machine was the abacus, a hand-operated computing machine using beads on wires. This machine is so old that its origin is only speculative. As technology developed, different calculating machines were invented, all of them mechanical. In 1946, however, the first large-scale electronic calculating machine was produced, the ENIAC. It contained over eighteen thousand vacuum tubes (filling a large room) and required a power supply half the size of the computer itself. The cost was prohibitive, too: over a million dollars to build and hundreds of thousands of dollars to use. Obviously, it was only used to solve very important problems. Since that time, a series of inventions have dramatically reduced the cost of building computer circuits. Today, a $10 computing device has essentially the same capabilities as the early ENIAC computer.

There are three generally recognized classes of computers today: the main-frame computer, the minicomputer, and the microcomputer. The differences among these classifications and the point where one class begins and the other leaves off are not well defined. In general, only price and packaging differentiate the three classifications. The minicomputer is less expensive and less powerful than the main frame, and the microcomputer is less expensive and less powerful than the minicomputer. At the same time, there is considerable overlap; the most powerful minicomputer is more powerful than the least powerful main frame, and so on. This overlapping situation is compounded by the fact that as technology improves, the capabilities of a lower computer classification exceed those of the past generation of the classification above it. Basically, a main frame is a main frame and a minicomputer is a minicomputer because that is what the manufacturer calls it.

There are very few products whose continued development has led to vast improvements in itself and at the same time a vast reduction in its cost; however, the microcomputer is one. It has been said that if the automotive industry had had a similar history, we would now have an automobile that would cost about $1.00 and travel at the speed of sound!

Because of these microcomputer advances, it is not hard to see why there are so many new microcomputer applications. This chapter will discuss some of those applications and related topics such as microcomputer structure, architecture, and families. The microprocessor is becoming increasingly important in industrial electronics as a topic unto itself. Thus, we no longer have the luxury of just knowing something about the microprocessor; now, we must become proficient in its operation and application.

Computer Structure

The structure of a basic computer system is illustrated in Figure 11.1. The basic system includes five fundamental units: arithmetic-logic unit, control unit (together, these comprise the central-processing unit), input, output, and memory (storage).

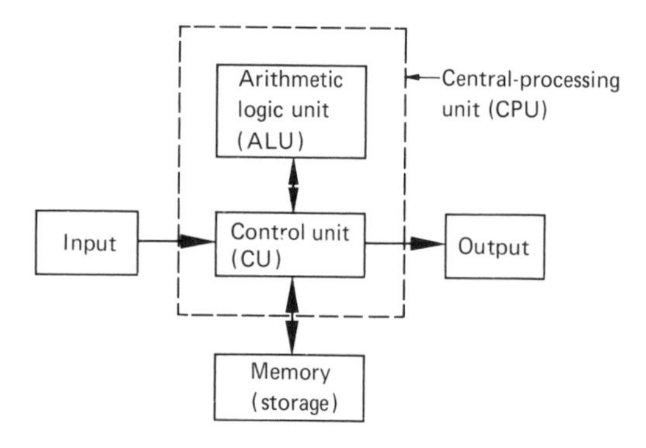

Figure 11.1 Five Fundamental Units of a Computer

Central-processing Unit (CPU)

At the center of the computer and the illustration is the *central-processing unit (CPU)*, which is generally defined by its composition. The CPU is composed of two units, the control unit (CU) and the arithmetic-logic unit (ALU).

The function of the *arithmetic-logic unit* is to perform arithmetic and logic operations on data passing through it. Typical arithmetic functions include addition and subtraction. Logic operations include the logical and, or, exclusive or, not, and shift operations.

The *control unit* is the center of control for the computer and has the responsibility of sequencing the operation of the entire computer system. The CU is generally physically associated with the ALU that it controls. And, as mentioned above, the combination of the CU and the ALU is called the central-processing unit. A *microprocessor* is basically a CPU on a single chip. A microprocessor is shown in Figure 11.2; the protective cap of the IC has been removed in this figure, and the circuit inside is exposed.

The CPU does not need to be implemented as a single component, however, and the CU can be separated from the ALU. Components called *bit-slices* implement the ALU section of a traditional computer, exclusive of the control section. The CU then must be designed and assembled separately. The bit-slice approach is best used in applications where commercially available microprocessors are inadequate.

Memory

Below the control unit in Figure 11.1 is the *storage*, or *memory*, section. The main purpose of the memory module is to store a list of instructions, called a program, and to store other information called data.

Instructions are binary codes that are recognized by the CU and that cause the microprocessor to perform some procedure. An example of an instruction is the ADD instruction. When the CU

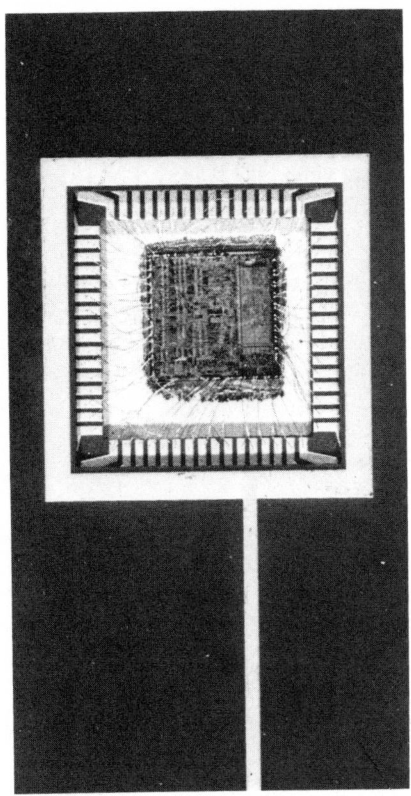

Figure 11.2 Top View of 64–Pin, Microprocessor IC

receives the code to add, it will cause the numbers to be added to be brought together in the ALU so that the ALU can add them.

A *program* is a sequence of instructions that has been written by the microprocessor user. A program is generally written in such a way as to implement an *algorithm*, which is a step-by-step specification of a sequence of operations that will solve some problem. An algorithm can be expressed in any form and in any language (computer language, English language, and so on). An example of an algorithm is a paragraph of directions for getting from my house to your house. A program is an algorithm written in a form the CU can recognize.

Data, which can also be contained in memory, are binary codes that represent numbers or characters. *Characters* are symbols that represent concepts such as the letter *A* or the symbol ?. A common code used to represent these characters (sometimes called alphanumerics) is the American Standard Code for Information Interchange (ASCII). This coding scheme uses eight bits, seven of which represent 128 standard ASCII characters. Bit 8 is a parity bit (discussed in Chapter 10) used for error checking.

Physically, several kinds of memories can be distinguished, depending on whether one can write information into the memory and then read it, or whether one can only read from the memory. Two of these types of memories are random-access memory and read-only memory.

Random-access memory (RAM) is a memory device in which information can be either written or read. The words *random access* refer to the fact that any location in the memory, called an *address*, can be accessed (made available for use). *Read-only memory* (*ROM*) is a memory that can only be read. The information in ROM was placed there only once, by the manufacturer. In actuality, ROM is also randomly accessible since any location can be accessed, but ROM was the name given to it when it was first produced, and the name has stayed with it ever since. More properly, ROM should be called read-mostly memory since the information has to be placed into the memory at some time, generally by the manufacturer.

Two more types of memory are *programmable read-only memory* (*PROM*) and *erasable programmable read-only memory* (*EPROM*). These types are subsets of ROM. The major difference between ROM and PROM is that the latter is not programmed at the manufacturer's facilities but at the user's. PROM may be written into only once; fusible links are melted in the device and prevent it from being reprogrammed. EPROM is also programmable by the user, but it can be reprogrammed because it uses different techniques and materials than PROM. An EPROM has a quartz window over the memory *chip*, the silicon material containing the electronic circuits in the integrated circuit. The quartz window allows ultraviolet light to erase the memory contents. At other times, when it is desired to preserve the program in EPROM, a cover is placed over the window.

Another important term associated with memory is *volatility*. A volatile memory is one that loses its information when power is removed from the device. RAMs are generally volatile. ROMs are nonvolatile; they do not lose their information when power is removed.

Input and Output

The *input module* is to the left of the CU in Figure 11.1; it supplies all information to the CU from any external source. A keyboard and a sensor (temperature sensor, pressure sensor, light sensor) are examples of input devices. A keyboard is generally the way the program is placed in memory, either by way of the CU or by direct access to the memory module. The process of storing or retrieving information by going directly to the memory and not through the control unit is called *direct memory access* (*DMA*). There may also be more than one input module to a system, such as a keyboard for entering the program and a sensor whose output is used in the program.

The *output module*, to the right of the CU in Figure 11.1, displays the information coming from the computer system. Examples of output devices are light-emitting diodes (LEDs), liquid crystal displays (LCDs), and cathode ray tubes (CRTs).

The combination of a CRT and a keyboard is referred to as a *terminal*. A terminal is used for both putting information into the computer and displaying the output from the computer system. An input-output function, like the terminal, is abbreviated I/O.

Buses

The remaining parts of Figure 11.1 to be discussed are the lines that connect the blocks. These lines, which are shown in Figure 11.3, are referred to as buses. A *bus* is a set of signals or wires grouped by

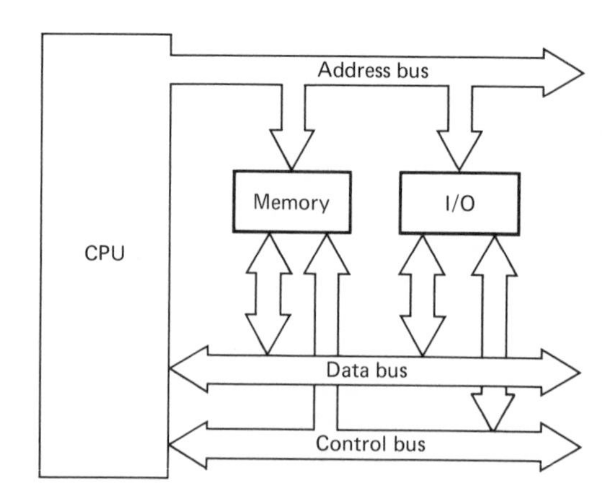

Figure 11.3 Typical Computer System Block Diagram

function. The modules of a microprocessor system are interconnected by means of three buses: the data bus, the address bus, and the control bus. An additional bus, not generally shown but which must be present, is the power bus.

The *data bus* transmits data between modules in Figure 11.3. An eight-bit microprocessor requires an eight-bit data bus (or eight separate paths, one for each bit) in order to transmit eight bits of information in parallel. The data bus is bidirectional. The term *bidirectional* refers to transmission of information in both directions on the bus (not at the same time). Bidirectional transmission is shown in Figure 11.3 by lines with arrowheads on both ends. Arrowheads on one end only indicate *unidirectional* (one direction) transmission.

The *address bus* of Figure 11.3 is used to select the origin or destination of signals transmitted on one of the other buses. A standard microprocessor address bus has 16 lines and can address 2^{16}, or 65,536, different locations. The number 65,536 is referred to as 64K in computer jargon. This name came about in the following way: Computer machine language is binary (based on powers of 2), and 2^{10}, which is equal to 1024, has been traditionally referred to as 1K (1000). Since 2^{16} is equal to 2^6 times 2^{10}, or 64 times 2^{10}, 2^{16} is incorrectly called 64K.

The *control bus* of Figure 11.3 is used for synchronizing the computer system. It carries both *status information*—signals indicating what operation is in progress—and control information to and from the microprocessor unit. (The standard abbreviation for the microprocessor unit has become MPU.) There are no fixed number of lines required for a control bus.

The *power bus* is seldom shown in schematics or block diagrams, but it is very necessary. It provides power to all the components of the system.

The preceding discussion has focused on the computer's *hardware*, the physical components of the system. In contrast with the hardware is the computer *software*, the programs in memory or on paper and all other instructions and manuals associated with the computer.

Microprocessor Architecture

Now that we have discussed the structure of the computer, we can take a closer look at microprocessor *architecture*, the internal construction of the unit, both physical and electronic. Knowledge of the architecture of a microprocessor is helpful in understanding the internal logic of the device.

The functional block diagram of Figure 11.4 is the architecture of Intel's 8080A microprocessor. (The pin connections shown in the figure are not in the correct order for the 8080A.) Notice the parts of the computer system we have previously discussed: the 16-line address bus, the 8-line data bus, the control bus (represented by the inputs and outputs of the timing and control block), the power bus (listed as power supplies), and the arithmetic-logic unit.

Associated with the address and data buses in Figure 11.4 are buffer/latches. In general, a *buffer* is another name for an interface circuit. The buffer allows signals to pass from one circuit to another when these circuits may be incompatible without the buffers. In this case, the buffer is used to provide enough current to the bus to operate the devices attached to that bus. If there are too many devices attached to the bus, the microprocessor buffers may require assistance from additional external buffers. In that application, the buffers are sometimes called *drivers. Latches* are devices that hold the information on their outputs as long as it is needed, even after the signals have disappeared from their inputs.

To the right in Figure 11.4 are a number of registers. A *register* can be thought of as a box that stores data. Figure 11.5 shows a simplified diagram of a register. It contains some means to store one or more bits of data, an input line, an output line, and a control signal to specify when the register is to be loaded from the input. A flip-flop is an example of a single-bit register.

Most computers provide a set of registers to be used by the programmer. *Data manipulation registers* are used for data manipulation; *control registers* are devoted to the control of the program

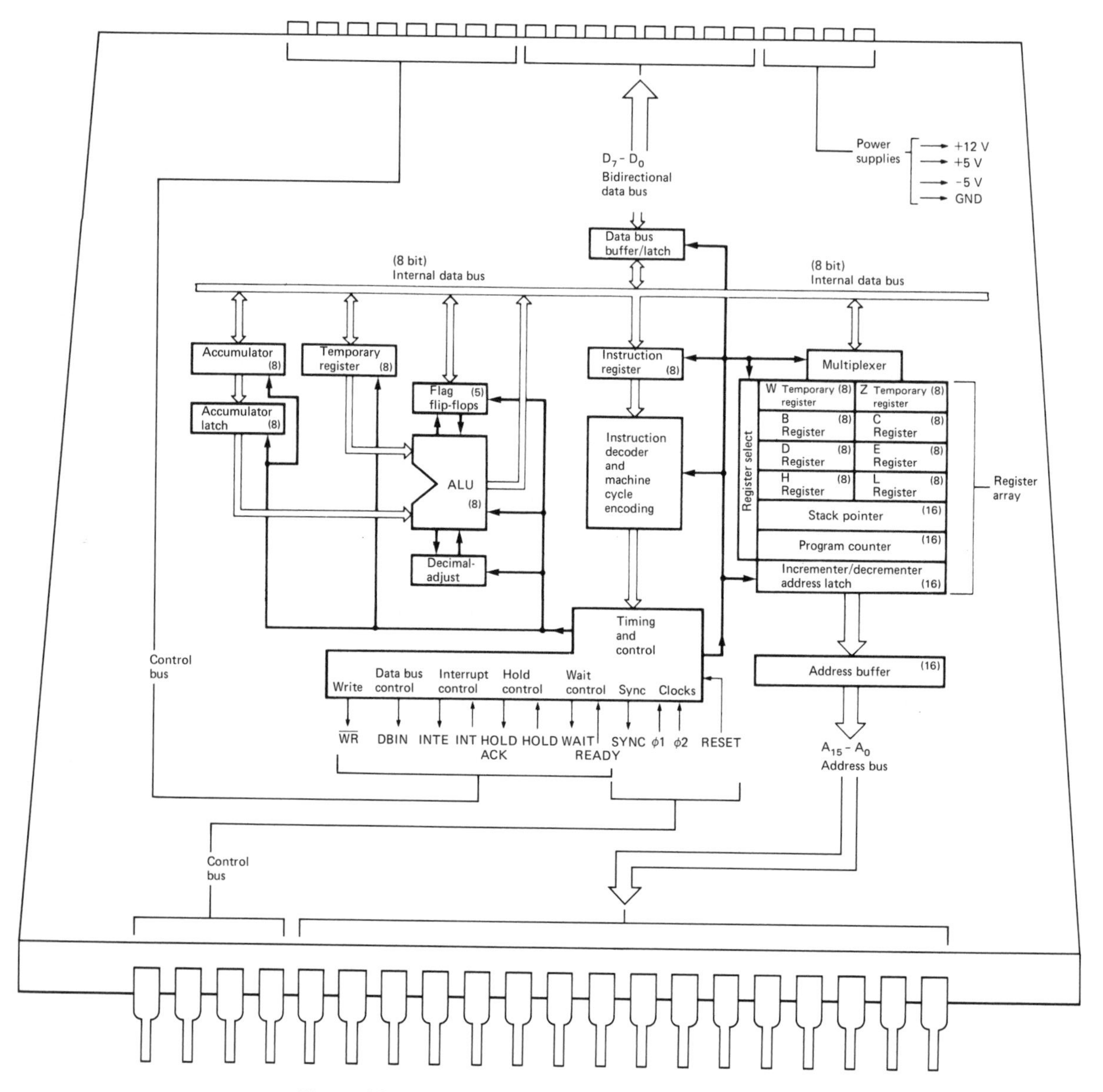

Figure 11.4 8080A MPU Functional Block Diagram

itself. Data manipulation registers in the 8080A of Figure 11.4 are the B, C, D, E, H, L registers and the accumulator (A) register. All these registers hold eight bits of information. The W, Z, and temporary registers are also data manipulation registers, but they are not directly accessible to the programmer. The control registers of the 8080A are the instruction register (IR, eight bits), the program counter (PC, 16 bits), and the stack pointer (SP, 16 bits).

The *instruction register* receives the instructions as they come in, one at a time, from the

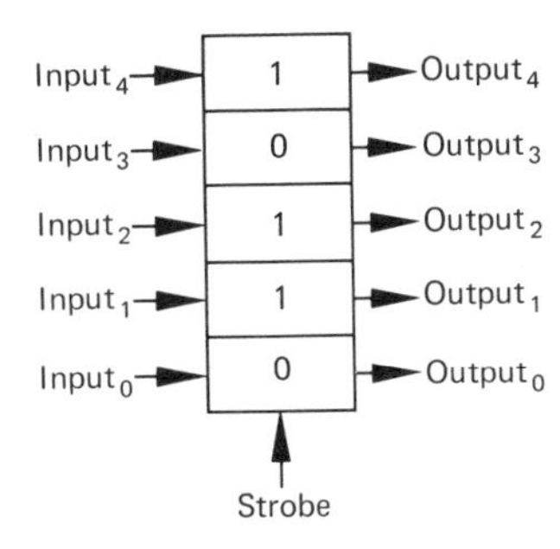

Figure 11.5 Simplified Register

program contained in memory. The instruction then passes to the instruction-decoding section. Here, the instruction is decoded by a complex set of logic gates that provide the timing and control to the microprocessor in order for it to execute (complete) the instruction decoded. The complex logic contained in the instruction-decoding block is called a *microprogram*. This program is a sequencing program for the control unit of any processor, and it is an example of *firmware*, a program that cannot be changed. A program in ROM is another example of firmware.

The *program counter* is the register that contains the address of the instruction in memory that is to be executed next. The program counter, then, determines the order in which the instructions are to be executed.

The *stack pointer* contains an address of a location in memory referred to as the stack. The *stack* generally contains an address that is a reference point in the memory when executing a subroutine. A *subroutine* is a program within the main program and is used when a repetitive operation is desired. Suppose a program has a subroutine in it. When the subroutine is executed, program control jumps to the location of the subroutine. The stack contains the address of the place to return to in the main program when the subroutine is completed.

The *flag flip-flops* in Figure 11.4 are associated with the arithmetic-logic unit and indicate the results of the most recent ALU operation. The *decimal-adjust* is also associated with the ALU and is used when the accumulator is working with binary-coded decimal (BCD) numbers. The BCD system is a method of representing the decimal numbers 0 through 9 by using four bits. For example, 0000_2 in the binary system represents 0_{10} in the decimal system; 0001_2 in binary represents 1_{10} in decimal; and so on. The subscript of each number represents its respective base (the number of symbols in that number system).

The program sequence in this microprocessor is as follows: Instructions are sequentially fetched (brought) from the memory to the microprocessor and executed. At the start of the instruction cycle, the address in the program counter is output on the address bus. This address accesses memory, and the instruction code at that address is brought into the microprocessor through the data bus. The instruction goes into the instruction register and is decoded into signals that cause the microprocessor to perform the operation the instruction required through another series of register transfers. The instruction cycle is complete when all operations pertaining to that instruction are complete. The next instruction cycle begins with an instruction fetch, and the process repeats itself again. As each instruction is decoded and executed, the program is completed.

Microprocessor Differences

The above discussion has shown the basic process of microprocessor and microcomputer operation. A *microcomputer* is a computer that uses a microprocessor as its CPU. If all microprocessors worked exactly the same way, there would be very few brand names of microprocessors, and choosing one to fit your application would be relatively easy.

That is not the case, however. There are numerous ways of implementing the basic microprocessor operations. As a case in point, consider the previously mentioned process of fetching the instruction to the microprocessor, executing it, and then fetching the next instruction. Since microprocessor manufacturers are very interested in speed and efficiency, some have developed a process whereby the next instruction is brought to the microprocessor before the previous one is completed. This next instruction is placed in a *queue* (waiting line) in the microprocessor and is immediately ready to execute, or it may even overlap the execution of the previous instruction. This process is called *pipelining*.

Many other differences exist among the microprocessor manufacturers' products, and sometimes even among a single manufacturer's microprocessors. Many companies, however, try to maintain upward compatibility among their microprocessors. That is, as a new and more powerful microprocessor is developed, it is designed so that the new microprocessor can directly replace an older model with little or no changes in the older system.

Microprocessor Families

Those microprocessors that have similar bus structures, hardware architectures, and instruction sets (all the basic instructions that a microprocessor recognizes) are considered to be in the same family. The family also includes the microprocessor support devices, such as clock generator chips, and the system support components, such as bus, direct memory access, and interrupt controllers. Many *peripheral devices* (closely associated devices) and memory devices are compatible with more than one family.

Some manufacturers who have developed and produced products that no one else has (called OEMs, original equipment manufacturers, or prime sources) will support only their own products. Other manufacturers may develop and produce their own products and, in addition, may manufacture products developed by OEMs. These companies that manufacture products for OEMs are called second sources. Many of the microprocessors on the market today are second-sourced. In rare cases, some second-sourced devices are not identical to the OEM's product. Of course, this situation causes great confusion and many problems.

In most cases, microprocessor users should stay with one family of products. It is easier to remember just one instruction set, architecture, and bus structure. At the same time, most manufacturers of any microprocessor family will have enough support devices to provide for all applications. It is also more economical to support one set of replacement parts. Furthermore, in the process of learning to use a microprocessor, more detail can be gleaned from one family than from many families.

Microprocessor/ Microcomputer Categories

A diagram listing most American-produced microprocessors and microcomputers is shown in Figure 11.6. In the figure, solid lines connect chips with relatively strong family relationships; dashed lines connect chips with weaker ties. Lines connecting chips in the different application areas show which families have the broadest application flexibility.

Besides the American firms listed in Figure 11.6, there are also many Japanese firms (NEC, Hitachi, and Fujitsu) and European firms (Philips and Siemens) that produce these products. Furthermore, Figure 11.6 does not represent all American microprocessors (μPs) and microcomputers (μCs); only those that have a relatively high volume of production are shown. The listings in the columns in Figure 11.6 are basically ordered according to volume of production, with the highest volume at the top. Because of market fluctuations and other factors, this diagram is not extremely accurate. The purpose of Figure 11.6 is to show the major categories and approximate production activity in the United States μP/μC industry.

The following discussion lists some observations about the various microprocessor/microcomputer categories shown in Figure 11.6. Additional information about rapidly developing technologies such as microprocessors can be obtained from technical journals and magazines. Three important

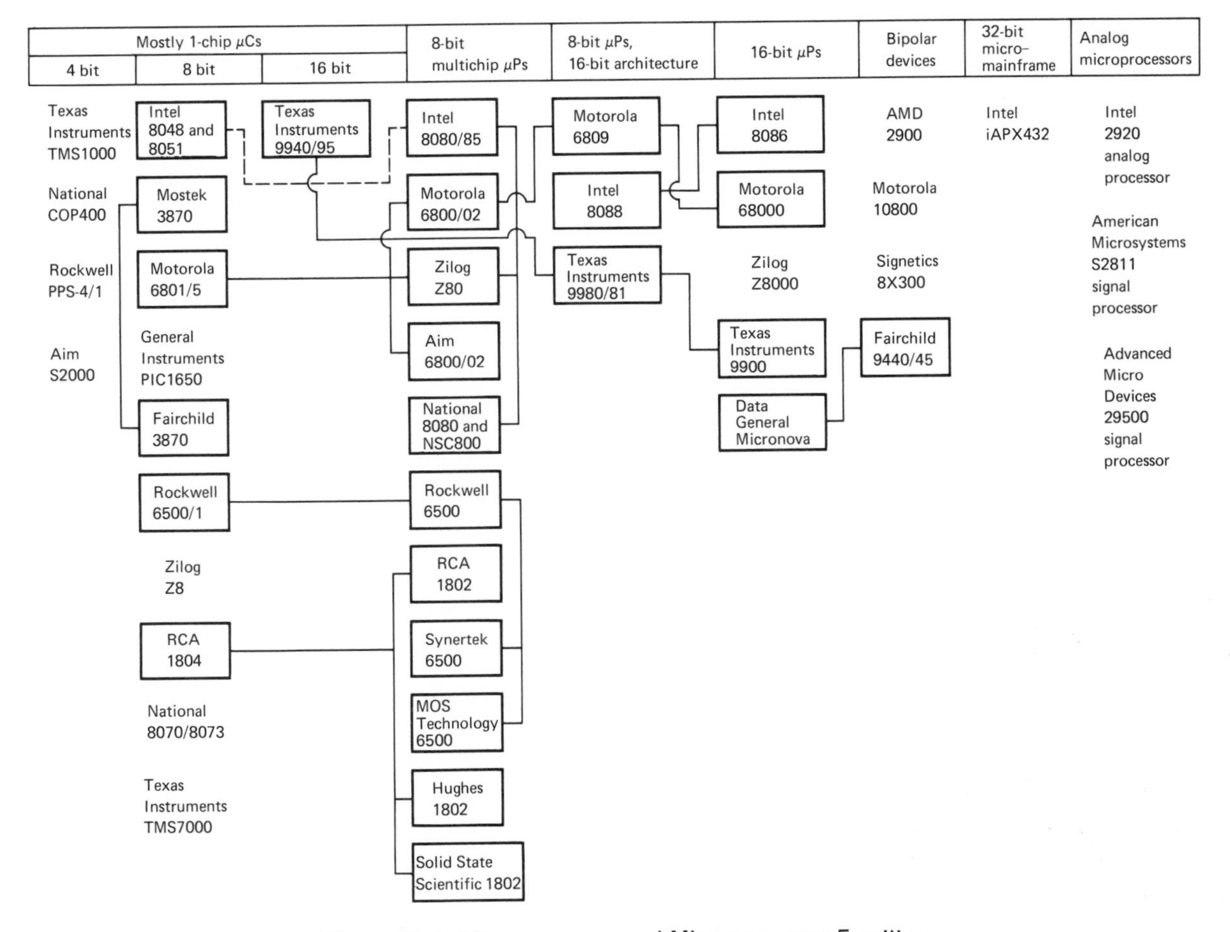

Figure 11.6 Microprocessor and Microcomputer Families

articles are noted in the bibliography for this chapter located at the back of the book.

One-Chip Microcomputers

To the left in Figure 11.6 are the one-chip μCs. These devices are available in 4-bit, 8-bit, or 16-bit data buses and split-memory (Harvard) or unified-memory (Von Neumann) architectures. The split-memory systems are predominantly in the 4-bit data bus types. In most cases, the race for highest-volume production is relatively close, except for one. Texas Instruments' TMS1000, 4-bit μC is by far the undisputed leader in its group, with an excess of 45 million units produced a year. The next closest competitor is the Japanese firm NEC (μCom-42, not shown in Figure 11.6) with about 30 million units per year.

Eight-Bit Multichip Microprocessors

Even with all the activity in the one-chip μC area, 8-bit multichip μPs are still very popular. There are at least two reasons for their popularity: (1) The large group of support chips and software developments allow great flexibility of application, and (2) some of these μPs are the building blocks for popular consumer products. Three of these familiar products are the Z80-based TRS-80 and the 6502-based Apple and PET computers.

Eight-Bit-Plus Microprocessors

The next group, the 8-bit μPs with 16-bit architecture, claims to achieve the best of two worlds. These devices have the ability to use the plentiful 8-bit support chips, on the one hand, while working with a more powerful 16-bit CPU, on the other. At any rate, for those who wish to learn 16-bit μPs, this group seems the logical group with which to start.

Sixteen-Bit and Thirty-Two-Bit Microprocessors

The full 16-bit and 32-bit μPs represent a different design philosophy from that of the 4-bit and 8-bit devices. The one-chip μCs intend to minimize hardware costs, while the 16-bit and 32-bit μPs intend to provide the lowest possible software cost. The 16-bit and 32-bit devices are also based on minicomputer and main-frame designs instead of calculator principles.

This change in design philosophy for the larger machines requires users not only to understand the new architecture but also to become proficient in one or more high-level languages such as PASCAL. Because of the inefficiency, at the human level, in programming in *machine language* (the ones and zeros that the computer understands), high-level languages have been developed. *High-level languages* are closer to the functional representation of algorithms and are independent of the internal architecture of the microprocessor. Thus, a program written in a high-level language should be transferable from one computer to another. High-level language is much easier for people to use and understand than machine language is.

In addition to the change in design philosophy of the 16-bit and 32-bit machines, for best results, the applications should also be different. If an 8-bit program and application were adapted to the larger machine, the major advantage obtained would be speed. Applications for the larger machines should be more ambitious programs only possible through the use of high-level language.

Bipolar Devices

The advantage of the bipolar (transistor-based) devices is speed. There are still many applications for bipolar devices, but these areas are shrinking owing to the steadily increasing capabilities of MOS (metal-oxide semiconductor) devices.

Analog Microprocessors

Just as the digital microprocessor, Intel's 4004, in 1972 triggered far-reaching changes in electronic design philosophy, so, too, have the analog microprocessors today. Intel's 2920, American Microsystems' S2811, NEC's μPD7720 (not listed in Figure 11.6), and Advanced Micro Devices' 29500 promise to bring about equally important changes in analog design approaches.

These devices are intended for use in the real-time digital signal processing (DSP) of analog signals. In particular, they are designed to implement the sum-of-products equations found in sampled-data analog systems. (Sampled-data analog systems were discussed in Chapter 9.) One of the largest markets for DSP is telecommunications voice filtering. All four of the analog μPs have the 8-bit dynamic range and 3 kHz bandwidth needed for telephone line voice processing, as discussed in Chapter 9.

Because analog μPs are only in their first generation, their progress is not predictable. Who would have thought the digital microprocessor would be where it is today!

Microprocessor Application Areas

Because of the relative ease with which a microprocessor system can be assembled and made operational, the only limitations to applying μPs to new applications are the competence and imagination of the user. For this reason, no list of applications will ever be complete. We can, however, distinguish four major microprocessor application areas. They are computer systems, industrial systems, consumer applications, and special-purpose systems. These areas are discussed in detail in this section.

Computer Systems

Microprocessors were first used in computer applications because the microprocessor's use initially required a significant amount of hardware and software expertise. These skills were most available in the computer business. And the microprocessor manufacturers were quick to see the application of the microprocessor as a device controller. Examples are paper tape readers, punchers, printers, keyboards, CRT controllers, and floppy disc controllers (FDC). (A *floppy disc* is a thin, flexible disc coated with a magnetic material and used as memory storage.) Microprocessors are now used for the control of virtually every computer peripheral that does not require bipolar speeds.

Besides peripheral device control, the microprocessor is also used as the basic building block for a small, rather simple computer. This simple computer was made available to the public by small (generally one-person or two-person) companies. No one could have anticipated the amount of sales resulting from these early ventures. Advertising was accomplished through electronics magazines, and business boomed. This market became known as the hobby market or the personal computing market. Because of this market, many microcomputer-related parts became standards merely because of their popularity and widespread use. (A standard is a measure against which a manufacturer should conform in order to be compatible with other manufacturers.)

Industrial Systems

Industrial microprocessor applications are generally in the replacement of minicomputers or complex hardware logic with low-cost, flexible microprocessors. Most industrial applications require analog inputs and outputs. The microprocessor system is the equivalent of an analog controller with a number of control loops. In this case, a *control loop* is the application of an algorithm that will regulate an output as a function of one or more inputs. The major impact of microprocessors in industrial control has been to provide a number of new functions, making process control simpler and more powerful.

Many industrial applications employ costly sensors and control mechanisms. Here, the cost of the microprocessor system is far overshadowed by the cost of the sensors and controls. Thus, the microprocessor's advantage is that of software and programming to replace costly hardware. Programming permits the use of functions of unlimited complexity, which could not previously be achieved. Changes are fast and simple, and algorithms can be improved or replaced with minimal hardware changes.

The following discussions describe some applications and software techniques made possible by the microprocessor in the industrial setting.

Data Logging

When a microprocessor is used for process control, an additional advantage results. It now becomes easy to add a bulk memory system, such as a cassette recorder or a floppy disc, which can record data continuously. This process is called *data logging*. During all idle times, or at regular intervals, the microprocessor can record all the process parameters in bulk memory for future reference. It may also use these recorded parameters to try to improve the process performance. This task is called *dynamic optimization*. For instance, the system could look up previous values of the control parameters that were found to be successful in improving the operation of the system and apply an algorithm to these values to further improve the process performance.

Status Feedback

Software techniques that once were reserved for minicomputer use are now being applied to microprocessors. One such technique is *status feedback*, which is the observation of the condition of a controlled device in order to determine its state. For example, suppose a relay was being controlled by a microprocessor, and the microprocessor commanded the relay to close. Status feedback would allow the microprocessor to verify that the operation of the relay was correct. If, however, the relay failed to close, the microprocessor could give the close command a second or third time. If closure now occurred, the first malfunction could be disregarded and an alarm sounded for preventive maintenance. If the relay failed to close after a number of tries, a backup system could be activated, if it were available. This relay failure, with a backup system, is called a *soft failure* because it degrades the operation of the system but does not stop it. Every controlled device should be monitored in this manner. Status feedback can also be used to periodically monitor the entire system's functioning ability. This process is a *system self-check.*

Reasonableness Testing

Status feedback is used with outputs, while reasonableness testing is used with inputs. *Reasonableness testing* determines if an input value is reasonable or not. Normally, a *bracket* (a range) is provided for every input parameter at any given time in the system. If the input value is within this bracket, it is accepted; if it is outside the bracket, it is rejected. For example, if a temperature reading in a steam system suddenly dropped to 0°, the reading would be disregarded. In effect, the sensor would be disconnected from the system. However, the microprocessor could continue to monitor the sensor. If the sensor reading became reasonable again for a certain length of time, the sensor output could be connected back into the system. This situation is an example of a temporary malfunction.

Connecting and disconnecting inputs can be accomplished in software with a technique called *confidence weighing,* where every sensor has a confidence weight assigned to it. A measurement is obtained from several input sensors, and each input is multiplied by its weight in order to arrive at the averaged input value. As an example, suppose two temperature sensors are used. If each has a weight of 50%, each sensor value will be multiplied by 0.5, and these weighted values will be added to arrive at the temperature. However, if one sensor value becomes unreasonable, it will be assigned a coefficient, or weight, of 0, and the second sensor will have a weight of 1. Now, the second sensor alone determines the temperature. If the first sensor again becomes reasonable, it could be assigned a new confidence weight, possibly dependent on the length of time it was reasonable.

Programmed Filtering

Whenever a microprocessor system samples a large number of inputs over a period of time, the input values should be filtered in order to reduce noise (spurious indications). A technique called *programmed filtering* is then employed. The simplest programmed filtering is averaging. The input values are added together and then divided by the number of inputs. The resulting value is the filtered input value. Simple averaging may not be the best filter to use since it is unduly influenced by extremes. Other algorithms could be employed that are not so adversely affected.

Consumer Applications

Consumer applications are characterized by large volume and lowest possible price. In this application area, one-chip and two-chip microcomputers are used. Advantages are that the microprocessor eliminates electromechanical or hardware logic, it provides more functions, and it provides reasonableness testing, such as not operating but, instead, flashing an alarm when the wrong controls have been punched. Typical examples of microprocessor-equipped consumer products are washing machines, sewing machines, microwave ovens, televisions, and electronic games. Microprocessors in the office are found in word-processing machines, copying machines, and telephone switchboards.

Specialized Applications

Specialized applications are characterized by some special overwhelming constraint, such as small size, low power, or high speed. Many times, the users of the devices have little regard for cost. The main areas for these applications are in government (military, avionic, aerospace) and the medical field. Examples are automated flight controls, navigation, communication, instrumentation, radar, electronic countermeasures, electronic warfare, pacemakers, and body function monitors.

Industrial Controllers

When computers first came on the scene, many managers thought that the computer could be used to control an entire industrial plant. Some tried and were successful—until the computer failed. Mass confusion and loss of many dollars resulted. Some managers then tried another plan: using a second computer as a backup. This plan worked but was very costly. Various schemes for running an industrial plant entirely by computer have been tried since, with varying degrees of success.

The advent of the microprocessor, as discussed previously, has had a dramatic effect on industry because of its compact size, high speed of operation, low cost, and high reliability. With the microprocessor, computing power could be decentralized. Each microcomputer could operate a small portion of the plant, maintaining interplant communication, if desired. Now, a computer failure would not be so devastating since microcomputers are inexpensive enough to have redundancy (duplication). The duplicate system would be ready to take over if needed. Examples of the application of the microprocessor to various industrial systems are shown in the following products.

Dedicated Single-Loop Controllers

A block diagram of a *dedicated single-loop controller* using a microprocessor is illustrated in Figure 11.7. The sensor(s)–ADC–microprocessor–DAC–value loop is the single loop. This controller contains a single microprocessor whose memory is preprogrammed at the factory, except for tuning constants, and it operates as any analog controller would. It also contains RAM, ROM, and PROM memories, process I/O modules, and a serial data transceiver for communicating with another computer. The ADC and DAC operate in much the same way as those devices discussed in Chapter 9.

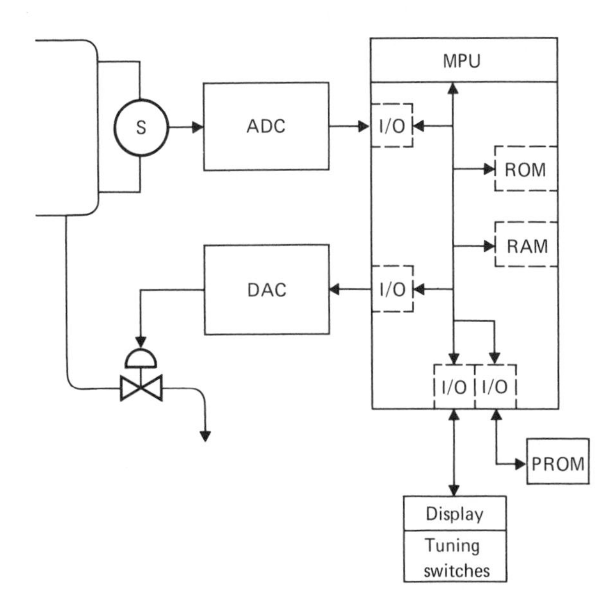

Figure 11.7 Dedicated Single-Loop Controller Using a Microprocessor

Figure 11.8 Taylor Instrument's Micro-Scan 1300 Single-Loop Controller

The single-loop controller is a controller that has one or more inputs and a single control output. The single-loop controller is one of the simplest types of control application for the microprocessor.

An example of a single-loop controller is the Micro-Scan 1300 Controller offered by Taylor Instrument; it is shown in Figure 11.8. This controller has many preprogrammed algorithms. Some of its standard features are remote analog adaptive gain (confidence weighting), input for start-up or sequential gain changes, digital 12-bit accuracy for control algorithms, and digital display of key loop parameters. The Micro-Scan 1300 Controller is contained in a relatively small package whose front dimensions are about 8 cm by 15 cm. Figure 11.9 shows a close-up view of the entry keyboard for the Micro-Scan 1300, which is located on the side of the controller in Figure 11.8.

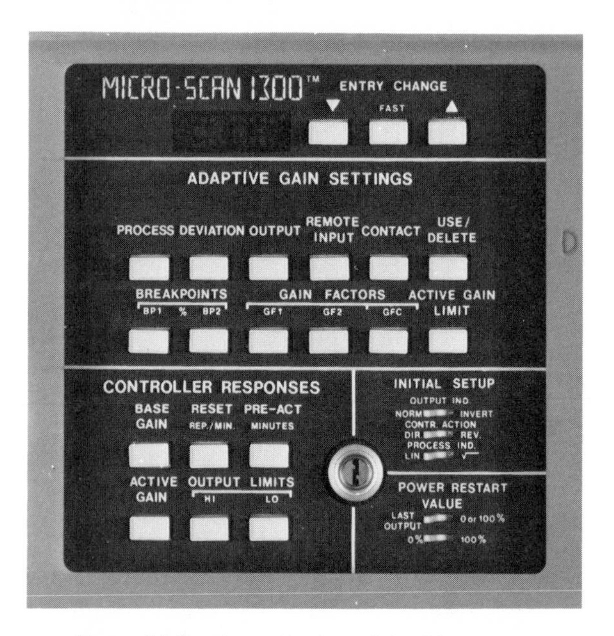

Figure 11.9 Entry Keyboard for Micro-Scan 1300

Two types of single-loop controllers are single input and output, and multiple inputs and outputs with a single control output. Both types that use microprocessors are limited to specific applications because they are preprogrammed. To adapt the unit to a new application, the manufacturer must reprogram the unit and possibly also make hardware changes.

Multiloop Controllers

The *multiloop controller* is a controller with multiple inputs and multiple control outputs. It has many more options to choose from, but it is still limited in the same manner as the single-loop controller.

Distributed Control

In *distributed control*, a number of remotely located microprocessors are controlled by a central microprocessor. These units are much larger and, of course, more costly, but they reduce the possibility of a complete system failure. Examples of this type of microprocessor system are Honeywell's TDC-2000 Total Distributed Control, Fischer's ProVos, Foxboro's Videospec, and XOMOX's System 1100.

In-House Design

If none of the commercially available systems are adequate for a specific application, an alternative is an in-house system, a system designed by the people who will be using it. A good way to design such a system is to review the microprocessor manufacturing data specifications and the preassembled printed circuit boards available from suppliers. Often, it is only a matter of purchasing preassembled boards and making the necessary interconnections to make the microprocessor ready for programming.

Conclusion

This discussion on computers and microprocessors has been necessarily short because of the large number of topics covered in this text. Nevertheless, the topic is an important one.

We have shown in this chapter that microprocessors are extensively used in industry. All levels of industrial control are amenable to microprocessor control. Thus, in the future, microprocessors will become as common as electric motors are today. Therefore, microprocessors should be well understood by everyone in the technical fields today.

Questions

1. The three classes of computers are ______, ______, and ______.
2. A CPU on a single chip is called a ______.
3. An ______ is a step-by-step specification of a sequence of operations that will solve some problem.
4. A software technique used to observe the condition of a controlled device in order to determine its state is called ______ ______.
5. A software technique used to determine if a measurement is within the correct range is called a ______ test.
6. List the five parts of a computer system and describe each part.
7. Define the following: random access, volatile, DMA, bus, microprocessor architecture, microprocessor family.

Programmable Controllers

CHAPTER

12

Objectives

On completion of this chapter, you should be able to:

- Define sequential control;
- Recognize a state description and a ladder diagram for an industrial process;
- Define programmable controller;
- List the differences among programmable controllers, numerical and position controllers, and computer controllers.

Introduction

In the 1960s, computers were considered by many in industry as the ultimate way of increasing efficiency, reliability, productivity, and automation of industrial processes. Computers possessed the ability to acquire and analyze data at extremely high speeds, make decisions, and then disseminate information back to the control process. However, there were disadvantages associated with computer control, such as high cost, complexity of programs, hesitancy on the part of industry personnel to rely on a machine, and lack of personnel trained in computer technology. Thus, computer applications through the 1960s were mostly in the area of data collection, on-line monitoring, and open-loop advisory control.

During the mid 1960s, though, a new concept of electronic controllers evolved, the *programmable controllers* (*PCs*). This concept developed from a mix of solid-state computer technology and traditional sequential controllers, such as the stepping drum (a mechanical rotating switching device) and the solid-state programmer with plug-in modules. This new device first came about as a result of problems faced by the auto industry, which had to scrap costly assembly line controls each time a new model went into production. The first PCs were installed in 1969 as electronic replacements of electromechanical relay controls. The PC presented the best compromise of existing relay ladder schematic techniques (a topic that will be discussed later in this chapter) and expanding solid-state technology. It increased the efficiency of the auto industry's system by eliminating the costly job of rewiring relay controls used in the assembly line process. The PC reduced the changeover downtime, increased flexibility, and considerably reduced the space requirements formerly used by the relay controls.

In this chapter, we will first discuss the basic concepts of sequential control. Next, we will describe the programmable controller. We end the chapter with a description of a specific PC, the

Allen-Bradley Bulletin 1772. The PC has gained wide acceptance in industry because of its ability to reconfigure processes economically and to be programmed easily. However, the 1772 is not the only PC on the market. Currently, there are about thirty-eight companies in the United States that manufacture PCs.

Sequential Control

Traditionally, industrial processes and control systems used relays, timers, and counters. These devices constitute a class of control systems used to control processes known as sequential control processes. A *sequential process* is a process in which one event follows another until the job is completed.

In this section, we first present an example of a sequential process. This process could be controlled either by the traditional electromechanical devices or by a PC. Following this presentation, we will illustrate why the PC is the better choice.

Automatic Mixer

Figure 12.1A is a simplified representation of an automatic mixing system. The tank in the diagram is filled with a fluid, agitated for a length of time, and then emptied. Figure 12.1B is a *state description* of the process, which gives the order of operation of the process. A state description is similar to a flowchart in computer programming. This sequential process is the kind of process that can easily be handled by a programmable controller.

The control logic for the process shown in Figure 12.1 has traditionally been depicted in the language of ladder diagrams. A *ladder diagram* is a diagram with a vertical line (the power line) on each side. All the components are placed between these two lines, connecting the two power lines with what look like rungs of a ladder—thus the name, ladder diagram. A ladder diagram for the process of Figure 12.1 is shown in Figure 12.2. The letter symbols in the diagram are defined in succeeding paragraphs.

Relay ladder diagrams are universally understood in industry, whether in the process industry, in manufacturing, on the assembly line, or inside electric appliances and products. Any new product increases its chances of success if it capitalizes on widely held concepts. Thus, the PC's ladder diagram language was a logical choice.

Some mention should be made at this point about electrical and electronics symbol designations. In general, there is a difference between electrical and electronics symbols and symbol

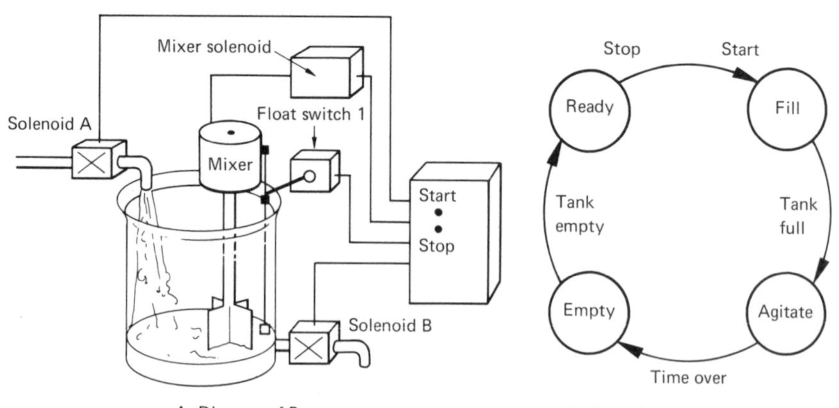

Figure 12.1 Sequential Control Process

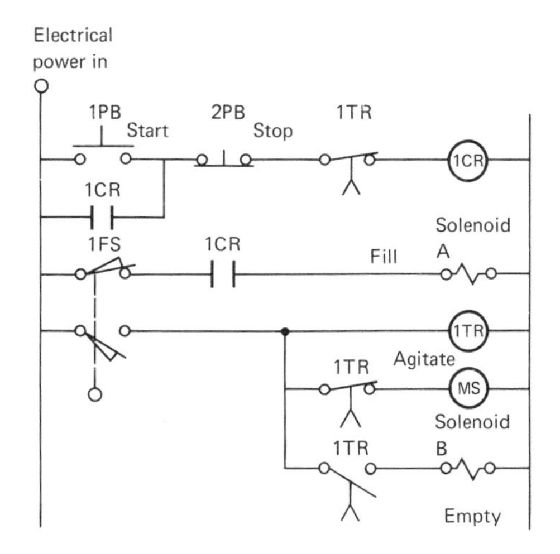

Figure 12.2 Ladder Diagram for Control of Automatic Mixer of Figure 12.1

designations. These two industries grew up somewhat independent of each other, and therefore differences exist. For example, the electronics symbol for a resistor is a zigzag line with a symbol designation of R_1. The same symbol in the electrical or industrial world is a rectangle with lines out the ends and with a symbol designation of 1R. These differences can sometimes be confusing. In this chapter, we will use the industrial symbols and symbol designations because the programmable controller developed as an industrial machine.

Now, let us follow the series of events for the full control cycle of the automatic mixer process. We first refer to Figure 12.1 and then to Figure 12.2. At the start of the process, the start push button (1PB) is pressed. The start button energizes a control relay (1CR) located in the start/stop switch box (not shown in Figure 12.1).

In Figure 12.2, the circle with 1CR in it is the coil portion of the control relay. It is located in the first line, or *rung*, of the ladder diagram. The vertical lines in Figure 12.2 with 1CR above them are not the symbol for capacitors but the industrial symbol for the relay contacts of the 1CR relay. They are shown in the normally open position (abbreviated NO). The same symbol with a slash drawn through it represents the normally closed (NC) relay contact.

When the relay (1CR) is energized (or pulled in or picked up), these relay contacts change state; in this case, they close. When the 1CR contact under the 1PB switch closes, it allows current to continue through the coil of the 1CR relay, even though the start push button (1PB) is released. This circuit holds the 1CR relay in as long as the power line power is applied, the stop button (2PB) is not pushed, and the timing relay (1TR) has not timed out.

Another 1CR contact is located in the second rung of the ladder diagram. When this 1CR closes, current can flow through solenoid A. Solenoid A is an electromechanical device that is electrically activated to mechanically open a valve, which allows fluid to flow into the tank. Fluid flows because the float switch (1FS) in rung 2 is closed. This situation is indicated in Figure 12.1 by the empty position for the float switch.

When the tank has filled, the float switch (1FS) changes to the filled position. This change de-energizes solenoid A, starts the timer relay (1TR in rung 3 of Figure 12.2, also in the start/stop box of Figure 12.1), and operates the mixer solenoid (MS).

After the timer has timed out, relay 1TR switches off the mixer and energizes solenoid B, which empties the tank. When the tank is empty, float switch (1FS) shuts off solenoid B and places the system in the ready position for the next manual start.

Notice that pressing the start switch (1PB) again once the cycle has started will have no adverse effect on the cycle. This protective logic should be designed into all processes, whether a PC or a computer is used.

Programmable Controller

In 1978, the National Electrical Manufacturers Association (NEMA) released a standard for programmable controllers. This standard was the result of four years of work by a committee made up of representatives from PC manufacturers. NEMA Standard ICS3–1978, part ICS3–304, defines a programmable controller as "a digitally operating electronic apparatus which uses a programmable memory for the internal storage of instructions for implementing specific functions such as logic, sequencing, timing, counting, and arithmetic to control, through digital or analog input/output modules, various types of machines or processes. A digital computer which is used to perform the functions of a programmable controller is considered to be within this scope. Excluded are drum and similar mechanical type sequencing controllers."

There is a tendency to confuse PCs with computers and programmable process controllers that are used for numerical control and for position control. Numerical and position control are used where a very large number of incremental positions are needed to complete a task. Examples are a lathe and a drilling machine. These tasks are not normally handled well by a PC, although that situation is changing.

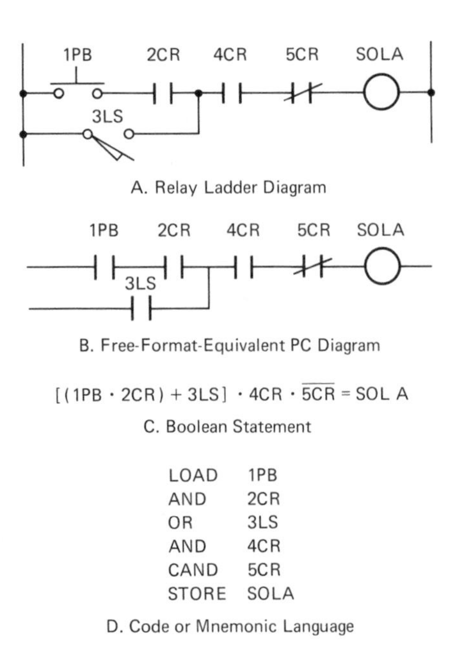

Figure 12.3 Comparison of Programmable Languages Used with Various PCs

What is the difference between a PC and a computer? To start with, all PCs are computers. The PC's block diagram structure is the same as that given in Chapter 11 for the computer. However, not all computers are PCs. The major differences that distinguish a PC from a computer are the PC's ability to operate in harsh environments, its different programming language, and its ease of troubleshooting and maintenance.

Programmable controllers are designed to operate in industrial environments that are dirty, are electrically noisy, have a wide fluctuation in temperatures (0°–60°C), and have relative humidities of from 0% to 95%. Air conditioning, which is generally required for computers, is not required for PCs.

The PC's programming language has been, by popular demand, the ladder diagram with standard relay symbology; see Figure 12.3A. The reason for the ladder diagram's popularity is that plant personnel are very familiar with relay logic from their previous experience with sequential controls. There are other PC languages in use, however. One language involves Boolean statements relating logical inputs such as and, or, and invert to a single statement output. Figure 12.3C is an example of a Boolean statement. Another type of PC language is the code or mnemonic (pronounced "nee-*mon*-ic") language, shown in Figure 12.3D. This language uses instructions such as AND, OR, LOAD, STORE, and so on. This type of language is very similar to computer assembly language.

The last major difference between PCs and computers is that of troubleshooting and maintenance. The PC can be maintained by the plant electrician or technician with minimal training. Most of the maintenance is done by replacing modules rather than components. Many times, the PC has a diagnostic program that assists the technician in locating bad modules. Computers, on the other hand, require highly trained electronics specialists to maintain and troubleshoot them.

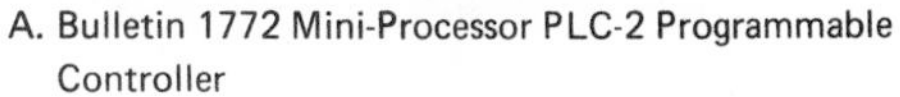

A. Bulletin 1772 Mini-Processor PLC-2 Programmable Controller

B. Bulletin 1770 Industrial Terminal System

Figure 12.4 Allen-Bradley's Programmable Controller and Terminal

Example of a Small System

In this section, we will describe a small Allen-Bradley programmable controller. It is discussed here because it is a representative small PC system and can easily be used for instructional purposes. Our main purpose in this chapter is to acquaint you with a PC and how it can be used.

The PC shown in Figure 12.4A is the Bulletin 1772 Mini-Processor PLC-2 Programmable Controller; it has small space requirements, powerful capabilities, compact design, and low cost. The PC has a Zilog Z80 microprocessor unit (MPU) and an Intel 8251 interface integrated circuit. The programming terminal shown in Figure 12.4B is the Bulletin 1770 Industrial Terminal system (catalog number 1770-TA). It represents the latest design in terminals by Allen-Bradley.

The PLC-2 contains a programmable read/write memory to store user-programmed instructions that implement specific functions, such as program logic, timing, counting, internal data storage and manipulation, and two-function arithmetic. Editing, or modification of the program, can be carried out at the installation.

The PLC-2 controller can be programmed to control functions that involve many types of cyclical or repetitive operations. It is designed to perform functions similar to those of relay or solid-state systems. The PLC-2 controller may interface with input and output devices such as limit, float, thumbwheel, and pressure switches; alarm, indicator, and annunciator (sound alarm) panels; push buttons and solenoids; motors and motor starters; and various solid-state devices.

Applications in industry include manufacturing of discrete parts and machine tools as well as uses in the areas of petrochemicals, pulp and paper products, metals, food and beverages, lumber, and materials handling. This controller is best suited for small applications where cost and compact installation are primary concerns. The controller allows interfacing with a maximum of 128 input/output (I/O) devices, although as few as 8 I/O may be used. The chassis is self-contained and requires about a ninth of a cubic meter of installation space (assuming a 32 I/O configuration). If the application later demands a more powerful processor, hardware expansion is available.

Figure 12.5 shows the 1770 terminal as a portable device. Figure 12.6 shows the 1770 in more detail. The terminal has a 23 cm cathode ray tube (CRT) for the display (Figure 12.6A). The terminal weighs less than 16 kg and is sufficiently rugged to be adaptable to most industrial environments.

The detachable, sealed touchpad keyboard is shown in Figure 12.6C. Figure 12.6D shows the keyboard overlay being placed on the keyboard. Figure 12.7 shows two choices for keyboard overlays: the ladder diagram overlay and the alphanumeric overlay.

Conclusion

Though programmable controllers are not designed to replace computers, they are useful and cost-effective for small- to medium-sized control systems. With the capability of operating in a distributed control system as a local controller, PCs will also retain their application in large, plantwide control systems. Within the next few years, PCs should become as valuable to industry as hand-held calculators have become to education.

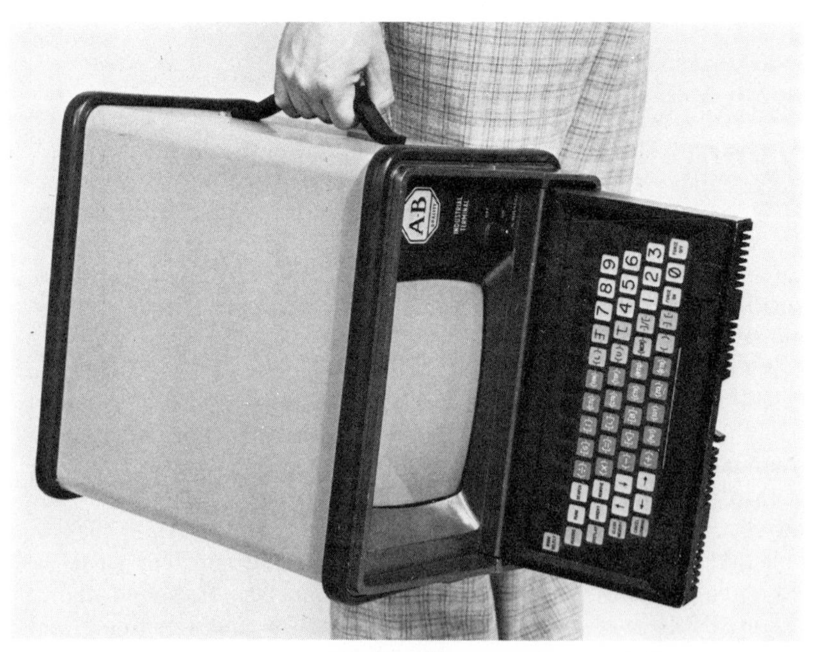

A. Portability

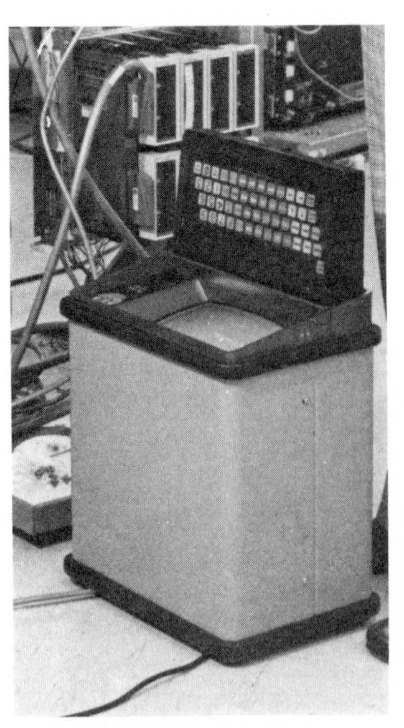
B. Operation in Any Position

Figure 12.5 Allen-Bradley 1770 Industrial Terminal

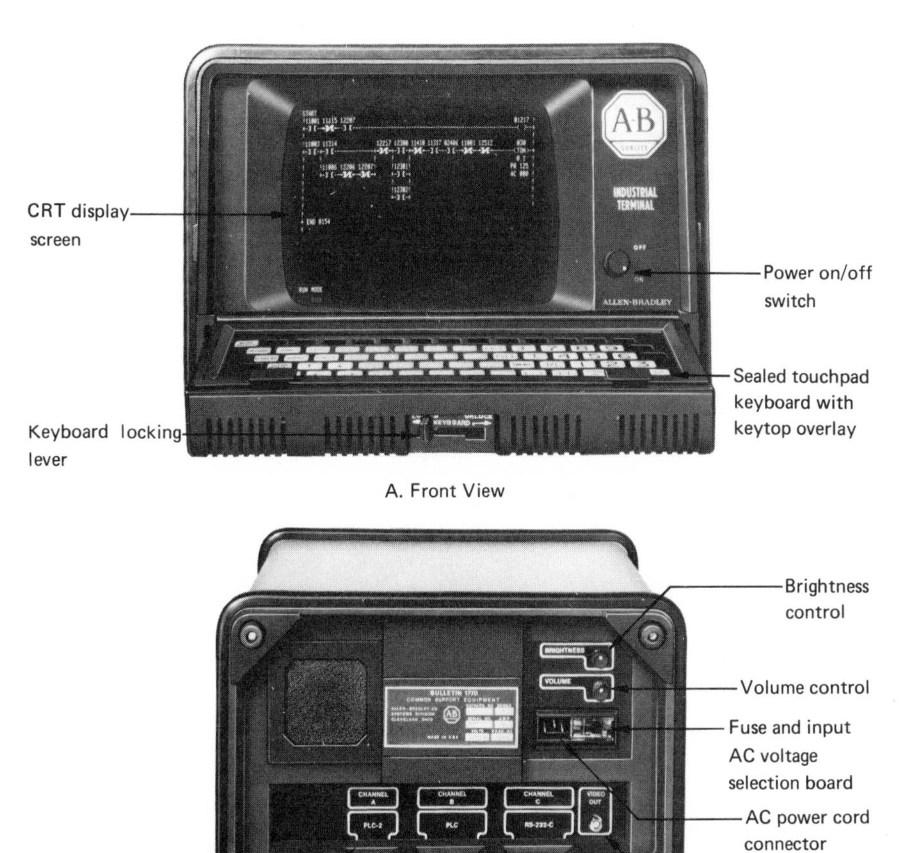

A. Front View

B. Rear View

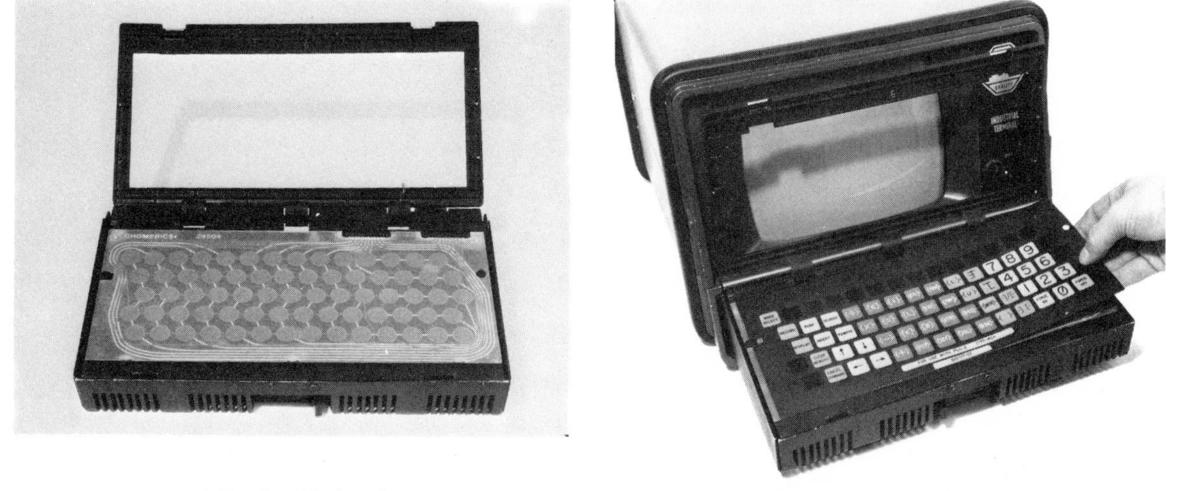

C. Touchpad Keyboard

D. Installing Keytop Overlay

Figure 12.6 Close-ups of Allen-Bradley 1770 Industrial Terminal

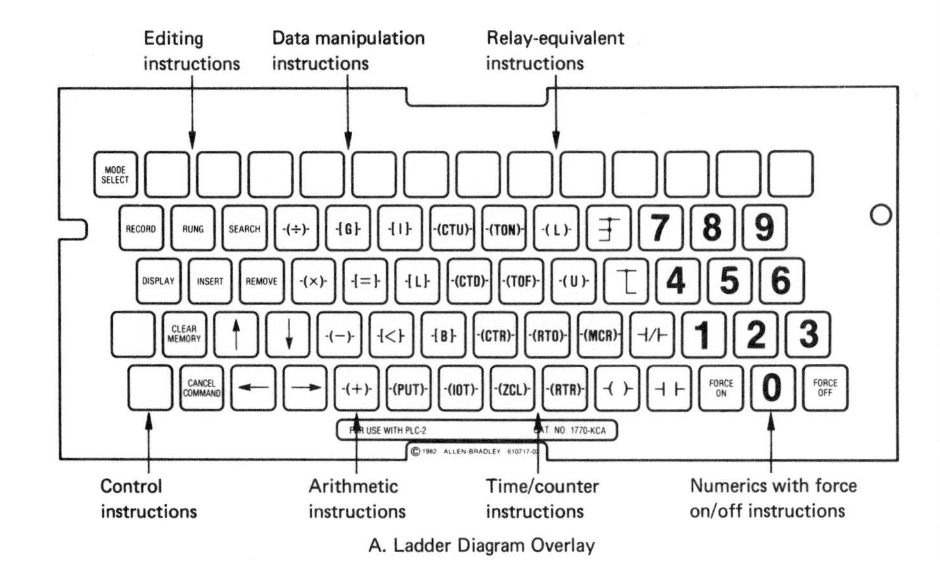

A. Ladder Diagram Overlay

B. Alphanumeric Overlay

Figure 12.7 Keytop Overlays for Allen-Bradley 1770 Terminal

Questions

1. A ______ process is a process in which one event follows another until the job is completed.
2. A diagram representing the order of operation of a process is called a ______ ______.
3. A digitally operating electronic apparatus that uses a programmable memory for the internal storage of instructions for implementing specific functions is called a ______ ______.
4. Draw the ladder diagram for an appliance such as a toaster.
5. List the differences between a PC and a computer.
6. After appropriate research, make a table showing industrial schematic symbols and the corresponding electronic symbols.

Bibliography

Chapter 1: Operational Amplifiers

*Berlin, H. W. *Design of Op-Amp Circuits with Experiments.* Indianapolis: H. W. Sams, 1977.

*Cirovic, M. M. *Integrated Circuit: A User's Handbook.* Reston, Va.: Reston, 1977.

Clayton, G. B. *Operational Amplifiers.* London: Butterworth, 1971.

*Coughlin, R. F., and Driscoll, F. F. *Operational Amplifiers and Linear Integrated Circuits.* 2nd ed. Englewood Cliffs, N.J.: Prentice-Hall, 1982.

Faulkenberry, L. M. *An Introduction to Operational Amplifiers.* New York: Wiley, 1977.

Graeme, J. G. *Designing with Operational Amplifiers.* New York: McGraw-Hill, 1977.

*Hughes, F. W. *Op Amp Handbook.* Englewood Cliffs, N.J.: Prentice-Hall, 1981.

Jung, W. C. *IC Op Amp Cookbook.* Indianapolis: H. W. Sams, 1974.

Melen, R., and Garland, H. *Understanding IC Operational Amplifiers.* 2nd ed. Indianapolis: H. W. Sams, 1975.

National Semiconductor Corporation. *Databook.* Santa Clara, Calif., 1980.

*Rutkowski, G. B. *Handbook of Integrated-Circuit Operational Amplifiers.* Englewood Cliffs, N.J.: Prentice-Hall, 1975.

Wait, J. V. *Introduction to Operational Amplifier Theory and Applications.* New York: McGraw-Hill, 1975.

Chapter 2: Applications of Op Amps

*Berlin, H. W. *Design of Active Filters with Experiments.* Indianapolis: H. W. Sams, 1977.

*———. *Design of Op-Amp Circuits with Experiments.* Indianapolis: H. W. Sams, 1977.

Burr-Brown Research Corporation. *Handbook of Operational Amplifier Applications.* Tucson, 1963.

Faulkenberry, L. M. *An Introduction to Operational Amplifiers.* New York: Wiley, 1977.

Garrett, P. H. *Analog I/O Design.* Reston, Va.: Reston, 1981.

Hnatek, E. R. *Applications of Linear Integrated Circuits.* New York: Wiley, 1975.

Jacob, J. M. *Applications and Designs with Analog Integrated Circuits.* Reston, Va.: Reston, 1982.

*Jung, W. C. *IC Op Amp Cookbook.* Indianapolis: H. W. Sams, 1974.

*———. *Audio IC Op-Amp Applications.* 2nd ed. Indianapolis: H. W. Sams, 1978.

Lancaster, D. *Active Filter Cookbook.* Indianapolis: H. W. Sams, 1975.

National Semiconductor Corporation. *Linear Applications.* Santa Clara, Calif., 1976.

———. *Application Note No. AN-211.* Santa Clara, Calif., 1979.

———. *LM-10 Datasheet.* Santa Clara, Calif., 1979.

RCA Corporation. *Linear Integrated Circuits.* Sommerville, N.J., 1978.

*Rutkowski, G. B. *Handbook of Integrated-Circuit Operational Amplifiers.* Englewood Cliffs, N.J.: Prentice-Hall, 1975

Wait, J. V. *Introduction to Operational Amplifier Theory and Applications.* New York: McGraw-Hill, 1975.

Chapter 3: DC Motors and Generators

Anderson, L. R. *Electric Machines and Transformers.* Reston, Va.: Reston, 1981.

* Books that have a technical rather than an engineering emphasis.

Fisher, F. "Convenient Comparison Charts Aid in Motor Selection." *EDN* 23 (August 5, 1978): 97–99.
Headquarters, Department of the Army. *Electric Motor and Generator Repair.* Washington, D.C.: U.S. Government Printing Office, 1972.
Kosow, I. L. *Electric Machinery and Transformers.* Englewood Cliffs, N.J.: Prentice-Hall, 1972.
*Lister, E. C. *Electric Circuits and Machines.* 4th ed. New York: McGraw-Hill, 1968.
Lloyd, T. C. *Electric Motors and Their Applications.* New York: Wiley, 1969.
*Naval Education and Training Support Command. *Electrician's Mate 3 & 2.* Washington, D.C.: U.S. Government Printing Office, 1974.
Richardson, D. V. *Rotating Electric Machinery and Transformer Technology.* Reston, Va.: Reston, 1978.
Wildi, T. *Electric Power Technology.* New York: Wiley, 1981.

Chapter 4: AC Motors

Anderson, R. *Electric Machines and Transformers.* Reston, Va.: Reston, 1981.
Kosow, I. L. *Electric Machinery and Transformers.* Englewood Cliffs, N.J.: Prentice-Hall, 1972.
*Lister, E. C. *Electric Circuits and Machines.* 4th ed. New York: McGraw-Hill, 1968.
Lloyd, T. C. *Electric Motors and Their Applications.* New York: Wiley, 1969.
Richardson, D. V. *Rotating Electric Machinery and Transformer Technology*, Reston, Va.: Reston, 1978.
Wildi, T. *Electric Power Technology.* New York: Wiley, 1981.

Chapter 5: Control Devices

*Bell, D. A. *Electronic Devices and Circuits.* 2nd ed. Reston, Va.: Reston, 1980.
*Coughlin, R. F., and Driscoll, F. F. *Solid State Devices and Applications.* Englewood Cliffs, N.J.: Prentice-Hall, 1975.
*Deboo, G. J., and Burrous, C. N. *Integrated Circuits and Semiconductor Devices.* New York: McGraw-Hill, 1977.
General Electric Company. *Transistor Manual.* 2nd ed. Auburn, N.Y., 1969.
———. *SCR Manual.* 6th ed. Auburn, N.Y., 1980.
Gottlieb, I. M. *Solid-State Power Electronics.* Indianapolis: H. W. Sams, 1979.
*Lurch, E. N. *Fundamentals of Electronics.* New York: Wiley, 1981.
*Maloney, T. J. *Industrial Solid-State Electronics.* Englewood Cliffs, N.J.: Prentice-Hall, 1979.
*Malvino, A. P. *Transistor Circuit Approximations.* 3rd ed. New York: McGraw-Hill, 1980.
RCA Corporation. *Thyristor and Rectifier Manual.* Somerville, N.J., 1975.
Rutkowski, G. B. *Solid-State Electronics.* 2nd ed. Indianapolis: Bobbs-Merrill, 1980.
Texas Instruments. *The Power Data Book.* Dallas.

Chapter 6: Power Control Circuits

*Bell, D. A. *Electronic Devices and Circuits.* 2nd ed. Reston, Va.: Reston, 1980.
*Coughlin, R. F., and Driscoll, F. F. *Solid State Devices and Applications.* Englewood Cliffs, N.J.: Prentice-Hall, 1975.
*Deboo, G. J., and Burrous, C. N. *Integrated Circuits and Semiconductor Devices.* New York: McGraw-Hill, 1977.
General Electric Company. *Transistor Manual.* 2nd ed. Auburn, N.Y., 1969.
———. *SCR Manual.* 6th ed. Auburn, N.Y., 1980.
Gottlieb, I. M. *Solid-State Power Electronics.* Indianapolis: H. W. Sams, 1979.
*Lurch, E. N. *Fundamentals of Electronics.* New York: Wiley, 1981.
*Maloney, T. J. *Industrial Solid-State Electronics.* Englewood Cliffs, N.J.: Prentice-Hall, 1979.
*Malvino, A. P. *Transistor Circuit Approximations.* 3rd ed. New York: McGraw-Hill, 1980.
RCA Corporation. *Thyristor and Rectifier Manual.* Somerville, N.J., 1975.
Rutkowski, G. B. *Solid-State Electronics.* 2nd ed. Indianapolis: Bobbs-Merrill, 1980.
Texas Instruments. *The Power Data Book.* Dallas.

Chapter 7: Transducers

Anderson, N. A. *Instrumentation for Process Measurement and Control.* 2nd ed. Radnor, Pa.: Chilton, 1972.
Andrew, W. G., and Williams, H. B. *Applied Instru-*

mentation in the Process Industries. 2nd ed. Houston: Gulf, 1979.

Deboo, G. J., and Burrous, C. N. *Integrated Circuits and Semiconductor Devices.* New York: McGraw-Hill, 1977.

*Driscoll, E. F. *Industrial Electronics: Devices, Circuits and Applications.* Chicago: American Technical Society, 1976.

*Fribance, A. E. *Industrial Instrumentation Fundamentals.* New York: McGraw-Hill, 1962.

Herceg, E. E. *Shaevitz Handbook of Measurement and Control.* Camden, N.J.: Shaevitz Engineering, 1976.

*Johnson, C. D. *Process Control Instrumentation Technology.* 2nd ed. New York: Wiley, 1982.

*Lenk, J. D. *Handbook of Controls and Instrumentation.* Englewood Cliffs, N.J.: Prentice-Hall, 1980.

Mansfield, P. H. *Electrical Transducers for Industrial Measurement.* London: Butterworth, 1973.

Micro Switch. *Solid State Sensors.* Freeport, Ill.: Honeywell.

Minnar, E. J., ed. *ISA Transducer Compendium.* Pittsburgh: Instrument Society of America, 1963.

Moore, R. L., ed. *Basic Instrumentation Lecture Notes and Study Guide.* 2nd ed. Pittsburgh: Instrument Society of America, 1976.

National Semiconductor Corporation. *Pressure Transducer Handbook.* Santa Clara, Calif., 1980.

*Norton, H. M. *Handbook of Transducers for Electronic Measuring Systems.* Englewood Cliffs, N.J.: Prentice-Hall, 1969.

O'Higgins, P. J. *Basic Instrumentation.* New York: McGraw-Hill, 1966.

Sheingold, D. H. *Transducer Interfacing Handbook.* Norwood, Mass.: Analog Devices, 1980.

Chapter 8: Process Control

Brewer, J. W. *Control Systems: Analysis, Design, and Simulation.* Englewood Cliffs, N.J.: Prentice-Hall, 1974.

Hougen, J. O. *Measurements and Control Applications.* Research Triangle Park, N.C.: Instrument Society of America, 1979.

*Hunter, R. P. *Automated Process Control Systems: Concepts and Hardware.* Englewood Cliffs, N.J.: Prentice-Hall, 1978.

*Johnson, C. D. *Process Control Instrumentation Technology.* 2nd ed. New York: Wiley, 1982.

Kuo, B. C. *Automatic Control Systems.* Englewood Cliffs, N.J.: Prentice-Hall, 1975.

Pericles, E., and Leff, E. *Introduction to Feedback Control Systems.* New York: McGraw-Hill, 1979.

Weyrick, R. C. *Fundamentals of Automatic Control.* New York: McGraw-Hill, 1975.

Chapter 9: Pulse Modulation

*Hnatek, E. R. *Applications of Linear Integrated Circuits.* New York: Wiley, 1975.

Journal Ministere Russie Defense 47, section 7, 25 (1845).

*Kennedy, G. *Electronic Communication Systems.* 2nd ed. New York: McGraw-Hill, 1977.

Miller, G. M. *Modern Electronic Communication.* Englewood Cliffs, N.J.: Prentice-Hall, 1978.

*National Semiconductor Corporation. *Special Functions Databook.* Santa Clara, Calif., 1979.

Shannon, C. E., and Weaver, W. *The Mathematical Theory of Communications.* Urbana: University of Illinois Press, 1949.

Taub, H., and Schilling, D. L. *Principles of Communication Systems.* New York: McGraw-Hill, 1971.

Chapter 10: Telemetry

Fisher, H. F. *Telemetry Transducer Handbook.* Technical Report no. WADD-TR-61-67, vol. I, rev. I. Wright-Patterson Air Force Base, Ohio: Air Force Systems Command, 1963.

*Gruenberg, E. L., ed. *Handbook of Telemetry and Remote Control.* New York: McGraw-Hill, 1967.

Inter-Range Instrumentation Group. *Telemetry Standards.* Document no. 106-73, rev. White Sands Missile Range, N.M.: Secretariat, Range Commanders Council, 1973.

Kennedy, G. *Electronic Communication Systems.* 2nd ed. New York: McGraw-Hill, 1977.

Martin, J. *Telecommunications and the Computer.* Englewood Cliffs, N.J.: Prentice-Hall, 1969.

Miller, G. M. *Modern Electronic Communication.* Englewood Cliffs, N.J.: Prentice-Hall, 1978.

*Zanger, H. *Electronic Systems Theory and Applications.* Englewood Cliffs, N.J.: Prentice-Hall, 1977.

Chapter 11: Microprocessors

*Cushman, R. H. "EDN's Sixth Annual μP/μC Chip Directory." *EDN* 24, no. 19 (October 20, 1979): 133–240.

———. "The Promise of Analog μPs: Low-cost Digital Signal Handling." *EDN* 25, no. 1 (January 5, 1980): 127–132.

*———. "EDN's Eighth Annual μP/μC Chip Directory." *EDN* 26, no. 22 (November 11, 1981): 101–216.

*Cushman, R. H., and Backler, J. "EDN's Seventh Annual μP/μC Chip Directory." *EDN* 25, no. 20 (November 5, 1980): 94–210.

Osborne, A. *An Introduction to Microcomputers.* Vol. O. Berkeley, Calif.: Osborne & Associates, 1979.

*Zaks, R. *Microprocessors from Chips to Systems.* Berkeley, Calif.: Sybex, 1977.

Chapter 12: Programmable Controllers

*Allen-Bradley Company. *Bulletin 1770 Industrial Terminal Systems User's Manual.* Cleveland.

*———. *Bulletin* 1772 *Mini-PLC-2 Programmable Controller.* Cleveland.

Andrew, W. G., and Williams, H. B. *Applied Instrumentation in the Process Industries.* Houston: Gulf, 1979.

Deltano, D. "Programming Your PC." *Instruments & Control Systems* 53 (July 1980): 37–40.

Hickey, J. "Programmable Controller Roundup." *Instruments & Control Systems* 54 (July 1981): 57–64.

Jannotta, K. "What is a PC?" *Instruments & Control Systems* 53 (February 1980): 21–25.

Data Sheets

The following data sheets have been included for your reference. They present detailed information about some of the IC chips used in this text. We hope that you will use this information to gain greater understanding of the IC chips and circuits we have used as examples. You may want to use these same IC chips in projects of your own design, and these data sheets should help you.

- 741 Operational Amplifier
- 3900 Current Differencing Amplifier
- SCR
- Triac
- 335 Temperature Sensor
- 555 Timer
- 565 Phase Locked Loop
- ADC 0801 Analog-to-digital Converter
- DAC 0808 Digital-to-analog Converter

The data sheets have been reproduced courtesy of National Semiconductor Corporation (© 1982) and Texas Instruments Incorporated.

Operational Amplifiers/Buffers

LM741/LM741A/LM741C/LM741E Operational Amplifier

General Description

The LM741 series are general purpose operational amplifiers which feature improved performance over industry standards like the LM709. They are direct, plug-in replacements for the 709C, LM201, MC1439 and 748 in most applications.

The amplifiers offer many features which make their application nearly foolproof: overload protection on the input and output, no latch-up when the common mode range is exceeded, as well as freedom from oscillations.

The LM741C/LM741E are identical to the LM741/LM741A except that the LM741C/LM741E have their performance guaranteed over a 0°C to +70°C temperature range, instead of −55°C to +125°C.

Schematic and Connection Diagrams (Top Views)

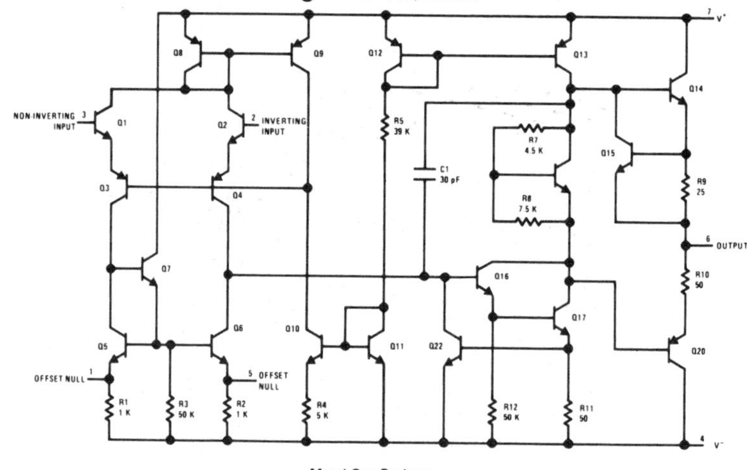

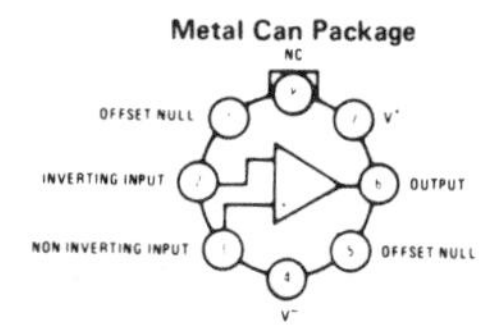

Order Number LM741H, LM741AH, LM741CH or LM741EH
See NS Package H08C

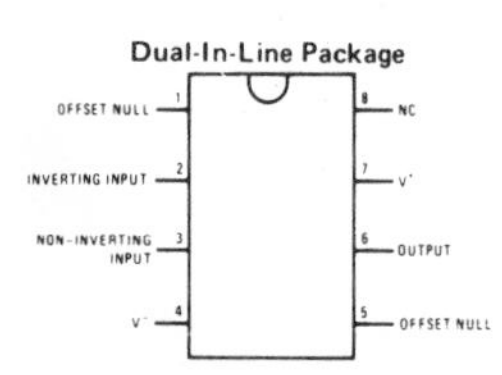

Order Number LM741CN or LM741EN
See NS Package N08B
Order Number LM741CJ
See NS Package J08A

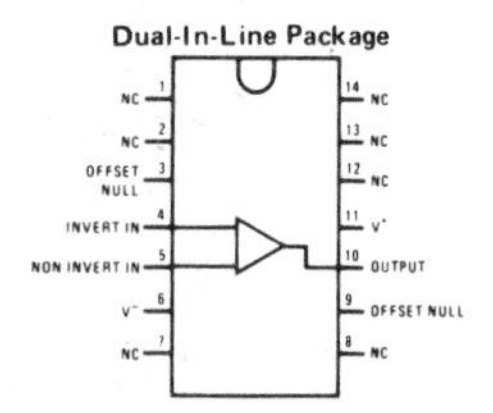

Order Number LM741CN-14
See NS Package N14A
Order Number LM741J-14, LM741AJ-14 or LM741CJ-14
See NS Package J14A

Absolute Maximum Ratings

	LM741A	LM741E	LM741	LM741C
Supply Voltage	±22V	±22V	±22V	±18V
Power Dissipation (Note 1)	500 mW	500 mW	500 mW	500 mW
Differential Input Voltage	±30V	±30V	±30V	±30V
Input Voltage (Note 2)	±15V	±15V	±15V	±15V
Output Short Circuit Duration	Indefinite	Indefinite	Indefinite	Indefinite
Operating Temperature Range	−55°C to +125°C	0°C to +70°C	−55°C to +125°C	0°C to +70°C
Storage Temperature Range	−65°C to +150°C	−65°C to +150°C	−65°C to +150°C	−65°C to +150°C
Lead Temperature (Soldering, 10 seconds)	300°C	300°C	300°C	300°C

Electrical Characteristics (Note 3)

PARAMETER	CONDITIONS	LM741A/LM741E			LM741			LM741C			UNITS
		MIN	TYP	MAX	MIN	TYP	MAX	MIN	TYP	MAX	
Input Offset Voltage	$T_A = 25°C$										
	$R_S \leq 10\ k\Omega$					1.0	5.0		2.0	6.0	mV
	$R_S \leq 50\Omega$		0.8	3.0							mV
	$T_{AMIN} \leq T_A \leq T_{AMAX}$										
	$R_S \leq 50\Omega$			4.0							mV
	$R_S \leq 10\ k\Omega$						6.0			7.5	mV
Average Input Offset Voltage Drift				15							μV/°C
Input Offset Voltage Adjustment Range	$T_A = 25°C$, $V_S = \pm 20V$	±10				±15			±15		mV
Input Offset Current	$T_A = 25°C$		3.0	30		20	200		20	200	nA
	$T_{AMIN} \leq T_A \leq T_{AMAX}$			70		85	500			300	nA
Average Input Offset Current Drift				0.5							nA/°C
Input Bias Current	$T_A = 25°C$		30	80		80	500		80	500	nA
	$T_{AMIN} \leq T_A \leq T_{AMAX}$			0.210			1.5			0.8	μA
Input Resistance	$T_A = 25°C$, $V_S = \pm 20V$	1.0	6.0		0.3	2.0		0.3	2.0		MΩ
	$T_{AMIN} \leq T_A \leq T_{AMAX}$, $V_S = \pm 20V$	0.5									MΩ
Input Voltage Range	$T_A = 25°C$							±12	±13		V
	$T_{AMIN} \leq T_A \leq T_{AMAX}$				±12	±13					V
Large Signal Voltage Gain	$T_A = 25°C$, $R_L \geq 2\ k\Omega$										
	$V_S = \pm 20V$, $V_O = \pm 15V$	50									V/mV
	$V_S = \pm 15V$, $V_O = \pm 10V$				50	200		20	200		V/mV
	$T_{AMIN} \leq T_A \leq T_{AMAX}$, $R_L \geq 2\ k\Omega$,										
	$V_S = \pm 20V$, $V_O = \pm 15V$	32									V/mV
	$V_S = \pm 15V$, $V_O = \pm 10V$				25			15			V/mV
	$V_S = \pm 5V$, $V_O = \pm 2V$	10									V/mV
Output Voltage Swing	$V_S = \pm 20V$										
	$R_L \geq 10\ k\Omega$	±16									V
	$R_L \geq 2\ k\Omega$	±15									V
	$V_S = \pm 15V$										
	$R_L \geq 10\ k\Omega$				±12	±14		±12	±14		V
	$R_L \geq 2\ k\Omega$				±10	±13		±10	±13		V
Output Short Circuit Current	$T_A = 25°C$	10	25	35		25			25		mA
	$T_{AMIN} < T_A \leq T_{AMAX}$	10		40							mA
Common-Mode Rejection Ratio	$T_{AMIN} \leq T_A \leq T_{AMAX}$										
	$R_S \leq 10\ k\Omega$, $V_{CM} = \pm 12V$				70	90		70	90		dB
	$R_S \leq 50\ k\Omega$, $V_{CM} = \pm 12V$	80	95								dB

Electrical Characteristics (Continued)

PARAMETER	CONDITIONS	LM741A/LM741E			LM741			LM741C			UNITS
		MIN	TYP	MAX	MIN	TYP	MAX	MIN	TYP	MAX	
Supply Voltage Rejection Ratio	$T_{AMIN} \le T_A \le T_{AMAX}$, V_S = ±20V to V_S = ±5V										
	$R_S \le 50\Omega$	86	96								dB
	$R_S \le 10\ k\Omega$				77	96		77	96		dB
Transient Response	T_A = 25°C, Unity Gain										
Rise Time			0.25	0.8		0.3			0.3		μs
Overshoot			6.0	20		5			5		%
Bandwidth (Note 4)	T_A = 25°C	0.437	1.5								MHz
Slew Rate	T_A = 25°C, Unity Gain	0.3	0.7			0.5			0.5		V/μs
Supply Current	T_A = 25°C					1.7	2.8		1.7	2.8	mA
Power Consumption	T_A = 25°C										
	V_S = ±20V		80	150							mW
	V_S = ±15V					50	85		50	85	mW
LM741A	V_S = ±20V										
	$T_A = T_{AMIN}$			165							mW
	$T_A = T_{AMAX}$			135							mW
LM741E	V_S = ±20V			150							mW
	$T_A = T_{AMIN}$			150							mW
	$T_A = T_{AMAX}$			150							mW
LM741	V_S = ±15V										
	$T_A = T_{AMIN}$					60	100				mW
	$T_A = T_{AMAX}$					45	75				mW

Note 1: The maximum junction temperature of the LM741/LM741A is 150°C, while that of the LM741C/LM741E is 100°C. For operation at elevated temperatures, devices in the TO-5 package must be derated based on a thermal resistance of 150°C/W junction to ambient, or 45°C/W junction to case. The thermal resistance of the dual-in-line package is 100°C/W junction to ambient.

Note 2: For supply voltages less than ±15V, the absolute maximum input voltage is equal to the supply voltage.

Note 3: Unless otherwise specified, these specifications apply for V_S = ±15V, $-55°C \le T_A \le +125°C$ (LM741/LM741A). For the LM741C/LM741E, these specifications are limited to $0°C \le T_A \le +70°C$.

Note 4: Calculated value from: BW (MHz) = 0.35/Rise Time(μs).

Operational Amplifiers/Buffers

LM2900/LM3900, LM3301, LM3401 Quad Amplifiers

General Description

The LM2900 series consists of four independent, dual input, internally compensated amplifiers which were designed specifically to operate off of a single power supply voltage and to provide a large output voltage swing. These amplifiers make use of a current mirror to achieve the non-inverting input function. Application areas include: ac amplifiers, RC active filters, low frequency triangle, squarewave and pulse waveform generation circuits, tachometers and low speed, high voltage digital logic gates.

Features

- Wide single supply voltage range or dual supplies — $4\ V_{DC}$ to $36\ V_{DC}$; $\pm 2\ V_{DC}$ to $\pm 18\ V_{DC}$
- Supply current drain independent of supply voltage
- Low input biasing current — 30 nA
- High open-loop gain — 70 dB
- Wide bandwidth — 2.5 MHz (Unity Gain)
- Large output voltage swing — $(V^+ - 1)$ Vp-p
- Internally frequency compensated for unity gain
- Output short-circuit protection

Schematic and Connection Diagrams

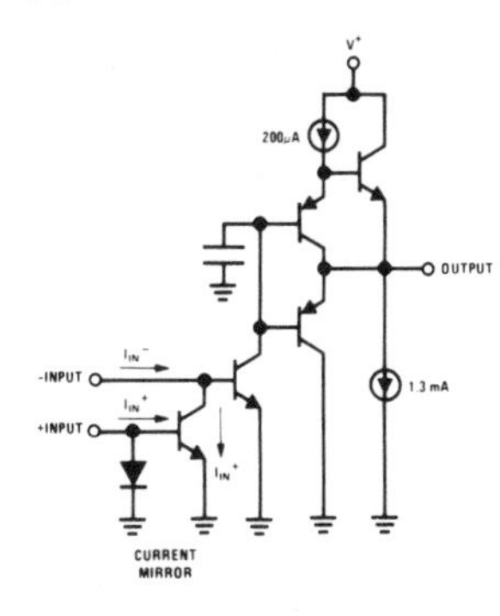

Order Number LM2900J
See NS Package J14A
Order Number LM2900N, LM3900N, LM3301N or LM3401N
See NS Package N14A

Dual-In-Line and Flat Package

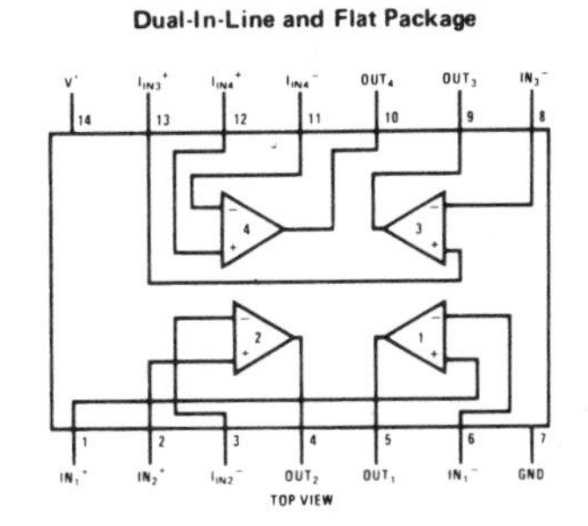

Typical Applications ($V^+ = 15\ V_{DC}$)

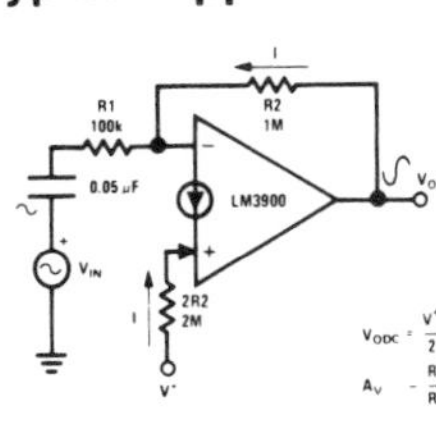

Inverting Amplifier

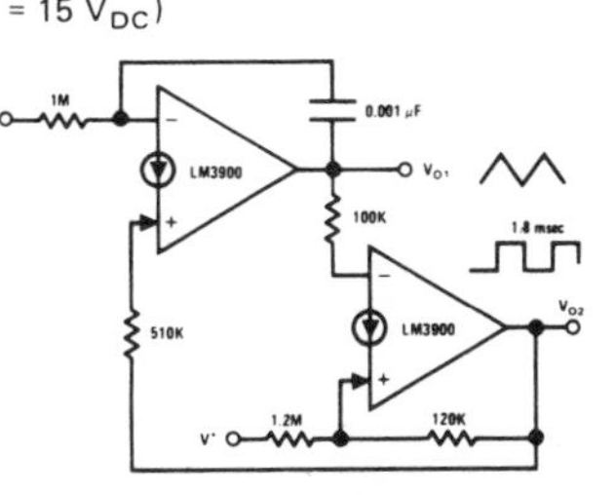

Triangle/Square Generator

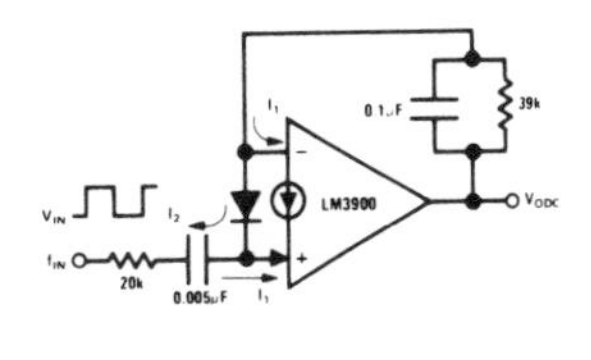

Frequency-Doubling Tachometer

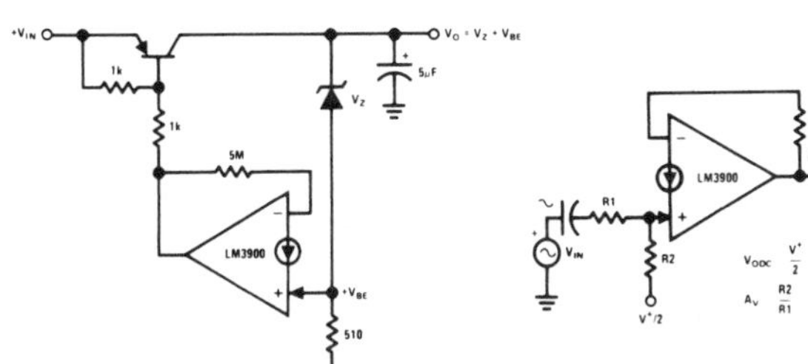

Low $V_{IN} - V_{OUT}$ Voltage Regulator

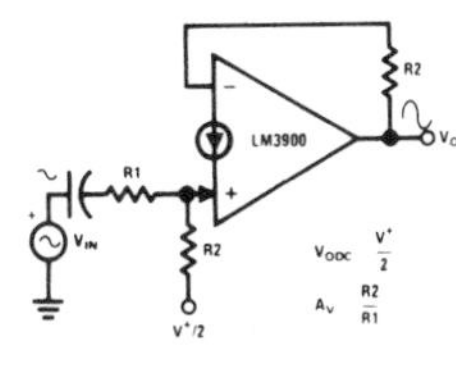

Non-Inverting Amplifier

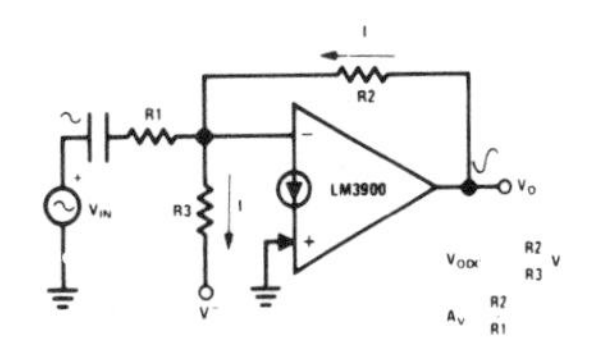

Negative Supply Biasing

Absolute Maximum Ratings

	LM2900/LM3900	LM3301	LM3401
Supply Voltage	32 V_{DC}	28 V_{DC}	18 V_{DC}
	±16 V_{DC}	±14 V_{DC}	±9 V_{DC}
Power Dissipation (T_A = 25°C) (Note 1)			
Cavity DIP	900 mW		
Flat Pack	800 mW		
Molded DIP	570 mW	570 mW	570 mW
Input Currents, I_{IN}^+ or I_{IN}^-	20 mA_{DC}	20 mA_{DC}	20 mA_{DC}
Output Short-Circuit Duration – One Amplifier T_A = 25°C (See Application Hints)	Continuous	Continuous	Continuous
Operating Temperature Range		−40°C to +85°C	0°C to +75°C
LM2900	−40°C to +85°C		
LM3900	0°C to +70°C		
Storage Temperature Range	−65°C to +150°C	−65°C to +150°C	−65°C to +150°C
Lead Temperature (Soldering, 10 seconds)	300°C	300°C	300°C

Electrical Characteristics (Note 6)

PARAMETER	CONDITIONS	LM2900			LM3900			LM3301			LM3401			UNITS
		MIN	TYP	MAX	MIN	TYP	MAX	MIN	TYP	MAX	MIN	TYP	MAX	
Open Loop														
Voltage Gain											800			V/mV
Voltage Gain	T_A = 25°C, f = 100 Hz	1.2	2.8		1.2	2.8		1.2	2.8		1.2	2.8		V/mV
Input Resistance	T_A = 25°C, Inverting Input		1			1			1		0.1	1		MΩ
Output Resistance			8			8			8			8		kΩ
Unity Gain Bandwidth	T_A = 25°C, Inverting Input		2.5			2.5			2.5			2.5		MHz
Input Bias Current	T_A = 25°C, Inverting Input		30	200		30	200		30	300		30	300	nA
	Inverting Input												500	nA
Slew Rate	T_A = 25°C, Positive Output Swing		0.5			0.5			0.5			0.5		V/μs
	T_A = 25°C, Negative Output Swing		20			20			20			20		V/μs
Supply Current	T_A = 25°C, R_L = ∞ On All Amplifiers		6.2	10		6.2	10		6.2	10		6.2	10	mA_{DC}
Output Voltage Swing	T_A = 25°C, R_L = 2k, V_{CC} = 15.0 V_{DC}													
V_{OUT} High	I_{IN}^- = 0, I_{IN}^+ = 0	13.5			13.5			13.5			13.5			V_{DC}
V_{OUT} Low	I_{IN}^- = 10μA, I_{IN}^+ = 0		0.09	0.2		0.09	0.2		0.09	0.2		0.09	0.2	V_{DC}
V_{OUT} High	I_{IN}^- = 0, I_{IN}^+ = 0 R_L = ∞, V_{CC} = Absolute Maximum Ratings		29.5			29.5			25.5			15.5		V_{DC}
Output Current Capability	T_A = 25°C													
Source		6	18		6	10		5	18		5	10		mA_{DC}
Sink	(Note 2)	0.5	1.3		0.5	1.3		0.5	1.3		0.5	1.3		mA_{DC}
I_{SINK}	V_{OL} = 1V, I_{IN} = 5μA		5			5			5			5		mA_{DC}

Electrical Characteristics (Continued) (Note 6)

PARAMETER	CONDITIONS	LM2900			LM3900			LM3301			LM3401			UNITS
		MIN	TYP	MAX	MIN	TYP	MAX	MIN	TYP	MAX	MIN	TYP	MAX	
Power Supply Rejection	T_A = 25°C, f = 100 Hz		70			70			70			70		dB
Mirror Gain	@ 20μA (Note 3)	0.90	1.0	1.1	0.90	1.0	1.1	0.90	1	1.10	0.90	1	1.10	μA/μA
	@ 200μA (Note 3)	0.90	1.0	1.1	0.90	1.0	1.1	0.90	1	1.10	0.90	1	1.10	μA/μA
ΔMirror Gain	@ 20μA To 200μA (Note 3)		2	5		2	5		2	5		2	5	%
Mirror Current	(Note 4)		10	500		10	500		10	500		10	500	μA_{DC}
Negative Input Current	T_A = 25°C (Note 5)		1.0			1.0			1.0			1.0		mA_{DC}
Input Bias Current	Inverting Input		300			300								nA

Note 1: For operating at high temperatures, the device must be derated based on a 125°C maximum junction temperature and a thermal resistance of 175°C/W which applies for the device soldered in a printed circuit board, operating in a still air ambient.

Note 2: The output current sink capability can be increased for large signal conditions by overdriving the inverting input. This is shown in the section on Typical Characteristics.

Note 3: This spec indicates the current gain of the current mirror which is used as the non-inverting input.

Note 4: Input V_{BE} match between the non-inverting and the inverting inputs occurs for a mirror current (non-inverting input current) of approximately 10μA. This is therefore a typical design center for many of the application circuits.

Note 5: Clamp transistors are included on the IC to prevent the input voltages from swinging below ground more than approximately −0.3 V_{DS}. The negative input currents which may result from large signal overdrive with capacitance input coupling need to be externally limited to values of approximately 1 mA. Negative input currents in excess of 4 mA will cause the output voltage to drop to a low voltage. This maximum current applies to any one of the input terminals. If more than one of the input terminals are simultaneously driven negative smaller maximum currents are allowed. Common-mode current biasing can be used to prevent negative input voltages; see for example, the "Differentiator Circuit" in the applications section.

Note 6: These specs apply for $-55°C \leq T_A \leq +125°C$, unless otherwise stated.

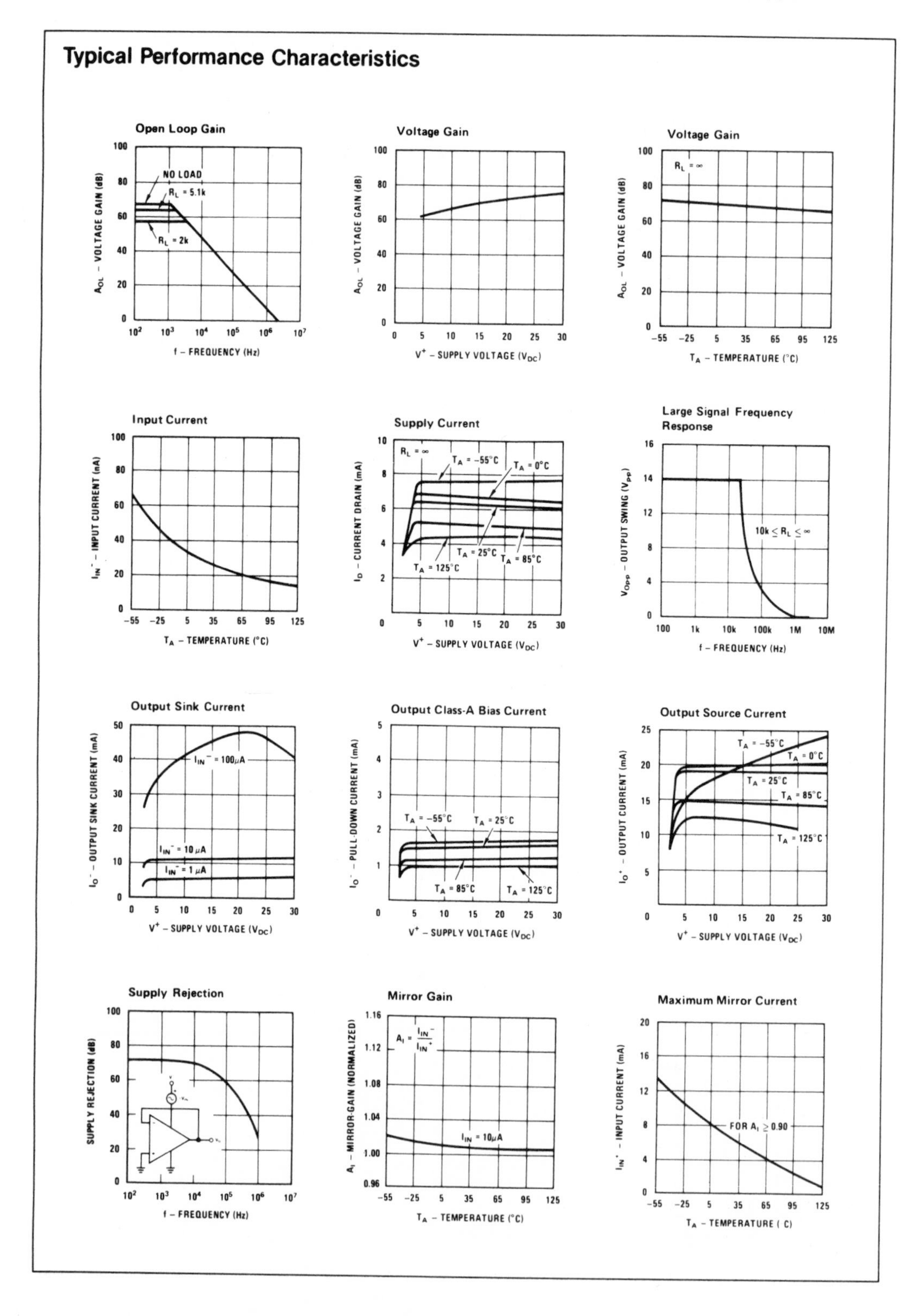
Typical Performance Characteristics
Open Loop Gain
NO LOAD
RL = 5.1k
RL = 2k
AOL – VOLTAGE GAIN (dB)
f – FREQUENCY (Hz)
Voltage Gain
AOL – VOLTAGE GAIN (dB)
V+ – SUPPLY VOLTAGE (VDC)
Voltage Gain
RL = ∞
AOL – VOLTAGE GAIN (dB)
TA – TEMPERATURE (°C)
Input Current
IIN– – INPUT CURRENT (nA)
TA – TEMPERATURE (°C)
Supply Current
RL = ∞
TA = –55°C
TA = 0°C
TA = 25°C
TA = 85°C
TA = 125°C
ID – CURRENT DRAIN (mA)
V+ – SUPPLY VOLTAGE (VDC)
Large Signal Frequency Response
10k ≤ RL ≤ ∞
VOp-p – OUTPUT SWING (Vpp)
f – FREQUENCY (Hz)
Output Sink Current
IIN– = 100μA
IIN– = 10 μA
IIN– = 1 μA
IO– – OUTPUT SINK CURRENT (mA)
V+ – SUPPLY VOLTAGE (VDC)
Output Class-A Bias Current
TA = –55°C
TA = 25°C
TA = 85°C
TA = 125°C
IO– – PULL-DOWN CURRENT (mA)
V+ – SUPPLY VOLTAGE (VDC)
Output Source Current
TA = –55°C
TA = 0°C
TA = 25°C
TA = 85°C
TA = 125°C
IO+ – OUTPUT CURRENT (mA)
V+ – SUPPLY VOLTAGE (VDC)
Supply Rejection
SUPPLY REJECTION (dB)
f – FREQUENCY (Hz)
Mirror Gain
AI = IIN– / IIN+
IIN = 10μA
AI – MIRROR-GAIN (NORMALIZED)
TA – TEMPERATURE (°C)
Maximum Mirror Current
FOR AI ≥ 0.90
IIN+ – INPUT CURRENT (mA)
TA – TEMPERATURE (°C)

Application Hints

When driving either input from a low-impedance source, a limiting resistor should be placed in series with the input lead to limit the peak input current. Currents as large as 20 mA will not damage the device, but the current mirror on the non-inverting input will saturate and cause a loss of mirror gain at mA current levels—especially at high operating temperatures.

Precautions should be taken to insure that the power supply for the integrated circuit never becomes reversed in polarity or that the unit is not inadvertently installed backwards in a test socket as an unlimited current surge through the resulting forward diode within the IC could cause fuzing of the internal conductors and result in a destroyed unit.

Output short circuits either to ground or to the positive power supply should be of short time duration. Units can be destroyed, not as a result of the short circuit current causing metal fuzing, but rather due to the large increase in IC chip dissipation which will cause eventual failure due to excessive junction temperatures. For example, when operating from a well-regulated +5 V_{DC} power supply at T_A = 25°C with a 100 kΩ shunt-feedback resistor (from the output to the inverting input) a short directly to the power supply will not cause catastrophic failure but the current magnitude will be approximately 50 mA and the junction temperature will be above T_J max. Larger feedback resistors will reduce the current, 11 MΩ provides approximately 30 mA, an open circuit provides 1.3 mA, and a direct connection from the output to the non-inverting input will result in catastrophic failure when the output is shorted to V^+ as this then places the base-emitter junction of the input transistor directly across the power supply. Short-circuits to ground will have magnitudes of approximately 30 mA and will not cause catastrophic failure at T_A = 25°C.

Unintentional signal coupling from the output to the non-inverting input can cause oscillations. This is likely only in breadboard hook-ups with long component leads and can be prevented by a more careful lead dress or by locating the non-inverting input biasing resistor close to the IC. A quick check of this condition is to bypass the non-inverting input to ground with a capacitor. High impedance biasing resistors used in the non-inverting input circuit make this input lead highly susceptible to unintentional ac signal pickup.

Operation of this amplifier can be best understood by noticing that input currents are differenced at the inverting-input terminal and this difference current then flows through the external feedback resistor to produce the output voltage. Common-mode current biasing is generally useful to allow operating with signal levels near ground or even negative as this maintains the inputs biased at $+V_{BE}$. Internal clamp transistors (see note 5) catch negative input voltages at approximately −0.3 V_{DC} but the magnitude of current flow has to be limited by the external input network. For operation at high temperature, this limit should be approximately 100μA.

This new "Norton" current-differencing amplifier can be used in most of the applications of a standard IC op amp. Performance as a dc amplifier using only a single supply is not as precise as a standard IC op amp operating with split supplies but is adequate in many less critical applications. New functions are made possible with this amplifier which are useful in single power supply systems. For example, biasing can be designed separately from the ac gain as was shown in the "inverting amplifier," the "difference integrator" allows controlling the charging and the discharging of the integrating capacitor both with positive voltages, and the "frequency doubling tachometer" provides a simple circuit which reduces the ripple voltage on a tachometer output dc voltage.

Typical Applications (Continued)

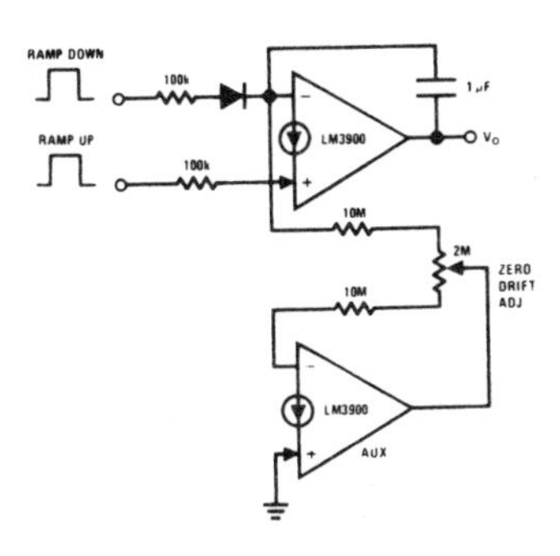

Low-Drift Ramp and Hold Circuit

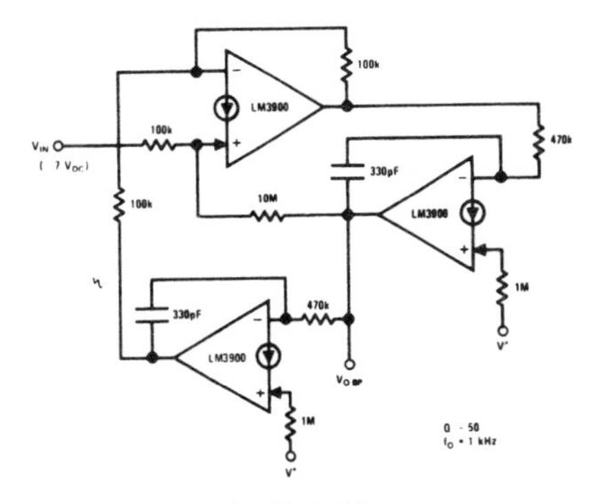

Bi-Quad Active Filter
(2nd Degree State-Variable Network)

Typical Applications (Continued)

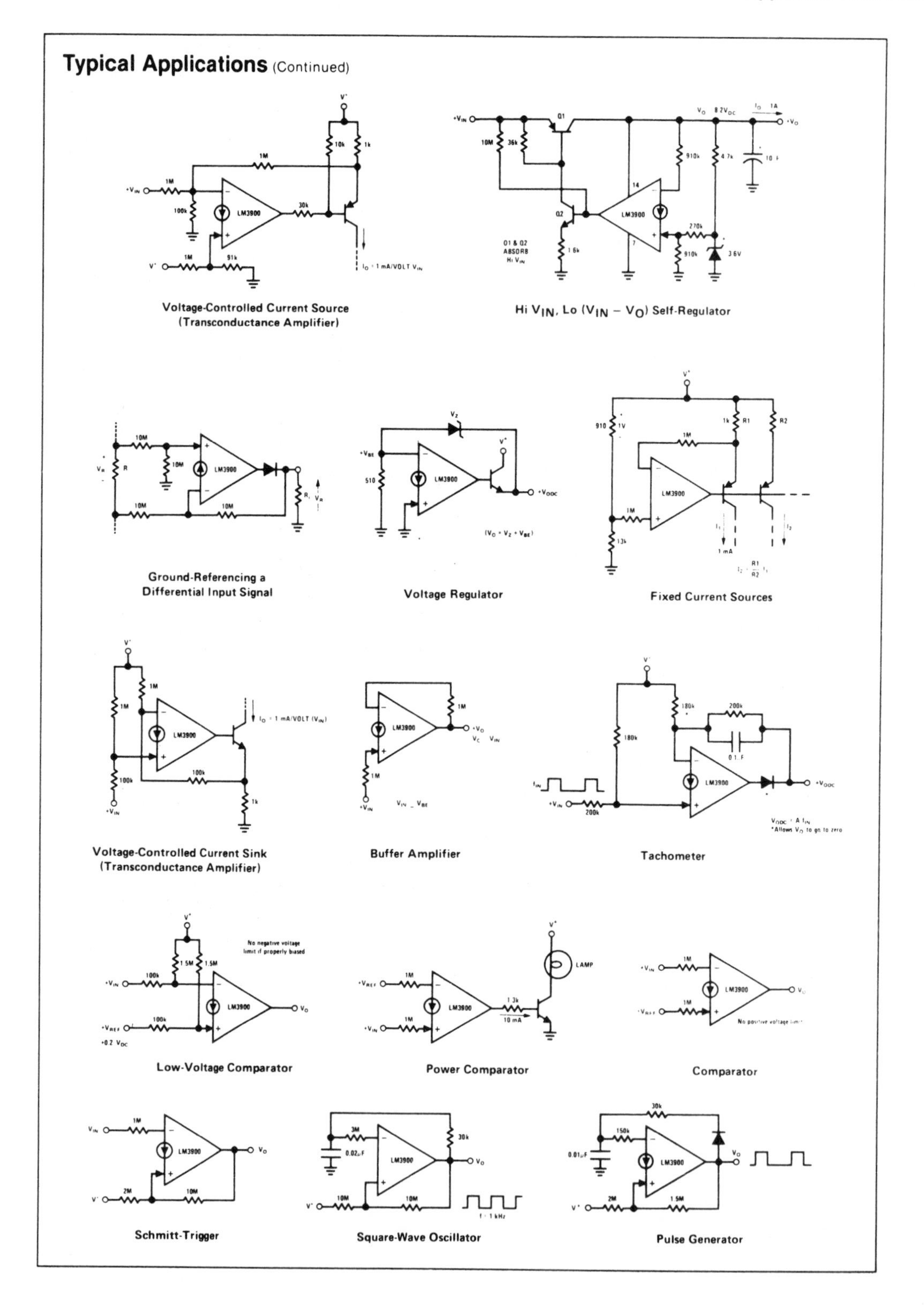

Voltage-Controlled Current Source (Transconductance Amplifier)

Hi V_{IN}, Lo (V_{IN} − V_O) Self-Regulator

Ground-Referencing a Differential Input Signal

Voltage Regulator

Fixed Current Sources

Voltage-Controlled Current Sink (Transconductance Amplifier)

Buffer Amplifier

Tachometer

Low-Voltage Comparator

Power Comparator

Comparator

Schmitt-Trigger

Square-Wave Oscillator

Pulse Generator

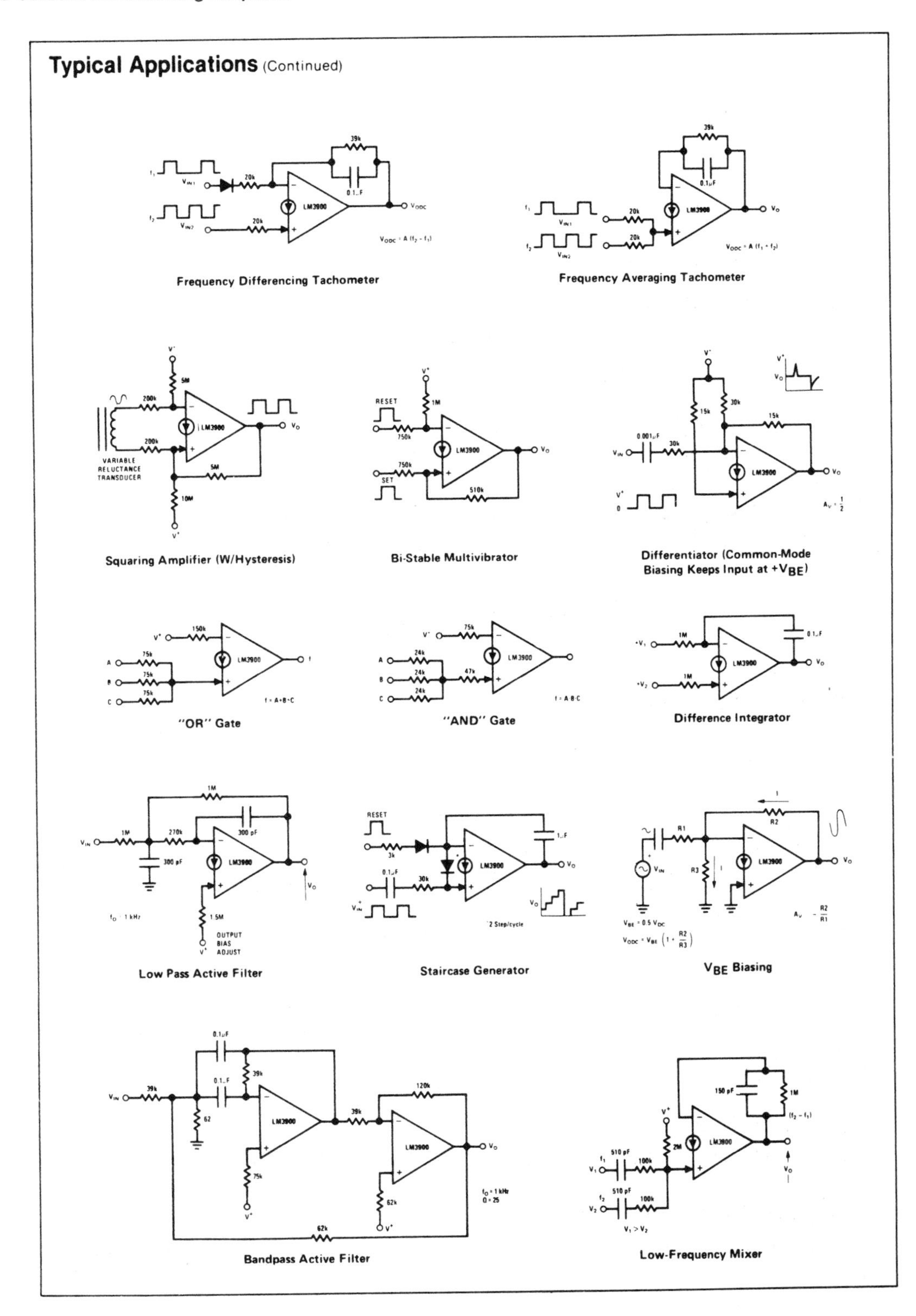

Frequency Differencing Tachometer

Frequency Averaging Tachometer

Squaring Amplifier (W/Hysteresis)

Bi-Stable Multivibrator

Differentiator (Common-Mode Biasing Keeps Input at $+V_{BE}$)

"OR" Gate

"AND" Gate

Difference Integrator

Low Pass Active Filter

Staircase Generator

V_{BE} Biasing

Bandpass Active Filter

Low-Frequency Mixer

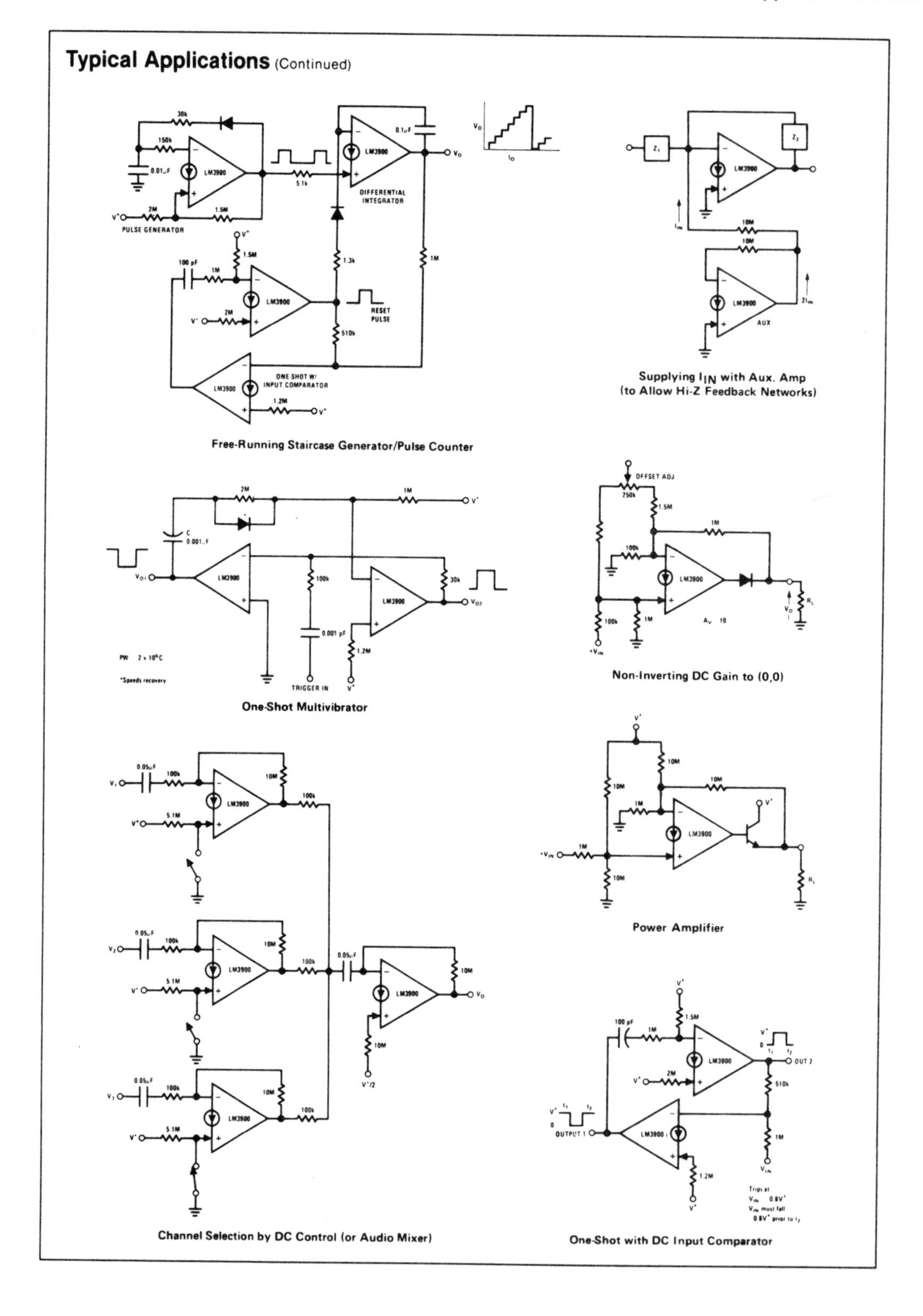

Free-Running Staircase Generator/Pulse Counter

Supplying I_{IN} with Aux. Amp (to Allow Hi-Z Feedback Networks)

One-Shot Multivibrator

Non-Inverting DC Gain to (0,0)

Power Amplifier

Channel Selection by DC Control (or Audio Mixer)

One-Shot with DC Input Comparator

Typical Applications (Continued)

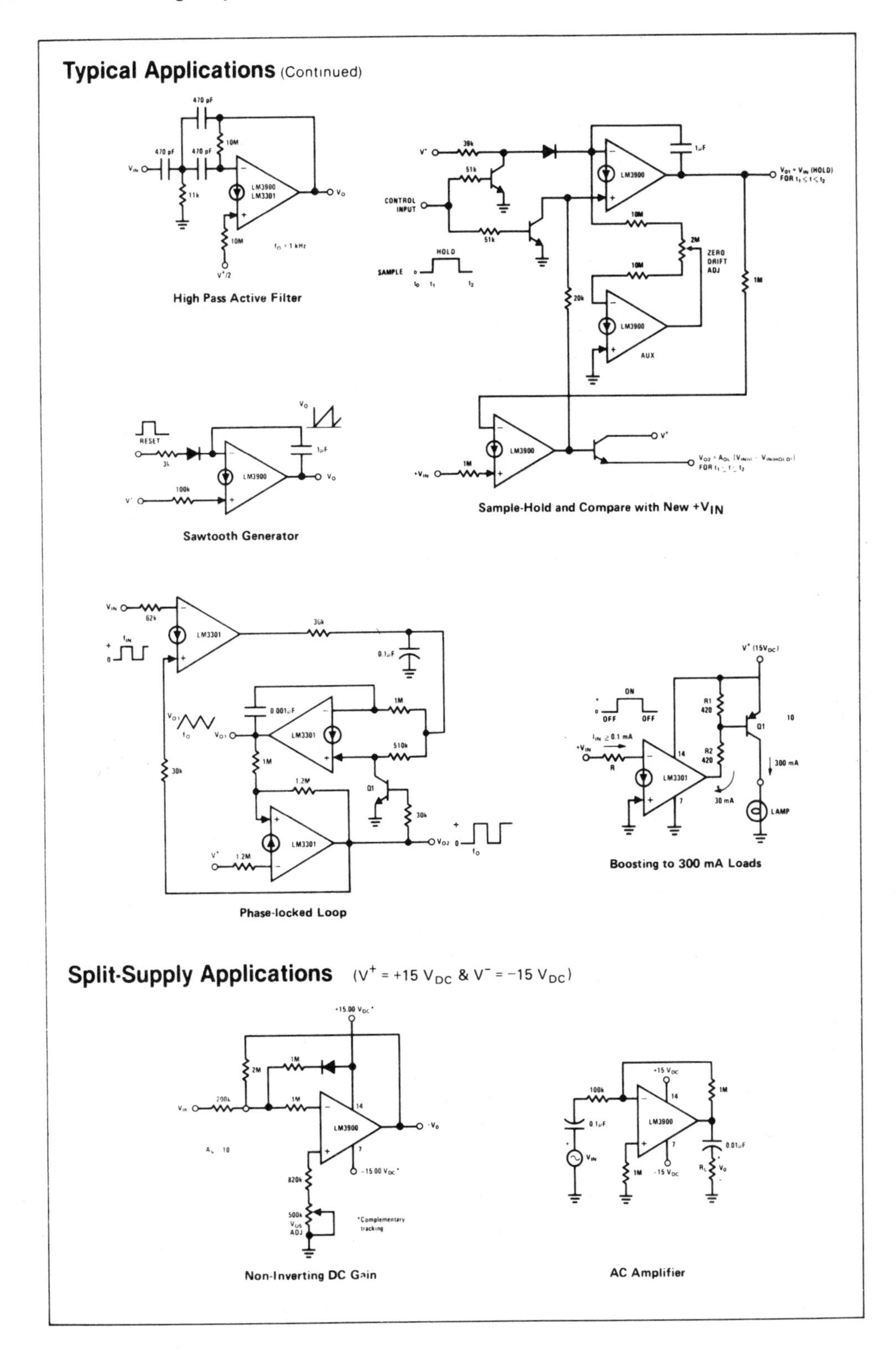

High Pass Active Filter

Sample-Hold and Compare with New $+V_{IN}$

Sawtooth Generator

Phase-locked Loop

Boosting to 300 mA Loads

Split-Supply Applications ($V^+ = +15\ V_{DC}$ & $V^- = -15\ V_{DC}$)

Non-Inverting DC Gain

AC Amplifier

TYPES TIC35, TIC36
P-N-P-N PLANAR EPITAXIAL SILICON REVERSE-BLOCKING TRIODE THYRISTORS

RADIATION-TOLERANT THYRISTORS

400 mA DC • 15 and 30 VOLTS

- Max I_{GT} of 5 mA after 1 x 10^{14} Fast Neutrons/cm^2
- Max V_{TM} of 1.6 V at I_{TM} of 1 A after 1 x 10^{14} Fast Neutrons/cm^2

description

The TIC35, TIC36 thyristors offer a significant advance in radiation-tolerant-device technology. Unique construction techniques produce thyristors which maintain useful characteristics after fast-neutron radiation fluences through 10^{15} n/cm^2.

mechanical data

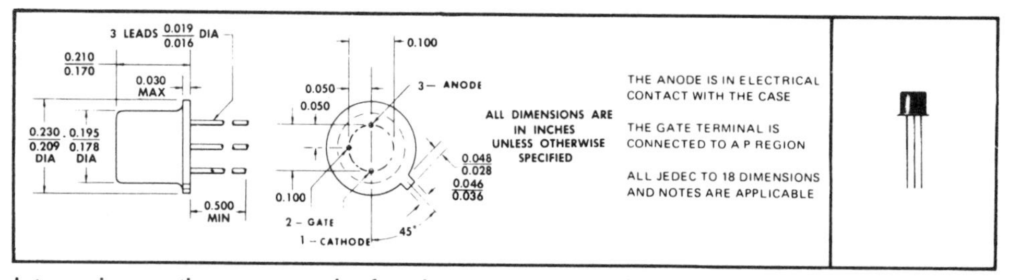

absolute maximum ratings over operating free-air temperature range (unless otherwise noted)

	TIC35	TIC36	UNIT
Continuous Off-State Voltage, V_D (See Note 1)	15	30	V
Repetitive Peak Off-State Voltage, V_{DRM} (See Note 1)	15	30	V
Continuous Reverse Voltage, V_R	5		V
Repetitive Peak Reverse Voltage, V_{RRM}	5		V
Nonrepetitive Peak Reverse Voltage, V_{RSM} (See Note 2)	5		V
Continuous On-State Current at (or below) 55°C Case Temperature (See Note 3)	400		mA
Continuous On-State Current at (or below) 25°C Free-Air Temperature (See Note 4)	225		mA
Average On-State Current (180° Conduction Angle) at (or below) 55°C Case Temperature (See Note 5)	320		mA
Surge On-State Current (See Note 6)	3		A
Peak Negative Gate Voltage	−4		V
Peak Positive Gate Current (Pulse Width ≤300 μs)	250		mA
Peak Gate Power Dissipation (Pulse Width ≤300 μs)	500		mW
Average Gate Power Dissipation	10		mW
Operating Free-Air or Case Temperature Range	−55 to 125		°C
Storage Temperature Range	−65 to 200		°C
Lead Temperature 1/16 Inch from Case for 10 Seconds	260		°C

NOTES:
1. These values apply when the gate-cathode resistance R_{GK} = 1 kΩ.
2. This value applies for a 5-ms rectangular pulse when the device is operating at (or below) rated values of peak reverse voltage and on-state current. Surge may be repeated after the device has returned to original thermal equilibrium.
3. These values apply for continuous d-c operation with resistive load. Above 55°C derate according to Figure 2.
4. These values apply for continuous d-c operation with resistive load. Above 25°C derate according to Figure 3.
5. This value may be applied continuously under single-phase, 60-Hz, half-sine-wave operation with resistive load. Above 55°C derate according to Figure 2.
6. This value applies for one 60-Hz half sine wave when the device is operating at (or below) rated values of peak reverse voltage and on-state current. Surge may be repeated after the device has returned to original thermal equilibrium.

TYPES TIC35, TIC36
P-N-P-N PLANAR EPITAXIAL SILICON REVERSE-BLOCKING TRIODE THYRISTORS

electrical characteristics at 25°C free-air temperature (unless otherwise noted)

PARAMETER		TEST CONDITIONS	MIN	TYP	MAX	UNIT
I_D	Static Off-State Current	V_D = Rated V_D, R_{GK} = 1 kΩ, T_A = 125°C			20	μA
I_R	Static Reverse Current	V_R = 5 V, R_{GK} = 1 kΩ, T_A = 125°C			100	μA
I_{GR}	Gate Reverse Current	V_{KG} = 4 V, I_A = 0			5	μA
I_{GT}	Gate Trigger Current	V_{AA} = 6 V, R_L = 100 Ω, V_{GG} = 6 V, $R_{G(source)} \geqslant$ 10 kΩ, $t_{p(g)} \geqslant$ 20 μs, T_A = −55°C			100	μA
		V_{AA} = 6 V, R_L = 100 Ω, V_{GG} = 6 V, $R_{G(source)} \geqslant$ 10 kΩ, $t_{p(g)} \geqslant$ 20 μs			20	
V_{GT}	Gate Trigger Voltage	V_{AA} = 6 V, R_L = 100 Ω, R_{GK} = 1 kΩ, $t_{p(g)} \geqslant$ 20 μs, T_A = −55°C			0.9	V
		V_{AA} = Rated V_D, R_L = 100 Ω, R_{GK} = 1 kΩ, $t_{p(g)} \geqslant$ 20 μs, T_A = 125°C	0.2			
		V_{AA} = 6 V, R_L = 100 Ω, R_{GK} = 1 kΩ, $t_{p(g)} \geqslant$ 20 μs			0.75	
I_H	Holding Current	V_{AA} = 6 V, R_{GK} = 1 kΩ, Initiating I_T = 10 mA, T_A = −55°C			4	mA
		V_{AA} = 6 V, R_{GK} = 1 kΩ, Initiating I_T = 10 mA			2	
V_{TM}	Peak On-State Voltage	I_{TM} = 1 A, See Note 7			1.6	V
dv/dt	Critical Rate of Rise of Off-State Voltage	V_D = Rated V_D, R_{GK} = 1 kΩ		12		V/μs

post-irradiation electrical characteristics at 25°C free-air temperature

PARAMETER		TEST CONDITIONS	RADIATION FLUENCE†	MIN	TYP	MAX	UNIT
I_{GT}	Gate Trigger Current	V_{AA} = 6 V, R_L = 100 Ω	1×10^{14} n/cm²			5	mA
V_{TM}	Peak On-State Voltage	I_{TM} = 1 A, See Note 7	1×10^{14} n/cm²			1.6	V

† Radiation is fast neutrons (n) at E $\geqslant$ 10 keV (reactor spectrum).

thermal characteristics

PARAMETER		MIN	TYP	MAX	UNIT
$\theta_{J\text{-}C}$	Junction-to-Case Thermal Resistance			124	°C/W
$\theta_{J\text{-}A}$	Junction-to-Free-Air Thermal Resistance			345	°C/W

NOTE: 7. These parameters must be measured using pulse techniques. t_w = 300 μs, duty cycle $\leqslant$ 2%. Voltage-sensing contacts, separate from the current-carrying contacts, are used.

TYPES TIC226B, TIC226D
SILICON BIDIRECTIONAL TRIODE THYRISTORS

8 A RMS • 200 V and 400 V
TRIACS
for
HIGH-TEMPERATURE, HIGH-CURRENT, and HIGH-VOLTAGE APPLICATIONS
• Typ dv/dt of 500 V/μs at 25°C

description

These devices are bidirectional triode thyristors (triacs) which may be triggered from the off-state to the cn-state by either polarity of gate signal with Main Terminal 2 at either polarity.

mechanical data

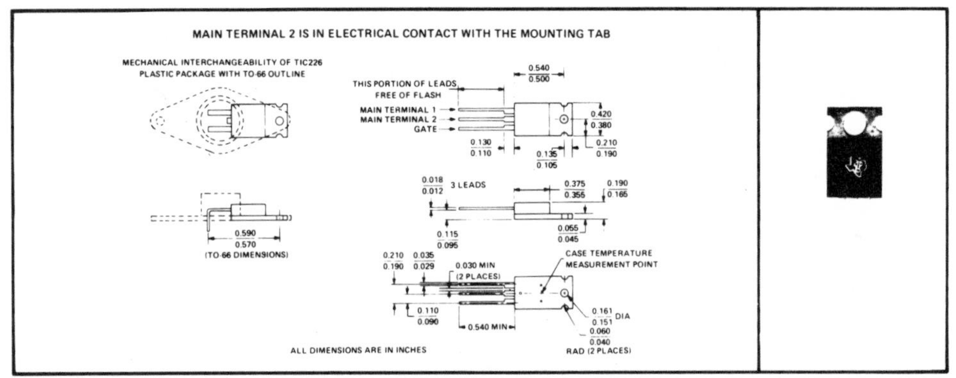

absolute maximum ratings over operating case temperature range (unless otherwise noted)†

			UNIT
Repetitive Peak Off-State Voltage, V_{DRM} (See Note 1)	TIC226B	200	V
	TIC226D	400	V
Full-Cycle RMS On-State Current at (or below) 85°C Case Temperature, $I_{T(RMS)}$ (See Note 2)		8	A
Peak On-State Surge Current, Full-Sine-Wave, I_{TSM} (See Note 3)		70	A
Peak On-State Surge Current, Half-Sine-Wave, I_{TSM} (See Note 4)		80	A
Peak Gate Current, I_{GM}		1	A
Peak Gate Power Dissipation, P_{GM}, at (or below) 85°C Case Temperature (Pulse Width ⩽ 200 μs)		2.2	W
Average Gate Power Dissipation, $P_{G(av)}$, at (or below) 85°C Case Temperature (See Note 5)		0.9	W
Operating Case Temperature Range		–40 to 110	°C
Storage Temperature Range		–40 to 125	°C
Lead Temperature 1/16 Inch from Case for 10 Seconds		230	°C

NOTES: 1. These values apply bidirectionally for any value of resistance between the gate and Main Terminal 1.
2. This value applies for 50-Hz to 60-Hz full-sine-wave operation with resistive load. Above 85°C derate according to Figure 2.
3. This value applies for one 60-Hz full sine wave when the device is operating at (or below) the rated value of on-state current. Surge may be repeated after the device has returned to original thermal equilibrium. During the surge, gate control may be lost.
4. This value applies for one 60-Hz half sine wave when the device is operating at (or below) the rated value of on-state current. Surge may be repeated after the device has returned to original thermal equilibrium. During the surge, gate control may be lost.
5. This value applies for a maximum averaging time of 16.6 ms.

†All voltage values are with respect to Main Terminal 1.

TYPES TIC226B, TIC226D
SILICON BIDIRECTIONAL TRIODE THYRISTORS

electrical characteristics at 25°C case temperature (unless otherwise noted)†

PARAMETER		TEST CONDITIONS	MIN	TYP	MAX	UNIT
I_{DRM}	Repetitive Peak Off-State Current	V_{DRM} = Rated V_{DRM}, $I_G = 0$, $T_C = 110°C$			±2	mA
I_{GTM}	Peak Gate Trigger Current	V_{supply} = +12 V†, $R_L = 10\ \Omega$, $t_{p(g)} \geq 20\ \mu s$		15	50	mA
		V_{supply} = +12 V†, $R_L = 10\ \Omega$, $t_{p(g)} \geq 20\ \mu s$		–25	–50	
		V_{supply} = –12 V†, $R_L = 10\ \Omega$, $t_{p(g)} \geq 20\ \mu s$		–30	–50	
		V_{supply} = –12 V†, $R_L = 10\ \Omega$, $t_{p(g)} \geq 20\ \mu s$		75		
V_{GTM}	Peak Gate Trigger Voltage	V_{supply} = +12 V†, $R_L = 10\ \Omega$, $t_{p(g)} \geq 20\ \mu s$		0.9	2.5	V
		V_{supply} = +12 V†, $R_L = 10\ \Omega$, $t_{p(g)} \geq 20\ \mu s$		–1.2	–2.5	
		V_{supply} = –12 V†, $R_L = 10\ \Omega$, $t_{p(g)} \geq 20\ \mu s$		–1.2	–2.5	
		V_{supply} = –12 V†, $R_L = 10\ \Omega$, $t_{p(g)} \geq 20\ \mu s$		1.2		
V_{TM}	Peak On-State Voltage	I_{TM} = ±12 A, I_G = 100 mA, See Note 6			±2.1	V
I_H	Holding Current	V_{supply} = +12 V†, $I_G = 0$, Initiating I_{TM} = 500 mA		20	60	mA
		V_{supply} = –12 V†, $I_G = 0$, Initiating I_{TM} = –500 mA		–30	–60	
I_L	Latching Current	V_{supply} = +12 V†, See Note 7		30	70	mA
		V_{supply} = –12 V†, See Note 7		–40	–70	
dv/dt	Critical Rate of Rise of Off-State Voltage	V_{DRM} = Rated V_{DRM}, $I_G = 0$, $T_C = 110°C$		500		V/μs
dv/dt	Critical Rate of Rise of Commutation Voltage	V_{DRM} = Rated V_{DRM}, I_{TRM} = ±12 A, $T_C = 85°C$, See Figure 3	5			V/μs

†All voltage values are with respect to Main Terminal 1.

NOTES: 6. This parameter must be measured using pulse techniques. $t_w \leq 1$ ms, duty cycle $\leq 2\%$. Voltage-sensing contacts, separate from the current-carrying contacts, are located within 0.125 inch from the device body.

7. The triacs are triggered by a 15-V (open-circuit amplitude) pulse supplied by a generator with the following characteristics: $R_G = 100\ \Omega$, $t_w = 20\ \mu s$, $t_r \leq 15$ ns, $t_f \leq 15$ ns, f = 1 kHz.

thermal characteristics

PARAMETER		MAX	UNIT
$R_{\theta JC}$	Junction-to-Case Thermal Resistance	1.8	°C/W
$R_{\theta JA}$	Junction-to-Free-Air Thermal Resistance	62.5	

Industrial Blocks

LM135/LM235/LM335, LM135A/LM235A/LM335A Precision Temperature Sensors

General Description

The LM135 series are precision, easily-calibrated, integrated circuit temperature sensors. Operating as a 2-terminal zener, the LM135 has a breakdown voltage directly proportional to absolute temperature at +10 mV/°K. With less than 1Ω dynamic impedance the device operates over a current range of 400 μA to 5 mA with virtually no change in performance. When calibrated at 25°C the LM135 has typically less than 1°C error over a 100°C temperature range. Unlike other sensors the LM135 has a linear output.

Applications for the LM135 include almost any type of temperature sensing over a −55°C to +150°C temperature range. The low impedance and linear output make interfacing to readout or control circuitry especially easy.

The LM135 operates over a −55°C to +150°C temperature range while the LM235 operates over a −40°C to +125°C temperature range. The LM335 operates from −40°C to +100°C. The LM135/LM235/LM335 are available packaged in hermetic TO-46 transistor packages while the LM335 is also available in plastic TO-92 packages.

Features

- Directly calibrated in °Kelvin
- 1°C initial accuracy available
- Operates from 400 μA to 5 mA
- Less than 1Ω dynamic impedance
- Easily calibrated
- Wide operating temperature range
- 200°C overrange
- Low cost

Schematic Diagram

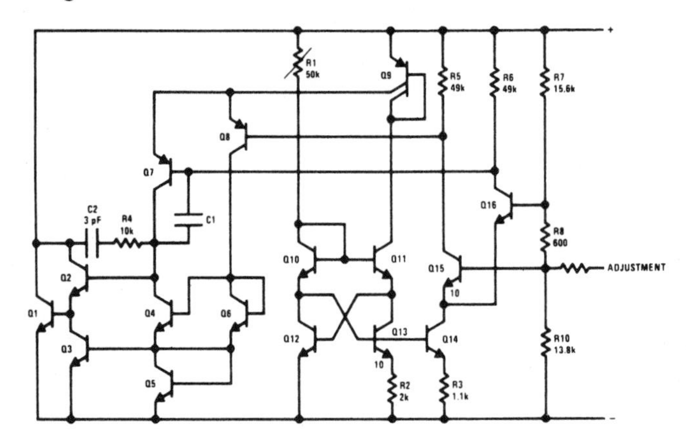

Typical Applications

Basic Temperature Sensor

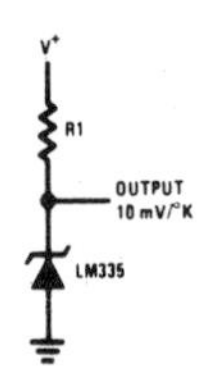

Calibrated Sensor

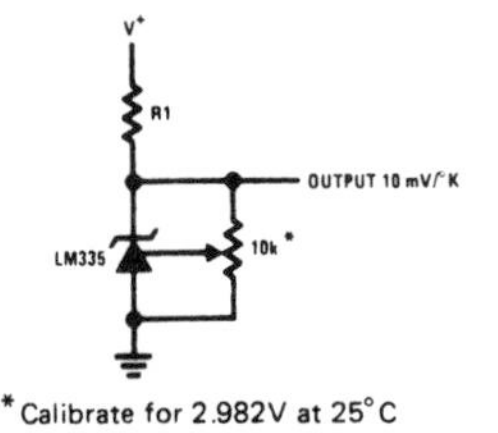

*Calibrate for 2.982V at 25°C

Wide Operating Supply

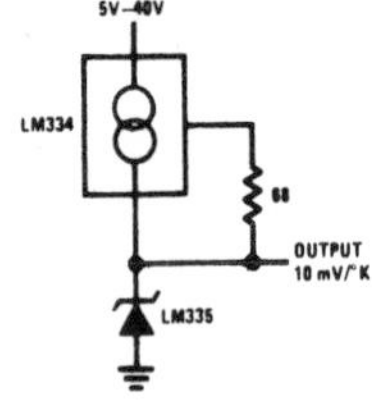

Absolute Maximum Ratings

Reverse Current		15 mA
Forward Current		10 mA
Storage Temperature		
TO-46 Package		−60°C to +180°C
TO-92 Package		−60°C to +150°C
Specified Operating Temperature Range		
	Continuous	**Intermittent (Note 2)**
LM135, LM135A	−55°C to +150°C	150°C to 200°C
LM235, LM235A	−40°C to +125°C	125°C to 150°C
LM335, LM335A	−40°C to +100°C	100°C to 125°C
Lead Temperature (Soldering, 10 seconds)		300°C

Temperature Accuracy LM135/LM235, LM135A/LM235A (Note 1)

PARAMETER	CONDITIONS	LM135A/LM235A			LM135/LM235			UNITS
		MIN	TYP	MAX	MIN	TYP	MAX	
Operating Output Voltage	$T_C = 25°C$, $I_R = 1$ mA	2.97	2.98	2.99	2.95	2.98	3.01	V
Uncalibrated Temperature Error	$T_C = 25°C$, $I_R = 1$ mA		0.5	1		1	3	°C
Uncalibrated Temperature Error	$T_{MIN} < T_C < T_{MAX}$, $I_R = 1$ mA		1.3	2.7		2	5	°C
Temperature Error with 25°C Calibration	$T_{MIN} < T_C < T_{MAX}$, $I_R = 1$ mA		0.3	1		0.5	1.5	°C
Calibrated Error at Extended Temperatures	$T_C = T_{MAX}$ (Intermittent)		2			2		°C
Non-Linearity	$I_R = 1$ mA		0.3	0.5		0.3	1	°C

Temperature Accuracy LM335, LM335A (Note 1)

PARAMETER	CONDITIONS	LM335A			LM335			UNITS
		MIN	TYP	MAX	MIN	TYP	MAX	
Operating Output Voltage	$T_C = 25°C$, $I_R = 1$ mA	2.95	2.98	3.01	2.92	2.98	3.04	V
Uncalibrated Temperature Error	$T_C = 25°C$, $I_R = 1$ mA		1	3		2	6	°C
Uncalibrated Temperature Error	$T_{MIN} < T_C < T_{MAX}$, $I_R = 1$ mA		2	5		4	9	°C
Temperature Error with 25°C Calibration	$T_{MIN} < T_C < T_{MAX}$, $I_R = 1$ mA		0.5	1		1	2	°C
Calibrated Error at Extended Temperatures	$T_C = T_{MAX}$ (Intermittent)		2			2		°C
Non-Linearity	$I_R = 1$ mA		0.3	1.5		0.3	1.5	°C

Electrical Characteristics (Note 1)

PARAMETER	CONDITIONS	LM135/LM235 LM135A/LM235A			LM335 LM335A			UNITS
		MIN	TYP	MAX	MIN	TYP	MAX	
Operating Output Voltage Change with Current	400 μA $< I_R <$ 5 mA At Constant Temperature		2.5	10		3	14	mV
Dynamic Impedance	$I_R = 1$ mA		0.5			0.6		Ω
Output Voltage Temperature Drift			+10			+10		mV/°C
Time Constant	Still Air		80			80		sec
	100 ft/Min Air		10			10		sec
	Stirred Oil		1			1		sec
Time Stability	$T_C = 125°C$		0.2			0.2		°C/khr

Note 1: Accuracy measurements are made in a well-stirred oil bath. For other conditions, self heating must be considered.

Note 2: Continuous operation at these temperatures for 10,000 hours for H package and 5,000 hours for Z package may decrease life expectancy of the device.

Typical Performance Characteristics

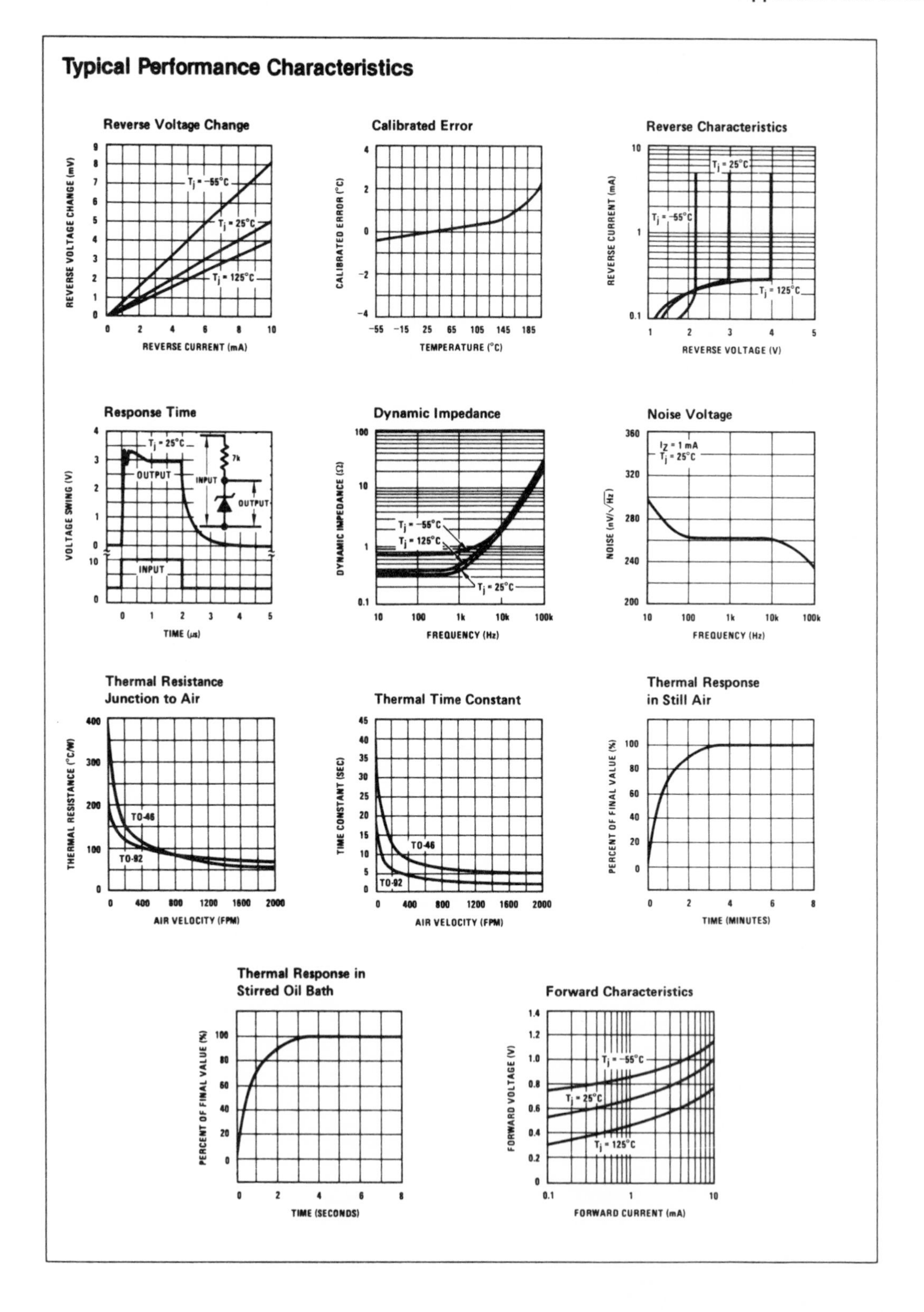

Industrial Blocks

LM555/LM555C Timer

General Description

The LM555 is a highly stable device for generating accurate time delays or oscillation. Additional terminals are provided for triggering or resetting if desired. In the time delay mode of operation, the time is precisely controlled by one external resistor and capacitor. For astable operation as an oscillator, the free running frequency and duty cycle are accurately controlled with two external resistors and one capacitor. The circuit may be triggered and reset on falling waveforms, and the output circuit can source or sink up to 200 mA or drive TTL circuits.

Features

- Direct replacement for SE555/NE555
- Timing from microseconds through hours
- Operates in both astable and monostable modes
- Adjustable duty cycle
- Output can source or sink 200 mA
- Output and supply TTL compatible
- Temperature stability better than 0.005% per °C
- Normally on and normally off output

Applications

- Precision timing
- Pulse generation
- Sequential timing
- Time delay generation
- Pulse width modulation
- Pulse position modulation
- Linear ramp generator

Schematic Diagram

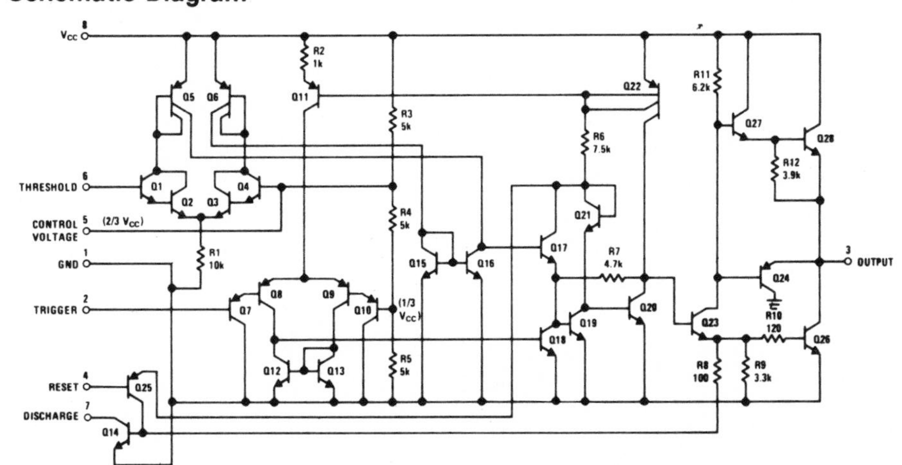

Connection Diagrams

Metal Can Package

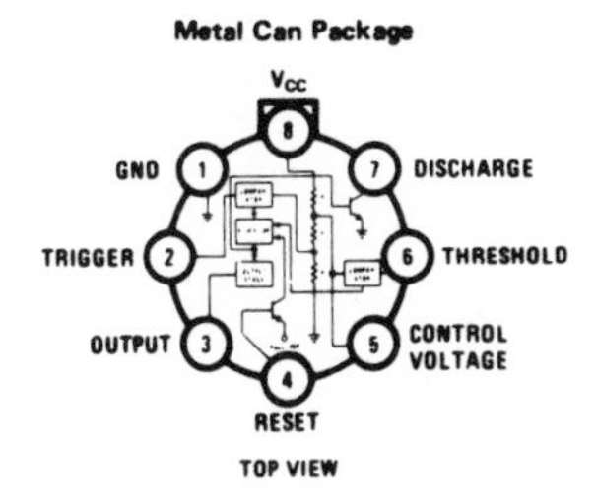

TOP VIEW

Order Number LM555H, LM555CH
See NS Package H08C

Dual-In-Line Package

GND 1 — 8 +V_{CC}
TRIGGER 2 — 7 DISCHARGE
OUTPUT 3 — 6 THRESHOLD
RESET 4 — 5 CONTROL VOLTAGE

TOP VIEW

Order Number LM555CN
See NS Package N08B
Order Number LM555J or LM555CJ
See NS Package J08A

Absolute Maximum Ratings

Supply Voltage	+18V
Power Dissipation (Note 1)	600 mW
Operating Temperature Ranges	
LM555C	0°C to +70°C
LM555	–55°C to +125°C
Storage Temperature Range	–65°C to +150°C
Lead Temperature (Soldering, 10 seconds)	300°C

Electrical Characteristics (T_A = 25°C, V_{CC} = +5V to +15V, unless otherwise specified)

PARAMETER	CONDITIONS	LIMITS						UNITS
		LM555			LM555C			
		MIN	TYP	MAX	MIN	TYP	MAX	
Supply Voltage		4.5		18	4.5		16	V
Supply Current	V_{CC} = 5V, R_L = ∞		3	5		3	6	mA
	V_{CC} = 15V, R_L = ∞ (Low State) (Note 2)		10	12		10	15	mA
Timing Error, Monostable								
Initial Accuracy			0.5			1		%
Drift with Temperature	R_A, R_B = 1k to 100 k, C = 0.1μF, (Note 3)		30			50		ppm/°C
Accuracy over Temperature			1.5			1.5		%
Drift with Supply			0.05			0.1		%/V
Timing Error, Astable								
Initial Accuracy			1.5			2.25		%
Drift with Temperature			90			150		ppm/°C
Accuracy over Temperature			2.5			3.0		%
Drift with Supply			0.15			0.30		%/V
Threshold Voltage			0.667			0.667		x V_{CC}
Trigger Voltage	V_{CC} = 15V	4.8	5	5.2		5		V
	V_{CC} = 5V	1.45	1.67	1.9		1.67		V
Trigger Current			0.01	0.5		0.5	0.9	μA
Reset Voltage		0.4	0.5	1	0.4	0.5	1	V
Reset Current			0.1	0.4		0.1	0.4	mA
Threshold Current	(Note 4)		0.1	0.25		0.1	0.25	μA
Control Voltage Level	V_{CC} = 15V	9.6	10	10.4	9	10	11	V
	V_{CC} = 5V	2.9	3.33	3.8	2.6	3.33	4	V
Pin 7 Leakage Output High			1	100		1	100	nA
Pin 7 Sat (Note 5)								
Output Low	V_{CC} = 15V, I_7 = 15 mA		150			180		mV
Output Low	V_{CC} = 4.5V, I_7 = 4.5 mA		70	100		80	200	mV
Output Voltage Drop (Low)	V_{CC} = 15V							
	I_{SINK} = 10 mA		0.1	0.15		0.1	0.25	V
	I_{SINK} = 50 mA		0.4	0.5		0.4	0.75	V
	I_{SINK} = 100 mA		2	2.2		2	2.5	V
	I_{SINK} = 200 mA		2.5			2.5		V
	V_{CC} = 5V							
	I_{SINK} = 8 mA		0.1	0.25				V
	I_{SINK} = 5 mA					0.25	0.35	V
Output Voltage Drop (High)	I_{SOURCE} = 200 mA, V_{CC} = 15V		12.5			12.5		V
	I_{SOURCE} = 100 mA, V_{CC} = 15V	13	13.3		12.75	13.3		V
	V_{CC} = 5V	3	3.3		2.75	3.3		V
Rise Time of Output			100			100		ns
Fall Time of Output			100			100		ns

Note 1: For operating at elevated temperatures the device must be derated based on a +150°C maximum junction temperature and a thermal resistance of +45°C/W junction to case for TO-5 and +150°C/W junction to ambient for both packages.

Note 2: Supply current when output high typically 1 mA less at V_{CC} = 5V.

Note 3: Tested at V_{CC} = 5V and V_{CC} = 15V.

Note 4: This will determine the maximum value of $R_A + R_B$ for 15V operation. The maximum total ($R_A + R_B$) is 20 MΩ.

Note 5: No protection against excessive pin 7 current is necessary providing the package dissipation rating will not be exceeded.

Typical Performance Characteristics

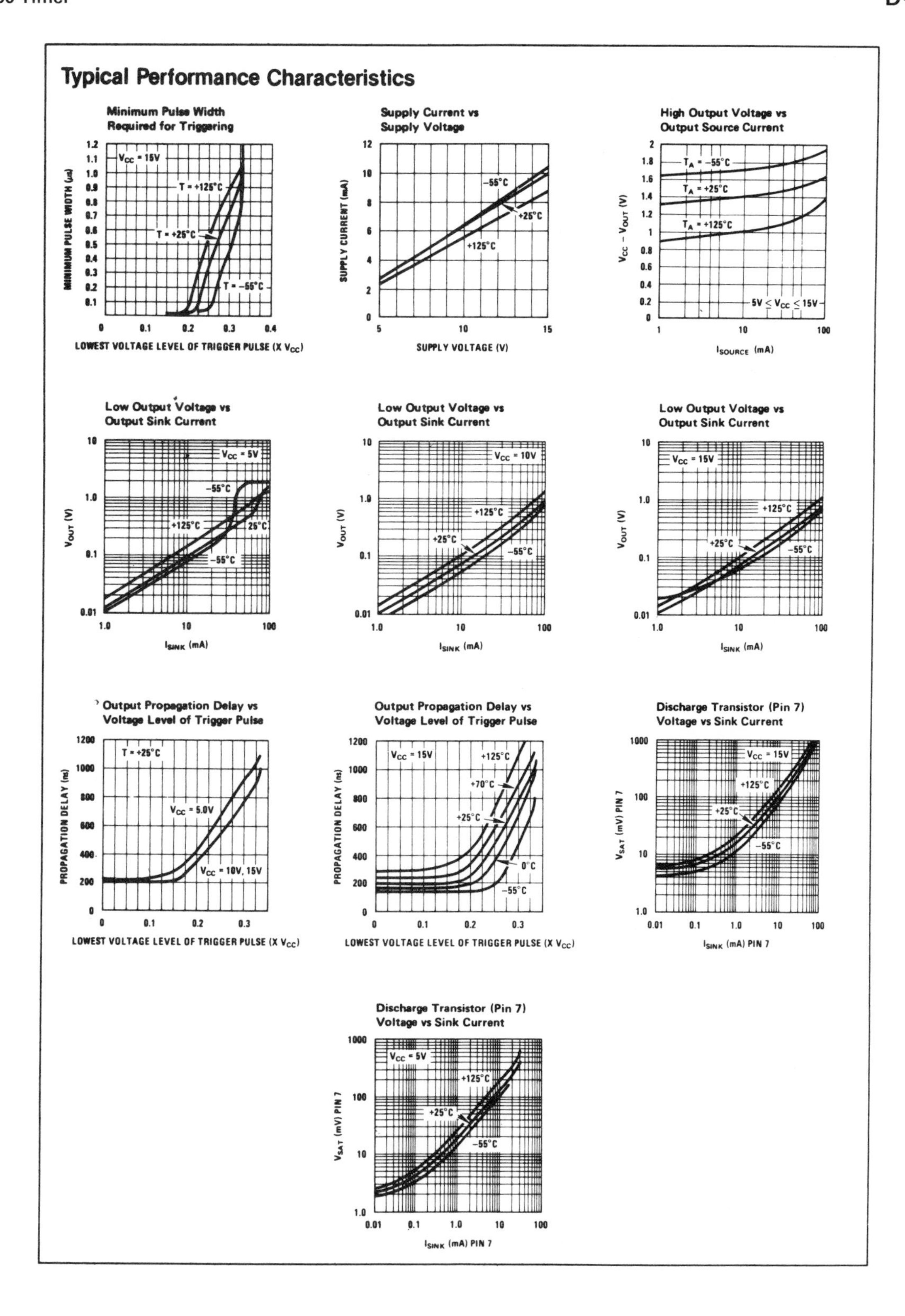

Applications Information

MONOSTABLE OPERATION

In this mode of operation, the timer functions as a one-shot (*Figure 1*). The external capacitor is initially held discharged by a transistor inside the timer. Upon application of a negative trigger pulse of less than 1/3 V_{CC} to pin 2, the flip-flop is set which both releases the short circuit across the capacitor and drives the output high.

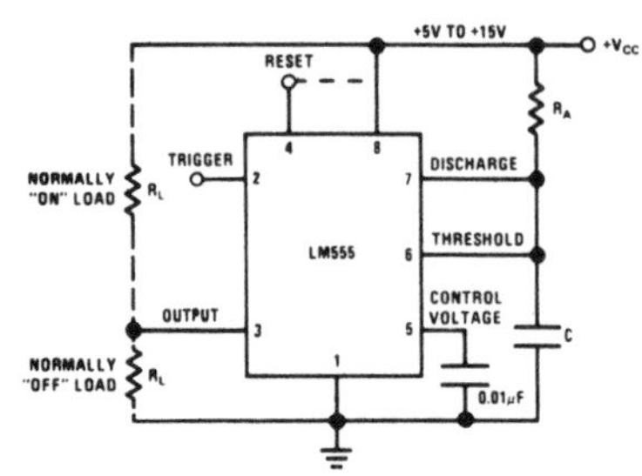

FIGURE 1. Monostable

The voltage across the capacitor then increases exponentially for a period of $t = 1.1\ R_A C$, at the end of which time the voltage equals 2/3 V_{CC}. The comparator then resets the flip-flop which in turn discharges the capacitor and drives the output to its low state. *Figure 2* shows the waveforms generated in this mode of operation. Since the charge and the threshold level of the comparator are both directly proportional to supply voltage, the timing internal is independent of supply.

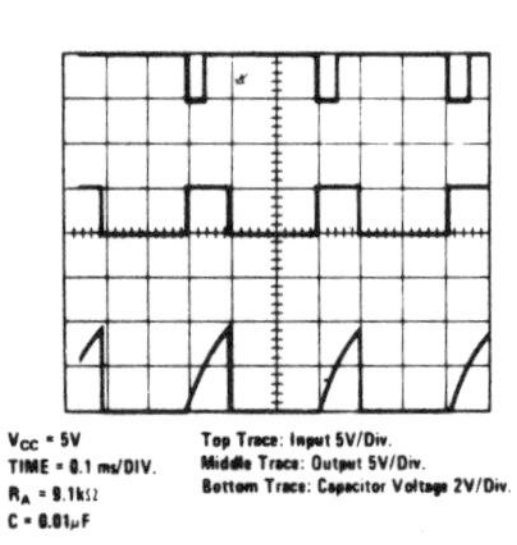

FIGURE 2. Monostable Waveforms

During the timing cycle when the output is high, the further application of a trigger pulse will not effect the circuit. However the circuit can be reset during this time by the application of a negative pulse to the reset terminal (pin 4). The output will then remain in the low state until a trigger pulse is again applied.

When the reset function is not in use, it is recommended that it be connected to V_{CC} to avoid any possibility of false triggering.

Figure 3 is a nomograph for easy determination of R, C values for various time delays.

NOTE: In monostable operation, the trigger should be driven high before the end of timing cycle.

ASTABLE OPERATION

If the circuit is connected as shown in *Figure 4* (pins 2 and 6 connected) it will trigger itself and free run as a multivibrator. The external capacitor charges through $R_A + R_B$ and discharges through R_B. Thus the duty cycle may be precisely set by the ratio of these two resistors.

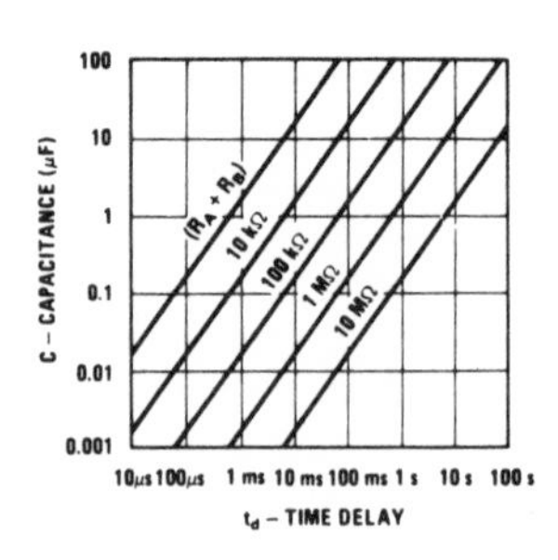

FIGURE 3. Time Delay

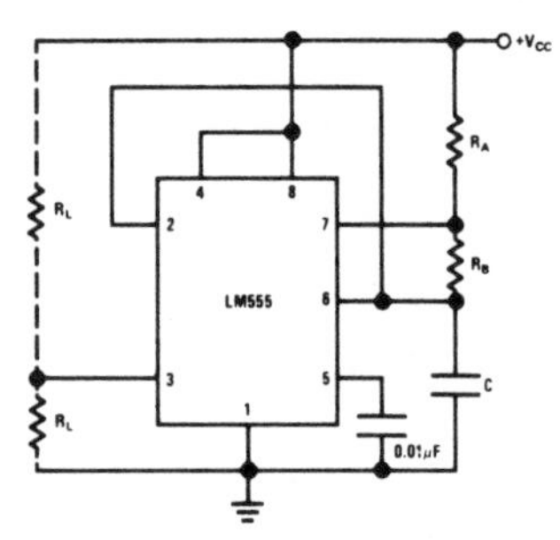

FIGURE 4. Astable

In this mode of operation, the capacitor charges and discharges between 1/3 V_{CC} and 2/3 V_{CC}. As in the triggered mode, the charge and discharge times, and therefore the frequency are independent of the supply voltage.

Figure 5 shows the waveforms generated in this mode of operation.

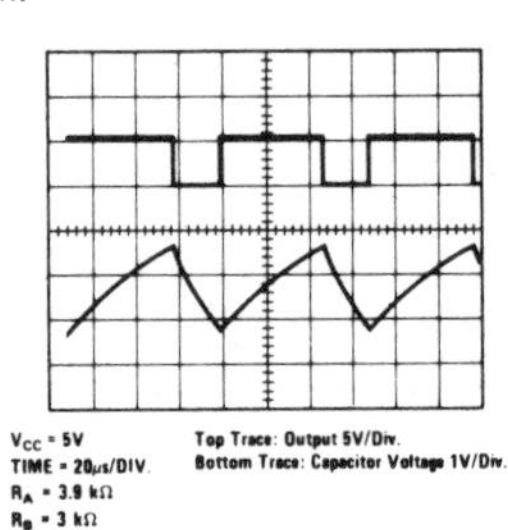

FIGURE 5. Astable Waveforms

The charge time (output high) is given by:

$$t_1 = 0.693\ (R_A + R_B)\ C$$

And the discharge time (output low) by:

$$t_2 = 0.693\ (R_B)\ C$$

Thus the total period is:

$$T = t_1 + t_2 = 0.693\ (R_A + 2R_B)\ C$$

Applications Information (Continued)

The frequency of oscillation is:

$$f = \frac{1}{T} = \frac{1.44}{(R_A + 2\,R_B)\,C}$$

Figure 6 may be used for quick determination of these RC values.

The duty cycle is: $D = \dfrac{R_B}{R_A + 2R_B}$

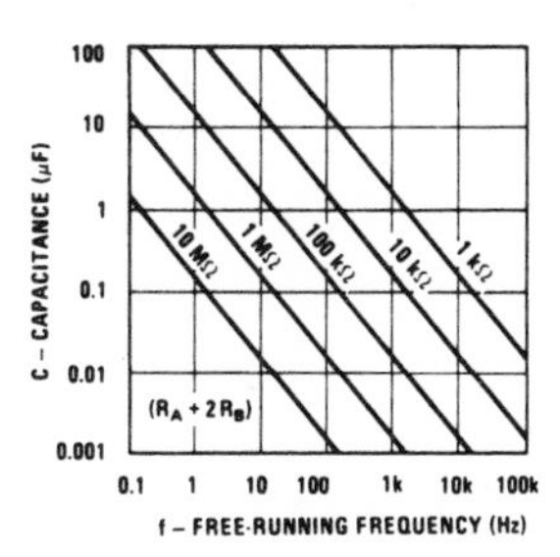

FIGURE 6. Free Running Frequency

FREQUENCY DIVIDER

The monostable circuit of *Figure 1* can be used as a frequency divider by adjusting the length of the timing cycle. *Figure 7* shows the waveforms generated in a divide by three circuit.

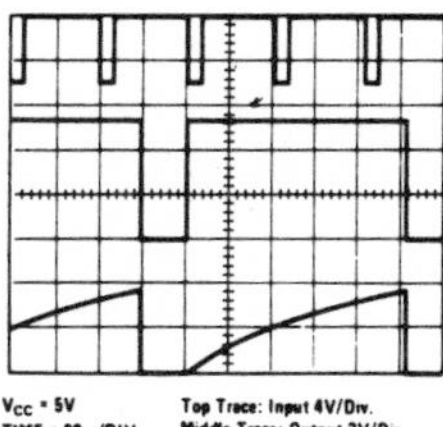

FIGURE 7. Frequency Divider

PULSE WIDTH MODULATOR

When the timer is connected in the monostable mode and triggered with a continuous pulse train, the output pulse width can be modulated by a signal applied to pin 5. *Figure 8* shows the circuit, and in *Figure 9* are some waveform examples.

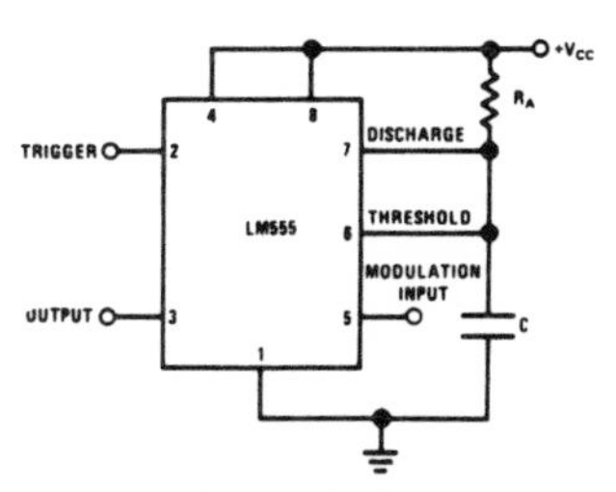

FIGURE 8. Pulse Width Modulator

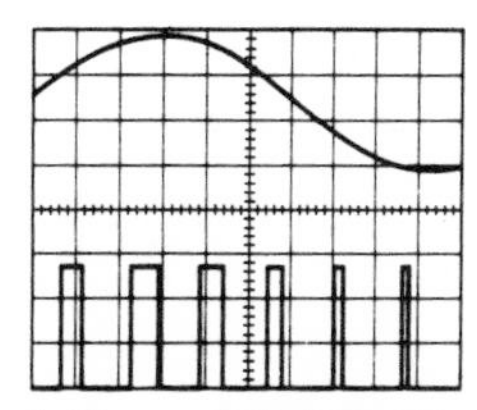

FIGURE 9. Pulse Width Modulator

PULSE POSITION MODULATOR

This application uses the timer connected for astable operation, as in *Figure 10*, with a modulating signal again applied to the control voltage terminal. The pulse position varies with the modulating signal, since the threshold voltage and hence the time delay is varied. *Figure 11* shows the waveforms generated for a triangle wave modulation signal.

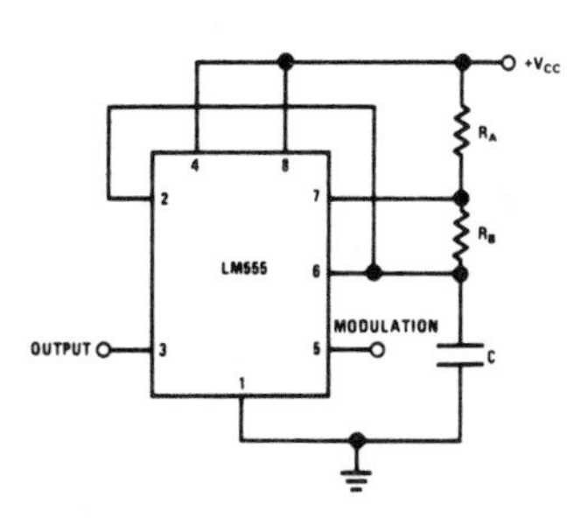

FIGURE 10. Pulse Position Modulator

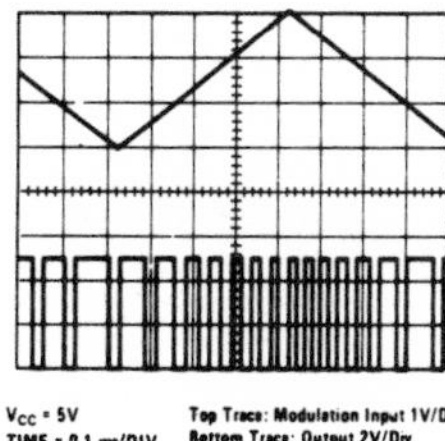

FIGURE 11. Pulse Position Modulator

LINEAR RAMP

When the pullup resistor, R_A, in the monostable circuit is replaced by a constant current source, a linear ramp is

Applications Information (Continued)

generated. *Figure 12* shows a circuit configuration that will perform this function.

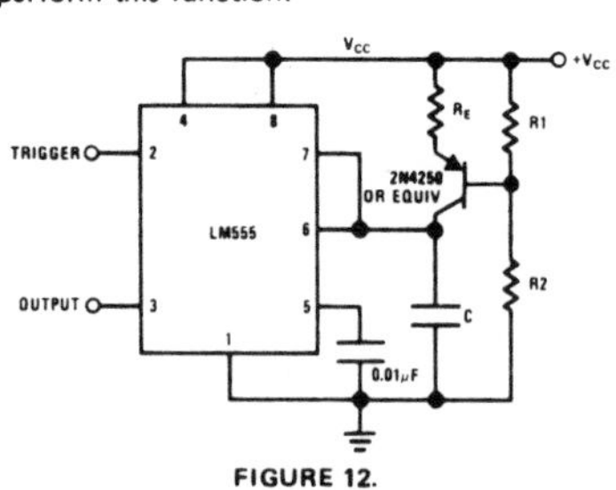

FIGURE 12.

Figure 13 shows waveforms generated by the linear ramp.

The time interval is given by:

$$T = \frac{2/3\ V_{CC}\ R_E\ (R_1 + R_2)\ C}{R_1\ V_{CC} - V_{BE}\ (R_1 + R_2)}$$

$$V_{BE} \simeq 0.6V$$

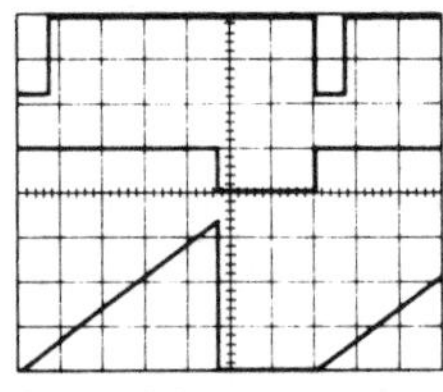

$V_{CC} = 5V$
TIME = 20μs/DIV.
$R_1 = 47\ k\Omega$
$R_2 = 100\ k\Omega$
$R_E = 2.7\ k\Omega$
$C = 0.01\mu F$

Top Trace: Input 3V/Div.
Middle Trace: Output 5V/Div.
Bottom Trace: Capacitor Voltage 1V/Div.

FIGURE 13. Linear Ramp

50% DUTY CYCLE OSCILLATOR

For a 50% duty cycle, the resistors R_A and R_B may be connected as in *Figure 14*. The time period for the output high is the same as previous, $t_1 = 0.693\ R_A\ C$. For the output low it is t_2 =

$$[(R_A\ R_B)/(R_A + R_B)]\ C\ Ln \left[\frac{R_B - 2R_A}{2R_B - R_A}\right]$$

Thus the frequency of oscillation is $f = \frac{1}{t_1 + t_2}$

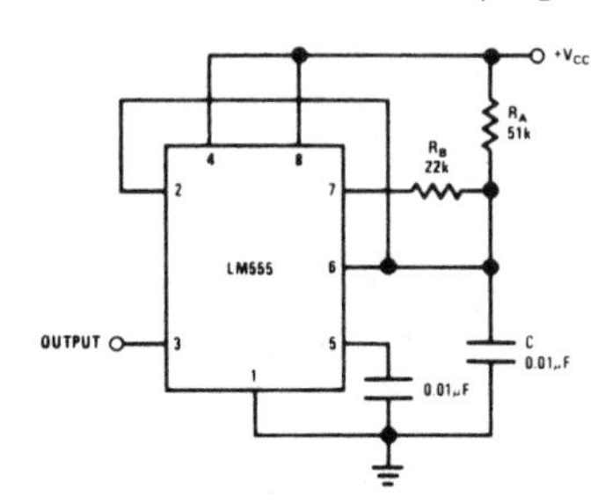

FIGURE 14. 50% Duty Cycle Oscillator

Note that this circuit will not oscillate if R_B is greater than 1/2 R_A because the junction of R_A and R_B cannot bring pin 2 down to 1/3 V_{CC} and trigger the lower comparator.

ADDITIONAL INFORMATION

Adequate power supply bypassing is necessary to protect associated circuitry. Minimum recommended is 0.1μF in parallel with 1μF electrolytic.

Lower comparator storage time can be as long as 10μs when pin 2 is driven fully to ground fpr triggering. This limits the monostable pulse width to 10μs minimum.

Delay time reset to output is 0.47μs typical. Minimum reset pulse width must be 0.3μs, typical.

Pin 7 current switches within 30 ns of the output (pin 3) voltage.

Industrial Blocks

LM565/LM565C Phase Locked Loop

General Description

The LM565 and LM565C are general purpose phase locked loops containing a stable, highly linear voltage controlled oscillator for low distortion FM demodulation, and a double balanced phase detector with good carrier suppression. The VCO frequency is set with an external resistor and capacitor, and a tuning range of 10:1 can be obtained with the same capacitor. The characteristics of the closed loop system—bandwidth, response speed, capture and pull in range—may be adjusted over a wide range with an external resistor and capacitor. The loop may be broken between the VCO and the phase detector for insertion of a digital frequency divider to obtain frequency multiplication.

The LM565H is specified for operation over the −55°C to +125°C military temperature range. The LM565CH and LM565CN are specified for operation over the 0°C to +70°C temperature range.

Features

- 200 ppm/°C frequency stability of the VCO
- Power supply range of ±5 to ±12 volts with 100 ppm/% typical
- 0.2% linearity of demodulated output
- Linear triangle wave with in phase zero crossings available
- TTL and DTL compatible phase detector input and square wave output
- Adjustable hold in range from ±1% to > ±60%.

Applications

- Data and tape synchronization
- Modems
- FSK demodulation
- FM demodulation
- Frequency synthesizer
- Tone decoding
- Frequency multiplication and division
- SCA demodulators
- Telemetry receivers
- Signal regeneration
- Coherent demodulators.

Schematic and Connection Diagrams

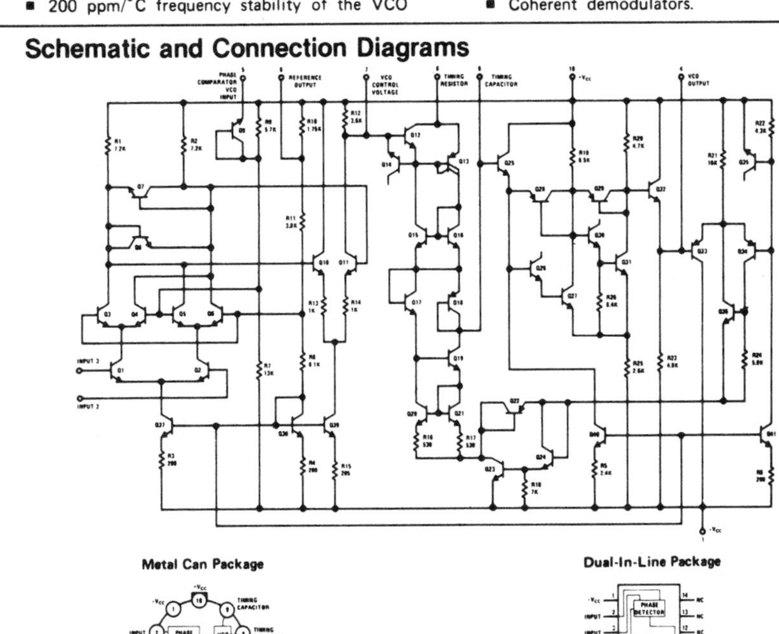

Metal Can Package

TOP VIEW

Order Number LM565H or LM565CH
See NS Package H10C

Dual-In-Line Package

TOP VIEW

Order Number LM565CN
See NS Package N14A

Absolute Maximum Ratings

Supply Voltage		±12V
Power Dissipation (Note 1)		300 mW
Differential Input Voltage		±1V
Operating Temperature Range	LM565H	−55°C to +125°C
	LM565CH, LM565CN	0°C to 70°C
Storage Temperature Range		−65°C to +150°C
Lead Temperature (Soldering, 10 sec)		300°C

Electrical Characteristics (AC Test Circuit, T_A = 25°C, V_C = ±6V)

PARAMETER	CONDITIONS	LM565			LM565C			UNITS
		MIN	TYP	MAX	MIN	TYP	MAX	
Power Supply Current			8.0	12.5		8.0	12.5	mA
Input Impedance (Pins 2, 3)	$-4V < V_2, V_3 < 0V$	7	10			5		kΩ
VCO Maximum Operating Frequency	C_o = 2.7 pF	300	500		250	500		kHz
Operating Frequency Temperature Coefficient			−100	300		−200	500	ppm/°C
Frequency Drift with Supply Voltage			0.01	0.1		0.05	0.2	%/V
Triangle Wave Output Voltage		2	2.4	3	2	2.4	3	$V_{p\text{-}p}$
Triangle Wave Output Linearity			0.2	0.75		0.5	1	%
Square Wave Output Level		4.7	5.4		4.7	5.4		$V_{p\text{-}p}$
Output Impedance (Pin 4)			5			5		kΩ
Square Wave Duty Cycle		45	50	55	40	50	60	%
Square Wave Rise Time			20	100		20		ns
Square Wave Fall Time			50	200		50		ns
Output Current Sink (Pin 4)		0.6	1		0.6	1		mA
VCO Sensitivity	f_o = 10 kHz	6400	6600	6800	6000	6600	7200	Hz/V
Demodulated Output Voltage (Pin 7)	±10% Frequency Deviation	250	300	350	200	300	400	mV_{pp}
Total Harmonic Distortion	±10% Frequency Deviation		0.2	0.75		0.2	1.5	%
Output Impedance (Pin 7)			3.5			3.5		kΩ
DC Level (Pin 7)		4.25	4.5	4.75	4.0	4.5	5.0	V
Output Offset Voltage $\|V_7 - V_6\|$			30	100		50	200	mV
Temperature Drift of $\|V_7 - V_6\|$			500			500		μV/°C
AM Rejection		30	40			40		dB
Phase Detector Sensitivity K_D		0.6	.68	0.9	0.55	.68	0.95	V/radian

Note 1: The maximum junction temperature of the LM565 is 150°C, while that of the LM565C and LM565CN is 100°C. For operation at elevated temperatures, devices in the TO-5 package must be derated based on a thermal resistance of 150°C/W junction to ambient or 45°C/W junction to case. Thermal resistance of the dual-in-line package is 100°C/W.

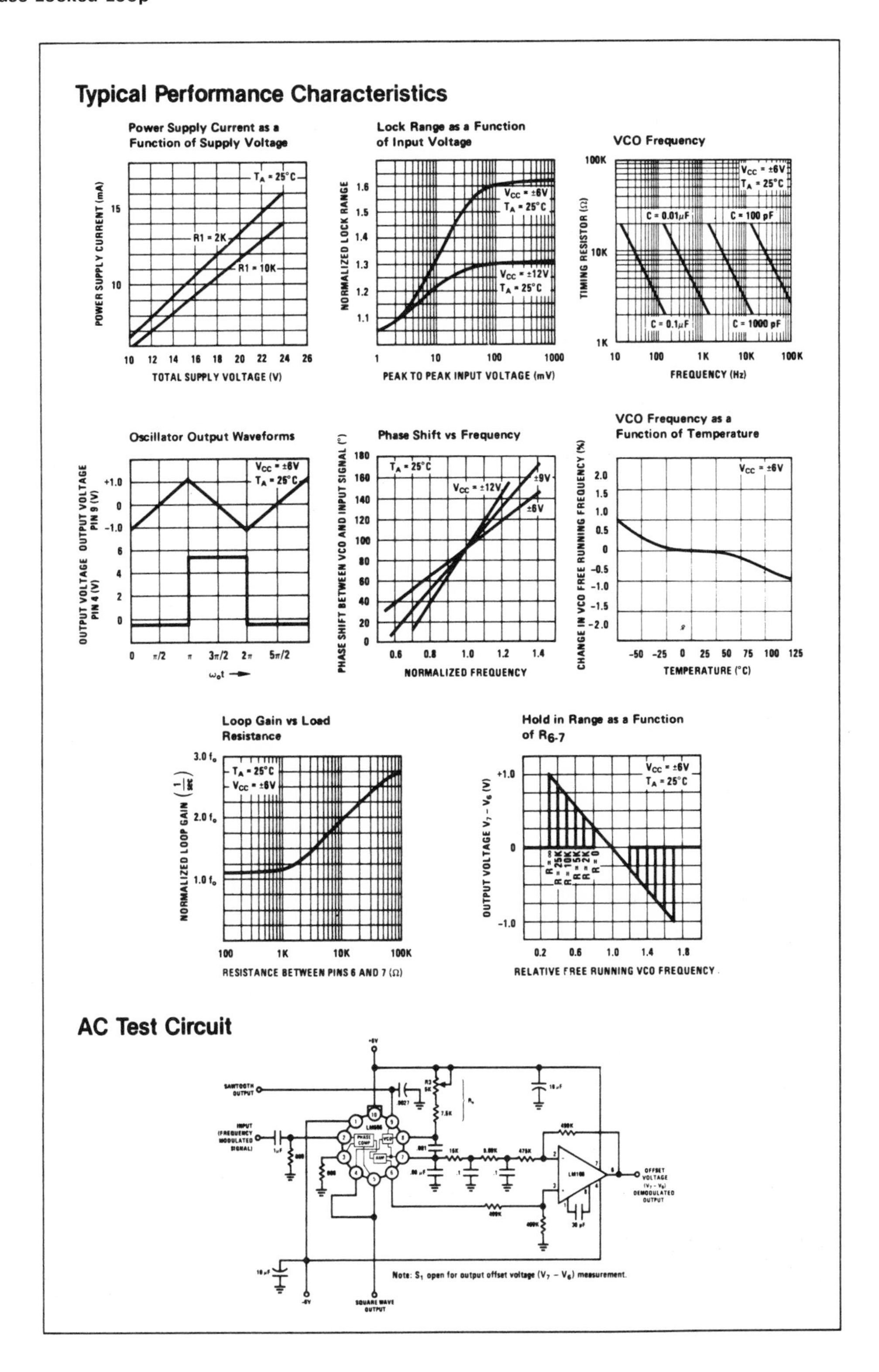
Typical Performance Characteristics
Power Supply Current as a Function of Supply Voltage
POWER SUPPLY CURRENT (mA)
TOTAL SUPPLY VOLTAGE (V)
T_A = 25°C
R1 = 2K
R1 = 10K
Lock Range as a Function of Input Voltage
NORMALIZED LOCK RANGE
PEAK TO PEAK INPUT VOLTAGE (mV)
V_CC = ±6V
T_A = 25°C
V_CC = ±12V
VCO Frequency
TIMING RESISTOR (Ω)
FREQUENCY (Hz)
C = 0.01μF
C = 100 pF
C = 0.1μF
C = 1000 pF
Oscillator Output Waveforms
OUTPUT VOLTAGE PIN 9 (V)
OUTPUT VOLTAGE PIN 4 (V)
ω_o t
Phase Shift vs Frequency
PHASE SHIFT BETWEEN VCO AND INPUT SIGNAL (°)
NORMALIZED FREQUENCY
V_CC = ±12V
±9V
±6V
VCO Frequency as a Function of Temperature
CHANGE IN VCO FREE RUNNING FREQUENCY (%)
TEMPERATURE (°C)
Loop Gain vs Load Resistance
NORMALIZED LOOP GAIN (1/sec)
RESISTANCE BETWEEN PINS 6 AND 7 (Ω)
Hold in Range as a Function of R6-7
OUTPUT VOLTAGE V7 – V6 (V)
RELATIVE FREE RUNNING VCO FREQUENCY
R = ∞
R = 25K
R = 10K
R = 5K
R = 2K
R = 0
AC Test Circuit
SAWTOOTH OUTPUT
INPUT (FREQUENCY MODULATED SIGNAL)
PHASE COMP
VCO
AMP
OFFSET VOLTAGE (V7 – V6) DEMODULATED OUTPUT
SQUARE WAVE OUTPUT
Note: S1 open for output offset voltage (V7 – V6) measurement.

Typical Applications

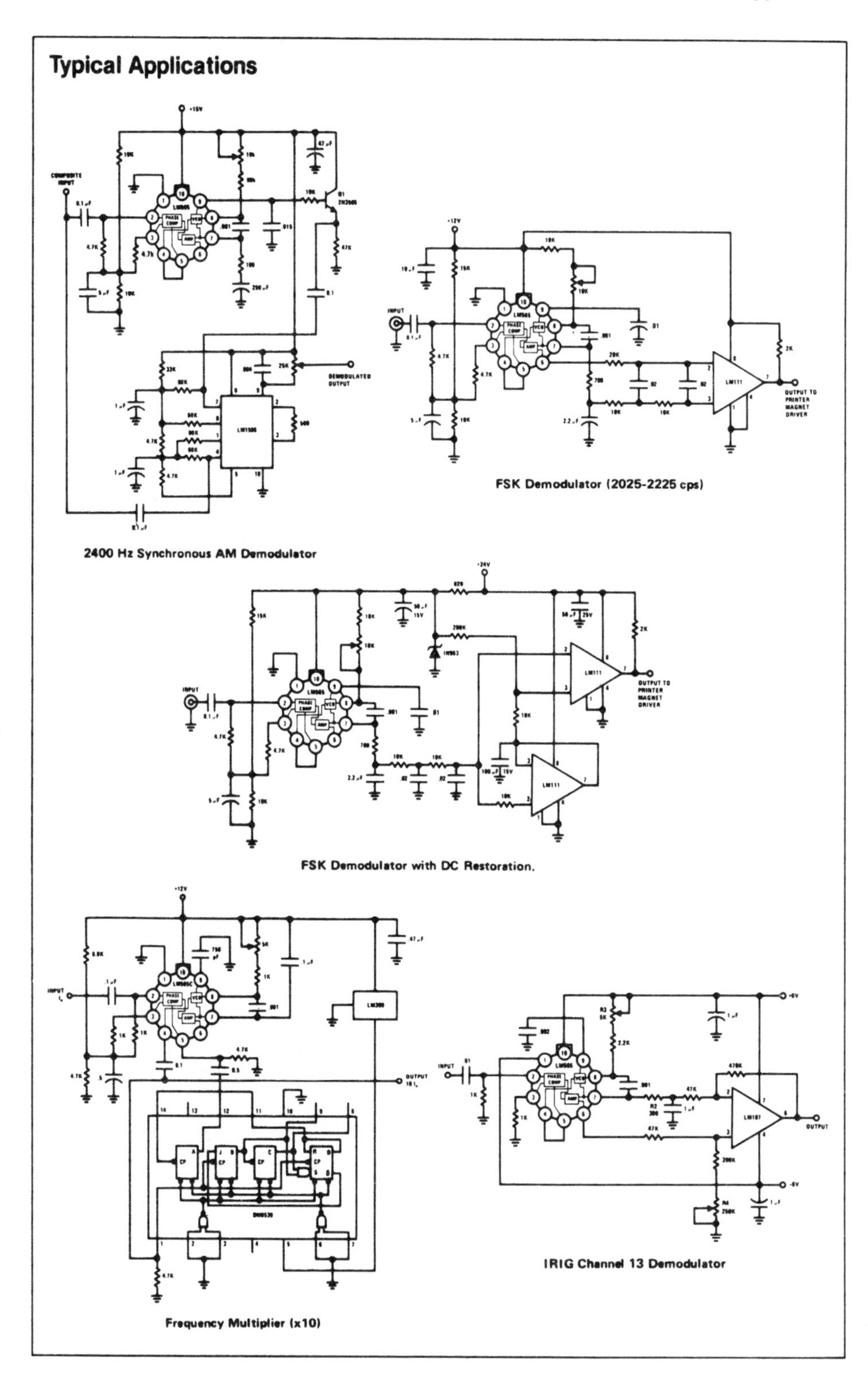

2400 Hz Synchronous AM Demodulator

FSK Demodulator (2025-2225 cps)

FSK Demodulator with DC Restoration.

Frequency Multiplier (x10)

IRIG Channel 13 Demodulator

Applications Information

In designing with phase locked loops such as the LM565, the important parameters of interest are:

FREE RUNNING FREQUENCY

$$f_o \cong \frac{1}{3.7\, R_0 C_0}$$

LOOP GAIN: relates the amount of phase change between the input signal and the VCO signal for a shift in input signal frequency (assuming the loop remains in lock). In servo theory, this is called the "velocity error coefficient".

$$\text{Loop gain} = K_o K_D \left(\frac{1}{\text{sec}}\right)$$

$$K_o = \text{oscillator sensitivity} \left(\frac{\text{radians/sec}}{\text{volt}}\right)$$

$$K_D = \text{phase detector sensitivity} \left(\frac{\text{volts}}{\text{radian}}\right)$$

The loop gain of the LM565 is dependent on supply voltage, and may be found from:

$$K_o K_D = \frac{33.6\, f_o}{V_c}$$

f_o = VCO frequency in Hz

V_c = total supply voltage to circuit.

Loop gain may be reduced by connecting a resistor between pins 6 and 7; this reduces the load impedance on the output amplifier and hence the loop gain.

HOLD IN RANGE: the range of frequencies that the loop will remain in lock after initially being locked.

$$f_H = \pm \frac{8\, f_o}{V_c}$$

f_o = free running frequency of VCO

V_c = total supply voltage to the circuit.

THE LOOP FILTER

In almost all applications, it will be desirable to filter the signal at the output of the phase detector (pin 7) this filter may take one of two forms:

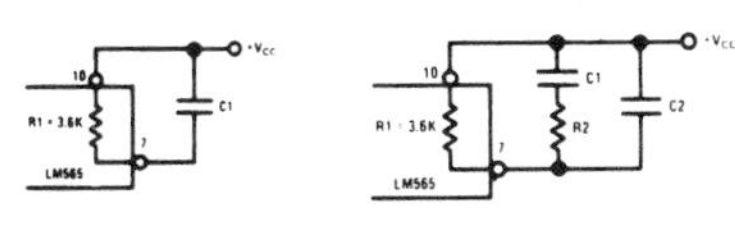

Simple Lag Filter   Lag-Lead Filter

A simple lag filter may be used for wide closed loop bandwidth applications such as modulation following where the frequency deviation of the carrier is fairly high (greater than 10%), or where wideband modulating signals must be followed.

The natural bandwidth of the closed loop response may be found from:

$$f_n = \frac{1}{2\pi}\sqrt{\frac{K_o K_D}{R_1 C_1}}$$

Associated with this is a damping factor:

$$\delta = \frac{1}{2}\sqrt{\frac{1}{R_1 C_1 K_o K_D}}$$

For narrow band applications where a narrow noise bandwidth is desired, such as applications involving tracking a slowly varying carrier, a lead lag filter should be used. In general, if $1/R_1C_1 < K_oK_d$, the damping factor for the loop becomes quite small resulting in large overshoot and possible instability in the transient response of the loop. In this case, the natural frequency of the loop may be found from

$$f_n = \frac{1}{2\pi}\sqrt{\frac{K_o K_D}{\tau_1 + \tau_2}}$$

$$\tau_1 + \tau_2 = (R_1 + R_2)\, C_1$$

R_2 is selected to produce a desired damping factor δ, usually between 0.5 and 1.0. The damping factor is found from the approximation:

$$\delta \simeq \pi \tau_2 f_n$$

These two equations are plotted for convenience.

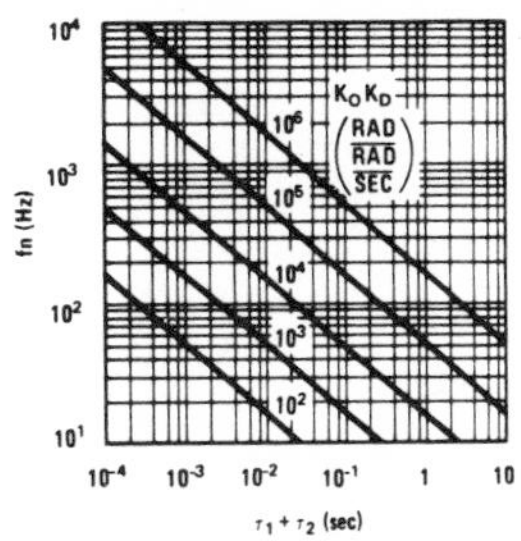

Filter Time Constant vs Natural Frequency

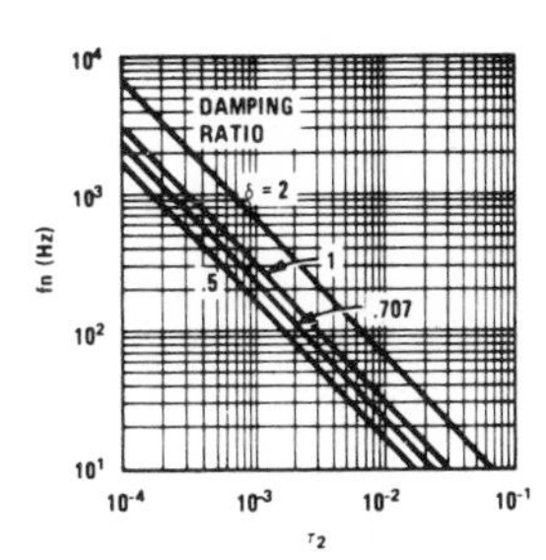

Damping Time Constant vs Natural Frequency

Capacitor C_2 should be much smaller than C_1 since its function is to provide filtering of carrier. In general $C_2 \leq 0.1\, C_1$.

A to D, D to A

ADC0801, ADC0802, ADC0803, ADC0804, ADC0805 8-Bit μP Compatible A/D Converters

General Description

The ADC0801, ADC0802, ADC0803, ADC0804 and ADC0805 are CMOS 8-bit successive approximation A/D converters which use a differential potentiometric ladder—similar to the 256R products. These converters are designed to allow operation with the NSC800 and INS8080A derivative control bus, and TRI-STATE® output latches directly drive the data bus. These A/Ds appear like memory locations or I/O ports to the microprocessor and no interfacing logic is needed.

A new differential analog voltage input allows increasing the common-mode rejection and offsetting the analog zero input voltage value. In addition, the voltage reference input can be adjusted to allow encoding any smaller analog voltage span to the full 8 bits of resolution.

Features

- Compatible with 8080 μP derivatives—no interfacing logic needed – access time – 135 ns
- Easy interface to all microprocessors, or operates "stand alone"
- Differential analog voltage inputs
- Logic inputs and outputs meet both MOS and T^2L voltage level specifications
- Works with 2.5V (LM336) voltage reference
- On-chip clock generator
- 0V to 5V analog input voltage range with single 5V supply
- No zero adjust required
- 0.3" standard width 20-pin DIP package
- Operates ratiometrically or with 5 V_{DC}, 2.5 V_{DC}, or analog span adjusted voltage reference

Key Specifications

- Resolution 8 bits
- Total error ±1/4 LSB, ±1/2 LSB and ±1 LSB
- Conversion time 100 μs

Typical Applications

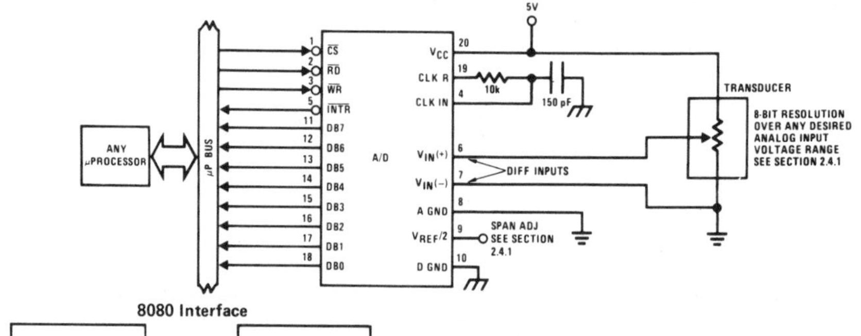

8080 Interface

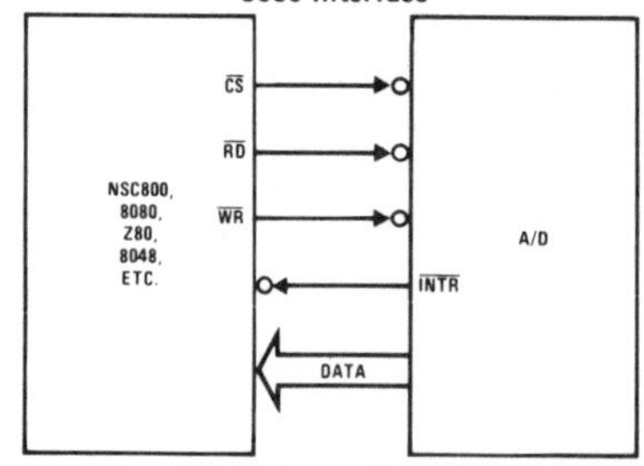

ERROR SPECIFICATION (INCLUDES FULL-SCALE, ZERO ERROR, AND NON-LINEARITY)			
PART NUMBER	FULL-SCALE ADJUSTED	$V_{REF}/2 = 2.500\ V_{DC}$ (NO ADJUSTMENTS)	$V_{REF}/2$ = NO CONNECTION (NO ADJUSTMENTS)
ADC0801	±1/4 LSB		
ADC0802		±1/2 LSB	
ADC0803	±1/2 LSB		
ADC0804		±1 LSB	
ADC0805			±1 LSB

Absolute Maximum Ratings (Notes 1 and 2)

Supply Voltage (V_{CC}) (Note 3)	6.5V
Voltage	
Logic Control Inputs	−0.3V to +18V
At Other Input and Outputs	−0.3V to (V_{CC} + 0.3V)
Storage Temperature Range	−65°C to +150°C
Package Dissipation at T_A = 25°C	875 mW
Lead Temperature (Soldering, 10 seconds)	300°C

Operating Ratings (Notes 1 and 2)

Temperature Range	$T_{MIN} \le T_A \le T_{MAX}$
ADC0801/02LD	$-55°C \le T_A \le +125°C$
ADC0801/02/03/04LCD	$-40°C \le T_A \le +85°C$
ADC0801/02/03/05LCN	$-40°C \le T_A \le +85°C$
ADC0804LCN	$0°C \le T_A \le +70°C$
Range of V_{CC}	4.5 V_{DC} to 6.3 V_{DC}

Electrical Characteristics

The following specifications apply for V_{CC} = 5 V_{DC}, $T_{MIN} \le T_A \le T_{MAX}$ and f_{CLK} = 640 kHz unless otherwise specified.

PARAMETER	CONDITIONS	MIN	TYP	MAX	UNITS
ADC0801: Total Adjusted Error (Note 8)	With Full-Scale Adj. (See Section 2.5.2)			±1/4	LSB
ADC0802: Total Unadjusted Error (Note 8)	$V_{REF}/2$ = 2.500 V_{DC}			±1/2	LSB
ADC0803: Total Adjusted Error (Note 8)	With Full-Scale Adj. (See Section 2.5.2)			±1/2	LSB
ADC0804: Total Unadjusted Error (Note 8)	$V_{REF}/2$ = 2.500 V_{DC}			±1	LSB
ADC0805: Total Unadjusted Error (Note 8)	$V_{REF}/2$ – No Connection			±1	LSB
$V_{REF}/2$ Input Resistance (Pin 9)	ADC0801/02/03/05	2.5	8.0		kΩ
	ADC0804 (Note 9)	1.0	1.3		kΩ
Analog Input Voltage Range	(Note 4) V(+) or V(−)	Gnd−0.05		V_{CC}+0.05	V_{DC}
DC Common-Mode Error	Over Analog Input Voltage Range		±1/16	±1/8	LSB
Power Supply Sensitivity	V_{CC} = 5 V_{DC} ±10% Over Allowed $V_{IN}(+)$ and $V_{IN}(-)$ Voltage Range (Note 4)		±1/16	±1/8	LSB

AC Electrical Characteristics

The following specifications apply for V_{CC} = 5 V_{DC} and T_A = 25°C unless otherwise specified.

PARAMETER		CONDITIONS	MIN	TYP	MAX	UNITS
T_c	Conversion Time	f_{CLK} = 640 kHz (Note 6)	103		114	μs
T_c	Conversion Time	(Note 5, 6)	66		73	$1/f_{CLK}$
f_{CLK}	Clock Frequency	V_{CC} = 5V, (Note 5)	100	640	1460	kHz
	Clock Duty Cycle	(Note 5)	40		60	%
CR	Conversion Rate In Free-Running Mode	$\overline{INTR}$ tied to $\overline{WR}$ with $\overline{CS}$ = 0 V_{DC}, f_{CLK} = 640 kHz			8770	conv/s
$t_{W(\overline{WR})L}$	Width of $\overline{WR}$ Input (Start Pulse Width)	$\overline{CS}$ = 0 V_{DC} (Note 7)	100			ns
t_{ACC}	Access Time (Delay from Falling Edge of $\overline{RD}$ to Output Data Valid)	C_L = 100 pF		135	200	ns
t_{1H}, t_{0H}	TRI-STATE Control (Delay from Rising Edge of $\overline{RD}$ to Hi-Z State)	C_L = 10 pF, R_L = 10k (See TRI-STATE Test Circuits)		125	200	ns
t_{WI}, t_{RI}	Delay from Falling Edge of $\overline{WR}$ or $\overline{RD}$ to Reset of $\overline{INTR}$			300	450	ns
C_{IN}	Input Capacitance of Logic Control Inputs			5	7.5	pF
C_{OUT}	TRI-STATE Output Capacitance (Data Buffers)			5	7.5	pF

Electrical Characteristics

The following specifications apply for V_{CC} = 5 V_{DC} and $T_{MIN} \leq T_A \leq T_{MAX}$, unless otherwise specified.

PARAMETER		CONDITIONS	MIN	TYP	MAX	UNITS
CONTROL INPUTS [Note: CLK IN (Pin 4) is the input of a Schmitt trigger circuit and is therefore specified separately]						
V_{IN} (1)	Logical "1" Input Voltage (Except Pin 4 CLK IN)	V_{CC} = 5.25 V_{DC}	2.0		15	V_{DC}
V_{IN} (0)	Logical "0" Input Voltage (Except Pin 4 CLK IN)	V_{CC} = 4.75 V_{DC}			0.8	V_{DC}
I_{IN} (1)	Logical "1" Input Current (All Inputs)	V_{IN} = 5 V_{DC}		0.005	1	μA_{DC}
I_{IN} (0)	Logical "0" Input Current (All Inputs)	V_{IN} = 0 V_{DC}	−1	−0.005		μA_{DC}
CLOCK IN AND CLOCK R						
V_T+	CLK IN (Pin 4) Positive Going Threshold Voltage		2.7	3.1	3.5	V_{DC}
V_T-	CLK IN (Pin 4) Negative Going Threshold Voltage		1.5	1.8	2.1	V_{DC}
V_H	CLK IN (Pin 4) Hysteresis $(V_T+) - (V_T-)$		0.6	1.3	2.0	V_{DC}
V_{OUT} (0)	Logical "0" CLK R Output Voltage	I_O = 360 μA V_{CC} = 4.75 V_{DC}			0.4	V_{DC}
V_{OUT} (1)	Logical "1" CLK R Output Voltage	I_O = −360 μA V_{CC} = 4.75 V_{DC}	2.4			V_{DC}
DATA OUTPUTS AND $\overline{INTR}$						
V_{OUT}(0)	Logical "0" Output Voltage Data Outputs $\overline{INTR}$ Output	 I_{OUT} = 1.6 mA, V_{CC} = 4.75 V_{DC} I_{OUT} = 1.0 mA, V_{CC} = 4.75 V_{DC}			 0.4 0.4	 V_{DC} V_{DC}
V_{OUT} (1)	Logical "1" Output Voltage	I_O = −360 μA, V_{CC} = 4.75 V_{DC}	2.4			V_{DC}
V_{OUT} (1)	Logical "1" Output Voltage	I_O = −10 μA, V_{CC} = 4.75 V_{DC}	4.5			V_{DC}
I_{OUT}	TRI-STATE Disabled Output Leakage (All Data Buffers)	V_{OUT} = 0 V_{DC} V_{OUT} = 5 V_{DC}	−3		 3	μA_{DC} μA_{DC}
I_{SOURCE}		V_{OUT} Short to Gnd, T_A = 25°C	4.5	6		mA_{DC}
I_{SINK}		V_{OUT} Short to V_{CC}, T_A = 25°C	9.0	16		mA_{DC}
POWER SUPPLY						
I_{CC}	Supply Current (Includes Ladder Current)	f_{CLK} = 640 kHz, $V_{REF}/2$ = NC, T_A = 25°C and $\overline{CS}$ = "1" ADC0801/02/03/05 ADC0804 (Note 9)		 1.1 1.9	 1.8 2.5	 mA mA

Note 1: Absolute maximum ratings are those values beyond which the life of the device may be impaired.

Note 2: All voltages are measured with respect to Gnd, unless otherwise specified. The separate A Gnd point should always be wired to the D Gnd.

Note 3: A zener diode exists, internally, from V_{CC} to Gnd and has a typical breakdown voltage of 7 V_{DC}.

Note 4: For $V_{IN}(-) \geq V_{IN}(+)$ the digital output code will be 0000 0000. Two on-chip diodes are tied to each analog input (see block diagram) which will forward conduct for analog input voltages one diode drop below ground or one diode drop greater than the V_{CC} supply. Be careful, during testing at low V_{CC} levels (4.5V), as high level analog inputs (5V) can cause this input diode to conduct—especially at elevated temperatures, and cause errors for analog inputs near full-scale. The spec allows 50 mV forward bias of either diode. This means that as long as the analog V_{IN} does not exceed the supply voltage by more than 50 mV, the output code will be correct. To achieve an absolute 0 V_{DC} to 5 V_{DC} input voltage range will therefore require a minimum supply voltage of 4.950 V_{DC} over temperature variations, initial tolerance and loading.

Note 5: Accuracy is guaranteed at f_{CLK} = 640 kHz. At higher clock frequencies accuracy can degrade. For lower clock frequencies, the duty cycle limits can be extended so long as the minimum clock high time interval or minimum clock low time interval is no less than 275 ns.

Note 6: With an asynchronous start pulse, up to 8 clock periods may be required before the internal clock phases are proper to start the conversion process. The start request is internally latched, see *Figure 2* and section 2.0.

Note 7: The $\overline{CS}$ input is assumed to bracket the $\overline{WR}$ strobe input and therefore timing is dependent on the $\overline{WR}$ pulse width. An arbitrarily wide pulse width will hold the converter in a reset mode and the start of conversion is initiated by the low to high transition of the $\overline{WR}$ pulse (see timing diagrams).

Note 8: None of these A/Ds requires a zero adjust (see section 2.5.1). To obtain zero code at other analog input voltages see section 2.5 and *Figure 5.*

Note 9: For ADC0804LCD typical value of $V_{REF}/2$ input resistance is 8 kΩ and of I_{CC} is 1.1 mA.

1.0 UNDERSTANDING A/D ERROR SPECS

A perfect A/D transfer characteristic (staircase waveform) is shown in *Figure 1a.* The horizontal scale is analog input voltage and the particular points labeled are in steps of 1 LSB (19.53 mV with 2.5V tied to the $V_{REF}/2$ pin). The digital output codes which correspond to these inputs are shown as D–1, D, and D+1. For the perfect A/D, not only will center-value (A–1, A, A+1, . . .) analog inputs produce the correct output digital codes, but also each riser (the transitions between adjacent output codes) will be located ±1/2 LSB away from each center-value. As shown, the risers are ideal and have no width. Correct digital output codes will be provided for a range of analog input voltages which extend ±1/2 LSB from the ideal center-values. Each tread (the range of analog input voltage which provides the same digital output code) is therefore 1 LSB wide.

Figure 1b shows a worst case error plot for the ADC0801. All center-valued inputs are guaranteed to produce the correct output codes and the adjacent risers are guaranteed to be no closer to the center-value points than ±1/4 LSB. In other words, if we apply an analog input equal to the center-value ±1/4 LSB, *we guarantee* that the A/D will produce the correct digital code. The maximum range of the position of the code transition is indicated by the horizontal arrow and it is guaranteed to be no more than 1/2 LSB.

The error curve of *Figure 1c* shows a worst case error plot for the ADC0802. Here we guarantee that if we apply an analog input equal to the LSB analog voltage center-value the A/D will produce the correct digital code.

Next to each transfer function is shown the corresponding error plot. Many people may be more familiar with error plots than transfer functions. The analog input voltage to the A/D is provided by either a linear ramp or by the discrete output steps of a high resolution DAC. Notice that the error is continuously displayed and includes the quantization uncertainty of the A/D. For example the error at point 1 of *Figure 1a* is +1/2 LSB because the digital code appeared 1/2 LSB in advance of the center-value of the tread. The error plots always have a constant negative slope and the abrupt upside steps are always 1 LSB in magnitude.

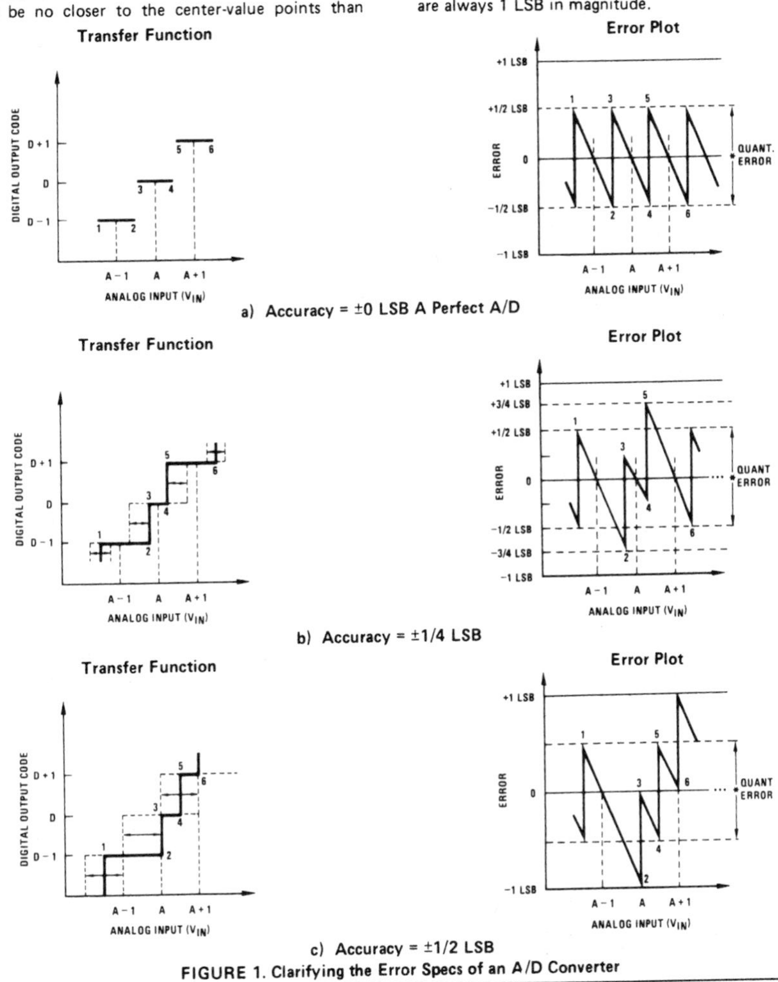

FIGURE 1. Clarifying the Error Specs of an A/D Converter

2.0 FUNCTIONAL DESCRIPTION

The ADC0801 series contains a circuit equivalent of the 256R network. Analog switches are sequenced by successive approximation logic to match the analog difference input voltage [$V_{IN}(+) - V_{IN}(-)$] to a corresponding tap on the R network. The most significant bit is tested first and after 8 comparisons (64 clock cycles) a digital 8-bit binary code (1111 1111 = full-scale) is transferred to an output latch and then an interrupt is asserted ($\overline{INTR}$ makes a high-to-low transition). A conversion in process can be interrupted by issuing a second start command. The device may be operated in the free-running mode by connecting $\overline{INTR}$ to the $\overline{WR}$ input with $\overline{CS}$ = 0. To insure start-up under all possible conditions, an external $\overline{WR}$ pulse is required during the first power-up cycle.

On the high-to-low transition of the $\overline{WR}$ input the internal SAR latches and the shift register stages are reset. As long as the $\overline{CS}$ input and $\overline{WR}$ input remain low, the A/D will remain in a reset state. *Conversion will start from 1 to 8 clock periods after at least one of these inputs makes a low-to-high transition.*

A functional diagram of the A/D converter is shown in *Figure 2.* All of the package pinouts are shown and the major logic control paths are drawn in heavier weight lines.

The converter is started by having $\overline{CS}$ and $\overline{WR}$ simultaneously low. This sets the start flip-flop (F/F) and the resulting "1" level resets the 8-bit shift register, resets the Interrupt (INTR) F/F and inputs a "1" to the D flop, F/F1, which is at the input end of the 8-bit shift register. Internal clock signals then transfer this "1" to the Q output of F/F1. The AND gate, G1, combines this "1" output with a clock signal to provide a reset signal to the start F/F. If the set signal is no longer present (either $\overline{WR}$ or $\overline{CS}$ is a "1") the start F/F is reset and the 8-bit shift register then can have the "1" clocked in, which starts the conversion process. If the set signal were to still be present, this reset pulse would have no effect (both outputs of the start F/F would momentarily be at a "1" level) and the 8-bit shift register would continue to be held in the reset mode. This logic therefore allows for wide $\overline{CS}$ and $\overline{WR}$ signals and the converter will start after at least one of these signals returns high and the internal clocks again provide a reset signal for the start F/F.

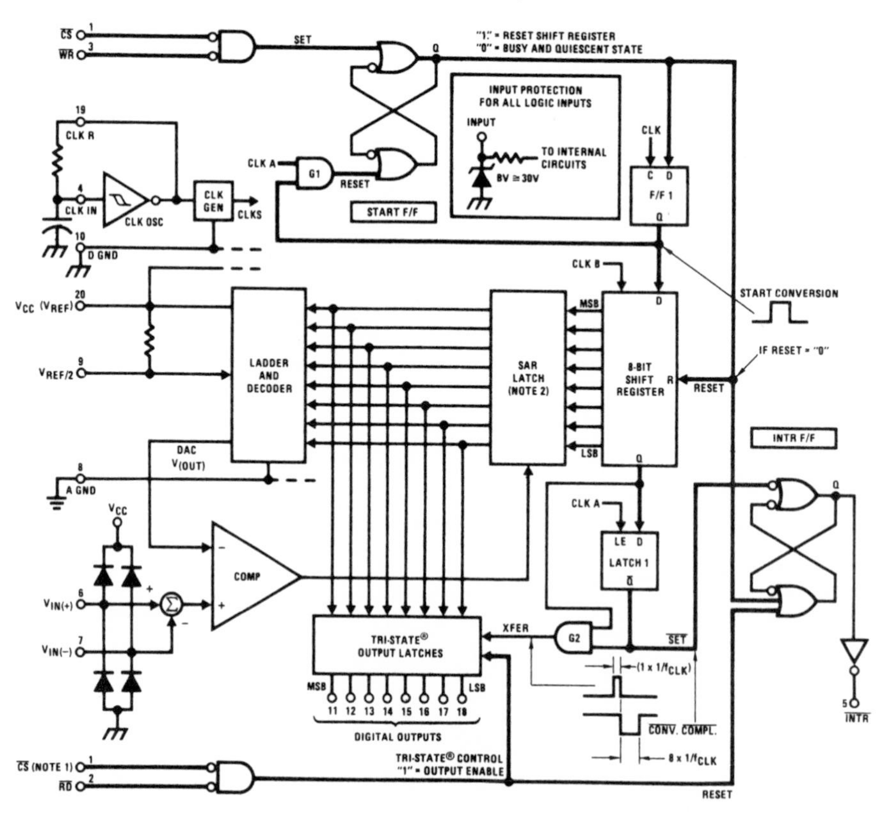

Note 1: $\overline{CS}$ shown twice for clarity.

Note 2: SAR = Successive Approximation Register.

FIGURE 2. Block Diagram

After the "1" is clocked through the 8-bit shift register (which completes the SAR search) it appears as the input to the D-type latch, LATCH 1. As soon as this "1" is output from the shift register, the AND gate, G2, causes the new digital word to transfer to the TRI-STATE output latches. When LATCH 1 is subsequently enabled, the Q output makes a high-to-low transition which causes the INTR F/F to set. An inverting buffer then supplies the $\overline{\text{INTR}}$ output signal.

Note that this $\overline{\text{SET}}$ control of the INTR F/F remains low for 8 of the external clock periods (as the internal clocks run at 1/8 of the frequency of the external clock). If the data output is continuously enabled ($\overline{\text{CS}}$ and $\overline{\text{RD}}$ both held low), the $\overline{\text{INTR}}$ output will still signal the end of conversion (by a high-to-low transition), because the $\overline{\text{SET}}$ input can control the Q output of the INTR F/F even though the RESET input is constantly at a "1" level in this operating mode. This $\overline{\text{INTR}}$ output will therefore stay low for the duration of the $\overline{\text{SET}}$ signal, which is 8 periods of the external clock frequency (assuming the A/D is not started during this interval).

When operating in the free-running or continuous conversion mode ($\overline{\text{INTR}}$ pin tied to $\overline{\text{WR}}$ and $\overline{\text{CS}}$ wired low—see also section 2.8), the START F/F is SET by the high-to-low transition of the $\overline{\text{INTR}}$ signal. This resets the SHIFT REGISTER which causes the input to the D-type latch, LATCH 1, to go low. As the latch enable input is still present, the $\overline{\text{Q}}$ output will go high, which then allows the INTR F/F to be RESET. This reduces the width of the resulting $\overline{\text{INTR}}$ output pulse to only a few propagation delays (approximately 300 ns).

When data is to be read, the combination of both $\overline{\text{CS}}$ and $\overline{\text{RD}}$ being low will cause the INTR F/F to be reset and the TRI-STATE output latches will be enabled to provide the 8-bit digital outputs.

2.1 Digital Control Inputs

The digital control inputs ($\overline{\text{CS}}$, $\overline{\text{RD}}$, and $\overline{\text{WR}}$) meet standard T^2L logic voltage levels. These signals have been renamed when compared to the standard A/D Start and Output Enable labels. In addition, these inputs are active low to allow an easy interface to microprocessor control busses. For non-microprocessor based applications, the $\overline{\text{CS}}$ input (pin 1) can be grounded and the standard A/D Start function is obtained by an active low pulse applied at the $\overline{\text{WR}}$ input (pin 3) and the Output Enable function is caused by an active low pulse at the $\overline{\text{RD}}$ input (pin 2).

2.2 Analog Differential Voltage Inputs and Common-Mode Rejection

This A/D has additional applications flexibility due to the analog differential voltage input. The $V_{IN}(-)$ input (pin 7) can be used to automatically subtract a fixed voltage value from the input reading (tare correction). This is also useful in 4 mA–20 mA current loop conversion. In addition, common-mode noise can be reduced by use of the differential input.

The time interval between sampling $V_{IN}(+)$ and $V_{IN}(-)$ is 4-1/2 clock periods. The maximum error voltage due to this slight time difference between the input voltage samples is given by:

$$\Delta V_e(\text{MAX}) = (V_P)\,(2\pi f_{cm})\left(\frac{4.5}{f_{CLK}}\right)$$

where:

ΔV_e is the error voltage due to sampling delay

V_P is the peak value of the common-mode voltage

f_{cm} is the common-mode frequency

As an example, to keep this error to 1/4 LSB (~5 mV) when operating with a 60 Hz common-mode frequency, f_{cm}, and using a 640 kHz A/D clock, f_{CLK}, would allow a peak value of the common-mode voltage, V_P, which is given by:

$$V_P = \frac{[\Delta V_{e(\text{MAX})}\,(f_{CLK})]}{(2\pi f_{cm})\,(4.5)}$$

or

$$V_P = \frac{(5 \times 10^{-3})\,(640 \times 10^3)}{(6.28)\,(60)\,(4.5)}$$

which gives

$$V_P \cong 1.9\text{V}.$$

The allowed range of analog input voltages usually places more severe restrictions on input common-mode noise levels.

An analog input voltage with a reduced span and a relatively large zero offset can be easily handled by making use of the differential input (see section 2.4 Reference Voltage).

2.3 Analog Inputs

2.3.1 Input Current

Normal Mode

Due to the internal switching action, displacement currents will flow at the analog inputs. This is due to on-chip stray capacitance to ground as shown in *Figure 3.*

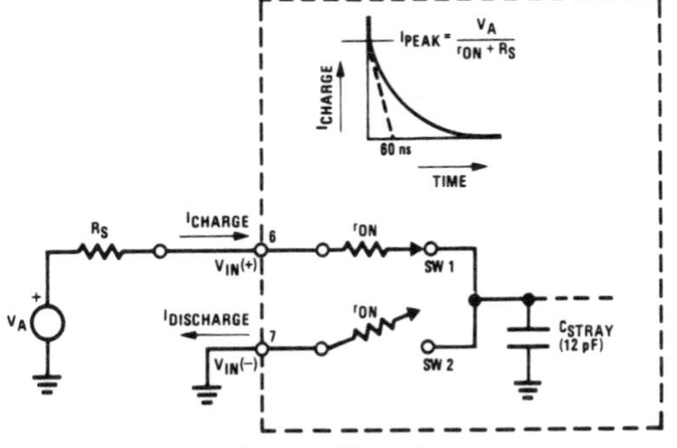

r_{ON} of SW 1 and SW 2 ≅ 5 kΩ

$\tau = r_{ON}\,C_{STRAY} \cong 5\text{ k}\Omega \times 12\text{ pF} = 60\text{ ns}$

FIGURE 3. Analog Input Impedance

The voltage on this capacitance is switched and will result in currents entering the $V_{IN}(+)$ input pin and leaving the $V_{IN}(-)$ input which will depend on the analog differential input voltage levels. These current transients occur at the leading edge of the internal clocks. They rapidly decay and *do not cause errors* as the on-chip comparator is strobed at the end of the clock period.

Fault Mode

If the voltage source which is applied to the $V_{IN}(+)$ pin exceeds the allowed operating range of V_{CC} + 50 mV, large input currents can flow through a parasitic diode to the V_{CC} pin. If these currents could exceed the 1 mA max allowed spec, an external diode (1N914) should be added to bypass this current to the V_{CC} pin (with the current bypassed with this diode, the voltage at the $V_{IN}(+)$ pin can exceed the V_{CC} voltage by the forward voltage of this diode).

2.3.2 Input Bypass Capacitors

Bypass capacitors at the inputs will average these charges and cause a DC current to flow through the output resistances of the analog signal sources. This charge pumping action is worse for continuous conversions with the $V_{IN}(+)$ input voltage at full-scale. For continuous conversions with a 640 kHz clock frequency with the $V_{IN}(+)$ input at 5V, this DC current is at a maximum of approximately 5 µA. Therefore, ***bypass capacitors should not be used at the analog inputs or the $V_{REF}/2$ pin*** for high resistance sources (> 1 kΩ). If input bypass capacitors are necessary for noise filtering and high source resistance is desirable to minimize capacitor size, the detrimental effects of the voltage drop across this input resistance, which is due to the average value of the input current, can be eliminated with a full-scale adjustment while the given source resistor and input bypass capacitor are both in place. This is possible because the average value of the input current is a precise linear function of the differential input voltage.

2.3.3 Input Source Resistance

Large values of source resistance where an input bypass capacitor is not used, ***will not cause errors*** as the input currents settle out prior to the comparison time. If a low pass filter is required in the system, use a low valued series resistor (≤ 1 kΩ) for a passive RC section or add an op amp RC active low pass filter. For low source resistance applications, (≤ 1 kΩ), a 0.1 µF bypass capacitor at the inputs will prevent pickup due to series lead inductance of a long wire. A 100Ω series resistor can be used to isolate this capacitor—both the R and C are placed outside the feedback loop—from the output of an op amp, if used.

2.3.4 Noise

The leads to the analog inputs (pins 6 and 7) should be kept as short as possible to minimize input noise coupling. Both noise and undesired digital clock coupling to these inputs can cause system errors. The source resistance for these inputs should, in general, be kept below 5 kΩ. Larger values of source resistance can cause undesired system noise pickup. Input bypass capacitors, placed from the analog inputs to ground, will eliminate system noise pickup but can create analog scale errors as these capacitors will average the transient input switching currents of the A/D (see section 2.3.1). This scale error depends on both a large source resistance and the use of an input bypass capacitor. This error can be eliminated by doing a full-scale adjustment of the A/D (adjust $V_{REF}/2$ for a proper full-scale reading—see section 2.5.2 on Full-Scale Adjustment) with the source resistance and input bypass capacitor in place.

2.4 Reference Voltage

2.4.1 Span Adjust

For maximum applications flexibility, these A/Ds have been designed to accommodate a 5 V_{DC}, 2.5 V_{DC} or an adjusted voltage reference. This has been achieved in the design of the IC as shown in *Figure 4*.

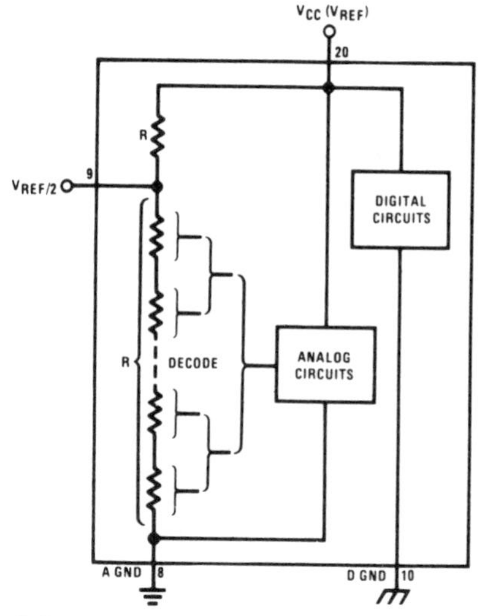

FIGURE 4. The $V_{REFERENCE}$ Design on the IC

Notice that the reference voltage for the IC is either 1/2 of the voltage which is applied to the V_{CC} supply pin, or is equal to the voltage which is externally forced at the $V_{REF}/2$ pin. This allows for a ratiometric voltage reference using the V_{CC} supply, a 5 V_{DC} reference voltage can be used for the V_{CC} supply or a voltage less than 2.5 V_{DC} can be applied to the $V_{REF}/2$ input for increased application flexibility. The internal gain to the $V_{REF}/2$ input is 2 making the full-scale differential input voltage twice the voltage at pin 9.

An example of the use of an adjusted reference voltage is to accommodate a reduced span—or dynamic voltage range of the analog input voltage. If the analog input voltage were to range from 0.5 V_{DC} to 3.5 V_{DC}, instead of 0V to 5 V_{DC}, the span would be 3V as shown in *Figure 5*. With 0.5 V_{DC} applied to the $V_{IN}(-)$ pin to absorb the offset, the reference voltage can be made equal to 1/2 of the 3V span or 1.5 V_{DC}. The A/D now will encode the $V_{IN}(+)$ signal from 0.5V to 3.5V with the 0.5V input corresponding to zero and the 3.5 V_{DC} input corresponding to full-scale. The full 8 bits of resolution are therefore applied over this reduced analog input voltage range.

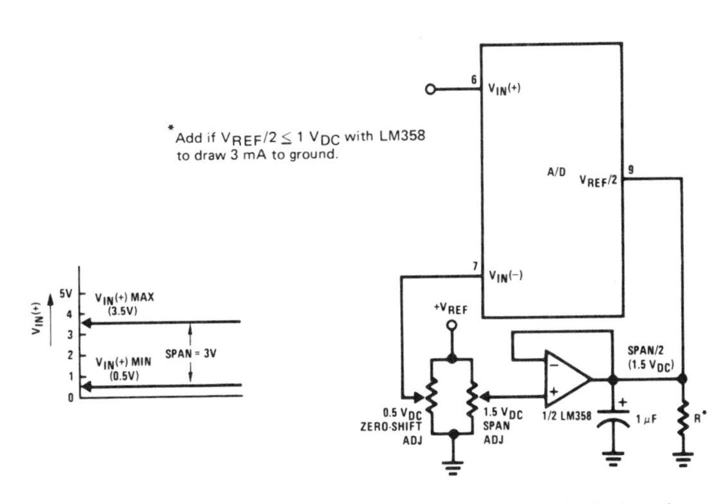

a) Analog Input Signal Example

b) Accommodating an Analog Input from 0.5V (Digital Out = 00_{HEX}) to 3.5V (Digital Out = FF_{HEX})

FIGURE 5. Adapting the A/D Analog Input Voltages to Match an Arbitrary Input Signal Range

2.4.2 Reference Accuracy Requirements

The converter can be operated in a ratiometric mode or an absolute mode. In ratiometric converter applications, the magnitude of the reference voltage is a factor in both the output of the source transducer and the output of the A/D converter and therefore cancels out in the final digital output code. The ADC0805 is specified particularly for use in ratiometric applications with no adjustments required. In absolute conversion applications, both the initial value and the temperature stability of the reference voltage are important accuracy factors in the operation of the A/D converter. For $V_{REF}/2$ voltages of 2.5 V_{DC} nominal value, initial errors of ± 10 mV_{DC} will cause conversion errors of ± 1 LSB due to the gain of 2 of the $V_{REF}/2$ input. In reduced span applications, the initial value and the stability of the $V_{REF}/2$ input voltage become even more important. For example, if the span is reduced to 2.5V, the analog input LSB voltage value is correspondingly reduced from 20 mV (5V span) to 10 mV and 1 LSB at the $V_{REF}/2$ input becomes 5 mV. As can be seen, this reduces the allowed initial tolerance of the reference voltage and requires correspondingly less absolute change with temperature variations. Note that spans smaller than 2.5V place even tighter requirements on the initial accuracy and stability of the reference source.

In general, the magnitude of the reference voltage will require an initial adjustment. Errors due to an improper value of reference voltage appear as full-scale errors in the A/D transfer function. IC voltage regulators may be used for references if the ambient temperature changes are not excessive. The LM336B 2.5V IC reference diode (from National Semiconductor) is available which has a temperature stability of 1.8 mV typ (6 mV max) over $0^\circ C \leq T_A \leq +70^\circ C$. Other temperature range parts are also available.

2.5 Errors and Reference Voltage Adjustments

2.5.1 Zero Error

The zero of the A/D does not require adjustment. If the minimum analog input voltage value, $V_{IN(MIN)}$, is not ground, a zero offset can be done. The converter can be made to output 0000 0000 digital code for this minimum input voltage by biasing the A/D V_{IN} (−) input at this $V_{IN(MIN)}$ value (see Applications section). This utilizes the differential mode operation of the A/D.

The zero error of the A/D converter relates to the location of the first riser of the transfer function and can be measured by grounding the V (−) input and applying a small magnitude positive voltage to the V (+) input. Zero error is the difference between the actual DC input voltage which is necessary to just cause an output digital code transition from 0000 0000 to 0000 0001 and the ideal 1/2 LSB value (1/2 LSB = 9.8 mV for $V_{REF}/2 = 2.500$ V_{DC}).

2.5.2 Full-Scale

The full-scale adjustment can be made by applying a differential input voltage which is 1-1/2 LSB down from the desired analog full-scale voltage range and then adjusting the magnitude of the $V_{REF}/2$ input (pin 9 or the V_{CC} supply if pin 9 is not used) for a digital output code which is just changing from 1111 1110 to 1111 1111.

2.5.3 Adjusting for an Arbitrary Analog Input Voltage Range

If the analog zero voltage of the A/D is shifted away from ground (for example, to accommodate an analog input signal which does not go to ground) this new zero reference should be properly adjusted first: A $V_{IN}(+)$ voltage which equals this desired zero reference plus 1/2 LSB (where the LSB is calculated for the desired analog span, 1 LSB = analog span/256) is applied to pin 6 and the zero reference voltage at pin 7 should then be adjusted to just obtain the 00_{HEX} to 01_{HEX} code transition.

The full-scale adjustment should then be made (with the proper $V_{IN}(-)$ voltage applied) by forcing a voltage to the $V_{IN}(+)$ input which is given by:

$$V_{IN}(+)\text{ fs adj} = V_{MAX} - 1.5\left[\frac{(V_{MAX} - V_{MIN})}{256}\right],$$

where:

V_{MAX} = The high end of the analog input range

and

V_{MIN} = the low end (the offset zero) of the analog range. (Both are ground referenced.)

The $V_{REF}/2$ (or V_{CC}) voltage is then adjusted to provide a code change from FE_{HEX} to FF_{HEX}. This completes the adjustment procedure.

2.6 Clocking Option

The clock for the A/D can be derived from the CPU clock or an external RC can be added to provide self-clocking. The CLK IN (pin 4) makes use of a Schmitt trigger as shown in *Figure 6*.

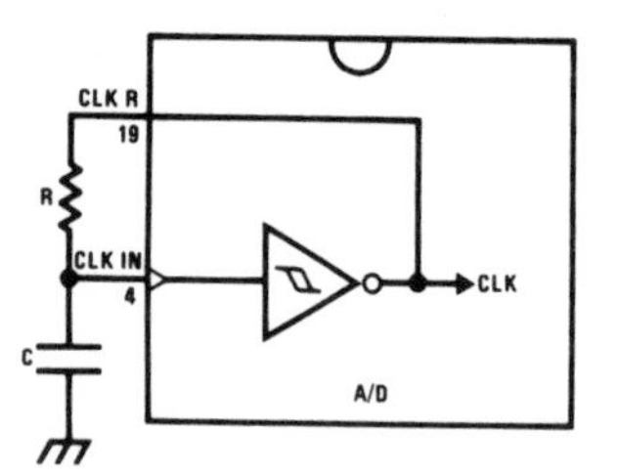

$$f_{CLK} \cong \frac{1}{1.1\ RC}$$

$R \cong 10\ k\Omega$

FIGURE 6. Self-Clocking the A/D

Heavy capacitive or DC loading of the clock R pin should be avoided as this will disturb normal converter operation. Loads less than 50 pF, such as driving up to 7 A/D converter clock inputs from a single clock R pin of 1 converter, are allowed. For larger clock line loading, a CMOS or low power T^2L buffer or PNP input logic should be used to minimize the loading on the clock R pin (do not use a standard T^2L buffer).

2.7 Restart During a Conversion

If the A/D is restarted ($\overline{CS}$ and $\overline{WR}$ go low and return high) during a conversion, the converter is reset and a new conversion is started. The output data latch is not updated if the conversion in process is not allowed to be completed, therefore the data of the previous conversion remains in this latch. The $\overline{INTR}$ output also simply remains at the "1" level.

2.8 Continuous Conversions

For operation in the free-running mode an initializing pulse should be used, following power-up, to insure circuit operation. In this application, the $\overline{CS}$ input is grounded and the $\overline{WR}$ input is tied to the $\overline{INTR}$ output. This $\overline{WR}$ and $\overline{INTR}$ node should be momentarily forced to logic low following a power-up cycle to guarantee operation.

2.9 Driving the Data Bus

This MOS A/D, like MOS microprocessors and memories, will require a bus driver when the total capacitance of the data bus gets large. Other circuitry, which is tied to the data bus, will add to the total capacitive loading, even in TRI-STATE (high impedance mode). Backplane bussing also greatly adds to the stray capacitance of the data bus.

There are some alternatives available to the designer to handle this problem. Basically, the capacitive loading of the data bus slows down the response time, even though DC specifications are still met. For systems operating with a relatively slow CPU clock frequency, more time is available in which to establish proper logic levels on the bus and therefore higher capacitive loads can be driven (see typical characteristics curves).

At higher CPU clock frequencies time can be extended for I/O reads (and/or writes) by inserting wait states (8080) or using clock extending circuits (6800).

Finally, if time is short and capacitive loading is high, external bus drivers must be used. These can be TRI-STATE buffers (low power Schottky is recommended such as the DM74LS240 series) or special higher drive current products which are designed as bus drivers. High current bipolar bus drivers with PNP inputs are recommended.

2.10 Power Supplies

Noise spikes on the V_{CC} supply line can cause conversion errors as the comparator will respond to this noise. A low inductance tantalum filter capacitor should be used close to the converter V_{CC} pin and values of 1 μF or greater are recommended. If an unregulated voltage is available in the system, a separate LM340LAZ-5.0, TO-92, 5V voltage regulator for the converter (and other analog circuitry) will greatly reduce digital noise on the V_{CC} supply.

2.11 Wiring and Hook-Up Precautions

Standard digital wire wrap sockets are not satisfactory for breadboarding this A/D converter. Sockets on PC boards can be used and all logic signal wires and leads should be grouped and kept as far away as possible from the analog signal leads. Exposed leads to the analog inputs can cause undesired digital noise and hum pickup, therefore shielded leads may be necessary in many applications.

A single point analog ground should be used which is separate from the logic ground points. The power supply bypass capacitor and the self-clocking capacitor (if used) should both be returned to digital ground. Any $V_{REF}/2$ bypass capacitors, analog input filter capacitors, or input signal shielding should be returned to the analog ground point. A test for proper grounding is to measure the zero error of the A/D converter. Zero errors in excess of 1/4 LSB can usually be traced to improper board layout and wiring (see section 2.5.1 for measuring the zero error).

3.0 TESTING THE A/D CONVERTER

There are many degrees of complexity associated with testing an A/D converter. One of the simplest tests is to apply a known analog input voltage to the converter and use LEDs to display the resulting digital output code as shown in *Figure 7*.

For ease of testing, the $V_{REF}/2$ (pin 9) should be supplied with 2.560 V_{DC} and a V_{CC} supply voltage of 5.12 V_{DC} should be used. This provides an LSB value of 20 mV.

If a full-scale adjustment is to be made, an analog input voltage of 5.090 V_{DC} (5.120 − 1 1/2 LSB) should be applied to the $V_{IN}(+)$ pin with the $V_{IN}(-)$ pin grounded. The value of the $V_{REF}/2$ input voltage should then be adjusted until the digital output code is just changing from 1111 1110 to 1111 1111. This value of $V_{REF}/2$ should then be used for all the tests.

The digital output LED display can be decoded by dividing the 8 bits into 2 hex characters, the 4 most significant (MS) and the 4 least significant (LS). Table I shows the fractional binary equivalent of these two 4-bit groups. By adding the decoded voltages which are obtained from the column: Input voltage value for a 2.560 $V_{REF}/2$ of both the MS and the LS groups, the value of

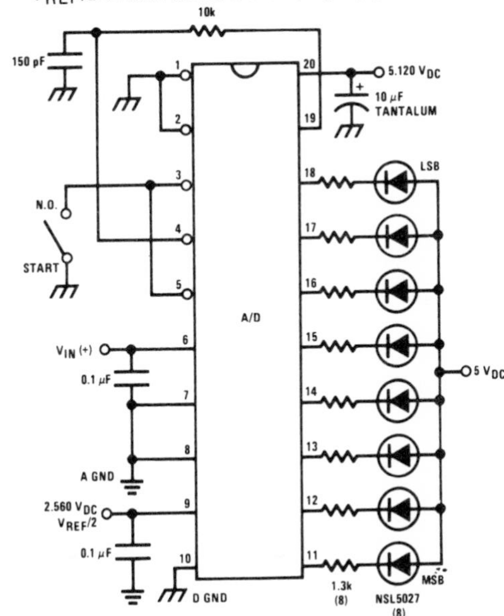

FIGURE 7. Basic A/D Tester

the digital display can be determined. For example, for an output LED display of 1011 0110 or B6 (in hex), the voltage values from the table are 3.520 + 0.120 or 3.640 V_{DC}. These voltage values represent the center-values of a perfect A/D converter. The effects of quantization error have to be accounted for in the interpretation of the test results.

For a higher speed test system, or to obtain plotted data, a digital-to-analog converter is needed for the test set-up. An accurate 10-bit DAC can serve as the precision voltage source for the A/D. Errors of the A/D under test can be provided as either analog voltages or differences in 2 digital words.

A basic A/D tester which uses a DAC and provides the error as an analog output voltage is shown in *Figure 8*. The 2 op amps can be eliminated if a lab DVM with a numerical subtraction feature is available to directly readout the difference voltage, "A–C". The analog input voltage can be supplied by a low frequency ramp generator and an X-Y plotter can be used to provide analog error (Y axis) versus analog input (X axis). The construction details of a tester of this type are provided in the NSC application note AN-179, "Analog-to-Digital Converter Testing".

For operation with a microprocessor or a computer-based test system, it is more convenient to present the errors digitally. This can be done with the circuit of *Figure 9*, where the output code transitions can be detected as the 10-bit DAC is incremented. This provides 1/4 LSB steps for the 8-bit A/D under test. If the results of this test are automatically plotted with the analog input on the X axis and the error (in LSB's) as the Y axis, a useful transfer function of the A/D under test results. For acceptance testing, the plot is not necessary and the testing speed can be increased by establishing internal limits on the allowed error for each code.

4.0 MICROPROCESSOR INTERFACING

To discuss the interface with 8080A and 6800 microprocessors, a common sample subroutine structure is used. The microprocessor starts the A/D, reads and stores the results of 16 successive conversions, then returns to the user's program. The 16 data bytes are stored in 16 successive memory locations. All Data and Addresses will be given in hexadecimal form. Software and hardware details are provided separately for each type of microprocessor.

4.1 Interfacing 8080 Microprocessor Derivatives (8048, 8085)

This converter has been designed to directly interface with derivatives of the 8080 microprocessor. The A/D can be mapped into memory space (using standard memory address decoding for $\overline{CS}$ and the $\overline{MEMR}$ and $\overline{MEMW}$ strobes) or it can be controlled as an I/O device by using the $\overline{I/O\ R}$ and $\overline{I/O\ W}$ strobes and decoding the address bits A0 → A7 (or address bits A8 → A15 as they will contain the same 8-bit address information) to obtain the $\overline{CS}$ input. Using the I/O space provides 256 additional addresses and may allow a simpler 8-bit address decoder but the data can only be input to the accumulator. To make use of the additional memory reference instructions, the A/D should be mapped into memory space. An example of an A/D in I/O space is shown in *Figure 10*.

A to D, D to A

DAC0808, DAC0807, DAC0806 8-Bit D/A Converters

General Description

The DAC0808 series is an 8-bit monolithic digital-to-analog converter (DAC) featuring a full scale output current settling time of 150 ns while dissipating only 33 mW with ±5V supplies. No reference current (I_{REF}) trimming is required for most applications since the full scale output current is typically ±1 LSB of 255 I_{REF}/256. Relative accuracies of better than ±0.19% assure 8-bit monotonicity and linearity while zero level output current of less than 4 μA provides 8-bit zero accuracy for $I_{REF} \geq 2$ mA. The power supply currents of the DAC0808 series are independent of bit codes, and exhibits essentially constant device characteristics over the entire supply voltage range.

The DAC0808 will interface directly with popular TTL, DTL or CMOS logic levels, and is a direct replacement for the MC1508/MC1408. For higher speed applications, see DAC0800 data sheet.

Features

- Relative accuracy: ±0.19% error maximum (DAC0808)
- Full scale current match: ±1 LSB typ
- 7 and 6-bit accuracy available (DAC0807, DAC0806)
- Fast settling time: 150 ns typ
- Noninverting digital inputs are TTL and CMOS compatible
- High speed multiplying input slew rate: 8 mA/μs
- Power supply voltage range: ±4.5V to ±18V
- Low power consumption: 33 mW @ ±5V

Block and Connection Diagrams

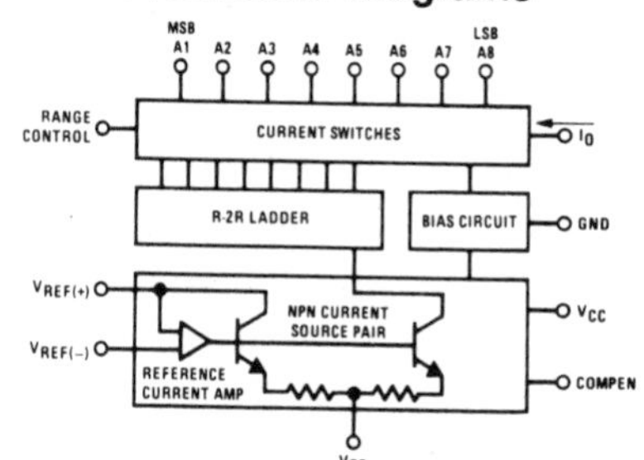

Dual-In-Line Package

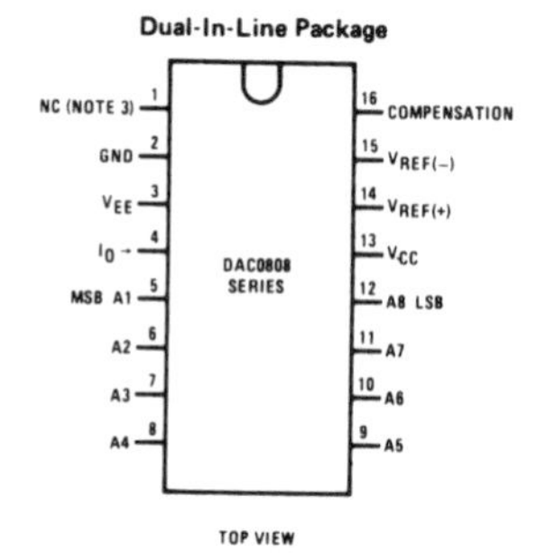

Typical Application

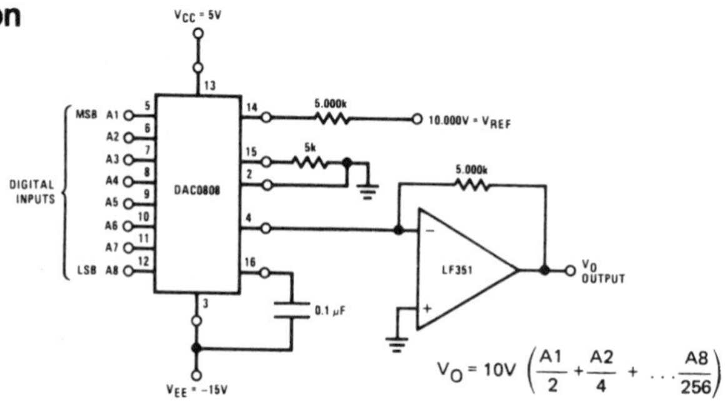

FIGURE 1. +10V Output Digital to Analog Converter

Ordering Information

ACCURACY	OPERATING TEMPERATURE RANGE	ORDER NUMBERS*					
		D PACKAGE (D16C)		J PACKAGE (J16A)		N PACKAGE (N16A)	
8-bit	$-55°C \leq T_A \leq +125°C$	DAC0808LD	MC1508L8				
8-bit	$0°C \leq T_A \leq +75°C$			DAC0808LCJ	MC1408L8	DAC0808LCN	MC1408P8
7-bit	$0°C \leq T_A \leq +75°C$			DAC0807LCJ	MC1408L7	DAC0807LCN	MC1408P7
6-bit	$0°C \leq T_A \leq +75°C$			DAC0806LCJ	MC1408L6	DAC0806LCN	MC1408P6

*Note. Devices may be ordered by using either order number.

Absolute Maximum Ratings

Power Supply Voltage	
V_{CC}	+18 V_{DC}
V_{EE}	−18 V_{DC}
Digital Input Voltage, V5–V12	−10 V_{DC} to +18 V_{DC}
Applied Output Voltage, V_O	−11 V_{DC} to +18 V_{DC}
Reference Current, I_{14}	5 mA
Reference Amplifier Inputs, V14, V15	V_{CC}, V_{EE}
Power Dissipation (Package Limitation)	1000 mW
Derate above T_A = 25°C	6.7 mW/°C
Operating Temperature Range	
DAC0808L	−55°C ≤ T_A ≤ +125°C
DAC0808LC Series	0 ≤ T_A ≤ +75°C
Storage Temperature Range	−65°C to +150°C

Electrical Characteristics

(V_{CC} = 5V, V_{EE} = −15 V_{DC}, V_{REF}/R14 = 2 mA, DAC0808: T_A = −55°C to +125°C, DAC0808C, DAC0807C, DAC0806C, T_A = 0°C to +75°C, and all digital inputs at high logic level unless otherwise noted.)

	PARAMETER	CONDITIONS	MIN	TYP	MAX	UNITS
E_r	Relative Accuracy (Error Relative to Full Scale I_O)	*(Figure 4)*				%
	DAC0808L (LM1508-8), DAC0808LC (LM1408-8)				±0.19	%
	DAC0807LC (LM1408-7), (Note 1)				±0.39	%
	DAC0806LC (LM1408-6), (Note 1)				±0.78	%
	Settling Time to Within 1/2 LSB (Includes t_{PLH})	T_A = 25°C (Note 2), *(Figure 5)*		150		ns
t_{PLH}, t_{PHL}	Propagation Delay Time	T_A = 25°C, *(Figure 5)*		30	100	ns
TCI_O	Output Full Scale Current Drift			±20		ppm/°C
MSB	Digital Input Logic Levels	*(Figure 3)*				
V_{IH}	High Level, Logic "1"		2			V_{DC}
V_{IL}	Low Level, Logic "0"				0.8	V_{DC}
MSB	Digital Input Current	*(Figure 3)*				
	High Level	V_{IH} = 5V		0	0.040	mA
	Low Level	V_{IL} = 0.8V		−0.003	−0.8	mA
I_{15}	Reference Input Bias Current	*(Figure 3)*		−1	−3	μA
	Output Current Range	*(Figure 3)*				
		V_{EE} = −5V	0	2.0	2.1	mA
		V_{EE} = −15V, T_A = 25°C	0	2.0	4.2	mA
I_O	Output Current	V_{REF} = 2.000V, R14 = 1000Ω, *(Figure 3)*	1.9	1.99	2.1	mA
	Output Current, All Bits Low	*(Figure 3)*		0	4	μA
	Output Voltage Compliance	E_r ≤ 0.19%, T_A = 25°C				
	Pin 1 Grounded,				−0.55, +0.4	V_{DC}
	V_{EE} Below −10V				−5.0, +0.4	V_{DC}
SRI_{REF}	Reference Current Slew Rate	*(Figure 6)*	4	8		mA/μs
	Output Current Power Supply Sensitivity	−5V ≤ V_{EE} ≤ −16.5V		0.05	2.7	μA/V
	Power Supply Current (All Bits Low)	*(Figure 3)*				
I_{CC}				2.3	22	mA
I_{EE}				−4.3	−13	mA
	Power Supply Voltage Range	T_A = 25°C, *(Figure 3)*				
V_{CC}			4.5	5.0	5.5	V_{DC}
V_{EE}			−4.5	−15	−16.5	V_{DC}
	Power Dissipation					
	All Bits Low	V_{CC} = 5V, V_{EE} = −5V		33	170	mW
		V_{CC} = 5V, V_{EE} = −15V		106	305	mW
	All Bits High	V_{CC} = 15V, V_{EE} = −5V		90		mW
		V_{CC} = 15V, V_{EE} = −15V		160		mW

Note 1: All current switches are tested to guarantee at least 50% of rated current.

Note 2: All bits switched.

Note 3: Range control is not required.

Test Circuits

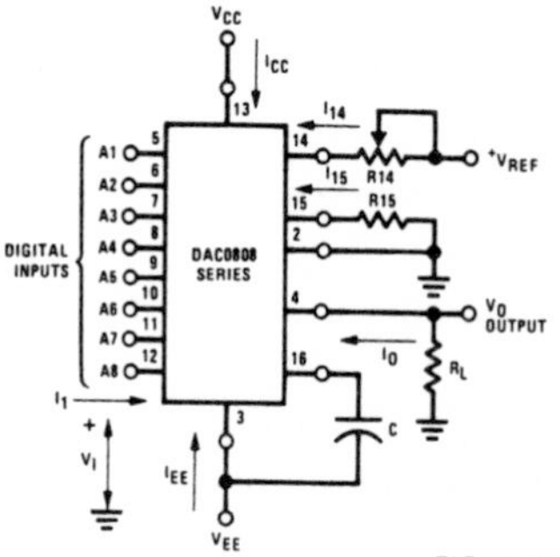

V_I and I_I apply to inputs A1–A8.

The resistor tied to pin 15 is to temperature compensate the bias current and may not be necessary for all applications.

$$I_O = K\left(\frac{A1}{2} + \frac{A2}{4} + \frac{A4}{16} + \frac{A5}{32} + \frac{A6}{64} + \frac{A7}{128} + \frac{A8}{256}\right)$$

where $K \cong \frac{V_{REF}}{R14}$

and A_N = "1" if A_N is at high level

A_N = "0" if A_N is at low level

FIGURE 3. Notation Definitions Test Circuit

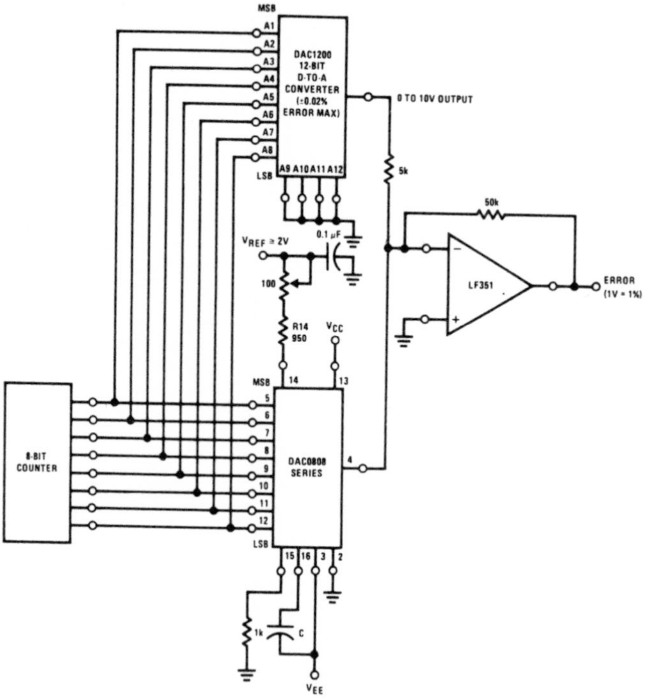

FIGURE 4. Relative Accuracy Test Circuit

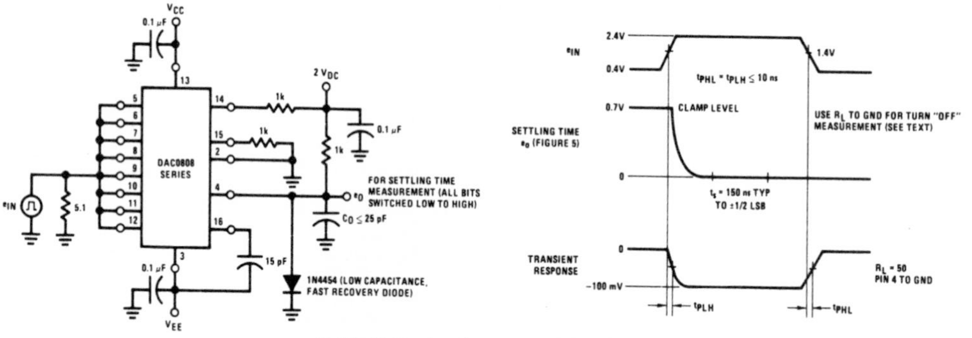

FIGURE 5. Transient Response and Settling Time

Test Circuits (Continued)

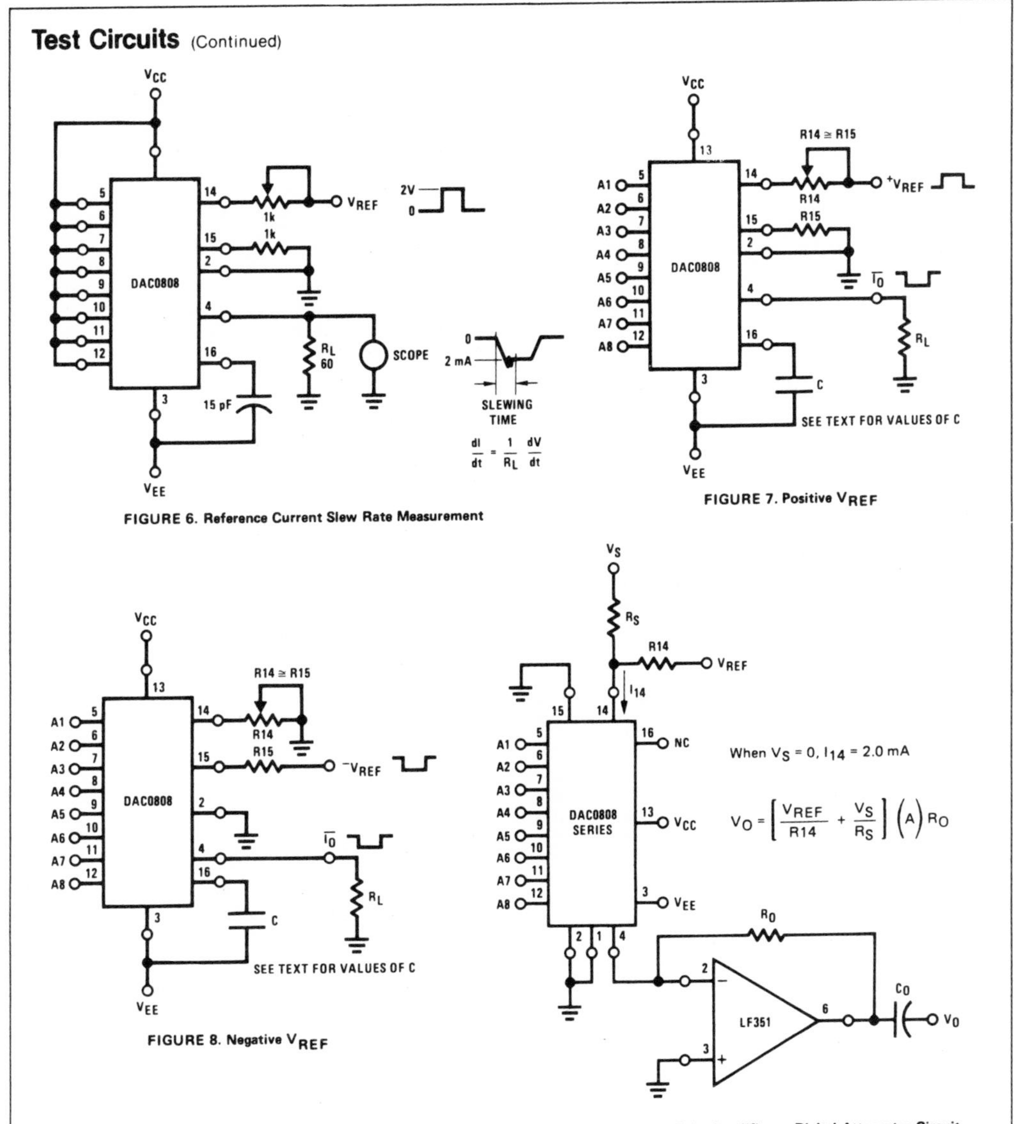

FIGURE 6. Reference Current Slew Rate Measurement

$$\frac{dI}{dt} = \frac{1}{R_L}\frac{dV}{dt}$$

FIGURE 7. Positive V_{REF}

FIGURE 8. Negative V_{REF}

$$V_O = \left[\frac{V_{REF}}{R14} + \frac{V_S}{R_S}\right](A) R_O$$

FIGURE 9. Programmable Gain Amplifier or Digital Attenuator Circuit

Application Hints

REFERENCE AMPLIFIER DRIVE AND COMPENSATION

The reference amplifier provides a voltage at pin 14 for converting the reference voltage to a current, and a turn-around circuit or current mirror for feeding the ladder. The reference amplifier input current, I_{14}, must always flow into pin 14, regardless of the set-up method or reference voltage polarity.

Connections for a positive voltage are shown in *Figure 7*. The reference voltage source supplies the full current I_{14}. For bipolar reference signals, as in the multiplying mode, R15 can be tied to a negative voltage corresponding to the minimum input level. It is possible to eliminate R15 with only a small sacrifice in accuracy and temperature drift.

The compensation capacitor value must be increased with increases in R14 to maintain proper phase margin; for R14 values of 1, 2.5 and 5 kΩ, minimum capacitor values are 15, 37 and 75 pF. The capacitor may be tied to either V_{EE} or ground, but using V_{EE} increases negative supply rejection.

Application Hints (Continued)

A negative reference voltage may be used if R14 is grounded and the reference voltage is applied to R15 as shown in *Figure 8*. A high input impedance is the main advantage of this method. Compensation involves a capacitor to V_{EE} on pin 16, using the values of the previous paragraph. The negative reference voltage must be at least 4V above the V_{EE} supply. Bipolar input signals may be handled by connecting R14 to a positive reference voltage equal to the peak positive input level at pin 15.

When a DC reference voltage is used, capacitive bypass to ground is recommended. The 5V logic supply is not recommended as a reference voltage. If a well regulated 5V supply which drives logic is to be used as the reference, R14 should be decoupled by connecting it to 5V through another resistor and bypassing the junction of the 2 resistors with 0.1 μF to ground. For reference voltages greater than 5V, a clamp diode is recommended between pin 14 and ground.

If pin 14 is driven by a high impedance such as a transistor current source, none of the above compensation methods apply and the amplifier must be heavily compensated, decreasing the overall bandwidth.

OUTPUT VOLTAGE RANGE

The voltage on pin 4 is restricted to a range of −0.6 to 0.5V when V_{EE} = −5V due to the current switching methods employed in the DAC0808.

The negative output voltage compliance of the DAC0808 is extended to −5V where the negative supply voltage is more negative than −10V. Using a full-scale current of 1.992 mA and load resistor of 2.5 kΩ between pin 4 and ground will yield a voltage output of 256 levels between 0 and −4.980V. Floating pin 1 does not affect the converter speed or power dissipation. However, the value of the load resistor determines the switching time due to increased voltage swing. Values of R_L up to 500Ω do not significantly affect performance, but a 2.5 kΩ load increases worst-case settling time to 1.2 μs (when all bits are switched ON). Refer to the subsequent text section on Settling Time for more details on output loading.

OUTPUT CURRENT RANGE

The output current maximum rating of 4.2 mA may be used only for negative supply voltages more negative than −7V, due to the increased voltage drop across the resistors in the reference current amplifier.

ACCURACY

Absolute accuracy is the measure of each output current level with respect to its intended value, and is dependent upon relative accuracy and full-scale current drift. Relative accuracy is the measure of each output current level as a fraction of the full-scale current. The relative accuracy of the DAC0808 is essentially constant with temperature due to the excellent temperature tracking of the monolithic resistor ladder. The reference current may drift with temperature, causing a change in the absolute accuracy of output current. However, the DAC0808 has a very low full-scale current drift with temperature.

The DAC0808 series is guaranteed accurate to within ±1/2 LSB at a full-scale output current of 1.992 mA. This corresponds to a reference amplifier output current drive to the ladder network of 2 mA, with the loss of 1 LSB (8 μA) which is the ladder remainder shunted to ground. The input current to pin 14 has a guaranteed value of between 1.9 and 2.1 mA, allowing some mismatch in the NPN current source pair. The accuracy test circuit is shown in *Figure 4*. The 12-bit converter is calibrated for a full-scale output current of 1.992 mA. This is an optional step since the DAC0808 accuracy is essentially the same between 1.5 and 2.5 mA. Then the DAC0808 circuits' full-scale current is trimmed to the same value with R14 so that a zero value appears at the error amplifier output. The counter is activated and the error band may be displayed on an oscilloscope, detected by comparators, or stored in a peak detector.

Two 8-bit D-to-A converters may not be used to construct a 16-bit accuracy D-to-A converter. 16-bit accuracy implies a total error of ±1/2 of one part in 65,536, or ±0.00076%, which is much more accurate than the ±0.019% specification provided by the DAC0808.

MULTIPLYING ACCURACY

The DAC0808 may be used in the multiplying mode with 8-bit accuracy when the reference current is varied over a range of 256:1. If the reference current in the multiplying mode ranges from 16 μA to 4 mA, the additional error contributions are less than 1.6 μA. This is well within 8-bit accuracy when referred to full-scale.

A monotonic converter is one which supplies an increase in current for each increment in the binary word. Typically, the DAC0808 is monotonic for all values of reference current above 0.5 mA. The recommended range for operation with a DC reference current is 0.5 to 4 mA.

SETTLING TIME

The worst-case switching condition occurs when all bits are switched ON, which corresponds to a low-to-high transition for all bits. This time is typically 150 ns for settling to within ±1/2 LSB, for 8-bit accuracy, and 100 ns to 1/2 LSB for 7 and 6-bit accuracy. The turn OFF is typically under 100 ns. These times apply when $R_L \leq 500\Omega$ and $C_O \leq 25$ pF.

Extra care must be taken in board layout since this is usually the dominant factor in satisfactoy test results when measuring settling time. Short leads, 100 μF supply bypassing for low frequencies, and minimum scope lead length are all mandatory.

Selected Answers

Chapter 1

Questions

1. high, operations
3. 0, offset
5. oscillation
7. inverting, noninverting

Problems

1. 20 V/μs
3. 250 kHz
5. 6.75 V
7. **a.** 4, **b.** −8 V, **c.** 0.4 mA

Chapter 2

Questions

1. noninverting, infinite
3. summing
5. log, multiplication
7. TTL
17. 0°

Problems

1. **a.** −5 V
3. 7.23 V peak
7. $V_{ref} = \pm 3$ V
9. **a.** 0.75, 250 mV

Chapter 3

Questions

1. D, S
3. I, D, D
5. I
9. lower

Problems

1. 2095 r/min
3. 972 Ω
5. 5 A, 0.5 Nm, 500 W
7. 2283 r/min
9. 0 V, 31.5 A

Chapter 4

Questions

1. series, DC
3. electromagnetic induction, electrical
5. synchronous, starting, rotor
19. synchronous speed

Problems

1. 300 r/min
3. 1800 r/min

Chapter 5

Questions

1. manual, mechanically operated, limit
3. positive, gate
5. triggering

Problems

1. **a.** yes, **b.** 24.5 mW
3. R ← 3.3 kΩ
5. 5.17 A
7. 976 Hz

Chapter 6

Questions

1. phase, gain
3. zero-voltage, motor
5. PLL, 0.001

Chapter 7

Questions

1. measurand, position
3. temperature, length
5. digital, photovoltaic
7. magnetoresistor
9. orifice plate, Bernoulli
15. resistance increases, negative, no

Problems

1. **a.** 130°C, **b.** 107°C
3. 0.86 cm
5. 0.256 s

Chapter 8

Questions

1. disturbance
3. Open loop, lags, loads
5. set point, gain
7. rate, rate
15. yes
21. no

Problems

1. 1, 2
3. 50%

Chapter 9

Questions

1. Telemetry
3. analog, digital
5. double polarity
7. position
9. quanta
11. aliasing
13. low-pass filter
15. no

Problems

1. 40 kHz
3. 625 Hz
5. 50 kHz

Chapter 10

Questions

1. transducer, transmission, display, interpretation
3. frequency, amplitude
5. smaller
7. trailing
17. channel 10
19. 5.4 kHz
21. channel 9

Problems

1. 100 kHz

Chapter 11

Questions

1. microcomputer, minicomputer, mainframe computer
3. algorithm
5. reasonableness

Chapter 12

Questions

1. sequential
3. programmable logic controller

Index